T0235331

Lecture Notes in Computer Science　　9473

Commenced Publication in 1973
Founding and Former Series Editors:
Gerhard Goos, Juris Hartmanis, and Jan van Leeuwen

More information about this series at http://www.springer.com/series/7410

Moti Yung · Liehuang Zhu
Yanjiang Yang (Eds.)

Trusted Systems

6th International Conference, INTRUST 2014
Beijing, China, December 16–17, 2014
Revised Selected Papers

 Springer

Editors
Moti Yung
Google
New York, NY
USA

Yanjiang Yang
Institute for Infocomm Research
Singapore
Singapore

Liehuang Zhu
Beijing Institute of Technology
Beijing
China

ISSN 0302-9743   ISSN 1611-3349 (electronic)
Lecture Notes in Computer Science
ISBN 978-3-319-27997-8   ISBN 978-3-319-27998-5 (eBook)
DOI 10.1007/978-3-319-27998-5

Library of Congress Control Number: 2015957794

LNCS Sublibrary: SL4 – Security and Cryptology

This Springer imprint is published by SpringerNature
The registered company is Springer International Publishing AG Switzerland

Preface

These proceedings contains 27 papers presented at the INTRUST (International Conference on Trustworthy Systems) 2014 conference, held in Beijing, China, in December 2014. INTRUST 2014 was the sixth international conference on the theory, technologies, and applications of trusted systems. It was devoted to all aspects of trusted computing systems, including trusted modules, platforms, networks, services, and applications, from their fundamental features and functionalities to design principles, architecture, and implementation technologies. The goal of the conference was to bring academic and industrial researchers, designers, and implementers together with end-users of trusted systems, in order to foster the exchange of ideas in this challenging and fruitful area.

INTRUST 2014 built on a series of highly successful international conferences, previously held as INTRUST 2013 (Graz, Austria), INTRUST 2012 (London, UK), INTRUST 2011 (Beijing, China), INTRUST 2010 (Beijing, China), and INTRUST 2009 (Beijing, China). The program of INTRUST 2014 consisted of five keynote speeches from Moti Yung (Columbia University and Google), Sheng Zhong (Nanjing University), Kui Ren (University at Buffalo, the State University of New York), Jinjun Chen (University of Technology, Sydney), and Rui Zhang (Institute of Information Engineering, Chinese Academy of Sciences).

All submissions were blind-reviewed, i.e., the Program Committee members provided reviews on anonymous submissions. Each submission was reviewed by at least two, and on average 3.2, Program Committee members. The individual reviewing phase was followed by in-depth discussions about the papers, which contributed greatly the quality of the final selection. A number of accepted papers were shepherded by some Program Committee members in order to make sure the review comments were addressed properly. We are very grateful to our hard-working and distinguished Program Committee for doing such an excellent job in a timely fashion.

For the proceedings, the papers have been divided into seven main categories, namely, signature and authentication, secure protocols and access control, cloud security, cryptographic aspects, software security, security analysis, and secure communication and privacy.

We would like to thank the conference General Chair Heyan Huang, the conference Honorary Chairs Liqun Chen and Yongfei Han, and the Publicity Chairs Xinyi Huang and Mingzhong Wang, for valuable guidance and assistance and for handling the arrangements in Beijing. We also would like to thank the program committee and the reviewers for their hard work. Thanks are also due to easyChair for providing the submission and review webserver.

On behalf of the conference organization and participants, we would like to express our appreciation to Beijing Institute of Technology, the National Nature Science Foundation of China, and the Technical Committee on Intelligent Information Network of the Chinese Association for Artificial Intelligence for their generous sponsorship of this event.

We would also like to thank all the authors who submitted their papers to the INTRUST 2014 conference, all external reviewers, and all the attendees of the conference. Authors of accepted papers are thanked again for revising their papers according to the feedback from the conference participants. The revised versions were not checked by the Program Committee, and thus the authors bear full responsibility for their contents. We thank the staff at Springer for their help in producing the proceedings.

July 2015

Moti Yung
Liehuang Zhu
Yanjiang Yang

Organization

Honorary Chairs

Liqun Chen   HP Laboratories, UK
Yongfei Han   BJUT and ONETS, China

General Chair

Heyan Huang   Beijing Institute of Technology, China

Program Chairs

Moti Yung   Google and Columbia University, USA
Liehuang Zhu   Beijing Institute of Technology, China
Yanjiang Yang   Institute for Infocomm Research, Singapore

Publicity Chairs

Xinyi Huang   Fujian Normal University, China
Mingzhong Wang   Beijing Institute of Technology, China

Program Committee

Endre Bangerter   Bern University of Applied Sciences, Switzerland
Zhen Chen   Tsinghua University, China
Zhong Chen   Peking University, China
Naccache David   Ecole Normale Superieure, France
Kurt Dietrich   NXP Semiconductors, Austria
Xuhua Ding   Singapore Management University
Dieter Gollmann   Hamburg University of Technology, Germany
Sigrid Guergens   Fraunhofer Institute for Secure Information
                                   Technology SIT, Germany
Weili Han   Fudan University, China
Jingyu Hua   Nanjing University, China
Xuejia Lai   Shanghai Jiaotong University, China
Jianxin Li   Beihang University, China
Shujun Li   University of Surrey, UK
Peter Lipp   Graz University of Technology, Austria

Jiqiang Liu	Beijing Jiaotong University, China
Javier Lopez	University of Malaga, Spain
Andrew Martin	University of Oxford, UK
Shin'Ichiro Matsuo	National Institute of Information and Communications Technology, Japan
Yi Mu	University of Wollongong, Australia
David Naccache	Ecole Normale Suprieure, France
Raphael C.-W. Phan	Loughborough University, UK
Bart Preneel	Katholieke Universiteit Leuven - COSIC, Belgium
Scott Rotondo	Oracle Corporation, USA
Kouichi Sakurai	Kyushu University, Japan
Willy Susilo	University of Wollongong, Australia
Qiang Tang	University of Luxembourg, Luxembourg
Claire Vishik	Intel Corporation (UK)
Qian Wang	Wuhan University, China
Jian Weng	Jinan University, China
Fan Wu	Shanghai Jiao Tong University, China
Yi Xie	Sun Yat-sen University, China
Chang Xu	Beijing Institute of Technology, China
Shouhuai Xu	University of Texas at San Antonio, USA
Rui Xue	Chinese Academy of Sciences, China
Huanguo Zhang	Wuhan University, China
Rui Zhang	Institute of Information Engineering, CAS, China
Xing Zhang	Beijing University of Technology, China
Xinwen Zhang	Huawei Research Center, China
Yuan Zhang	Nanjing University, China
Zijian Zhang	Beijing Institute of Technology, China
Sheng Zhong	Nanjing University, China
Yan Zhou	Beijing University, China
Yongbin Zhou	Institute of Information Engineering, Chinese Academy of Sciences, China

Steering Committee

Yongfei Han	BJUT and ONETS, China
Moti Yung	Google and Columbia University, USA
Liqun Chen	HP Laboratories, UK
Robert Deng	SMU, Singapore
Chris Mitchell	RHUL, UK

Additional Reviewers

Anada, Hiroaki
Dong, Wenyu
Fernandez, Carmen
Guo, Fuchun
Han, Jinguang
King-Lacroix, Justin
Ma, Ge
Mao, Xianping
Matsumoto, Shinichi
Mendel, Florian
Moyano, Francisco
Spreitzer, Raphael

Su, Chunhua
Wei, Yu
Weng, Jian
Xu, Chang
Xu, Kun
Xu, Lei
Xue, Weijia
Yang, Guomin
Yang, Rupeng
Yuan, Wei
Zhang, Su
Zheng, Qingji

Contents

Identity-Embedding Method
for Decentralized Public-Key Infrastructure

Hiroaki Anada[1]([envelope]), Junpei Kawamoto[2], Jian Weng[2,3], and Kouichi Sakurai[1,2]

[1] Institute of Systems,
Information Technologies and Nanotechnologies, Fukuoka, Japan
{anada,sakurai}@isit.or.jp
[2] Department of Informatics, Faculty of Information Science and Electrical
Engineering, Kyushu University, Fukuoka, Japan
{kawamoto,sakurai}@inf.kyushu-u.ac.jp
[3] Department of Computer Science, College of Information Science and Technology,
Jinan University, Guangzhou, China
cryptjweng@gmail.com

Abstract. A public key infrastructure (PKI) is for facilitating the authentication and distribution of public keys. Currently, the most commonly employed approach to PKI is to rely on certificate authorities (CAs), but recently there has been arising more need for decentralized peer-to-peer certification like Webs of Trust. In this paper, we propose an identity-embedding method suitable for decentralized PKI. By embedding not only ID of the candidate public-key owner itself but also IDs of his guarantors into PK, we can construct *Web of guarantors* on public keys. Here guarantors can be chosen arbitrarily by the candidate public-key owner. Our embedding method uses a combination of two public-key cryptosystems; the first cryptosystem is for PKI directly. Here we employ a technique to embed a string into a public key of the first cryptosystem. As such a string, we choose a concatenation of ID of a candidate public-key owner, IDs of his guarantors, and a public key of the second cryptosystem. This embedded public key of the second cryptosystem is used by the candidate public-key owner that he certainly knows the secret key that corresponds to the public key of the first cryptosystem. Then, with an aid of a broadcast mechanism of an updated public-key list on a peer-to-peer network, we can attain the decentralized PKI. Such an embedding method is concretely realized by the RSA encryption with the Lenstra's algorithm, which can be used as the first cryptosystem. As the second cryptosystem, we employ an elliptic curve encryption whose security is equivalent to the security of the RSA encryption, where the former achieves shorter key size than the latter. We write down concrete values of parameters for a realization of the embedding.

Keywords: Identity management · Public key infrastructure · Decentralized system · RSA · Elliptic curve

M. Yung et al. (Eds.): INTRUST 2014, LNCS 9473, pp. 1–14, 2016.
DOI: 10.1007/978-3-319-27998-5_1

1   Introduction

A network is for participants, so trusted identity management among participants is a must requirement. Transactions are certainly available when they are based on reliable identities of participants; we can communicate each other, even in broadcasting, only when one party recognize the other with certainty. Concerning a trusted identity management, a public key infrastructure (PKI) is responsible for facilitating the authentication and distribution of public keys. It maintains a database of (ID, PK) pairs, where ID represents an identity, and PK represents a corresponding public key.

Currently, the most commonly employed approaches to PKIs are classified into two categories: centralized PKI with certificate authorities (CAs) and decentralized PKI of peer-to-peer certification, often referred to as Webs of Trust. CA acts a trusted third party that is responsible for distributing and managing digital certificates for a network of users. A typical example is the use of CA in the Secure Sockets Layer protocol (SSL). In the use case, an identity of a server (or a client) is an IP address and it is assured by a certificate issued. Here the certificate has, as an evidence, a digital signature generated by the certificate authority. Then the identity of the certificate authority is assured by a certificate with a signature issued by another certificate authority. Hence, there arises a chain of signatures, which forms a hierarchy of certificates with a top; that is, a root certificate. Thus the use of CAs creates single points of failure in PKI. There have been numerous recent incidents showing that too much trust is being placed in CAs. CAs have been hacked, and have even accidentally issued subordinate root certificates to customers. Additionally, while the CA system is centralized enough to introduce single points of failure, it is not centralized enough to ensure consistency. Since there exist multiple CAs, they may certify different public keys corresponding to the same identity that may yield violation of identity retention [7][1].

In contrast, in the second major PKI, Web of Trust, authentication is entirely decentralized; users are able to designate others as trustworthy by signing their public keys. A user thus accumulates a certificate containing his public key and digital signatures from entities that have assured him as trustworthy. The certificate is trusted by another user if he is able to verify that the certificate contains the signature of someone he trusts. As for motivation, Web of Trust needs relatively lower fee (in some cases, free) compared to the fee for CA-based hierarchical (centralized) certification. Another motivation of decentralized certification comes from risk control mentioned above; in a general theory of risk control, it is better to have PKI with more than one root to avoid single point failure. Ultimately saying, it is more desirable to do identity management in a *flat* manner, where flat means that *any* participant can be a guarantor. However, PGP [21] does not offer identity retention, because much like in the case of CAs

---

[1] In the research of functional encryption, there is a notion of *multi-authority* (for instance, in Lewko and Waters [15]), which means there can be more than one authorities that issue private secret keys without violation of identity retention. But in this paper, we consider this type of decentralization as introducing multiple CAs.

there is no guarantee of consistency, and nothing prevents multiple users from creating public keys for the same identity illegally.

In this paper, we propose an identity-embedding method for decentralized PKI like Webs of Trust. Our identity-embedding method does not resolve the problem of identity retention directly, but can be used as a building block for decentralized PKI. This is because it does not need any issuing center of certificates of public keys. Instead, it needs for a participant to prove that he certainly knows the secret key to a participant, as follows.

Our decentralized PKI is on its underlying P2P network. When the network is initiated, our decentralized PKI assumes that the network has more than one participants. Each initiator generates a pair of public key and secret key and writes it into a public-key list. When a candidate public-key owner wants to take part in the underlying network, two processes are done in our decentralized PKI.

The first process is an authentication process. We use two public key cryptosystems in the first process; a candidate public-key owner generates a pair of public key and secret key of the first cryptosystem. Then, by using the second cryptosystem, the candidate public-key owner tries to prove that he certainly knows the corresponding secret key to verifiers who have been already participants. Here the verifiers are chosen as guarantors arbitrarily by the candidate public-key owner. Hence our PKI can be decentralized in a flat manner.

The second process is a broadcast process. One of guarantors adds the newly generated public key of the candidate public-key owner to the public-key list. Then the guarantor broadcasts the updated public-key list along the underlying P2P network.

The features of our decentralized PKI can be summarized in Table 1. Here, we also compare our decentralized PKI with *identity-based* PKI that has been realized with the invention of identity-based encryption [4], where any public string can be a public key.

Table 1. PKI: Centralized versus Identity-Based versus Decentralized.

Item	Centralized PKI	Identity-Based PKI	Our Decentralized PKI
Need of CA:	$\checkmark$	-	-
Need of KI Center:	-	$\checkmark$	-
PK is verified by:	Checking Certificate	PK itself	Cha.-Res. Protocol
Trust is made by:	Root CA	KI Center	Web of Guarantors

KI: Key Issuing; PK: Public Key; Cha.-Res.: Challenge-and-Response

As is stated in the above, our decentralized PKI does not need any CA. It is also notable that our decentralized PKI does not need any key issuing center, whereas identity-based PKI needs a private secret-key issuing center that yields the key-escrow problem [11, 13].

1.1 Previous Work

It is well known that there has been a history on decentralized identity management and PKI. Zimmermann, who developed PGP [21], can be considered as the pioneers of decentralized trust management. Keeping the spirit of low cost management, PGP uses a concept of a Web of Trust to establish the authenticity of the binding between a public key and its owner. Its decentralized trust model is an alternative to the centralized trust model of a public key infrastructure (PKI), which relies exclusively on a certificate authority (or a hierarchy of such).

Following Zimmermann's work, there appeared a lot of work. Blaze et al. [3] proposed a trust management system which they call PolicyMaker. Sander and Ta-Shma [18] proposed an auditable anonymous electronic cash system, whose security relies on ability of an underlying network to maintain the integrity of a public database. Concerning a digital rights management (DRM), we can see recent years several work like Qiu et al. [17], which proposed a model of social trust between content sharers of DRM-related content.

As for the Lenstra's algorithm, Lenstra [14] proposed a more efficient algorithm that allows us to embed any string I into a modulus N of the RSA encryption. The length of embeddable string I is almost the half of the bit length of N. Following this work, Kitahara et al. [13] proposed a modified algorithm to produce the two factors ($N = pq$) of the same length. Upper bounds of the length of embeddable string I have been provided by Graham and Shparlinski [9], Meng [16] and Kitahara et al. [13]. Applications have been proposed, for example, in Kitahara et al. [12].

As for a broadcast mechanism of an updated public-key list, proof-of-work is an exciting area of research [2,7,8]. It originates from the work of Dwork and Naor of CRYPTO '92 paper [6]. We only rely on the result of Fromknecht, Velicanu and Yakoubov [7] and Andrychowicz and Dziembowski [2], and employ their broadcast mechanisms to update a public-key list.

1.2 Our Contributions

Following the traditional spirit [3,17,18,21], we contribute in two points. The first contribution is to propose an identity-embedding method for a decentralized PKI. The embedded string is taken as a concatenation of identity data of a candidate public-key owner, identity data of his guarantors, and a public key of the second cryptosystem (that is, we embed *a second public key into a first public key*). Like a digital signature, this embedding structure functions as a preventer of manipulations on public keys and hence our embedding method prevents falsification on the public-key list.

The second contribution is to provide the above identity-embedding method concretely. It is known that elliptic curve encryption with shorter public key achieves the same level of security as RSA encryption. Using this fact, we can embed a public key of an elliptic curve encryption into a modulus of an RSA encryption. we provide a security proof, for this concrete construction of public key, that our embedding method prevents impersonation that an adversary tries to pretend an honest public-key owner without the corresponding secret key. Actually we give a security proof based on the theory of key encapsulation mechanism (KEM): our KEM is secure against adaptive chosen-plaintext attacks on one-wayness based on the Gap-CDH assumption in the random oracle model.

We note that our ID-embedding method can be used, for example, in the RSA encryption and signature in SSL. As a result, we can exit the hierarchy of certificate authorities.

Parameter values of our concrete decentralized PKI can be as follows. Here IFP denotes the Integer Factorization Problem and ECDLP denotes the Elliptic Curve Discrete Logarithm Problem. λ is the security parameter against exhaustive search, N is a modulus of the RSA encryption and p_0 is a prime order of the employed cyclic group of ECDLP. Table 2 shows parameter values.

Table 2. Parameter Size (bit).

Sec. Param. against Exhaustive Search	λ	112	128	192
Equiv. Length of Modulus for IFP	$\lambda_{RSA} = \|N\|$	2048	3072	7680
Max. Length of Embed. Info	$\|I\| = \|N\|/2 - \log_2(\|N\|) - 1$	1012	1523	3826
Equiv. Length of Order for ECDLP	$\|p_0\| = 2\lambda$	224	256	384
Length of Expression for a Point on EC	$\|P\| = 2\|p_0\|$	448	512	768
Room for IDs to be Embedded	$\|I\| - \|P\|$	564	1011	3058

One notable thing is that an equivalent length of prime order p_0 of a cyclic group on a elliptic curve for ECDLP is shorter than the factor p of a modulus N of the RSA encryption. Therefore we can make room to embed a public key of the elliptic curve cryptosystem into the public key of the RSA encryption.

When we use e-mail address for identity data ID, 70 characters are available because in Table 2 there are 564 bits remaining for identity data. That is, 70 byte. On condition that 1 character needs 1 byte, we can use 23 characters for identity data of a candidate public-key owner and 23 characters for identity data of two guarantors, like: `alice@decentralized.com`, `bob@flat.com` and `charlie@lowfee.com`.

As for the efficiency of the embedding method, the work of Kitahara et al. [13] assures that it is as efficient as the usual RSA cryptosystem in the key generation as well as encryption and decryption.

1.3   Organization of This Paper

In Sect. 2, we explain required notations and notions. In Sect. 3, we state our generic decentralized PKI. In Sect. 4, we describe our concrete decentralized PKI in the RSA and elliptic curve encryption setting. In Sect. 5, we conclude our work.

2   Preliminaries

The security parameter against exhaustive search is denoted by λ. A multiplicative cyclic group of order p_0 is denoted by $\mathbb{G}_{p_0}$. The ring of the exponent domain of $\mathbb{G}_{p_0}$, which consists of integers from 0 to $p_0 - 1$ with modulo p_0 operation, is denoted by $\mathbb{Z}_{p_0}$. When an algorithm A with input a outputs z, we denote it as $z \leftarrow A(a)$.

2.1   Embedding Technique into a Modulus of RSA Encryption

As a variant of the RSA encryption, In 1995, Vanstone and Zuccherato [19] proposed an algorithm that allows us to embed any string I into a modulus N of the RSA encryption. But it has a trade off between the time to generate the modulus N and the bit length of I. This is because that the algorithm needs factorization of I as an integer. So, when we embed I whose bit length is a half of that of N, the algorithm needs quite long time both in theory and in practice.

After that, in 1998, Lenstra [14] proposed a more efficient algorithm that allows us to embed any string I into a modulus N of the RSA encryption. The length of embeddable string I is almost the half of the bit length of N. The time to generate the modulus N is almost the same as the time to generate the modulus N of the normal RSA. The following algorithm is a modified version by Kitahara et al. [13]. Here we denote the length of a modulus N of RSA encryption that has λ-bit security against exhaustive search as λ_{RSA}.

Lenstra's Algorithm (a Modified Version [13])

1. Put $N' = I \parallel 00 \cdots 0$ s.t. $|N'| = \lambda_{\mathrm{RSA}}$.
2. Choose a prime p s.t. $|p| = \lambda_{\mathrm{RSA}}/2$ at random.
3. Compute $q' = \lceil N'/p \rceil$.
4. Compute the minimum positive integer t s.t. $q' + t$ is a prime.
5. Put $q = q' + t$.
6. Compute $N = pq$.
7. If the higher bits of N is equal to I, then return (p, q, N) else go back to 2.

Note here that (1) and (2) can be swapped.

2.2   The CDH and the Gap-CDH Problems and Assumptions

A quadruple (g, X, Y, Z) of elements in $\mathbb{G}_{p_0}$ is called a Diffie-Hellman (DH) tuple if (g, X, Y, Z) is written as (g, g^x, g^y, g^{xy}) for some elements x and y in $\mathbb{Z}_{p_0}$. A CDH problem instance is a triple $(g, X = g^x, Y = g^y)$, where the exponents x and y are random and unknown to a solver. A DDH problem instance is a quadruple (g, X, Y, Z). The DDH oracle $\mathcal{DDH}$ is an oracle which, queried about a DDH problem instance (g, X, Y, Z), replies the correct boolean decision whether (g, X, Y, Z) is a DH-tuple or not.

A CDH problem solver $\mathcal{S}$ that is allowed to access $\mathcal{DDH}$ polynomially many times is called a Gap-CDH problem solver. We define the following experiment.

$$\mathbf{Exprmt}_{\mathcal{S},\mathbb{G}_{p_0}}^{\text{gap-cdh}}(1^\lambda)$$

$$x, y \leftarrow \mathbb{Z}_{p_0}, X := g^x, Y := g^y, Z \leftarrow \mathcal{S}^{\mathcal{DDH}}(g, X, Y)$$

If $Z = g^{xy}$ then return WIN else return LOSE.

We define the *Gap-CDH advantage of $\mathcal{S}$ over* **Grp** as:

$$\mathbf{Adv}_{\mathcal{S},\mathbb{G}_{p_0}}^{\text{gap-cdh}}(\lambda) \overset{\text{def}}{=} \Pr[\mathbf{Exprmt}_{\mathcal{S},\mathbb{G}_{p_0}}^{\text{gap-cdh}}(1^\lambda) \text{ returns WIN}].$$

We say that the Gap-CDH Assumption holds for **Grp** if, for any PPT algorithm $\mathcal{S}$, $\mathbf{Adv}_{\mathcal{S},\mathbb{G}_{p_0}}^{\text{gap-cdh}}(\lambda)$ is negligible in λ.

2.3 Key Encapsulation Mechanism [10, 1]

A *key encapsulation mechanism (KEM) KEM* is a triple of PPT algorithms (**KG**, **Enc**, **Dec**). **KG** is a key generator which returns a pair of a public key and a matching secret key (PK, SK) on an input λ. **Enc** is an encapsulation algorithm which, on an input PK, returns a pair (K, ψ), where K is a random string and ψ is an encapsulation of K. **Dec** is a decapsulation algorithm which, on an input (SK, ψ), returns the decapsulation $\widehat{K}$ of ψ. We require KEM to satisfy the completeness condition that the decapsulation $\widehat{K}$ of a consistently generated ciphertext ψ by **Enc** is equal to the original random string K with probability one. For this requirement, we simply force **Dec** deterministic.

2.4 Adaptive Chosen Ciphertext Attack on One-Wayness of KEM

An adversary $\mathcal{A}$ performs an *adaptive chosen ciphertext attack on one-wayness* of a KEM (called one-way-CCA2, for short) in the following way [1].

$$\mathbf{Exprmt}_{\mathcal{A},\text{KEM}}^{\text{ow-cca2}}(1^\lambda)$$

$$(pk, sk) \leftarrow \mathbf{KG}(1^\lambda), (K^*, \psi^*) \leftarrow \mathbf{Enc}(pk)$$

$$\widehat{K^*} \leftarrow \mathcal{A}^{\mathcal{DEC}(sk,\cdot)}(pk, \psi^*)$$

If $\widehat{K^*} = K^* \wedge \psi^* \notin \{\psi_i\}_{i=1}^{q_{dec}}$ then return WIN

else return LOSE.

In the above experiment, $\psi_i, i = 1, \ldots, q_{dec}$ mean ciphertexts for which $\mathcal{A}$ queries its decapsulation oracle $\mathcal{DEC}(sk, \cdot)$ for the answers. Here the number q_{dec} of queries is polynomial in k. Note that the challenge ciphertext ψ^* itself must not be queried to $\mathcal{DEC}(sk, \cdot)$, as is described $\psi^* \notin \{\psi_i\}_{i=1}^{q_{dec}}$ in the experiment.

We define *the one-way-CCA2 advantage of $\mathcal{A}$ over* KEM as:

$$\mathbf{Adv}_{\mathcal{A},\text{KEM}}^{\text{ow-cca2}}(\lambda) \overset{\text{def}}{=} \Pr[\mathbf{Exprmt}_{\mathcal{A},\text{KEM}}^{\text{ow-cca2}}(1^\lambda) \text{ returns WIN}].$$

We say that a KEM is secure against adaptive chosen ciphertext attacks against one-wayness (one-way-CCA2-secure, for short) if, for any PPT algorithm $\mathcal{A}$, $\mathbf{Adv}^{\mathrm{ow\text{-}cca2}}_{\mathcal{A},\mathrm{KEM}}(\lambda)$ is negligible in k.

Note that if a KEM is IND-CCA2 secure [5], then it is one-way-CCA2 secure. So IND-CCA2 security is a stronger notion than one-way-CCA2 security.

3   Our Generic Description of Embedding Method and Decentralized PKI

In this section, we describe our generic embedding method and a related decentralized PKI. We first state an assumption for the underlying network. Next, we explain a design principle of our embedding method. Then, we describe a related decentralized PKI.

Assumption for Underlying Network. Our decentralized PKI utilizes a public-key list that is public to the underlying network. The public-key list should be examined and maintained by all participants who are active in the network. The security of our decentralized PKI will partially rely on the ability of an underlying network to maintain the integrity of the public-key list.

3.1   Components and Procedures of Our Generic Decentralized PKI

Initiation. We start with at least n initiators. n must be equal to the number of guarantors for a candidate public-key owner. Each initiator generates a pair of public key and secret key and writes it to a public-key list.

Generation of Candidate Public-Key Owner's New Key. When a candidate public-key owner wants to join the underlying network, he executes the following.

1. Generate a secret key sk_0 by running $\mathbf{KG}(\lambda)$.
2. Compute a value of the one-way function at sk_0: $\overline{sk_0} := f(sk_0)$.
3. Put $I = \mathrm{ID}_{\mathrm{cand}} \parallel \mathrm{ID}_{\mathrm{grnt}_1} \parallel \cdots \parallel \mathrm{ID}_{\mathrm{grnt}_n} \parallel \overline{sk_0}$.
4. Put $pk' = I \parallel 00\cdots0$.
5. Apply the embedding algorithm to pk' to obtain (pk, sk).
6. Put $\mathrm{PK}_{\mathrm{cand}} = pk$, $\mathrm{SK}_{\mathrm{cand}} = sk$.

Identification of Candidate Public-Key Owner

1. A verifier generates a random challenge according to a challenge-and-response identification protocol and send it to the candidate public-key owner.
2. Receiving the random challenge, the candidate public-key owner generates a response according to the protocol, and send it to the verifier.
3. Receiving the response, the verifier verifies it and outputs accept or reject.

The above protocol is executed by all guarantors, $i = 1, \ldots, n$, and any participant who wants to verify the identity of other participant with whom the former communicates.

Local Update of Public-Key List. After finishing the above verification, one of the guarantors adds the newly generated public key PK_{cand} of the new participant to the public-key list.

Broadcast of the Updated Identity List. The same guarantor broadcasts the updated public-key list along the underlying network. Integrity of the public-key list is assured by the above assumption.

4   Instantiation

In this section, we instantiate our generic decentralized PKI by employing the RSA encryption as the first cryptosystem and the elliptic curve encryption as the second cryptosystem. We use the modified version [13] of the Lenstra's algorithm [14] as our main tool.

4.1   Components and Procedures of Our Decentralized PKI: Instantiation.

We assume that the Elliptic Curve Discrete-Logarithm problem for the employed cyclic group $\mathbb{G}_{p_0}$ of prime order p_0 and the Integer-Factorization problem for the employed RSA modulus N have almost the same difficulty ([20]).

Initiation. This phase is executed generically according to the description in Sect. 3.1.

Generation of Candidate Public-Key Owner's New Key.

1. Generate a prime p s.t. $|p| = \lambda_{RSA}/2$ at random.
2. Compute $P := g^p$ in $\mathbb{G}_{p_0}$.
3. Put $I = ID_{cand} \| ID_{grnt_1} \| \cdots \| ID_{grnt_n} \| P$.
4. Put $N' = I \| 00 \cdots 0$ s.t. $|N'| = \lambda_{RSA}$.
5. Apply the modified Lensra's algorithm to obtain (N, q) and (e, d) s.t. $ed = 1 \mod \phi(N)$.
6. Put $PK_{cand} = (N, e)$, $SK_{cand} = (q, d)$.

Identification of Candidate Public-Key Owner.

1. The verifier chooses a random exponent t from $\mathbb{Z}_{p_0}$, computes $h = P^t$ in $\mathbb{G}_{p_0}$, and send it to the candidate public-key owner.
2. Receiving the random challenge h, the candidate public-key owner computes $K' = h^q$ by using his secret key q, then compute its hash value $K = H_\mu(K)$ a response, and then send it to the verifier.

3. Receiving the response K, the verifier verifies it by examining the following equation, and outputs `accept` or `reject` accordingly.

$$K \stackrel{?}{=} H_\mu(g^{Nt}).$$

The above protocol is executed by all guarantors, $i = 1, \ldots, n$, and any participant who wants to verify the identity of other participant with whom the former communicates.

Correctness of the above protocol is assured by:

$$K = H_\mu(K') = H_\mu((P^t)^q) = H_\mu((P^q)^t) = H_\mu(((g^p)^q)^t) = H_\mu(g^{Nt}).$$

Note that we can view the above procedures as a hashed key encapsulation mechanism [1,10], h-EGKEM, via putting $g := P$, $x := q$ and $X = g^x = P^q := g^N$ in its algorithm described in Fig. 1.

Note that, for a prime q (a factor of the RSA modulus N) that is longer than p_0 (the equivalent prime order of the group $\mathbb{G}_{p_0}$ of the elliptic curve cryptography), *collision resistance of the secret keys* $x = q \bmod p_0$ is assured by *the Dirichlet's Theorem on Primes in Arithmetic Progressions*.

Note also that, if a candidate candidate public-key owner generate a modulus N illegally, say, $N = pqr$ (three factors), the candidate public-key owner merely weaken his security of RSA.

Key Generation
- **KG**: given λ as an input;
  - $(p_0, \mathbb{G}_{p_0}, g) \leftarrow \mathbf{Grp}(\lambda), x \leftarrow \mathbb{Z}_{p_0}, X := g^x$
  - $\mathrm{PK}_0 := (g, X), \mathrm{SK}_0 := (g, x)$, return $(\mathrm{PK}_0, \mathrm{SK}_0)$

Encapsulation
- **Enc**: given PK_0 as an input;
  - $a \leftarrow \mathbb{Z}_{p_0}, K' := X^a, K := H_\mu(K'), h := g^a, \psi := h$, return (K, ψ)

Decapsulation
- **Dec**: given SK_0 and $\psi = h$ as an input;
  - $\widehat{K'} := h^x$
  - $\widehat{K} := H_\mu(\widehat{K'})$, return $\widehat{K}$

Fig. 1. Our Hashed ElGamal KEM: h-EGKEM.

Local Update and Broadcast of Public-Key List. These phases are executed generically according to the description in Sect. 3.1.

4.2   Attack and Security in Our Instantiation

An attack to be considered on the above verification procedure is an impersonation by a cheating verifier. More precisely, in the phase that an honest public-key owner tries to prove that he knows a factoring of N to an honest verifier, a man-in-the-middle adversary can execute impersonation. In our case, the security against this attack is reduced to the one-way-CCA2 security of our h-EGKEM.

Theorem 1. *If the key encapsulation mechanism h-EGKEM is one-way-CCA2 secure, then the protocol of identification of a candidate public-key owner is secure against man-in-the-middle attacks of impersonation.*

Proof. Suppose that there is a PPT, man-in-the-middle adversary $\mathcal{M}$. Then, putting $g := P$ and $X := g^N$, we can make a PPT adversary that attacks on h-EGKEM and that has the same success probability.   □

Theorem 2. *The key encapsulation mechanism h-EGKEM is one-way-CCA2 secure based on the Gap-CDH assumption for* **Grp** *in the random oracle model. More precisely, for any PPT one-way-CCA2 adversary $\mathcal{A}$ on h-EGKEM, and assuming that $\mathcal{A}$ issues a hash query every time when $\mathcal{A}$ computes a hash value, there exist a PPT Gap-CDH problem solver $\mathcal{S}$ on* **Grp** *which satisfies the following tight reduction.*

$$\mathbf{Adv}^{ow\text{-}cca2}_{\mathcal{A},h\text{-}EGKEM}(\lambda) \leqslant \mathbf{Adv}^{gap\text{-}cdh}_{\mathcal{S},\mathbb{G}_{p_0}}(\lambda).$$

Proof. Employing any given adversary $\mathcal{A}$ on h-EGKEM as subroutine, we construct, in the random oracle model, a PPT Gap-CDH problem solver $\mathcal{S}$ as follows (see Fig. 2).

$\mathcal{S}$ is given a CDH problem instance (g, X, Y) as input. $\mathcal{S}$ initialize the hash table T, whose row consists of the format (h, K', K). $\mathcal{S}$ sets $\mathrm{PK}_0 := X$ and $\phi : * := Y$, where the latter is the challenge message that should be responded by $\mathcal{A}$. $\mathcal{S}$ invokes $\mathcal{A}$ on input $\mathrm{PK}_0 := X$ and $\phi : *$.

$\mathcal{S}$ must answer decapsulation queries and hash queries of $\mathcal{A}$. Those answers can be made as is in the Fig. 2.

Finally, when $\mathcal{A}$ responds an answer $\widehat{K}^*$, $\mathcal{S}$ can extract the answer $Z := K'$ of the CDH problem instance (g, X, Y).   □

4.3   Discussion

Multiple Identities. As we can see from our procedures, there is a possibility that more than one public key N are issued on the same ID_{cand}. That is, N_1 and N_2 ($N_1 \neq N_2$) both have the same ID_{cand}. It would be desirable if guarantors could control this phenomenon.

Given (g, X, Y) as input;

Initial Setting
- Initialize the hash table T
- $\mathrm{PK}_0 := (g, X)$, $\psi^* := Y$
- Invoke $\mathcal{A}$ on (PK_0, ψ^*)

Answering $\mathcal{A}$'s Queries and Extracting the Answer

- When $\mathcal{A}$ queries its decap. oracle $\mathcal{DEC}(\mathrm{SK}, \cdot)$ for the answer of $\psi = h$;
- If $\psi = \psi^*$, then $\widehat{K} := \bot$
- else if h is in T, then pick K in the same row, then $\widehat{K} := K$
- else if h is not in T, then search K' s.t. $\mathcal{DDH}(g, X, h, K') = 1$
- -- If there is such K', then pick K in the same row, $\widehat{K} := K$, $T := T \cup \{(h, K', K)\}$
- -- else $K' \xleftarrow{\$} \mathbb{G}_{p_0}$, $\widehat{K} := K := H(K')$, $T := T \cup \{(h, K', K)\}$
- Reply $\widehat{K}$ to $\mathcal{A}$

- When $\mathcal{A}$ queries its hash oracle $H(\cdot)$ for the hash value of K';
- If K' is in T, then pick K in the same row
- else if K' is not in T, search h s.t. $\mathcal{DDH}(g, X, h, K') = 1$
- -- If there is such h, then pick K in the same row, $T := T \cup \{(h, K', K)\}$
- -- else $K' \xleftarrow{\$} \mathbb{G}_{p_0}$, $K := H(K')$, $T := T \cup \{(h, K', K)\}$
- Reply K to $\mathcal{A}$

- When $\mathcal{A}$ responds $\widehat{K}^*$;
- Search K in T s.t. $K = \widehat{K}^*$, pick K' in the same row
- Return $Z := K'$

Fig. 2. A Gap-CDH problem solver $\mathcal{S}$ for the proof of Theorem 2.

Revocation. As is discussed for PGP and a web of trust, our decentralized PKI also has the problem revocation. A simple way to enable revocation is to maintain a revocation list as well as our public-key list. We have to rely on an assumption that the underlying network can maintain the integrity of the revocation list, too.

Anonymity of Guarantors. In the real world it might be better to anonymise guarantors. This is a matter to be pursued. Using an anonymous credential system can be considered to resolve this matter though it needs a (centralized) issuer of credentials.

5 Conclusions

We proposed an embedding method for a decentralized PKI; ID of a candidate public-key owner, IDs of his guarantors, and a public key of the second cryptosystem was embedded into a public key of the first cryptosystem. This embedding functions as a preventer of manipulations on public keys. But it prevents not only falsification but also impersonation. We realized our decentralized PKI concretely; we could embed a public key of an elliptic curve encryption into a modulus of an RSA encryption. We note that the resulting decentralized PKI can be used as an alternative of the RSA encryption and signature in SSL.

Acknowledgements. The third author was partially supported by the Invitation Programs for Foreign-based Researchers provided by the National Institute of Information and Communications Technology (NICT), Japan.

The first, second and forth authors were partially supported by the Bilateral Joint Research Projects/Seminars FY2014 by Japan Society for the Promotion of Science under the research project name "Computational Aspects of Mathematical Design and Analysis of Secure Communication Systems Based on Cryptographic Primitives", who appreciate sincere thanks for discussion with Sushmita Ruj in Indian Statistical Institute and Avishek Adhikari in University of Calcutta.

References

1. Anada, H., Arita, S.: Identification schemes from key encapsulation mechanisms. In: Nitaj, A., Pointcheval, D. (eds.) AFRICACRYPT 2011. LNCS, vol. 6737, pp. 59–76. Springer, Heidelberg (2011)

2. Andrychowicz, M., Dziembowski, S.: Distributed cryptography based on the proofs of work. Cryptology ePrint Archive, Report 2014/796 (2014). http://eprint.iacr.org/

3. Blaze, M., Feigenbaum, J., Lacy, J.: Decentralized Trust Management. In: Proceedings of the 1996 IEEE Symposium on Security and Privacy, pp. 164–173. IEEE Computer Society Press (1996)

4. Boneh, D., Franklin, M.: Identity-based encryption from the weil pairing. In: Kilian, J. (ed.) CRYPTO 2001. LNCS, vol. 2139, p. 213. Springer, Heidelberg (2001)

5. Cramer, R., Shoup, V.: A practical public key cryptosystem provably secure against adaptive chosen ciphertext attack. In: Krawczyk, H. (ed.) CRYPTO 1998. LNCS, vol. 1462, p. 13. Springer, Heidelberg (1998)

6. Dwork, C., Naor, M.: Pricing via processing or combatting junk mail. In: Brickell, E.F. (ed.) CRYPTO 1992. LNCS, vol. 740, pp. 139–147. Springer, Heidelberg (1993)

7. Fromknecht, C., Velicanu, D., Yakoubov, S.: A decentralized public key infrastructure with identity retention. Cryptology ePrint Archive, Report 2014/803 (2014). http://eprint.iacr.org/

8. Garman, C., Green, M., Miers, I.: Decentralized Anonymous Credentials. In: IACR Cryptology ePrint Archive vol. 2013, p. 622 (2013)

9. Graham, S.W., Shparlinski, I.E.: On RSA moduli with almost half of the bits prescribed. Discrete Appl. Math. **156**(16), 3150–3154 (2008)

10. Kiltz, E.: Chosen-ciphertext security from tag-based encryption. In: Halevi, S., Rabin, T. (eds.) TCC 2006. LNCS, vol. 3876, pp. 581–600. Springer, Heidelberg (2006)

11. Kitahara, M., Yasuda, T., Nishide, T., Sakurai, K.: Embedding method of owner's information into public key of RSA encryption and its application to digital rights management system. In: IPSJ SIG Technical report, vol. 2014-CSEC65, p.3. Information Processing Society of Japan (2014)

12. Kitahara, M., Nishide, T., Sakurai, K.: A method for embedding secret key information in RSA public key and its application. In: Proceedings of the Sixth International Conference on Innovative Mobile and Internet Services in Ubiquitous Computing, pp. 665–670. IEEE (2012)

13. Kitahara, M., Yasuda, T., Nishide, T., Sakurai, K.: Upper bound of the length of information embedd in RSA public key efficiently. In: AsiaPKC@AsiaCCS, pp. 33–38. ACM (2013)

14. Lenstra, A.K.: Generating RSA moduli with a predetermined portion. In: Ohta, K., Pei, D. (eds.) ASIACRYPT 1998. LNCS, vol. 1514, pp. 1–10. Springer, Heidelberg (1998)

15. Lewko, A., Waters, B.: Decentralizing attribute-based encryption. In: Paterson, K.G. (ed.) EUROCRYPT 2011. LNCS, vol. 6632, pp. 568–588. Springer, Heidelberg (2011)

16. Meng, X.: On RSA moduli with half of the bits prescribed. J. Number Theory **133**(1), 105–109 (2013)

17. Qiu, Q., Tang, Z., Li, F., Yu, Y.: A personal DRM scheme based on social trust. Chin. J. Electron. **21**(4), 719–724 (2012)

18. Sander, T., Ta-Shma, A.: Auditable, anonymous electronic cash (extended abstract). In: Wiener, M. (ed.) CRYPTO 1999. LNCS, vol. 1666, pp. 555–579. Springer, Heidelberg (1999)

19. Vanstone, S.A., Zuccherato, R.J.: Short RSA keys and their generation. J. Cryptology **8**(2), 101–114 (1995)

20. Yasuda, M., Shimoyama, T., Kogure, J., Izu, T.: On the strength comparison of the ECDLP and the IFP. In: Visconti, I., De Prisco, R. (eds.) SCN 2012. LNCS, vol. 7485, pp. 302–325. Springer, Heidelberg (2012)

21. Zimmermann, P., and associates llc (2014). http://www.philzimmermann.com/EN/background/index.html. Accessed 20 September 2014

Diversification of System Calls in Linux Binaries

Sampsa Rauti[(✉)], Samuel Laurén, Shohreh Hosseinzadeh,
Jari-Matti Mäkelä, Sami Hyrynsalmi, and Ville Leppänen

University of Turku, 20014 Turku, Finland
{sjprau,smrlau,shohos,jmjmak,sthyry,villep}@utu.fi

Abstract. This paper studies the idea of using large-scale diversification to protect operating systems and make malware ineffective. The idea is to first diversify the system call interface on a specific computer so that it becomes very challenging for a piece of malware to access resources, and to combine this with the recursive diversification of system library routines indirectly invoking system calls. Because of this unique diversification (i.e. a unique mapping of system call numbers), a large group of computers would have the same functionality but differently diversified software layers and user applications. A malicious program now becomes incompatible with its environment. The basic flaw of operating system monoculture – the vulnerability of all software to the same attacks – would be fixed this way.

Specifically, we analyze the presence of system calls in the ELF binaries. We study the locations of system calls in the software layers of Linux and examine how many binaries in the whole system use system calls. Additionally, we discuss the different ways system calls are coded in ELF binaries and the challenges this causes for the diversification process. Also, we present a diversification tool and suggest several solutions to overcome the difficulties faced in system call diversification. The amount of problematic system calls is small, and our diversification tool manages to diversify the clear majority of system calls present in standard-like Linux configurations. For diversifying all the remaining system calls, we consider several possible approaches.

1   Introduction

Malicious software, or malware, is one of the main security challenges in today's information security. Malware uses knowledge about the identical interfaces of operating systems to achieve its goals. To access resources on a computer, a malicious program has to know the interface that provides the resources. Because of the prevailing operating system monoculture, an adversary can create a single malicious program that works for hundreds of millions of computers that use the same operating system.

The operation of malware would become considerably more difficult if it could not issue system calls and successfully use resources on a computer. Therefore,

---

This research has been funded by MATINE project 3301.

our approach is to make malware ineffective by using large-scale system call diversification. All software on a certain computer can be diversified so that it becomes very challenging for a malicious program to access resources. As a result of this, a large set of computers would have exactly the same functionality but differently diversified software layers and user applications. Because of the diversification of software layers, a piece of malware no more knows the "language" used in the system and becomes incompatible with its environment.

Even if a piece of malware would be able to find out how the resources are accessed on one computer, large-scale attacks are still very difficult, as malware knows the secret of applied diversification on one computer only. A costly analysis needs to be separately performed on each host. In other words, the diversification can be seen as a computer-specific secret. Our diversification scheme does not affect the work of a software developer because it is done on the binary level. Only some problematic cases, often found in libraries, may have to be dealt on source code level. The diversification of binaries does not change the user experience in any way, because the semantics and functionality of programs are preserved. Changing the system call numbers or mangled names of library routines does not affect performance either.

One part of this diversification process is to diversify the system calls that are used to access resources in an operating system. The idea of this paper is to study the diversification of system calls in Linux (see Sect. 2). More specifically, we discuss the challenge of recognizing and diversifying the direct system calls in Linux ELF (Executable and Linkable Format) binaries. We present a method for API diversification and describe a concrete tool used to achieve diversification. Based on the tests performed with this tool, we also present an experimental study of presence and distribution of system calls in Linux ELF binaries. We also discuss several possible solutions for the challenges faced when recognizing the system calls from binary files. By using our tool and these methods together, we believe 100 % accuracy in system call diversification can be achieved.

1.1 Our Goal

In this paper, our first goal is to characterize where the system calls are applied, what can be said about their distribution in the different software layers of an operating system that will be diversified. For example, how many binaries perform system calls and how often system calls are used in libraries and user applications. It obvious that libraries perform the majority of system calls, but statistics of the role of direct system calls in application binaries is also important for considering schemes for securing a whole system. If applications often perform such calls, an automatic binary transformation tool seems necessary instead of just recompiling the diversified libraries from sources.

As the second goal, we want to find out how system calls are coded in ELF binaries. Understanding this is essential to successfully diversify all the system calls in binary files. We study the different ways system calls can be coded in ELF files. We chose to diversify applications and libraries on binary level rather than on source code level. This way, we do not need to have diversifiers for several high-level programming languages. It is also easier to handle updates that arrive in binary form. Moreover, commercial applications or device drivers are usually not available in source code form.

The third goal is to present different ways to diversify the system calls in ELF binaries. Because system calls are not always presented in binaries in the same straightforward way, we may need to make use of several approaches to ensure that the system call diversification scheme is as perfect as possible.

1.2   Contributions and Structure of the Paper

To the best of our knowledge, our work is the first detailed study of static system call diversification in Linux ELF binaries. We also provide a concrete implementation of an automatic diversifier tool. This paper makes the idea of static system call number remapping, previously presented by Chew and Song [3], more concrete. We apply the system call diversification in practice to solve a more general problem of rendering malware useless. Unlike some earlier work on system call diversification (see for example [10,13]), our tool performs the diversification statically after compilation and thus does not introduce any runtime performance loss. Compared to the earlier similar solutions, our approach is also more system-wide because it also diversifies the kernel. We implement a proof of concept diversifier and test our solution using two popular Linux distributions. We provide a detailed description of this tool and also present results on its accuracy.

One of the contributions of this paper is an empirical study of the presence of system calls in ELF binaries. We conclude that most system calls in two Linux distributions we tested are found in libraries like `libc`. Also, the vast majority of binaries do not contain direct system calls at all, which makes their diversification much easier. Another contribution of this paper is to present various solutions to diversify the differently coded system calls in ELF binaries. We believe that even though there are many challenges, diversifying 100 % of system calls is feasible by applying the different methods we consider in Sect. 6.

The rest of the paper is organized as follows. Section 2 presents an overview of our API diversification method for operating system protection. Section 3 covers some basic concepts needed to understand our diversification method and experimental tool in more concrete sense, like Linux layer structure and ELF binary files.

Section 4 discusses coding of system calls in ELF binaries and the challenges faced when trying to identify and diversify them. Section 5 presents our study of presence of system calls in the ELF binaries. We discuss our diversifier tool implementation and the experimental setting. We also present some results; where in the system are system calls located, and how many binaries contain direct system calls in the tested Linux distributions? Many possible solutions for these challenges and achieving 100 % diversification accuracy are covered in Sect. 6. Section 7 concludes the paper.

2   An Overview of Our API Diversification Scheme and Threat Scenarios

2.1   Our API Diversification Scheme

An operating system provides a variety of services that user application can utilize in a shared manner [25]. Therefore, the operating system and its system call interface can be thought as an abstraction between user applications and the services provided by hardware of a computer. System calls are a fundamental set of services in an operating system [23].

In order to interact with its environment, an application needs to use system calls. This can be achieved by either calling the system call interface directly or using libraries that provide wrappers for system calls. Therefore, in our view preventing a piece of malware from accessing the system call interface consists of two separate parts:

1. Diversify the system call numbers.
2. Diversify the system call implementations in the kernel and all the functions calling them directly or indirectly.

Diversification refers to meaning-preserving mapping in a programming language. That is, the program code is transformed to a different form, but its semantics are preserved so that the user of the program experiences no visible change when using the program. In this paper, system call diversification simply refers to changing the mapping of system call numbers.

The first part of our scheme means that we change the system call numbers defined in the operating system kernel. As a consequence, all code in libraries and user applications that call system calls directly using these numbers must be diversified accordingly as well, or they will stop working.

On the other hand, when a system call is not invoked directly, the call passes through several software layers before it reaches the system call implementation. In order to prevent a piece of malware from invoking system calls, we have to recursively diversify all the functions that make these system calls. We refer to the set of these functions directly or indirectly calling the system call implementations as the *transitive closure*. All these functions must be diversified (by changing their function signature, for example) to prevent a piece malware from using them to access a computer's resources. Trusted applications that are diversified correctly can still access the resources.

This paper deals with changing the system call numbers and diversifying all the direct system calls accordingly in user applications and libraries. Another part of our protection scheme, diversifying the transitive closure, has already been discussed in our other publication [17] and is not covered in this paper.

The trusted software layers and user applications are diversified with a secret diversification function which makes them compatible with new system call numbers defined in the kernel. As a result, the entry points that lead to the system calls are diversified in the whole system, preventing malware and any untrusted applications from using them to invoke system calls.

2.2   Threat Scenarios

Our solution is meant to protect computers from the malicious code that is either executed as its own process or as a part of another executing process. In the second case, the malicious code can observe the system calls its host program invokes and gradually learn the system call mappings used in the system. However, this would still require advanced analysis.

Without observing any program's actions, it is very difficult to guess correct system call numbers. Linux currently only uses around 320 system calls and their numbers are 32-bit, so there is a very high chance that our function can map calls so that malware cannot make valid system calls even by mistake. Of course, the mapping function should be designed so that it never maps an ID to itself. Illegal system calls could also be logged for further inspection. A program that randomly tries invoking large amount of system calls can be seen as suspicious.

It is important to note that in our approach we assume that a piece of malware has no access to the file system, for example, and has no way to analyze files or our secret diversification function that has been applied to binary files. File system access requires using system calls and thus an external malware does not have an easy access to the file system.

Our approach adopts a proactive view by preventing malware from harmfully interacting with its environment before its execution. As the number of malicious programs keeps growing and they keep transforming, traditional fingerprint-based antivirus software is becoming increasingly inefficient in the fight against malware [1]. Also, antivirus programs often only detect the threats they are already aware of. This is why complementary approaches are needed.

Together with diversifying the transitive closure, the system call number diversification should make it much harder for malicious programs from opening any resources on a computer. Untrusted programs do not know either the system call numbers nor the names of functions in other applications or libraries that lead to system calls.

3   Linux Layer Structure and ELF Binary Files

3.1   Linux Layer Structure

Linux system calls are implemented as named routines in the operating system kernel. In the user-space facing system call ABI, each call routine corresponds to a kernel-defined system call number. There are around 320 system calls in Linux, each with its own number [27].

The software layer structure of Linux is very roughly illustrated in Fig. 1. Linux contains wrappers in order to make it easier to issue system calls, which are implemented in different parts of kernel. However, as explained in Sect. 2, in this paper we are only interested in the cases where either libraries or user applications call the system call interface directly. We aim at recognizing and diversifying all these entry points in the binaries of the whole system.

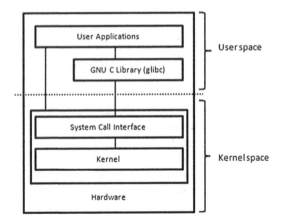

Fig. 1. Linux layer structure.

The most important library making the system calls in Linux is libc, which most user applications use to access operating system services. However, several other libraries, many command line tools and even some web browsers also make direct system calls. The distribution of system calls will be examined closer in two test environments in Sect. 4.2.

3.2 Structure of ELF Files

The Executable and Linkable Format (ELF) is a standard file format for executables, object code and shared libraries. It is used on many different platforms and in several operating systems. It is the standard binary file format for Unix-like systems.

Figure 2 shows the structure of an ELF file. An ELF header in the beginning of the ELF file describes the file's organization. The header also contains information on object file type and the instruction set architecture, for example.

Sections contain lots of miscellaneous information: instructions, data, symbol table, relocation information etc. Some sections are special, like sections for uninitialized and initialized data, sections holding debug information and comments, sections for read-only data and strings and sections for symbol tables [6].

A program header table is an optional part of ELF files. It tells the system how to create a process image and execute the program. Relocatable files do not need a process image. Information in a section header table describes the ELF file's sections. Each section has its own entry in the table. Every entry provides information such as the section name, the section size etc. Files used during linking always have a section header table, otherwise it is optional.

The sections in ELF files have a type attribute. For example, PROGBITS-type sections are reserved for program-specific data. This can be either executable code or other data. In order to diversify the system calls in ELF files, our diver-

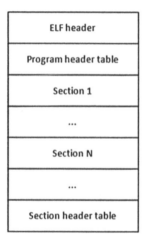

Fig. 2. ELF file structure.

sifier tool analyzes the PROGBITS-type sections that are marked the be executable with SHF_EXECINSTR flag.

4   On Coding of System Calls in ELF Files

For the purposes of this paper, a system call can be seen consisting of two separate phases. As we have already seen, the first phase puts the value of the system call into a predefined register. The second phase transfers control to the operating system's system call handler. The exact mechanism for this is architecture and operating system dependent. On x86-64 Linux system calls are made using the SYSCALL instruction. For instance, the following commands are used to invoke sys_write (system call number 1):

```
mov $1, %eax
syscall
```

However, there are several factors that make identifying the system calls more difficult. For example, there might be a jump command between the two phases:

```
cmp $1, %eax
je equal
mov $0, %eax
jmp over
equal:
 mov $1, %eax
over:
 syscall
```

Here, system call 1 is invoked if the value in EAX register is 1, otherwise system call 0 is invoked. It is much harder for the analyzer to deduce what system call will be invoked.

Also, when the analysis is restricted to simple mov commands that directly move a value to a register where a system number is stored, many problems arise. For example, this leaves out a complicated setting where a value is moved to a register indirectly through other registers. Also, it seems compilers often write the binary so that the register value is first put to the memory and then into an appropriate register. These kind of indirect approaches are difficult to analyze without tracing the control flow of the program. There might also be other commands affecting the register values before the system call is made, say incrementing EAX register, for example.

Of course, it is interesting to ask whether these different ways to code the system calls in binaries have any visible reasons in the source code. It is pretty clear some of them do. For example, the jump we saw in the example earlier is probably created as a result of a conditional statement, like an if-statement, in the source code. It also seems that loop structures in the source code often create coding in the binary where the two phases the system call consists of are not consecutive. For example, we noticed that setting register values indirectly through other registers can be a result of a loop structure in the source code. Use of function pointers in the source code probably also affects the coding of the binary file.

On the other hand, all the binary codings of system calls do not seem to have clear explanations in the source code. For example, the fact that register values are sometimes circulated through memory seems very arbitrary and apparently associated with optimization made by the compiler. We noticed the version of compiler may greatly affect the coding of system calls in the binaries it creates. The compiler configuration and compiling environment probably also have an effect on this. Many compilers have switches that can be used to configure the level of optimization.

One more noteworthy problem is alignment, that is, the way data is arranged and accessed. Because our tool disassembles the binary file in a straightforward manner by processing it from the beginning to the end, any excessive data or empty space between instructions (zero bytes) lead to failure. There is no reliable way to detect when this failure takes place. In the worst case scenario, an erroneous system call could be found from a binary file. In practice, however, compilers usually should not produce this kind of faulty program code. Programs that somehow determine the system call at runtime based on the user input would naturally also be problematic, as it is impossible to identify the correct system call number in this case with static analysis. Some commercial products may also use different obfuscation methods in their binaries, which ironically makes our diversification task much harder. Moreover, different kinds of self-modifying programs are always difficult to handle for diversifiers.

5   Experimental Study on the Presence of System Calls in Linux Binaries

To find out how system calls are distributed between different parts and binaries in Linux distributions, we performed an experimental study on the presence of system calls in all binaries in tested systems.

5.1   Settings of Studied Linux Environments

We conducted our study using 64-bit Fedora and Gentoo Linux distributions. Fedora Linux was selected because it provided a full-fledged desktop environment with all the associated software out of the box. In contrast, the Gentoo installation we used was fairly minimal with only a few packages outside of the `default/linux/amd64/13.0` profile. Additionally, we conducted separate tests with C standard library implementation `glibc`. We concentrated on `glibc` specifically because it is one of the main libraries containing system calls. Characterization of test environment was deemed to be especially important since there are multiple factors that can cause results to vary considerably. When analyzing binary files, the most obvious source of differentiation is the compiler used to create the said binaries. Using a different compiler and even different version of the same compiler can lead to differences in produced binaries, which in turn might affect the results our system call analysis. Aside from the compiler version, the compiler settings used to produce the binaries play a central role. For example, we noticed that our analysis tool performed radically worse when the binaries were compiled with no optimizations at all. Because our tools performance is highly dependent upon how the register allocation is done, all compiler settings that might affect this can potentially alter our results.

Because we are analyzing all the installed applications and shared libraries, precisely describing our experimental setting would require us to list all the specific versions of the installed packages including potential distribution specific changes. Also we would have to record detailed information about the build environment, including compiler versions and settings.

Because of how compiler dependent our analysis is, conducting experimental studies using Linux distributions with binary based packaging, presents us with certain challenges. As we are using precompiled binaries we cannot know how they were produced, let alone control the specific compiler settings. This might make precisely replicating our results more complicated, but at the same time, it means that our test environment resembles a real-world test scenario more closely, since we assume that a typical end-user does not have control over how their binaries have been produced.

We used 64-bit Fedora Linux version 20 based installation (kernel version 3.14.4) as our test platform. We had also installed various other applications and libraries that were needed during the development of our analysis software. Of course, knowing only the release number of the distribution leaves out many details about the system, since the software has received various updates during the release cycle. The Gentoo installation (kernel version 3.6.0) was considerably more minimal and contained relatively small number of packages. No desktop environment was installed for this system.

5.2 Distribution of System Calls in Binaries

We analyzed the direct system calls found in the binary files of two Linux distributions, Fedora and Gentoo. In addition to amount and distribution of system calls, we also wanted to see how well our diversification tool could identify the system calls in binaries.

In Fedora, 5649 binaries were analyzed. Only 18 of those contained any system calls. These binaries are shown in Table 1. For each binary, the amount of system calls successfully identified by our tool, the amount of unidentified calls and the total amount of system calls in that binary are shown. In this context, identifying refers to successfully recognizing the correct system call number. In unidentified cases, we find a SYSCALL command but cannot recognize a system call number associated with it.

Table 1. System calls found in binaries of Linux Fedora distribution.

Binary path	Identified	Not identified	Total calls
/lib64/libunwind-x86_64.so.8.0.1	1	0	1
/lib64/libcrypt-2.18.so	1	0	1
/lib64/librt-2.18.so	24	5	29
/lib64/libc-2.18.so	394	35	429
/lib64/libanl-2.18.so	3	3	6
/lib64/libnss_db-2.18.so	1	0	1
/lib64/libgomp.so.1.0.0	17	13	30
/lib64/libaio.so.1.0.0	5	0	5
/lib64/libaio.so.1.0.1	5	0	5
/lib64/ld-2.18.so	31	5	36
/lib64/libunwind.so.8.0.1	2	0	2
/lib64/rtkaio/librtkaio-2.18.so	45	14	59
/lib64/xulrunner/crashreporter	0	6	6
/lib64/xulrunner/libxul.so	3	56	59
/lib64/firefox/crashreporter	0	6	6
/lib64/firefox/libxul.so	3	56	59
/sbin/ldconfig	109	9	118
/sbin/sln	79	8	87
Total	**723**	**216**	**939**

We can see that large amount of system calls is located in `libc`, the C standard library. With this library, our diversifier performs well, recognizing over 90 % of calls. A few other libraries like `rtkaio` – a library used for asynchronous I/O – also contain direct system calls.

Some command line tools like `ld`, a dynamic linker, `ldconfig`, which is used configure dynamic linker run-time bindings, and `sln`, symbolic link creator also seem to make quite many system calls. Our tool performs well with all of these binaries.

There are also a few problematic binaries, like Mozilla's `libxul` library in this case. This binary encodes system calls in difficult ways, as it seems to favor using intermediate registers for passing values instead of direct assignments to appropriate registers to make a system call (see Sect. 4). Because of this problematic library, which appears in the system two times, our tool identifies about 70 % of the system calls in binaries in this system.

In the same way as with Fedora, binaries in Gentoo distribution were also analyzed. Only 9 of 569 binaries contained direct system calls. These binaries are shown in Table 2. There was no desktop environment installed in this system, which explains the smaller amount of binaries. Most system calls are in libraries, and about half of system calls are made in `libc`. Our tool performs well in this distribution, identifying 92 % of the system calls.

We can conclude from these results that even in rather large standard distributions, there are very few binaries with direct system calls. User application very rarely make direct system calls and use `libc` instead. This makes our diversification task easier. Especially in restricted environments with only a few user programs our diversifier would perform well.

However, even though our tool can recognize the system calls pretty well, there are still some problematic cases that were not identified correctly. We will look at some solutions to these problems in Sect. 6.

Table 2. System calls found in binaries of Linux Gentoo distribution.

Binary path	Identified	Not identified	Total calls
/lib64/libanl-2.17.so	1	5	6
/lib64/libc-2.17.so	411	19	430
/lib64/librt-2.17.so	24	5	29
/lib64/libnss_db-2.17.so	1	0	1
/lib64/ld-2.17.so	32	5	37
/lib64/libcrypt-2.17.so	3	0	3
/lib64/libpthread-2.17.so	144	23	167
/sbin/sln	84	5	89
/sbin/ldconfig	102	5	107
Total	**802**	**67**	**869**

5.3   A Closer Look at Diversification of System Calls in Libc

In the experiments, we also analyzed and diversified different versions of libc library. Because most of the system calls are located in standard libraries and not in the application's code (see Sect. 4.2), these libraries are a good target for analysis. User applications do not usually have any need to use the system call interface directly, and invoking the system calls indirectly using a standard library makes the application less dependent on certain operating system version by including an additional abstraction level.

We studied glibc version 686554bff63dff0f8b20c84e9bdca45e643f9d9c, which we compiled with gcc (GCC) 4.8.2 20131212 (Red Hat 4.8.2-7). This library was analyzed with our diversification tool both on Fedora 20 (64-bit) and on Gentoo. The results for both distributions are shown in Table 3.

Table 3. The system calls found in libc.

Distribution	Identified	Not identified	Total calls
Fedora	*380*	*35*	*415*
Gentoo	*398*	*18*	*416*

As, we can see, over 90 % of calls were successfully diversified in Fedora and over 95 % in Gentoo. Identifying the system calls succeeded well in our tests, because many routines in standard libraries are just simple wrappers for system calls. These routines simply take a set of parameters for system calls, put them into appropriate registers so that the system call can use them, and then invoke the system call with a predefined number.

However, some of the routines in libc include conditional execution of system calls. That is, these routines decide the system call to be invoked based on some external factor or invoke system calls as a part of a loop. These cases are of course more problematic. Additionally, standard libraries also usually contain routines that can be used to invoke an arbitrary system call. These routines take a system call number as their parameter, which makes it hard to rewrite them statically.

We also studied systems calls in musl, an implementation of C standard library. The version we used was 8a2d8719873a46d5cc5c54e688d47ea134c67c84. This library was compiled with several different optimization settings, which demonstrates well the effects that optimization performed by the compiler may have on our diversifier tool. musl was tested on Fedora and the same compiler was used as in previous tests with libc.

musl was compiled using several different optimization options. Results are shown in Table 4. Switch -O0 means no optimization. As shown in Table 4 results for the library with no optimization are really bad. However, with all optimized binaries, our tool performs well, identifying over 90 % of system calls. The switch

-Os means the binary is optimized for size. -O1, -O2 and -O3 refer to increasing level of optimization. Generally, it seems that optimization performed by a compiler is a big advantage for our tool.

We can also see that there are much less system calls in total in the binary that has not been optimized. It seems that leaving the optimization out results in more calls to the functions wrapping system calls in libc instead of inlining the syscall instruction in each function. This also explains why our tool does not perform that well with non-optimized binaries. This is because the wrappers circulate the system call number through the stack instead of putting it directly to a register, which causes problems for our diversifier tool.

It would be interesting to test more libc implementations with several versions of different compilers and see how well our tool performs when analyzing the compiled binaries.

Table 4. The effects of compiler optimization to diversification of musl library.

Optimization	Identified	Not identified	Total calls
-O0	7	291	298
-Os	372	27	399
-O1	379	27	406
-O2	373	29	402
-O3	375	31	406

6   Methods for System Call Diversification

Basically, the idea of system call diversification simply means that a system call number is replaced with another number. The easiest part in diversifying system calls is changing their numbers in the kernel code. How this is done depends on the architecture and kernel version. In x86-64 architecture, for example, the system call numbers are listed in arch/x86/syscalls/syscall_64.tbl. On compiling, definitions in this file are propagated to several header files.

The part that causes more problems is changing the system calls numbers in all binary files to correspond the new numbers we have set in the kernel. As we have seen, the system calls take place in several phases in the binary code, which causes several problems described in the previous section. In this section, we take a look at our own diversification tool and then present some solutions that would help to increase its accuracy to 100 %.

6.1   Our Tool and Recognizing the System Calls

To demonstrate the feasibility of system call diversification, we implemented an experimental diversification tool as a proof of concept. Our tool rewrites the system calls in x86-64 ELF-64 binaries by making use of a simple linear sweep algorithm [22]. This is a straightforward disassembly method that decodes everything appearing in sections of the executable that are typically reserved for machine code. We limit the analysis to executable PROGBITS-type sections in ELF binaries. The diversification is done after compile time before execution.

The tool tries to find system calls by walking through the program code sections linearly. It looks for SYSCALL commands used in x86-64 architecture. When such a command is found, it starts searching the system call number associated with this call. This is done by backtracking from the location of SYSCALL command and trying to find the command where the system call number is set. As the number of system call to be invoked is put into a register, our tool looks for commands that change values of RAX, EAX, AX, AH or AL registers.

Therefore, our diversifer tool uses the following two methods to identify the system calls:

1. *Recognize two consecutive phases.* As we have seen, in the simplest scenario we simply recognize two consecutive phases of the system call in the binary code. However, when there are other commands between these phases, this trivial approach will not work.
2. *Recognize two phases with a gap.* When the two phases of the system call are not consecutive, we have to find the command making the system call first and then backtrack to the call that puts the system call number into a register. Here, the potential jumps between the two phases should be somehow recognized and handled.

Table 5. Amount of gaps in system calls in Fedora and Gentoo.

Gap size	Fedora	Gentoo
0	722	736
1	34	29
2	15	19
3	17	8
4	42	6
5	4	7
Over 5	0	2
Total	834	807

Table 5 shows the amount of gaps found in Fedora and Gentoo distributions. In Fedora, 87 % of the system calls have no gaps and in Gentoo, 91 % of the system calls have no gaps. The vast majority of system calls are trivial in this sense. Fedora did not have any gaps bigger than 5 instructions. Gentoo has only two of these, the largest gap being 9 instructions.

When testing our tool, we used SysTap, a tool for real time analysis of running processes in user and kernel spaces. This way, we made sure that the programs that had been diversified with our tool worked correctly during this dynamic instrumentation – that is, they used the new system call numbers changed by our diversifier tool.

As seen from the results in Sect. 5.2, our tool still needs improvement. Next, we will take a look at many approaches that could be used to further improve our diversifier tool.

6.2 Challenges

During the development of our tool, we identified some problems in our approach to system call identification. Most of the challenges were linked to the use of a simple linear-sweep based disassembly algorithm. These problems are well known in literature and have for example been discussed by Schwarz et al. [22].

Our algorithm works by first disassembling the executable PROGBITS-type sections of the ELF files. After the initial disassembly we scan the binary for x86-64 specific SYSCALL instructions. If we find such instructions we stop the disassembly process and start backtracking. The backtracking process starts looking for preceding instructions that could assign a value to one of the accumulator registers RAX, EAX, AX, AH, AL. These registers are used for storing the system call number, and as our intention is to patch the system call numbers, we have to figure out what the original system call number was. The backtracking process might fail if it finds a control flow instruction or an instruction that might modify one of the accumulators in an unknown way. Also, we can only identify the system call numbers if the assignments assigning them use only immediate values for storing the numbers. This leaves out all cases where the system call number is assigned indirectly from memory or from another register.

There are various problems in this approach. First of all, the linear-sweep based disassembly process is susceptible to several hard to identify errors. If there is empty space or program data between instructions this might cause the disassembly to produce incorrect results. The fact that the malfunction might not be identified makes the situation even worse, the disassembly process might continue as if nothing unusual had happened producing false instructions or it might stop if the disassembler confronts an invalid instruction.

To solve this problem we would have to utilize a recursive disassembly algorithm. Such algorithm would first start the disassembly from a prespecified offset and continue until a control flow instruction is found. Then the algorithm would have to figure out the possible targets of the control flow instruction and continue the decoding process from there. This approach would solve some of the problems but increases the complexity of the tool considerably, because the recursive approach requires us to figure out the potential control flow paths. For example, if a jump target is specified to be in a certain register we have to figure out how that register gets its value. We would have to perform some form of data-flow analysis to be able to handle these kinds of indirect jumps. The situation is even more complicated. In order to build a control-flow graph we need a data-flow graph which in turn requires a control-flow graph to be in place. Henrik Theiling [26] refers to this as a chicken and egg problem.

Reconstruction of control-flow graphs from binaries has been widely studied. Theiling presented an bottom-up approach for the flow graph approximation [26]. Cooper et al. introduced an algorithm for building a control flow graph approximation and then refines it [7]. Kinder et al. devised an abstract interpretation-based framework that produces the most precise overapproximation of the control-flow graph with respect to the used abstract domain [12].

Performing a proper data-flow analysis would also help us figure out how the system call numbers are assigned. With a data-flow graph in place we could backtrack through the indirect assignments and find out how the registers' values are formed. The control-flow graph would also help us solve challenges like the ones presented in Sect. 6.2.

6.3   Methods to Improve System Call Diversification

There are several methods we can use to improve the system call diversifier so that it can handle the remaining problematic system calls:

1. *Include the diversification calculation in the binary.* We can embed the diversification calculation – that is, the calculation determining the new diversified system call number – somewhere in the binary. However, this might cause some relocation problems. This approach would also make a potential leakage of diversified code quite dangerous. As a result, the secret new system call mappings defined by the diversification function would be revealed. However, considering we assume a piece of malware should not be able to perform system calls and get access to the file system in order to analyze the diversified binary code, this approach should be pretty safe. If the malware finds some way to get into the memory space of an executing process, however, it can try to analyze the meanings of diversified system calls.

2. *Change compiler or compiler switch settings.* Sometimes the order of commands in the machine code can be changed as an optimization made by the compiler and system call numbers can be circulated through registers and memory before they are put in the appropriate registers in order to make a system call. This could probably often be prevented with correct compiler settings. This method naturally has some problems. For example, we cannot expect all software developers – like the major browser suppliers – to compile their binaries for us using a certain compiler or some specific configuration. Many open source applications could be compiled from source codes on the target machine using a specific compiler, though.

3. *Rewriting parts of the source code.* Many problems faced in the binary code diversification process can probably be traced back to the source code. As a consequence, rewriting some of the source code differently might solve the problem. In many systems, rewriting would cause too much work if it would be done for all user applications. However, it is a possible solution for example for many standard libraries like libc.

4. *Hard-code the diversification.* For some of the most problematic code sections, the diversification could be hard-coded in binaries. While not usually a preferable solution, this could be done for some standard parts of the Linux operating system.

The various methods for more accurate diversification we have discussed in this section all have some challenges. However, they can still be successfully used at least for some standard set of libraries and applications. Also, these methods are very feasible in some more or less restricted environments. Systems used in industry or military and embedded systems in general are easier to adapt this way, and security is often a major concern for these systems. In these systems, we believe we can reach 100 % diversification accuracy.

7   Related Work

To better position our work in the Linux based software ecosystem, we shortly discuss existing related technologies in this section and provide a summary of related research in the field.

7.1   Related Technologies

The traditional UNIX point of view to security is based on a discretionary user and group based restriction of file/process privileges to perform operations, with the exception of a superuser with access to all such resources. The system is binary in nature, i.e. an operation is either prohibited or allowed to full extent. It was later extended with more flexible *access control lists* (ACL) and policy based controlling mechanisms (PolKit) [9].

Another way to control actions is sandboxing. The *chroot* mechanism [9] provides an isolated view of the file system. As the superuser is allowed to break out from the chroot "jail", local root exploits pose a security threat. The chroot also has other attack vectors such as the ptrace system call. For mount points there is a noexec flag that prevents the execution of binaries from that file system, but will not prevent interpreting scripts from such locations.

Sandboxing is not limited to file systems. For example, Linux provides namespace isolation for process identifiers, network interfaces, firewall rules, routing, and inter-process communication and a related container framework (LXC) [11]. Other types of resource limits can be imposed via the ulimits, sysctl, and control group interfaces. The *Linux Secure Computing Mode* (seccomp) mechanism can be used to isolate a process from system on system call level with only a very limited interface to outside system via already-open file descriptors.

A more disciplined approach to security is mandatory access control. Frameworks such as SELinux and AppArmor introduce a policy based mechanism to security with modular hooks directly on kernel level. The policy is enforced by kernel, but its definition comes from userspace, which also deals with logging and informing about policy violations. The frameworks enable a fine-grained policy control with a small runtime overhead, and while the framework can be transparently set up on a system without changing the userspace applications, programs that are not designed for such a rigorous enforcement of permissions may trigger false warnings with careless resource usage patterns.

7.2   Related Research

In 1993, Cohen [4] introduced a general method of program diversity to protect operating systems. He proposed the exploitation of the evolutionary defenses to produce more complex and unique program instances. The higher complexity of the program increases the work an attacker has to do to understand the program's behavior in order to perform an attack. Moreover, with the uniqueness of the program, the attacker is no longer able to impact a substantial number of program versions with a single attack. This way, the attacker is forced to design individual attack versions for each of the program instances.

According to the classification of Collberg [5], there are various obfuscation techniques available: code obfuscation, data obfuscation, layout obfuscation, and preventing transformation. Based on the distribution format of the software, different techniques are applicable [15]. In [15] these techniques are used at the binary level.

Binary obfuscation makes reverse engineering the software significantly harder. In the reverse engineering process the machine code is disassembled into assembly code. The assembly code is then decompiled and the high-level code is recovered [14]. Linn and Debray [14] propose adding "junk bytes" into the instructions where the disassembler is expecting code. This method can disrupt the disassembly process to produce disassembly errors or at least make disassembled code more complex. The candidate instruction code should be incomplete (to confuse disassembler) and unreachable during the execution (to save the program's semantics). In [16], similar to [14], the goal is to make the disassembly of the machine code and thus the reverse engineering harder. They propose two different obfuscation techniques that make it more difficult for the disassembler to find the actual control flow of the binary code. One technique is to modify the control transfer instructions so that they cause traps and signals. The other technique is to add new bogus instructions (e.g., adding the conditional jumps that are disassembled but are never taken, or adding junk bytes that cause incorrect disassembly). Falcarin et al. [8] propose a novel binary obfuscation technique that is based on code mobility and code splitting at binary level. Their approach aims at obstructing the static and dynamic analysis and therefore the reverse engineering. Mimimorphism [28] is another binary obfuscation technique that the malware can use to hide itself from static and semantic analysis.

The idea of system call diversification was introduced by Chew and Song[3] for the first time, to mitigate the computer intrusions. In [3], the randomization is applied to operating system to defeat buffer overflows. One of their proposed methods is randomizing the system call mappings. Each system call is mapped to a corresponding numbers in a table. By altering (randomizing) the mappings, the original system call will no longer work.

System call diversification has also been studied in [13]. The authors propose it as a countermeasure against injection code attacks. This work is continued in [10], where the authors apply instruction set randomization and address space layout randomization simultaneously. These papers advocate randomization (diversfication) that happens dynamically at load time or run time, which causes some performance loss. They also require de-randomization, because the kernel is not diversified. We diversify binaries after they have been compiled so that the run-time performance is not affected. Unlike these earlier papers, our approach also provide system-wide protection by also diversifying the kernel.

Srivastava et al. [24] have designed an attack called Illusion. Illusion obfuscates the kernel's system calls that are used by the attacker to hide the actual operation of the malicious program. With the help of Illusion, attackers can stay invisible to the malware analyzers; since these analyzers rely on the standard system call interface to detect any changes and also the analyzers do not consider the actual execution behavior of the system call in the kernel. Moreover, Illusion is not detected by the tools checking the integrity of the kernel; because it does not make any alteration in data structure or code of the kernel. In addition, they have designed a detection system for detecting their attack.

The basic behavior of a program is recognizable by following its execution flow, i.e. by tracing the sequence of the system calls the program invokes in execution phase. Brusch et al. [2], proposed an obfuscator that works at kernel-level and randomizes the sequence of the invoked system calls. Randomization makes the program's execution flow unpredictable for attacks.

In our previous research we have studied the applicability of diversification techniques in different levels of software. In [17] we aim at concealing the system call interface in order to protect the operating system, while in [21] we focus on diversification at higher levels, i.e. Ajax applications. We propose a proxy-like obfuscator [21] to defeat the online banking Trojans. We implemented our approach in [19] and illustrated its efficiency. In two other papers [18,20] we consider the use of diversification techniques to mitigate the man-in-the-browser attacks.

8   Conclusions

In this paper, we have presented a scheme for large-scale system call diversification for operating system protection and also implemented a concrete diversifier tool to demonstrate feasibility of our approach. Our experiments show that a large majority of system calls is handled well by our tool, but there are still some challenges. Still, the numbers of unidentified calls were usually relatively small for analyzed binary files.

To overcome the challenges, we have also discussed several ways to increase the accuracy of our diversification scheme to 100 %. Based on this, we believe system call diversification is a feasible approach for protecting operating systems from malware. This is especially true for systems where a certain set of well-known libraries and applications is used and in many embedded systems that are more restricted in nature.

The small total amount of system calls also makes things easier. As we have seen, very few binaries in the tested Linux distributions contained direct system calls. Most direct system calls are in standard libraries and well-known command line tools, not in the ordinary applications.

There are still many open questions related to our diversification scheme. How and where do we store the system call number mapping as a secret? How would we invoke our diversification tool in an operating system? Would it run in the kernel or in user space? How is it protected? These details will be discussed in future work.

Only Linux has been covered in this paper. It would be interesting to also study our diversification scheme in other operating systems. Most likely, similar methods can be used and there are similar challenges present in the contexts of those systems, too.

References

1. Apvrille, A., Strazzere, T.: Reducing the window of opportunity for android malware gotta catch 'em all. Int. J. Ambient Comput. Intell. **8**(1–2), 61–71 (2012)
2. Bruschi, D., Cavallaro, L., Lanzi, A.: An efficient technique for preventing mimicry and impossible paths execution attacks. In: Performance, Computing, and Communications Conference, 2007, IPCCC 2007. IEEE Internationa, pp. 418–425, April 2007
3. Chew, M., Song, D.: Mitigating buffer overflows by operating system randomization (2002)
4. Cohen, F.B.: Operating system protection through program evolution. Comput. Secur. **12**(6), 565–584 (1993)
5. Collberg, C., Thomborson, C., Low, D.: A taxonomy of obfuscation tranformations. Technical report 148, The University of Auckland (1997)
6. TIS Committee: Tool Interface Standard. Executable and Linking Format (ELF) Specification. Version 1.2. Submitted to Journal of Information Security and Applications (Elsevier), under evaluation (1995)
7. Cooper, K.D., Harvey, T.J., Waterman, T.: Building a control-flow graph from scheduled assembly code. Technical report 02–399, Rice University (2002)
8. Falcarin, P., Carlo, S.D., Cabutto, A., Garazzino, N., Barberis, D.: Exploiting code mobility for dynamic binary obfuscation. In 2011 World Congress on Internet Security (WorldCIS), pp. 114–120, February 2011
9. Jang, M.H., Jang, M.: Security Strategies in Linux Platforms and Applications. Jones & Bartlett Publishers, Burlington (2010)
10. Jiang, X., Wang, H.J., Xu, D., Wang, Y.-M.: Randsys: thwarting code injection attacks with system service interface randomization. In: IEEE International Symposium on Reliable Distributed Systems, SRDS 2007, pp. 209–218 (2007)
11. Kerrisk, M.: The Linux Programming Interface. No Starch Press, San Francisco (2010)
12. Kinder, J., Zuleger, F., Veith, H.: An abstract interpretation-based framework for control flow reconstruction from binaries. In: Jones, N.D., Müller-Olm, M. (eds.) VMCAI 2009. LNCS, vol. 5403, pp. 214–228. Springer, Heidelberg (2009)

13. Liang, Z., Liang, B., Li, L.: A system call randomization based method for countering code injection attacks. In: International Conference on Networks Security, Wireless Communications and Trusted Computing, NSWCTC 2009, pp. 584–587 (2009)
14. Linn, C., Debray, S.: Obfuscation of executable code to improve resistance to static disassembly. In: Proceedings of the 10th ACM Conference on Computer and Communications Security, CCS 2003, pp. 290–299. ACM, New York, USA (2003)
15. Madou, M., Anckaert, B., De Bus, B., De Bosschere, K., Cappaert, J., Preneel, B.: On the effectiveness of source code transformations for binary obfuscation. In: Proceedings of the International Conference on Software Engineering Research and Practice (SERP06), pp. 527–533. CSREA Press (2006)
16. Popov, I.V., Debray, S.K., Andrews, G.R.: Binary obfuscation using signals. In: USENIX Security (2007)
17. S. Rauti, J. Holvitie, and V. Leppänen. Towards a Diversification Framework for Operating System Protection. In: Proceedings of International Conference on Computer Systems and Technologies, CompSysTech 2014 (2014)
18. Rauti, S., Leppänen, V.: Browser extension-based man-in-the-browser attacks against Ajax applications with countermeasures. In: Proceedings of International Conference on Computer Systems and Technologies, CompSysTech 2012, pp. 251–258. ACM Press (2012)
19. Rauti, S., Leppänen, V.: A proxy-like obfuscator for web application protection. Int. J. Inf. Technol. Secur. **5**(1) (2014)
20. Lee, J.W., Lee, Y.J., Kim, H.K., Hwang, B., Ryu, K.H.: Discovering temporal relation rules mining from interval data. In: Shafazand, H., Tjoa, A.M. (eds.) EurAsia-ICT 2002. LNCS, vol. 2510, pp. 57–66. Springer, Heidelberg (2002)
21. Rauti, S., Leppänen, V.: Resilient code protection by JavaScript and HTML obfuscation for Ajax applications against man-in-the-browser attacks. Submitted to Journal of Information Security and Applications (Elsevier), under evaluation (2014)
22. Schwarz, B., Debray, S., Andrews, G.: Disassembly of executable code revisited. In: Proceedings of Ninth Working Conference on Reverse Engineering, pp. 45–54 (2002)
23. Sobell, M.G.: A Practical Guide to Linux. Addison-Wesley, Boston (1999)
24. Srivastava, A., Lanzi, A., Giffin, J., Balzarotti, D.: Operating system interface obfuscation and the revealing of hidden operations. In: Holz, T., Bos, H. (eds.) DIMVA 2011. LNCS, vol. 6739, pp. 214–233. Springer, Heidelberg (2011)
25. Tanenbaum, A.S.: Modern Operating Systems, 3rd edn. Prentice Hall Press, Upper Saddle River (2007)
26. Theiling, H.: Extracting safe and precise control flow from binaries. In: Proceedings of Seventh International Conference on Real-Time Computing Systems and Applications, pp. 23–30. IEEE (2000)
27. Wang, S.P.: Mastering Linux. CRC Press, Boca Raton (2011)
28. Wu, Z., Gianvecchio, S., Xie, M., Wang, H.: Mimimorphism: a new approach to binary code obfuscation. In: Proceedings of the 17th ACM Conference on Computer and Communications Security, CCS 2010, pp. 536–546. ACM, New York, USA (2010)

Outsourced KP-ABE with Enhanced Security

Chao Li [(✉)], Bo Lang, and Jinmiao Wang

State Key Laboratory of Software Development Environment,
Beihang University, Beijing, China
{lichao,wangjinmiao}@nlsde.buaa.edu.cn, langbo@buaa.edu.cn

Abstract. Although Key-Policy Attribute-Based Encryption (KP-ABE) has been widely applied to protect data in cloud computing, it is always criticized for its inefficiency drawbacks, coming from both key-issuing and decryption. Recently, some papers proposed the outsourcing solutions. But adversaries in the attack model of these researches were divided into two categories, and it is assumed that the two cannot communicate with each other, which is obviously unrealistic. In this paper, we first proved that there are severe security vulnerabilities in these schemes for this assumption, and then proposed a security enhanced Chosen Plaintext Attack (SE-CPA) model, which eliminates the improper limitations. By utilizing Proxy Re-Encryption (PRE), we also constructed a concrete KP-ABE Outsourcing scheme (O-KP-ABE) and proved its security under SE-CPA model. Comparisons with existing schemes show that our construction has comprehensive advantages in security and efficiency.

Keywords: KP-ABE · Cloud computing · Computation outsourcing · Attack model · Proxy re-encryption

1 Introduction

As a new vision of public key based encryption, ABE [18] has attracted a great deal of attentions. For the first time, ABE enables efficient fine-grained access control on ciphertexts. In an ABE cryptosystem, the private key and the ciphertext are associated with an attribute group or an access policy respectively, and a particular key can decrypt a particular ciphertext only if the associated attribute group and access policy are matched. ABE can be classified as KP-ABE [11] and Ciphertext-Policy Attribute Based Encryption [4,13,19] (CP-ABE). In KP-ABE, the access policy is assigned to key and the attribute group is assigned to ciphertext. CP-ABE is just the opposite. Also, they are suitable for different application scenarios. The former is data-centered with data attributes; the latter is user-centered with user attributes.

Though ABE is a promising primitive to design fine-grained access control systems, it is being criticized for the inefficiency drawbacks, which is firstly reflected in decryption. The decryption of ABE is based on the expensive bilinear pairing operations whose number is in proportion to the complexity of access policy. This drawback appears more serious on resource-constrained equipment

© Springer International Publishing Switzerland 2015
M. Yung et al. (Eds.): INTRUST 2014, LNCS 9473, pp. 36–50, 2015.
DOI: 10.1007/978-3-319-27998-5_3

such as mobile devices and sensors. Beyond decryption, another cause for ABE's inefficiency is key-issuing, which requires a great quantity of modular exponentiations. When a large number of users request for their private keys, the Public Key Generator (PKG) may be overloaded. Moreover, in most of the existing ABE schemes the revocation of any single user requires key-update at PKG for all the remaining users who share common attributes with the revoked one. All of these heavy tasks centralized at PKG would make it a bottleneck of the whole system. Furthermore, user revocation is very common in a scalable ABE system. Since the policy is bound to key, in KP-ABE more tasks are needed in generating a key and it is also more difficult to recognize the users affected by the revocation. Thus the problems above are more serious to KP-ABE.

To address the aforementioned inefficiency drawbacks of ABE, Green et al. [12] firstly introduced the notion of outsourcing the decryption of ABE, which largely eliminates the decryption overhead from users. Based on the existing ABE schemes [11,19], Green et al. also constructed concrete schemes. In these schemes, a pair of keys (SK, TK) is generated during the key-issuing phase by utilizing the key-blinding technique, in which SK is the user Secret Key and TK is the Transformation Key. When decrypting a ciphertext CT, the user firstly provides his TK to an untrusted server, say a proxy maintained by a Cloud Service Provider (CSP) which can be called D-CSP (Decryption CSP)[1]. Then the D-CSP can translate the ciphertext CT into a simple ciphertext CT' with TK if the attribute group and access policy associated with CT and TK are matched. And then the user can restore the message from CT' with SK by only one ElGamal [10] decryption. We will henceforth refer to this paper as Green11.

However, Green et al. have not considered the key-issuing computation overhead at PKG. Li et al. [15] did that. They extended the outsourcing idea to key-issuing of KP-ABE and proposed a new scheme model called Outsourced ABE (OABE). The system model of OABE can be represented in Fig. 1. In OABE, three CSPs are involved, i.e. D-CSP, S-CSP (Store CSP), and KG-CSP (Key Generation CSP), among which the KG-CSP is a new added party compared to Green11. The function of KG-CSP is to complete the delegated key-issuing computation to relieve the PKG's load. When a user applies for his key pair, he first submits to PKG his AP (Access Policy), with which the PKG can work out the relative OK (Outsourcing Key) and sends it to the KG-CSP. Then the KG-CSP can use OK to compute TK and returns back TK to PKG. At last, the PKG calculates the final key pair (SK, TK) and sends it to the user. This is the whole process of key-issuing, and the decrypting process is the same to Green11. We will refer to this paper as LCLJ13. Although LCLJ13 can outsource decrypting and key-issuing simultaneously, it still requires the user to do two pairings during the decryption phase. In another paper [16] aiming at the checkability of outsourced ABE, the authors achieving the same decryption efficiency as Green11 by adopting the key-blinding technique. However, in this scheme the key-blinding work is done by PKG, thus it hasn't eased the PKG's burden compared to the traditional non-outsourced KP-ABE. We will henceforth refer to this paper as LHLC13 and it has the same system model with LCLJ13.

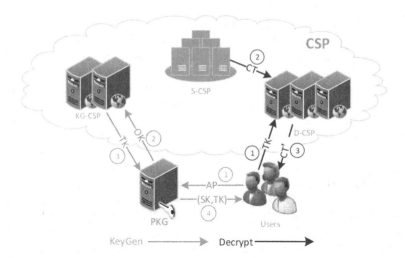

Fig. 1. The system model of OABE (The numbers on the arrows represent the execution sequence.)

LCLJ13 and LHLC13 also share the same attack model, in which the adversaries were classified into two types: a curious user colluding with D-CSP; a curious KG-CSP. But the two types of adversaries are not allowed to collude. In the system of OABE, all CSPs are from third parties, thus they cannot be completely trusted. So the above assumption is obviously unrealistic. Actually, in their schemes, by colluding with KG-CSP, a curious user can decrypt any ciphertext. We will give the proof in Sect. 4.1.

Our Contributions. In this paper, we focus on the KP-ABE outsourcing scheme which can outsource decryption and key-issuing simultaneously. Although LCLJ13 and LHLC13 can do that, their efficiency and security have yet to be improved, especially security. We will firstly prove that there are severe security vulnerabilities in these schemes for their assumption of no collusion between curious users and KG-CSP. Then, based on the analysis of the features of KP-ABE and all the possible attacks in real world to the outsourcing schemes, we propose a security enhanced CPA (SE-CPA) model. SE-CPA assumes all CSPs are honest but curious, which means they will follow our protocols, but try to find out as much secret information as possible based on their possessions. Thus, curious users may collude with any CSP, which eliminates the limitation in the attack model of LCLJ13 and LHLC13 where KG-CSP is not allowed to collude with users.

Then by utilizing the technique of PRE, we construct a new concrete KP-ABE outsourcing scheme (O-KP-ABE) with proved enhanced security under SE-CPA. In addition, compared with existing schemes, O-KP-ABE has the relatively highest comprehensive efficiency in key-issuing and decrypting.

Organization. The rest of the paper is organized as follows. In Sect. 2, we introduce the related work. We give the necessary background information in Sect. 3. In Sect. 4, we first give a detailed security analysis of existing schemes and then we describe our new KP-ABE outsourcing model and security enhanced CPA (SE-CPA) model. We present a new concrete KP-ABE outsourcing scheme (O-KP-ABE) and prove its security under SE-CPA in Sect. 5. In Sect. 6, we compare our scheme with the relevant schemes in security and efficiency. Finally, we conclude our work.

2   Related Work

ABE outsourcing. Green et al. [12] firstly proposed the idea of outsourcing ABE and constructed the concrete schemes with outsourced decryption. Then Zhou et al. [20] and Li et al. [17] proposed ABE schemes with outsourced encryption and decryption respectively. Green et al. and Zhou et al. both use the key-blinding technique to achieve the outsourcing of decryption. In key-blinding, the user firstly chooses a random value as the blind factor, and then he runs exponentiations on the original key components with that blind factor. However, they all didn't consider the computation overhead at PKG. Li et al. [15] firstly considered that issue and proposed a concrete scheme which can outsource decryption and key-issuing simultaneously. They proposed a different method by introducing a default attribute. The default attribute is appended to each data's attribute group and each user's access policy. In addition, there are also some papers [14,16] researching on the verification of outsourcing results. They generally did that by appending a redundancy with ciphertext.

Proxy Re-Encryption (PRE): Blaze et al. [5] first propose the notion of PRE and give a simple concrete scheme. PRE can be represented by the formula $D(\prod(E(m, e_A), \pi_{A \rightarrow B}), d_B) = m$, which means that the ciphertext encrypted by A's public key e_A after being re-encrypted by proxy key $\pi_{A \rightarrow B}$ can be decrypted by B's secret key d_B. $\pi_{A \rightarrow B}$ is public and the re-encryption work can be done by an untrusted proxy server without fearing the leakage of message m, and user secret keys d_A, d_B. Further research about PRE was made by Ateniese et al. [2], which concludes the features a PRE scheme should possess and its application scenarios. They also put forward an improved PRE scheme. Besides, some papers [7] search for the PRE technique with more security.

3   Background

3.1   Bilinear Maps

Let $\mathbb{G}_1$ and $\mathbb{G}_T$ be two multiplicative cyclic groups of prime order p and g be a generator of $\mathbb{G}_1$. $e : \mathbb{G}_1 \times \mathbb{G}_1 \rightarrow \mathbb{G}_T$ is a bilinear map with the properties:

- Bilinearity: for all $u, v \in \mathbb{G}_1$ and $a, b \in \mathbb{Z}_p$, we have $e(u^a, v^b) = e(u, v)^{ab}$.
- Non-degeneracy: $e(g, g) \neq 1$.

We say that $\mathbb{G}_1$ is a bilinear group if the group operation in $\mathbb{G}_1$ and the bilinear map $e : \mathbb{G}_1 \times \mathbb{G}_1 \rightarrow \mathbb{G}_T$ are both efficiently computable. Notice that the map e is symmetric since $e(g^a, g^b) = e(g, g)^{ab} = e(g^b, g^a)$.

3.2 DBDH Assumption

The security of our construction is based on the complexity of Decisional Bilinear Diffie-Hellman (DBDH) assumption [19] below.

Firstly, we define the DBDH problem as follows. A challenger chooses a group $\mathbb{G}$ of prime order p according to the security parameter. Let $a, b, c \in \mathbb{Z}_p$ be chosen at random and g be a generator of $\mathbb{G}$. The adversary when given (g, g^a, g^b, g^c) must distinguish a valid tuple $e(g, g)^{abc} \in \mathbb{G}_T$ from a random element R in $\mathbb{G}_T$.

Then we can get the definition of DBDH assumption:

Definition 1. *(DBDH Assumption) We say that the DBDH assumption holds if no polytime algorithm has a non-negligible advantage in solving the DBDH problem.*

3.3 Access Structure

Definition 2. *(Access Structure [3]) Let $\{P_1, P_2, \ldots, P_n\}$ be a set of parties. A collection $\mathbb{A} \subseteq 2^{\{P_1, P_2, \ldots, P_n\}}$ is monotone if $\forall B, C$: if $B \in \mathbb{A}$ and $B \subseteq C$ then $C \in \mathbb{A}$. An access structure (respectively, monotone access structure) is a collection (respectively, monotone collection) $\mathbb{A}$ of non-empty subsets of $\{P_1, P_2, \ldots, P_n\}$, i.e., $\mathbb{A} \subseteq 2^{\{P_1, P_2, \ldots, P_n\}} / \{\emptyset\}$. The sets in $\mathbb{A}$ are called the authorized sets, and the sets not in $\mathbb{A}$ are called the unauthorized sets.*

In the context of ABE, the role of the parties is taken by the attributes. Thus, the access structure $\mathbb{A}$ will contain the authorized sets of attributes.

4 New Models for KP-ABE with Outsourcing

4.1 Security Analysis of Existing Schemes

In Sect. 1, we have pointed out the security problems in LCLJ13 and LHLC13, but haven't given the detailed proofs. For the outline of these two schemes, please refer to the Sect. 1 section and Fig. 1. In this section, we take LCLJ13 for example to prove that if curious users collude with the KG-CSP, they can decrypt any ciphertext CT. LHLC13 has the same problem. The concrete process is as follows.

Assume a curious user will collude with KG-CSP, and his secret key is SK $=$ $(\text{SK}_1, \text{SK}_2)$, in which $\text{SK}_2 = (d_{\theta 0} = g_2^{x_2}(g_1 h)^{r_\theta}, d_{\theta 1} = g^{r_\theta})$. As KG-CSP computes all delegated key-issuing work, it may store the copies of all OKs, including the curious user's, and assume it is OK $= x_1$. Given a ciphertext CT$=(C_0 =$ $Me(g_1, g_2)^s, C_1 = g^s, E_\theta = (g_1 h)^s)$, the curious user performs the following steps:

Firstly, with public parameter g_2, OK and C_1 in CT, he can calculate $e(g, g_2)^{x_1 s}$; with SK_2 and E_θ in CT, he can calculate $\frac{e(C_1, d_{\theta 0})}{e(d_{\theta 1}, E_\theta)} = e(g_1, g_2)^{x_2 s}$.

Secondly, as the master key is $x = x_1 + x_2$, he is able to get $e(g, g_2)^{xs}$ through calculation.

Thirdly, as $C_0 = m.e(g, g_2)^{xs}$, the curious user can recover m.

As CT in the above process can be any ciphertext, the curious user can decrypt all ciphertexts by colluding with KG-CSP. This is obviously incorrect. Thus there exist severe security vulnerabilities in LCLJ13 and LHLC13.

4.2   Model of KP-ABE with Outsourcing

In this section, we give our KP-ABE outsourcing model by modifying the model of KP-ABE with outsourced decryption in Green11. Our model also supports the outsourcing of decryption and key-issuing simultaneously. The model is similar to LCLJ13 shown in Fig. 1, but it needs no subsequent processing of outsourced key-issuing, i.e. the PKG in our model needn't do any further computation after receiving TK from KG-CSP. By distinguishing OK and TK, on one hand the user could decrypt the ciphertext himself when the network is unavailable, on the other hand the D-CSP needn't generate TK from OK whenever it translates ciphertext. Our new KP-ABE outsourcing scheme consists of six algorithms, rather than seven like LCLJ13.

Setup(U). The setup algorithm takes the attributes universe U and the implicit security parameter as input. It is used to initialize the system and output the public parameter PK and master key MK. This algorithm is performed by PKG.

Encrypt(PK, M, S). The encryption algorithm takes as input the public parameters PK, a message M and a set of attributes S. It will encrypt M and produce a ciphertext CT. This algorithm is performed by Data Owner (DO).

Keygen_IN(𝔸, MK, PK). This algorithm is the first step of key-issuing and is performed by PKG. It takes as input the access structure 𝔸, the master key MK and the public parameters PK. It outputs the outsourcing key OK and user private key SK.

Keygen_OUT(OK, PK). This algorithm is the second step of key-issuing and is performed by KG-CSP. It takes as input the outsourcing key OK and public parameters PK. It will output the transformation key TK and return it to PKG.

Transform_OUT(TK, CT). This algorithm complete the preprocessing of ciphertext and is performed by D-CSP. It firstly checks whether the attribute set S in CT satisfies the access structure 𝔸 in TK. It outputs the partially decrypted ciphertext CT' if S ∈ 𝔸, otherwise it outputs ⊥.

Decrypt (CT, CT', SK). This algorithm takes as input the ciphertext CT, partially decrypted ciphertext CT' and user private key SK. It output the message M if S ∈ 𝔸, otherwise ⊥.

4.3   Enhanced Security Model

This section analyzes all possible attacks to the KP-ABE outsourcing model given in Sect. 4.2 under CPA and proposes a new Security Enhanced CPA model SE-CPA.

As the above outsourcing model will outsource the majority of work during key-issuing and decrypting to the third party which is not completely trusted, more information will be leaked. Even though the computations of key-issuing and decryption are outsourced to different parties, they may collude with each other. Thus, the outsourcing model above will face attacks different from any previous ones, which results in a different attack model.

Through careful analysis of the new outsourcing model, we find attackers under CPA may get the following information or services:

- Like the basic ABE schemes, the attacker is able to achieve the service of key-issuing, and thus get the key pair (SK, TK) corresponding to the specific access policy $\mathbb{A}$.
- Since KG-CSP is not trusted, it may save the copies of all OKs sent from PKG and the corresponding TKs. So the attacker may get all of the (OK, TK).
- Combining the above two points, the attacker can get the tuple of keys (OK, TK, SK) corresponding to the specific policy $\mathbb{A}$.
- As S-CSP and D-CSP are both untrusted and the TK corresponding to any access policy $\mathbb{A}$ can be achieved, the attacker is able to get the transforming service to all ciphertexts.

Based on these observations, we can propose the new CPA model SE-CPA, and the model is defined as follows:

Init. The adversary $\mathcal{A}$ declares the set of attributes S^* and submits it to the challenger $\mathcal{C}$.

Setup. The challenger $\mathcal{C}$ runs the Setup algorithm of KP-ABE outsourcing scheme and gives the public parameters PK to adversary $\mathcal{A}$.

Phase 1. The adversary $\mathcal{A}$ is allowed to make any of the following queries repeatedly:

i. Query for (SK,OK,TK) corresponding to the access structure $\mathbb{A}$ with the restriction that for all $x \in Y_\mathbb{A}, x \notin S^*$, in which $Y_\mathbb{A}$ is the collection of the attributes in $\mathbb{A}$.

ii. Query for (OK,TK) corresponding to the access structure $\mathbb{A}$, with the restriction that for all $x \in Y_\mathbb{A}, x \in S^*$, in which $Y_\mathbb{A}$ is the collection of the attributes in $\mathbb{A}$.

iii. Query for the transforming ciphertext CT' corresponding to CT encrypted with S^*.

Challenge. $\mathcal{A}$ sends to $\mathcal{C}$ two equal length messages m_0, m_1. Then $\mathcal{C}$ flips a random coin b, and encrypts m_b with S^*. The ciphertext CT will be sent to $\mathcal{A}$.

Phase 2. The adversary repeats Phase 1.

Guess. $\mathcal{A}$ outputs a guess b' of b.

Definition 3. *(SE-CPA Secure KP-ABE with Outsourcing) An KP-ABE outsourcing scheme is SE-CPA secure if all polynomial time adversaries have at most a negligible advantage in the game of SE-CPA.*

5 O-KP-ABE Scheme

In this section, we give the concrete construction of our new Outsourcing KP-ABE scheme (O-KP-ABE) based on the scheme model described in Sect. 4.2. The access structure of O-KP-ABE is represented by a tree. After the description of O-KP-ABE, we will prove its security under the enhanced security model SE-CPA.

5.1 Access Trees

Our construction uses tree-based access structure which is represented by $\mathcal{T}$. Each interior node of the tree is a threshold gate and the leaves are associated with attributes. This structure is very expressive. For example, we can represent a tree with "AND" and "OR" gates by using respectively 2 of 2 and 1 of 2 threshold gates. A user is able to decrypt a ciphertext if and only if the attributes in ciphertext satisfies the access structure in the user private key. The definitions of $\mathcal{T}$ and relative functions are identical to paper [11], for more information about $\mathcal{T}$ please refer to [11].

5.2 Construction of O-KP-ABE

Let $\mathbb{G}_1$ be a bilinear group of prime order p, and let g be a generator of $\mathbb{G}_1$. In addition, let $e : \mathbb{G}_1 \times \mathbb{G}_1 \to \mathbb{G}_T$ denote the bilinear map. We also define the Lagrange coefficient $\Delta_{i,S}$ for $i \in \mathbb{Z}_p$ and a set, S, of elements in $\mathbb{Z}_p$: $\Delta_{i,S}(x) = \prod_{j \in S, j \neq i} \frac{x-j}{i-j}$.

Our construction consists of 6 algorithms.

Setup(U). First, choose a bilinear group $\mathbb{G}_1$ of prime order p with a generator g and a bilinear map $e : \mathbb{G}_1 \times \mathbb{G}_1 \to \mathbb{G}_T$. Next, determine the universe of attributes according to the actual application situation $U = \{a_1, a_2, \ldots, a_n\}$, and let i represent the index of attribute a_i in U. Finally, choose $\alpha, t_i, \beta_i \in \mathbb{Z}_p, 1 \leq i \leq n$ and $g_2 \in \mathbb{G}_1$, in which t_i and β_i correspond to a_i. Then the master key is MK $= \{\alpha, t_i, \beta_i\}, 1 \leq i \leq n$, and the public parameters PK are

$$\text{PK} = \{U, g, g_1 = g^{\alpha}, g_2, T_i = g^{t_i}, P_i = g_2^{t_i^{-1}\beta_i}\}, 1 \leq i \leq n.$$

Encrypt(PK, M, S). To encrypt a message $M \in \mathbb{G}_T$ under a set of attributes S, choose a random value $s \in \mathbb{Z}_p$ and publish the ciphertext as:

CT= $\{S, C_0 = M.e(g_1, g_2)^s = M.e(g, g_2)^{\alpha s}, \{C_y = T_i^s\}_{y \in S}\}$, in which y represents an attribute and i is the index of y in the universe of attributes U.

Keygen_IN($\mathcal{T}$, MK, PK). The algorithm proceeds as follows. Firstly, choose a random value $z \in \mathbb{Z}_p$, and calculate $\delta = \alpha/z$. Then, choose a polynomial q_x for each node x (including the leaves) in the tree $\mathcal{T}$. These polynomials are chosen in the following way in a top-down manner, starting from the root node r.

For each node x in the tree, set the degree η_x of the polynomial q_x to be one less than the threshold value k_x of that node, that is $\eta_x = k_x - 1$. Then, for the root node r, set $q_r(0) = \delta$ and η_r other points of the polynomial q_r randomly to define it completely. For any other node x, set $q_x(0) = q_{\text{parent}(x)}(\text{index}(x))$ and choose η_x other points randomly to completely define q_x.

Once the polynomials have been decided, for each leaf node x, we can get the value of $q_x(0)$, and then calculate $d_x = q_x(0)/\beta_i$, in which i is the index of x in the universe of attributes.

Thus, the outsourcing key is OK= $\{\mathcal{T}, \{d_x\}_{x \in Y_T}\}$, in which Y_T is the attributes set of leaves in $\mathcal{T}$. And the user private key is $SK = z$.

Keygen_OUT(OK, PK). For each element d_x in OK calculate:

$$D_x = P_i^{d_x} = g_2^{t_i^{-1} \beta_i \frac{q_x(0)}{\beta_i}} = g_2^{t_i^{-1} q_x(0)},$$

in which i is the index of attribute x in the universe of attributes U. The transformation key is: TK= $\{\mathcal{T}, \{D_x\}_{x \in Y_T}\}$, in which Y_T is the attributes set of leaves in $\mathcal{T}$.

Transform_OUT(TK, CT). The transformation procedure is defined as a recursive algorithm TransformNode(x, TK, CT) which takes as input the ciphertext CT, the transformation key TK and a node x in the tree. This recursive algorithm outputs a group element of $\mathbb{G}_T$ or $\perp$.

If the node x is a leaf node then:

$$\text{TransformNode}(x, \text{TK}, \text{CT}) = \begin{cases} e(D_x, C_x) = e(g_2^{t_i^{-1} q_x(0)}, g^{t_i s}) \\ = e(g, g_2)^{q_x(0)s} \quad \text{if } x \in S \\ \\ \perp \quad \text{otherwise} \end{cases}$$

If x is not a leaf node, the algorithm TransformNode(x, TK, CT) proceeds as follows: for all nodes z that are children of x, it calls TransformNode (z, TK, CT) and stores the output as F_z. Let S_x be an arbitrary k_x sized set of child nodes z such that $F_z \neq \perp$. If no such set exists then the node was not satisfied and the function returns $\perp$.

Otherwise, we compute:

$$F_z = \text{TransformNode}(x, \text{TK}, \text{TK})$$

$$= \prod_{z \in S_x} F_z^{\Delta_{i,S'}(0)}, \text{ where } \begin{matrix} i = \text{index}(z) \\ S'_x = \{\text{index}(z) : z \in S_x\} \end{matrix}$$

$$= \prod_{z \in S_x} (e(g, g_2)^{sq_z(0)})^{\Delta_{i,S'}(0)}$$

$$= \prod_{z \in S_x} (e(g, g_2)^{sq_{\text{parent}(z)}(\text{index}(z))})^{\Delta_{i,S'}(0)}$$

$$= \prod_{z \in S_x} e(g, g_2)^{sq_x(i)\Delta_{i,S'}(0)}$$

$$= e(g, g_2)^{sq_x(0)}$$

Thus we can know that $\text{CT'}=\text{TransformNode}(r,\text{TK},\text{CT})=e(g, g_2)^{s\delta} = e(g, g_2)^{s\alpha/z}$ if and only if CT satisfies the TK.

Decrypt(CT, CT', SK). If the user has the privilege to access the data, then upon receiving CT' from D-CSP, the user completely decrypts the ciphertext and gets the message M$= C_0/CT'^z$.

5.3   Proof of Security Under SE-CPA

We prove the following theorem:

Theorem 1. *If an adversary can break the scheme of O-KP-ABE under the SE-CPA model, then a simulator can be constructed to solve the DBDH problem with a non-negligible advantage.*

Proof. Suppose there exists a polynomial adversary $\mathcal{A}$ who can attack our scheme under the SE-CPA model with the advantage ε, then we can build a simulator $\mathcal{S}$ who can win DBDH problem with a non-negligible advantage $\varepsilon/2$. The process of simulation is as follows:

Firstly, the challenger $\mathcal{C}$ set the groups $\mathbb{G}_1$ and $\mathbb{G}_T$ with an efficient bilinear map e, and a generator g. Then $\mathcal{C}$ flips a fair binary coin u, outside of $\mathcal{S}'s$ view. If $u = 0$, the challenger sets $(A, B, C, Z) = (g^a, g^b, g^c, e(g, g)^{abc})$; otherwise it sets $(A, B, C, Z) = (g^a, g^b, g^c, e(g, g)^z)$ for random a, b, c, z. We assume the universe of attributes, U is defined.

Init. $\mathcal{A}$ chooses the set of attributes S^* it wishes to be challenged upon and sents it to $\mathcal{S}$.

Setup. The simulator $\mathcal{S}$ sets $g_1 = A = g^\alpha$ (thus, $a = \alpha$) and $g_2 = B$. For each $a_i \in U$, it chooses random values $t'_i, \beta'_i \in \mathbb{Z}_p$. If $a_i \in S^*$, the simulator sets $T_i = g^{t'_i}$ and $P_i = B^{t'^{-1}_i \beta'_i} = g_2^{t'^{-1}_i \beta'_i}$ (thus, $t_i = t'_i, \beta_i = \beta'_i$); if $a_i \notin S^*, \mathcal{S}$ sets $T_i = A^{t'_i} = g^{at'_i}$ and $P_i = B^{t'^{-1}_i \beta'_i} = g_2^{t'^{-1}_i \beta'_i} = g_2^{(at'_i)^{-1}\alpha\beta'_i}$ (thus, $t_i = \alpha t'_i, \beta_i = \alpha\beta'_i$).

So the public parameters are PK= $\{U, g, g_1 = A, g_2 = B, T_i, P_i\}, 1 \le i \le n$ and they will be sent to $\mathcal{A}$.

Phase 1. The adversary $\mathcal{A}$ is allowed to make any of the following queries repeatedly:

i $\mathcal{A}$ submits an access tree $\mathcal{T}$ with the restriction that for all $x \in Y_{\mathcal{T}}, x \notin S^*$, in which $Y_{\mathcal{T}}$ is the attribute set of leaves in $\mathcal{T}$. And $\mathcal{S}$ must construct the corresponding key tuple (SK,OK,TK).
   $\mathcal{S}$ chooses a random value $z \in \mathbb{Z}_p$ and sets SK=z.
   Then, set $q_r(0) = z$ and calculate the value of $q_x(0)$ of each leaf node x in tree $\mathcal{T}$ following the steps of Keygen_IN. Next, let $Q_x(0) = \alpha \cdot q_x(0)$, thus $Q_r(0) = \alpha/z$. Since the simulator sets $\beta_i = \alpha\beta_i'$ for all $x \notin S^*$, we can calculate $d_x = Q_x(0)/\beta_i = q_x(0)/\beta_i'$. So, OK= $\{\mathcal{T}, \{d_x\}_{x \in Y_{\mathcal{T}}}\}$.
   For each element d_x in OK calculate $D_x = P_i^{d_x} = B^{t_i'^{-1}\beta_i'd_x} = B^{q_x(0)/t_i'}$, in which i is the index of attribute x in the universe of attributes U. The transformation key is TK= $\{\mathcal{T}, \{D_x\}_{x \in Y_{\mathcal{T}}}\}$, in which $Y_{\mathcal{T}}$ is the attributes set of leaves in $\mathcal{T}$.
   Finally, $\mathcal{S}$ sends (SK,OK,TK) to $\mathcal{A}$.
ii $\mathcal{A}$ submits an access tree $\mathcal{T}$ with the restriction that for all $x \in Y_{\mathcal{T}}, x \in S^*$, in which $Y_{\mathcal{T}}$ is the attributes set of leaves in $\mathcal{T}$. And $\mathcal{S}$ must construct the corresponding key tuple (OK,TK).
   $\mathcal{S}$ chooses a random value $z' \in \mathbb{Z}_p$, and sets $\delta = 1/z'$ (thus $z = \alpha z'$). Then, set $q_r(0) = 1/z'$ and calculate the value of $q_x(0)$ of each leaf node x in tree $\mathcal{T}$ following the steps of Keygen_IN. Since the simulator sets $\beta_i = \beta_i'$ for all $x \in S^*$, we can calculate $d_x = q_x(0)/\beta_i = q_x(0)/\beta_i'$. So, OK=$\{\mathcal{T}, \{d_x\}_{x \in Y_{\mathcal{T}}}\}$.
   Next, for each element d_x in OK calculate $D_x = P_i^{d_x} = B^{t_i'^{-1}\beta_i'd_x} = B^{q_x(0)/t_i'}$, and TK= $\{\mathcal{T}, \{d_x\}_{x \in Y_{\mathcal{T}}}\}$.
   Finally, $\mathcal{S}$ sends (OK,TK) to $\mathcal{A}$.
iii $\mathcal{A}$ submits a ciphertext CT encrypted by S^*, the simulator must transform it to CT'.
   Firstly, $\mathcal{S}$ should construct an access tree $\mathcal{T}$ which is satisfied by S^*. The simplest method may be choosing one attribute from S^* to construct such a tree $\mathcal{T}$. Then, by query ii $\mathcal{S}$ can get the corresponding transformation key TK which can be used to transform CT to CT'.

Challenge. The adversary $\mathcal{A}$ submits two challenge messages m_0, m_1 with equal length to $\mathcal{S}$. The simulator $\mathcal{S}$ will flip a fair binary coin b, and returns an encryption of m_b. The ciphertext is outputted as CT= $\{S^*, C_0 = m_b \cdot Z, \{C_y = C^{t_i'}\}_{y \in S^*}\}$.
   If $u = 0$ then $Z = e(g, g)^{abc}$. If we let $s = c$, then we have $C_0 = m_b \cdot e(g, g)^{abc} = m_b \cdot e(g^a, g^b)^c = m_b \cdot e(g_1, g_2)^s, C_y = C^{t_i'} = g^{t_i's} = T_i^s$. Therefore, the ciphertext is a valid random encryption of message m_b.

If $u = 1$, then $Z = e(g, g)^z$. Thus, $C_0 = m_b \cdot e(g, g)^z$. Since z is random, C_0 will be a random element of $\mathbb{G}_T$ from adversaries view and the message contains no information about m_b.

Phase 2. Repeat Phase 1.

Guess. $\mathcal{A}$ will submit a guess b' of b. If $b' = b$, the simulator will output $u' = 0$ to indicate it was given a valid BDH-tuple, otherwise it will output $u' = 1$ to indicate it was given a random 4-tuple.

In the case where $u = 1$ the adversary gains no information about b. Therefore, we have $\Pr(b \neq b' \mid u = 1) = \frac{1}{2}$. Since the simulator guess $u' = 1$ when $b' \neq b$, we have $\Pr(u' = u \mid u = 1) = \frac{1}{2}$.

If $u = 0$ then the adversary sees an encryption of m_b. The adversary's advantage in this situation is ε by definition. Therefore, we have $\Pr(b = b' \mid u = 0) = \frac{1}{2} + \varepsilon$. Since the simulator guess $u' = 0$ when $b' = b$, we have $\Pr(u' = u \mid u = 0) = \frac{1}{2} + \varepsilon$.

Thus, the overall advantage of simulator in the DBDH game is $\frac{1}{2}\Pr(u' = u \mid u = 0) + \frac{1}{2}\Pr(u' = u \mid u = 1) - \frac{1}{2} = \varepsilon/2$.

6 Analysis and Discussions

6.1 Analysis

This section compares our scheme with other existing KP-ABE outsourcing schemes in efficiency and security. The results are shown in Table 1.

Table 1. Comparison in efficiency and security between our scheme and others. G and P stand for the maximum time to compute an exponentiation in G and a pairing respectively. |Y| denotes the number of leaves in access tree. SS represents the time to share a secret in key-issuing phase.

Scheme	KG Ops	Dec Ops	Security level
Green11	$5|Y|G$	1G	RCCA
LCLJ13	3G	2P	CPA
LHLC13	$(2|Y|+5)G$	1G	CPA
O-KP-ABE	SS	1G	SE-CPA

As Green11 has not considered the outsourcing of key-issuing, the number of exponentiations that it must accomplish is proportional to the size of the access tree. Thus, its efficiency is relatively low. LCLJ13 can outsource decrypting and key-issuing simultaneously and PKG only needs to complete three exponentiations during the key-issuing phase. However, the user still has to complete two times pairings when decrypting ciphertexts. LHLC13 improves the decryption efficiency of LCLJ13, and the user only needs to complete one exponentiation in decryption. But its efficiency in key-issuing decreases sharply, even no better than the original scheme [11] without outsourcing. Thus, the outsourcing of LHLC13 seems meaningless.

Our O-KP-ABE scheme has the highest efficiency of decryption, in which the user only needs one exponentiation. And during key-issuing, PKG only needs to do the work of secret sharing. We can divide the process of key-issuing into two phases: secret sharing and key components calculating. The former consists of multiplication and division, and the latter consists of exponentiations. Experiments show that the former only accounts for a few portion of the latter.

In the aspect of security, Green11 can resist the Replayable Chosen Ciphertext Attack (RCCA)[8]. The traditional notion of security against chosen-ciphertext attacks (CCA) is a bit too strong, since it does not allow any bit of the ciphertext to be altered. However, there exist encryption schemes that are not CCA secure, but seem sufficiently secure "for most practical purposes". For these reasons, Canetti et al. proposed the notion of RCCA. On one hand, RCCA security accepts as secure some non-CCA schemes; on the other hand, it suffices for most of existing applications of CCA security. Thus, the security of RCCA lies between CPA and CCA. Although the authors of LCLJ13 and LHLC13 declared that their schemes are CPA secure, we have proved that they have severe security vulnerability when collusion is considered. We have proved the security of our scheme under the SE-CPA model, which means O-KP-ABE has remove the security vulnerability in LCLJ13 and LHLC13. Thus our scheme has a relatively higher security compared with LCLJ13 and LHLC13.

6.2   Discussions

Achieving Higher Security. In this work, we considered the selectively CPA security in which the adversary must declare the attribute sets he wishes to challenge on before starting the game. This may be a little bit unrealistic in practice, as this means no security can be guaranteed if the adversary chooses the attributes after he saw the public parameters. Thus we should consider higher security without that limitation, namely eliminating the Init stage in the security model [12]. Besides, we should also search for an ABE outsourcing scheme which can resist CCA. There are many techniques can be used to promote a CPA secure public key encryption to be CCA secure, for example, the one-time-signature technique [6,9]. Both of them can be our future work.

Verifiability. Although in our scheme the proxy servers, namely CSPs, cannot learn anything useful, there is no guarantee on the correctness of the outsourcing results. In some applications users or PKG often requests to check whether the outsourcing work is indeed done correctly. This is another important issue in outsourcing KP-ABE, and some approaches have been proposed. For example, Lai et al. [14] and Li et al. [16] addressed this problem by appending a redundancy with ciphertext. However, both of them only considered the verification of outsourced decrypting, they have not considered the same request for outsourced key-issuing. In our future work, we will consider the verification issue of our O-KP-ABE scheme.

7   Conclusion

Existing KP-ABE outsourcing schemes assume that KG-CSP and the curious user will not collude, which is obviously unrealistic. This paper first proposed a new security enhanced attack model SE-CPA. In SE-CPA, all CSPs are thought to be curious and allowed to collude with curious users, which is more practical. Then, we constructed a concrete outsourcing scheme O-KP-ABE and proved its security under SE-CPA. Except for the relative higher security, O-KP-ABE also has relative higher efficiency. Hence, our construction has a comprehensive advantage over existing schemes in security and efficiency.

Acknowledgments. This work was supported by the National Natural Science Foundation of China (Grant No.61170088) and Foundation of the State Key Laboratory of Software Development Environment (Grant No. SKLSDE-2014ZX-05).

References

1. Armbrust, M., Fox, A., Griffith, R., Joseph, A.D., Katz, R., Konwinski, A., Lee, G., Patterson, D., Rabkin, A., Stoica, I., et al.: A view of cloud computing. Commun. ACM **53**(4), 50–58 (2010)
2. Ateniese, G., Fu, K., Green, M., Hohenberger, S.: Improved proxy re-encryption schemes with applications to secure distributed storage. ACM Trans. Inf. Syst. Secur. (TISSEC) **9**(1), 1–30 (2006)
3. Beimel, A.: Secure schemes for secret sharing and key distribution. Ph.D. thesis, Technion-Israel Institute of Technology, Faculty of Computer Science (1996)
4. Bethencourt, J., Sahai, A., Waters, B.: Ciphertext-policy attribute-based encryption. In: IEEE Symposium on Security and Privacy, 2007, SP 2007, pp. 321–334. IEEE (2007)
5. Blaze, M., Bleumer, G., Strauss, M.J.: Divertible protocols and atomic proxy cryptography. In: Nyberg, K. (ed.) EUROCRYPT 1998. LNCS, vol. 1403, pp. 127–144. Springer, Heidelberg (1998)
6. Canetti, R., Halevi, S., Katz, J.: Chosen-ciphertext security from identity-based encryption. In: Cachin, C., Camenisch, J.L. (eds.) EUROCRYPT 2004. LNCS, vol. 3027, pp. 207–222. Springer, Heidelberg (2004)
7. Canetti, R., Hohenberger, S.: Chosen-ciphertext secure proxy re-encryption. In: Proceedings of the 14th ACM Conference on Computer and Communications Security, pp. 185–194. ACM (2007)
8. Canetti, R., Krawczyk, H., Nielsen, J.B.: Relaxing chosen-ciphertext security. In: Boneh, D. (ed.) CRYPTO 2003. LNCS, vol. 2729, pp. 565–582. Springer, Heidelberg (2003)
9. Cheung, L., Newport, C.: Provably secure ciphertext policy abe. In: Proceedings of the 14th ACM Conference on Computer and Communications Security, pp. 456–465. ACM (2007)

10. El Gamal, T.: A public key cryptosystem and a signature scheme based on discrete logarithms. In: Blakely, G.R., Chaum, D. (eds.) CRYPTO 1984. LNCS, vol. 196, pp. 10–18. Springer, Heidelberg (1985)

11. Goyal, V., Pandey, O., Sahai, A., Waters, B.: Attribute-based encryption for fine-grained access control of encrypted data. In: Proceedings of the 13th ACM Conference on Computer and Communications Security, pp. 89–98. ACM (2006)

12. Green, M., Hohenberger, S., Waters, B.: Outsourcing the decryption of abe ciphertexts. In: USENIX Security Symposium, p. 3 (2011)

13. Ibraimi, L., Tang, Q., Hartel, P., Jonker, W.: Efficient and provable secure ciphertext-policy attribute-based encryption schemes. In: Bao, F., Li, H., Wang, G. (eds.) ISPEC 2009. LNCS, vol. 5451, pp. 1–12. Springer, Heidelberg (2009)

14. Lai, J., Deng, R.H., Guan, C., Weng, J.: Attribute-based encryption with verifiable outsourced decryption. IEEE Trans. Inf. Forensics Secur. **8**(8), 1343–1354 (2013)

15. Li, J., Chen, X., Li, J., Jia, C., Ma, J., Lou, W.: Fine-grained access control system based on outsourced attribute-based encryption. In: Crampton, J., Jajodia, S., Mayes, K. (eds.) ESORICS 2013. LNCS, vol. 8134, pp. 592–609. Springer, Heidelberg (2013)

16. Li, J., Huang, X., Li, J., Chen, X., Xiang, Y.: Securely outsourcing attribute-based encryption with checkability. IEEE Trans. Parallel Distrib. Syst. **25**, 2201–2210 (2013)

17. Li, J., Jia, C., Li, J., Chen, X.: Outsourcing encryption of attribute-based encryption with mapreduce. In: Chim, T.W., Yuen, T.H. (eds.) ICICS 2012. LNCS, vol. 7618, pp. 191–201. Springer, Heidelberg (2012)

18. Sahai, A., Waters, B.: Fuzzy identity-based encryption. In: Cramer, R. (ed.) EUROCRYPT 2005. LNCS, vol. 3494, pp. 457–473. Springer, Heidelberg (2005)

19. Waters, Brent: Ciphertext-policy attribute-based encryption: an expressive, efficient, and provably secure realization. In: Catalano, Dario, Fazio, Nelly, Gennaro, Rosario, Nicolosi, Antonio (eds.) PKC 2011. LNCS, vol. 6571, pp. 53–70. Springer, Heidelberg (2011)

20. Zhou, Z., Huang, D.: Efficient and secure data storage operations for mobile cloud computing. In: Proceedings of the 8th International Conference on Network and Service Management, pp. 37–45. International Federation for Information Processing (2012)

A Simulated Annealing Algorithm for SVP Challenge Through y-Sparse Representations of Short Lattice Vectors

Dan Ding[1(✉)] and Guizhen Zhu[2]

[1] Department of Computer Science and Technology,
Tsinghua University, Beijing 100084, China
dingd09@mails.tsinghua.edu.cn
[2] Data Communication Science and Technology Research Institute,
Beijing 100191, China
zhugz08@gmail.com

Abstract. In this paper, we propose a novel simulated annealing algorithm for the shortest vector problem through y-sparse representations of short lattice vectors. A Markov analysis proves that the algorithm guarantees to converge to the shortest vector at a probability 1, under certain conditions to ensure strong ergodicity of its inhomogeneous Markov chain. After that, we propose a polynomial-time approximation version of our algorithm, and the experimental results under benchmarks in SVP challenge [27] show that the simulated annealing one outperforms the famous Kannan's algorithm in two aspects: it runs exponentially faster and it succeeds in searching the shortest vectors in lattices of higher dimensions. Therefore, our newly-proposed algorithm is a fast and efficient SVP solver and paves a completely new road for SVP algorithms.

Keywords: Lattice-based cryptography · Simulated annealing · Shortest vector problem · Inhomogeneous markov chain · Strong ergodicity

1 Introduction

A lattice is a discrete additive subgroup of a Euclidean space $\mathbb{R}^m$. The lattice amounts for the set of all the integral linear combinations of n linear independent vectors $\mathbf{b}_1, \ldots, \mathbf{b}_n \in \mathbb{R}^m$. The fact that lattice vectors are discrete implies that there exists a nonzero shortest vector in the lattice, which leads to the two most famous lattice problems: the shortest vector problem (SVP), which is, given a lattice, to find the shortest nonzero vector in the lattice, and the closet vector problem (CVP), which is, given a lattice and a target vector, to find the lattice vector closest to the target vector.

D. Ding—National Natural Science Foundation of China (Grant No. 61133013) and 973 Program (Grant No. 2013CB834205).
G. Zhu—National Development Foundation for Cryptological Research (No. MMJJ201401003).

M. Yung et al. (Eds.): INTRUST 2014, LNCS 9473, pp. 51–69, 2015.
DOI: 10.1007/978-3-319-27998-5_4

Both problems are of prime importance to the public-key cryptography in recent years, because, as a promising candidate for post-quantum cryptosystems, a variety of new public-key cryptography [4,12,26] and one-way functions [2,21, 22] are proposed based on the hardness of the two famous lattice problems and their variants. CVP has long been proved to be NP-hard by P. van Emde Baos in 1981 through classical Cook/Karp reduction [31], and the proof is refined by D. Miccancio et al. [20], while, at the same time, the hardness of other lattice problem SVP remains an open problem until SVP is proved to be NP-hard under a randomized reduction by M. Ajtai in 1998 [3]. Therefore, both CVP and SVP are hard enough to afford the security of lattice-based cryptography.

Due to the hardness of SVP, SVP algorithms for searching the exact or approximate shortest lattice vector are attracting more and more attention from the cryptology community. In the past 30 years, a number of algorithms are proposed to attempt the shortest (short) vectors in lattices, since the seminal paper by A. K. Lenstra et al. [18] presents the celebrated **LLL** algorithm for a short lattice basis in 1982. We divide the variety of SVP algorithms into two categories: the theoretically sound algorithms and the practically sound ones. The algorithms that are included in the former category enjoy rigorous theoretical proofs of time and space complexity, but they are mostly of exponential space complexity, which defies practical implementations for high-dimensional lattices due to shortage of computational resources. The sieve algorithms [5,25,32] and the Voronoi-cell computation-based algorithm [23] are of this category, which are both proved to be $2^{\mathcal{O}(n)}$ in time and space complexity. The algorithms in the latter category, though some of them might fail to have theoretical analysis of time complexity, are mostly of polynomial space complexity and amenable for practical realization and gains fabulous experimental results in searching the short vectors in lattices of high-density (SVP challnge). The **BKZ** algorithms [8,28], and enumeration algorithms (such as the Kannan-Helfrich algrotihm [15] and enumeration with extreme pruning [10]) fall into this category, all of which are of polynomial space complexity. A novel genetic algroithm for the shortest vector problem is proposed in 2014 [9], and it behaves well in experimental results and running time, though it is still an open problem to estimate theoretically its time complexity.

In this paper, we aims at propose a SVP algorithm of a completely new type using the ideas of *simulated annealing*. Annealing is a process in condensed matter physics: it, firstly, heats a solid up to a temperature at which all particles in the solid are free to rearrange themselves randomly, and, then, it cools down gradually the solid to make its particles forms into a regularly-arranged, crystallized structure. The simulated annealing apologizes the annealing of solids and simulation of solving large-scale NP-hard combinatorial optimization problems. For this reason, the name "simulated annealing" comes into being. Simulated annealing algorithms [7,16,19] have widely studied and applied to a variety of NP-hard problems for searching optimum solutions, such as the traveling salesman problem (TSP). The simulated annealing proceeds as follows: first, it generates a sequence of solutions by choosing randomly from the neighbours of current solution, and a control parameter was adopted to decide whether a newly-generated solution is accepted; at first, the new solution is always accepted no matter how good it is, and, the control parameter decreasing slowly (cooling down), the algorithm are more and more likely to accept better solutions, and finally it only accept better ones, and the best solution ever found is returned.

Our contributions in this paper are twofold: first, we propose a novel simulated annealing algorithm for the shortest vector problem, and prove theoretically that the algorithm will guarantee to converge to the shortest vector under certain conditions, by proving strong ergodicity of the Markov chain we build up for the algorithm; second, we implement a practical approximation algorithm with some implementation details (choosing suitable parameters for the algorithm to be of polynomial time complexity) and perform a plethora of experiments to show that the algorithm is efficient in searching the shortest vectors in a variety of lattices in SVP challenge [27]. Therefore, our simulated annealing algorithm for the shortest vector problem is a fast and efficient SVP solver, which paved a completely new direction for both practically and theoretically sound SVP algorithms.

The rest of the paper is organized as follows: Sect. 2 provides some necessary backgrounds, including lattices, LLL- and BKZ-reduced basis, y-sparse representations, and Markov processes; In Sect. 3, we describe in detail our simulated annealing algorithm for shortest vector problem; A Markov analysis is performed in Sect. 4, and a fast approximation algorithm of simulated annealing is proposed in Sect. 5. Experimental results are reported and compared in Sect. 6, and the conclusion and future work are proposed in Sects. 7 and 8 respectively.

2   Preliminaries

Let n be an integer, and let $\mathbb{R}^n$ be the n-dimensional Euclidean space with the inner product, denoted as $\langle \cdot, \cdot \rangle$, and the Euclidean norm of $\mathbf{v}$ is defined as $\|\mathbf{v}\| = \sqrt{\sum_{i=1}^{n} v_i^2}$, in which $\mathbf{v} = (v_1, \ldots, v_n) \in \mathbb{R}^n$. The closed sphere in $\mathbb{R}^n$ is denoted as $\mathcal{B}_n(\mathbf{O}, r)$ with $\mathbf{O}$ as its origin and r its radius. The linear space spanned by a set of vectors is denoted by $\mathrm{span}(\cdot)$ and its orthogonal complement $\mathrm{span}(\cdot)^\perp$, and $\mathbf{B}^T$ is the transpose of a matrix $\mathbf{B}$. We denote $\lfloor \cdot \rceil$ as the closest integer to a real number.

2.1   Lattices

A *lattice* $\mathcal{L}$ is defined as the set of all integral combinations of n linear independent vectors $\mathbf{b}_1, \mathbf{b}_2, \ldots, \mathbf{b}_n \in \mathbb{R}^m (m \geq n)$, where the vectors are referred to as the *basis* of the lattice, and n as its *rank*. If $m = n$, the lattice is called *full-rank*. All the lattices we discuss throughout this paper are full-rank lattices unless specified otherwise.

Conveniently, if we have a matrix $\mathbf{B} = [\mathbf{b}_1, \ldots, \mathbf{b}_n] \in \mathbb{R}^{n \times n}$ with the n linear independent vectors as its columns, we define the lattice $\mathcal{L}$ generated by the basis $\mathbf{B}$, denoted as $\mathcal{L}(\mathbf{B})$, as

$$\mathcal{L}(\mathbf{B}) = \{\mathbf{Bx}|\mathbf{x} \in \mathbb{Z}^n\} = \{\mathbf{v} \in \mathbb{R}^n \,|\, \mathbf{v} = \sum_{i=1}^{n} \mathbf{b}_i x_i, x_i \in \mathbb{Z}\}.$$

A single lattice can be generated by a series of equivalent bases, or, in other words, the basis of one lattice is not unique. For a specific basis $\mathbf{B}$, we define the fundamental parallelepiped $\mathcal{P}(\mathbf{B})$ of the lattice as

$$\mathcal{P}(\mathbf{B}) = \{\mathbf{Bx}|\mathbf{x} = (x_1, \ldots, x_n) \in \mathbb{R}^n, \text{ for all } 0 \le x_i < 1, i = 1, 2, \ldots, n\}.$$

The *determinant* $\det(\mathcal{L})$ of a lattice $\mathcal{L}$ is defined as the volume of the fundamental parallelepiped $\mathcal{P}(\mathbf{B})$ by selecting any basis $\mathbf{B}$. More precisely, for any basis $\mathbf{B}$ of a lattices $\mathcal{L}$, the determinant of $\mathcal{L}$ is:

$$\det(\mathcal{L}) = \sqrt{\det(\mathbf{B}^T \mathbf{B})} = \sqrt{\det(\langle \mathbf{b}_i, \mathbf{b}_j \rangle)_{0 \le i,j \le n}}.$$

The determinant of a lattice is well-defined in the sense that the determinant does not depend on the choice of the basis. The ith *successive minimum* $\lambda_i(\mathcal{L})$ of a lattice $\mathcal{L}$ is defined by the smallest radium of a sphere within which there are i linearly independent lattice points, or:

$$\lambda_i(\mathcal{L}) = \inf\{r \in \mathbb{R}^n \,|\, \dim\{\operatorname{span}(\mathcal{L} \cap \mathcal{B}_n(\mathbf{O}, r))\} = i\}.$$

For a basis $\mathbf{B} = [\mathbf{b}_1, \ldots, \mathbf{b}_n]$ of a lattice $\mathcal{L}(\mathbf{B}) \in \mathbb{R}^{n \times n}$, we define its *Gram-Schmidt Orthogonalization* $\mathbf{B}^* = [\mathbf{b}_1^*, \ldots, \mathbf{b}_n^*]$ as the Gram-Schmidt orthogonalization procedure:

$$\mathbf{b}_i^* = \mathbf{b}_i - \sum_{j=1}^{i-1} \mu_{i,j} \mathbf{b}_j^*,$$

where

$$\mu_{i,j} = \frac{\langle \mathbf{b}_i, \mathbf{b}_j^* \rangle}{\langle \mathbf{b}_j^*, \mathbf{b}_j^* \rangle}, \text{ for } 1 \le i < j \le n.$$

In other words, the Gram-Schmidt procedure is projecting $\mathbf{b}_i$ to the space orthogonal to the space spanned by $\mathbf{b}_1, \ldots, \mathbf{b}_{i-1}$.

Provided with a basis $\mathbf{B} = [\mathbf{b}_1, \ldots, \mathbf{b}_n]$, we have:

$$[\mathbf{b}_1, \mathbf{b}_2, \ldots, \mathbf{b}_n]_{n \times n} = [\mathbf{b}_1^*, \mathbf{b}_2^*, \ldots, \mathbf{b}_n^*]_{n \times n} \begin{bmatrix} 1 & \mu_{2,1} & \cdots & \mu_{n,1} \\ & 1 & \cdots & \mu_{n,2} \\ & & \ddots & \vdots \\ & & & 1 \end{bmatrix}_{n \times n} . \quad (1)$$

Thus we have

$$\mathbf{b}_i = \mathbf{b}_i^* + \sum_{j=1}^{i-1} \mu_{i,j} \mathbf{b}_j^*, \text{ for } 1 \le i \le n \quad (2)$$

and $\det(\mathbf{B}) = \prod_{i=1}^{n} \|\mathbf{b}_i^*\|$.

For more details about lattices, refer to [20].

2.2 Korkin-Zolotarev Basis and Blockwise Korkin-Zolotarev Basis

Let $\mathbf{B} = [\mathbf{b}_1, \ldots, \mathbf{b}_n]$ be a basis of a lattice $\mathcal{L} \in \mathbb{R}^n$. We define an operator $\pi_i : \mathbb{R}^n \mapsto \text{span}(\mathbf{b}_1, \ldots, \mathbf{b}_{i-1})^\perp$ as the projection on the orthogonal complement of the span of the first $i - 1$ bases of $\mathbf{B}$, for all $i \in \{1, 2, \ldots, n\}$. We define $\mathcal{L}_i^{(k)}$ as the lattice of rank k generated by the basis $[\pi_i(\mathbf{b}_i), \ldots, \pi_i(\mathbf{b}_{i+k-1})]$ in which $i + k < n + 1$. Clearly, it is true that $\mathcal{L}_i^{(n-i+1)} = \pi_i(\mathcal{L})$, which denotes the lattice of rank $n - i + 1$ generated by basis $[\pi_i(\mathbf{b}_i), \ldots, \pi_i(\mathbf{b}_n)]$. Hence, we have

$$
\left[\pi_i(\mathbf{b}_i), \pi_i(\mathbf{b}_{i+1}), \ldots, \pi_i(\mathbf{b}_n) \right]_{n \times (n-i+1)} =
$$
$$
\left[\mathbf{b}_i^*, \mathbf{b}_{i+1}^*, \ldots, \mathbf{b}_n^* \right]_{n \times (n-i+1)}
\begin{bmatrix}
1 & \mu_{i+1,i} & \cdots & \mu_{n,i} \\
 & 1 & \cdots & \mu_{n,i+1} \\
 & & \ddots & \vdots \\
 & & & \mu_{n,n-1} \\
 & & & 1
\end{bmatrix}_{(n-i+1) \times (n-i+1)},
$$
$$(3)$$

which yields that $\pi_i(\mathbf{b}_j) = \mathbf{b}_j^* + \sum_{k=i}^n \mu_{j,k} \mathbf{b}_k^*$ for $i < j$, and, in particular, we have $\pi_i(\mathbf{b}_i) = \mathbf{b}_i^*$ and $\pi_i(\mathbf{b}_j) = 0$ for $j < i$.

In terms of the denotations aforementioned, we define a basis $\mathbf{B} = [\mathbf{b}_1, \ldots, \mathbf{b}_n]$ as *reduced in the sense of Korkin and Zolatarev* or *Korkin-Zolatarev basis* if it satisfies that:

1. its $|\mu_{i,j}| \leq 1/2$, for $1 \leq j < i \leq n$;
2. $\pi_i(\mathbf{b}_i)$ is the shortest vector of the lattice $\mathcal{L}_i^{(n-i+1)}$ under the Euclidean norm, for $1 \leq i \leq n$.

Similarly, we can further define a basis $\mathbf{B} = [\mathbf{b}_1, \ldots, \mathbf{b}_n]$ as a *blockwise Korkin-Zolotarev basis* with block size β, or *BKZ-reduced* if the following holds:

1. its $|\mu_{i,j}| \leq 1/2$, for $1 \leq j < i \leq n$;
2. $\pi_i(\mathbf{b}_i)$ is the shortest vector of the lattice $\mathcal{L}_i^{(\min(\beta, n-i+1))}$ under the Euclidean norm, for $1 \leq i \leq n$.

2.3 y-Sparse Representations of Short Lattice Vectors: Lattice Vectors from Another Point of View

As defined in Subsect. 2.1, a lattice vector $\mathbf{v} \in \mathcal{L}(\mathbf{B})$ can be represented as $\mathbf{v} = \mathbf{B}\mathbf{x}$, in which $\mathbf{x}$ is an integer vector. Then $\mathbf{x}$ can corresponds to a specific lattice vector $\mathbf{v}$ under a basis $\mathbf{B}$. The y-sparse representation is to regards the lattice $\mathbf{v}$ from another point of view, which is endowed with some excellent properties.

Given a lattice basis $\mathbf{B} = [\mathbf{b}_1, \ldots, \mathbf{b}_n]$ and its Gram-Schmidt orthogonalization $\mathbf{B}^* = [\mathbf{b}_1^*, \ldots, \mathbf{b}_n^*]$ with its factor matrix $\mu = \{\mu_{ij}\}_{1 \leq i,j \leq n} \in \mathbb{R}^{n \times n}$ such that $\mathbf{B} = \mathbf{B}^* \mu^T$, for any vector $\mathbf{v} \in \mathcal{L}(\mathbf{B})$, or $\mathbf{v} = \mathbf{B}\mathbf{x}$, in which $\mathbf{x} = [x_1, \ldots, x_n] \in \mathbb{Z}^n$, we define another vector $\mathbf{t} = [t_1, \ldots, t_n] \in \mathbb{R}^n$ as, for $1 \leq i \leq n$,

$$
t_i = \begin{cases}
0 & \text{for } i = n, \\
\sum_{j=i+1}^n \mu_{j,i} x_j & \text{for } i < n.
\end{cases}
$$

and another vector $\mathbf{y} = (y_1, y_2, \ldots, y_n) \in \mathbb{Z}^n$ as, for $1 \leq i \leq n$,

$$y_i = \lfloor x_i + t_i \rceil.$$

Thereby, the definition establishes a one-to-one correspondence between a lattice vector $\mathbf{v}$ and its $\mathbf{y}$ as below:

$$\mathbf{y} \xleftarrow{\ \mathbf{y}=\mathbf{x}+\lfloor \mathbf{t} \rceil\ } \mathbf{x} \xleftarrow{\ \mathbf{v}=\mathbf{Bx}\ } \mathbf{v},$$

We call $\mathbf{v}$ is correspondent to $\mathbf{y}$, or, $\mathbf{v} \sim \mathbf{y}$.

We call such a representation as sparse because most of the elements in $\mathbf{y}$ corresponding to short lattice vectors under a **BKZ**-reduced basis are zero's. For example, the shortest vector of the 40-dimensional lattice (generated by $seed = 0$) in SVP challenge $\mathbf{v}$ is (-398 -305 -268 125 96 214 284 -108 37 -2 402 228 -243 -33 -76 -265 -3 558 323 552 -419 -408 217 2 440 375 -153 108 79 80 -299 -81 385 -80 -53 -294 -170 380 164 172), (and $\|\mathbf{v}\| = 1702$), and its corresponding $\mathbf{y}$ under its 5-**BKZ** reduced basis is (0 0 0 0 0 0 0 0 0 0 0 0 0 0 0 0 0 0 0 0 0 1 0 0 0 0 0 0 0 0 0 0 -1 -1 0 0 0 1). We can see that only 4 nonzero elements in $\mathbf{y}$ and they are all distributed in the second half with absolute value of 1.

Actually, most y-sparse representations of the short vectors in the lattice under a **BKZ**-reduced basis shares this excellent property. Therefore, for an integer vector $\mathbf{y} = (y_1, y_2, \ldots, y_n)$ corresponding to a short vector in a lattice under a **BKZ**-reduced basis, we have the following two heuristics as follows:

1. The first half integer elements $y_1, \ldots, y_{\lfloor n/2 \rfloor}$ in $\mathbf{y}$ are all zero's;
2. The absolute value $y_{\lfloor n/2 \rfloor+1}, \ldots, y_n$ of the second half integers in $\mathbf{y}$ is bounded by $\sqrt{\frac{\lambda_1}{\|\mathbf{b}_i^*\|}}$ instead of $\frac{\lambda_1}{\|\mathbf{b}_i^*\|}$, for $\lfloor n/2 \rfloor + 1 \leq i \leq n$;

Moreover, we call an lattice vector $\mathbf{v}$ is *feasible* if its corresponding $\mathbf{y} = (y_1, y_2, \ldots, y_n)$ satisfies the following condition:

$$\sum_{i=1}^{n} y_i^2 \|\mathbf{b}_i^*\|^2 \leq 4\lambda_1^2(\mathbf{B}). \tag{4}$$

Since, as discussed in [9], any lattice vector corresponding to $\mathbf{y}$ has a maximum squared norm of $\sum_{i=1}^{n} y_i^2 \|\mathbf{b}_i^*\|^2$, the vector can be the shortest vector under the condition that its maximum squared norm is smaller than the squared first minima $\lambda_1(\mathbf{B})^2$. Equivalently, that a vector is feasible is the same as that it lies in the hypersphere as expressed in Eq. 4.

For a more rigorous treatment of y-sparse representation, refer to [9].

3   Simulated Annealing: A Novel Algorithm for the Shortest Vector Problem

In this section, we will described in detail our simulated annealing algorithm for the shortest vector problem.

3.1 Motivations: The Annealing Process from Condensed Matter Physics

The simulated annealing algorithm, as its original form in [7,16], attempts to solve the large scale optimization problems by simulating the annealing process of solids. For this reason, the algorithm is called "simulated annealing". In condensed matter physics, annealing denotes a process as: a solid is heated up to the maximum temperature at which all the particles of the solid are free to rearrange themselves randomly in a liquid phase, followed by a cooling process in which the temperature is lowered such that the particles arrange themselves in the ground energy state into a regularly rearranged, crystallized phase, under the conditions that the maximum temperature is high and the cooling process is slow enough. To be more precise, the annealing starts off at the maximum temperature, say T_0, and the cooling phase of the annealing process can be described as follows: at each temperature $T(T < T_0)$, the probability that particles of the solid are at the energy state E (and the ground state is 0) is given by the *Boltzmann distribution*:

$$Pr\{\mathbf{E} = E\} = \frac{1}{N(T)} \cdot \exp(-\frac{E}{k_B T}),$$

in which $N(T)$ is the normalization factor, and k_B is Boltzmann constant. As the temperature T deceases, the Boltzmann distribution concentrates the on the state with lowest energy, or the ground energy state, and, finally, when the temperature $T = 0$, only the ground state gains a non-zero probability of occurrence. That is, the particle can only be at the ground state at last, and the process of annealing ends up.

As far as a combinatorial optimization problem is concerned, the energy state can be taken as the cost function C for each configuration and the control parameter c replaces the temperature T. Then, the simulated annealing algorithm can be viewed similarly as a procedure in which a sequence of configurations is generated as the control parameter c is decreasing slowly, and finally it generates the configuration with the smallest value of cost function, or the optimum solution for the combinatorial problem. The algorithm can be described as follows. First, the control parameter c is set as a high value, and, then, a sequence of configurations is generated as the control parameter c decreases: a generation mechanism is defined to, given a current configuration i, choose at random another configuration j from i's neighbourhood, corresponding to a small perturbation added to i. Let the ΔC_{ij} be the difference of the values cost function of the two configurations i and j, or $C(j) - C(i)$, then the probability that j is chosen as the next configuration in the sequence is given as follows: if $\Delta C_{ij} < 0$, or j is better than i, j is accepted at a probability 1, and, if $\Delta C_{ij} \geq 0$, or j is no good as i, i is accepted by the Boltzmann distribution at a probability of $\exp(-\frac{\Delta C_{ij}}{c})$, as shown as below

$$Pr\{j \text{ is chosen from } i\} = \begin{cases} 1 & \text{if } \Delta C_{ij} < 0, \\ \exp(-\frac{\Delta C_{ij}}{c}) & \text{if } \Delta C_{ij} \geq 0. \end{cases}$$

Table 1. The Simulated Annealing Algorithm for SVP The Simulated Annealing Algorithm for SVP

Input: A β-**BKZ** reduced basis $\mathbf{B} = [\mathbf{b}_1, \ldots, \mathbf{b}_n]$ of a lattice $\mathcal{L}$.
Output: The Shortest Nonzero Vector $\mathbf{v}'$ in the lattice $\mathcal{L}(\mathbf{B})$
1. Compute $\mathbf{B}$'s Gram-Schmidt Orthogonalizations $\mathbf{B}^* = [\mathbf{b}_1^*, \ldots, \mathbf{b}_n^*]$ and its factor matrix $\mu = \{\mu_{ij}\}_{1 \leq i,j \leq n}$;
2. Estimate the first minima λ_1 of the lattice by Gaussian Heuristic;
3. Initialize $\alpha = (\alpha_1, \ldots, \alpha_n)$ as $\alpha_i \leftarrow 0$, for $1 \leq i < \lfloor \frac{n}{2} \rfloor$, and $\alpha_i \leftarrow \sqrt{\frac{\lambda_1}{\|\mathbf{b}_i^*\|}}$, for $\lfloor \frac{n}{2} \rfloor \leq i \leq n$;
4. Choose a y-sparse representation of a starting vector $\mathbf{y}$ at random;
5. $\mathbf{v}'' = \text{SIMULATEDANNEALING}(\mathbf{B}, \mathbf{B}^*, \alpha, \mathbf{y})$;
6. **Return** $\mathbf{v}''$;

Clearly, as the control parameter c is decreasing, the worse configurations are accepted at a smaller and smaller probability, and, finally, while the control parameter c is approaching zero, only better ones are accepted (that is only configurations with smaller value of cost function enjoy a nonzero probability and all the worse ones a probability of zero). Therefore, the simulated annealing is reduced into a hill climbing, or local search, as the control parameter is zero.

In the following three subsections, we apply the simulated annealing to the shortest vector problem, or SVP, which is a classical combinatorial optimization problem, and the section that follows we will prove that the simulated annealing algorithm will be doomed to converge to the shortest lattice vector under the circumstances that the control parameter c is, initially, high, and decreases slowly enough.

3.2 Overview

In this subsection, we will discuss the main body of our simulated annealing algorithm. As shown in Table 1, given a β-**BKZ** reduced lattice basis $\mathbf{B}$, the algorithm first computes its Gram-Schmidt orthogonalization $\mathbf{B}^* = [\mathbf{b}_1^*, \ldots, \mathbf{b}_n^*]$, as well as its factor matrix $\mu = \{\mu_{i,j}\}_{1 \leq i,j \leq n}$. Second, it estimates the Euclidean length of the first minima, or the shortest vector, using the Gaussian heuristic [8]. Third, we bound the first half of the y-sparse representation of the short lattice vectors as 0 and the second half as $\sqrt{\lambda_1/\|\mathbf{b}_i^*\|}$, by the heuristic for random lattices from SVP challenge as in Sect. 2. After that, an initial vector (its y-sparse representation $\mathbf{y}$) in lattice $\mathcal{L}(\mathbf{B})$ is chosen at random, and the algorithm invokes the simulated annealing procedure SIMULATEDANNEALING() by giving $\mathbf{y}$ as the parameter to find the shortest vector $\mathbf{v}''$. Finally, the shortest vector is returned and the whole algorithm terminates.

3.3 Simulated Annealing

SIMULATEDANNEALING() is the main procedure of our simulated annealing algorithm for SVP, which, on an input of an initial lattice vector, it simulates the cooling process, or annealing, to find the shortest vector in the given lattice. Before devoting to describing the procedure, we should define the generation mechanism, the cooling schedule, which denotes the scheme how the control factor deceases, and the cost function for the simulated algorithm as aforementioned in the first subsection: the cost function is defined as the Euclidean norm, or the ℓ_2-norm, of the lattice vectors, and the optimum solution is, thence, the shortest lattice vector under the Euclidean norm (the first minimum λ_1); the cooling schedule as in [19] by multiplying the current control parameter c_k with a decreasing coefficient $\beta < 1$, i.e.,

$$c_{k+1} = \beta \cdot c_k, (k = 0, 1, \ldots),$$

and we have

$$c_k = \beta^k \cdot c_0;$$

the generation mechanism is defined to choose a new vector from the current vector from its neighbourhood by adding a small perturbation to the current vector, which we implement by PERTURB() in the next subsection.

As in Table 2, the procedure proceeds as follows. Initially, it set the initial control parameter c_0 and the decreasing coefficient β, and, second, it sets k as 0 and the final value of the control parameter c_{final} is set as a small real number approximate to 0. Note that the initial control parameter c_0 must be high enough to accept all the neighbour vectors enjoys a good accept probability, and the decreasing coefficient β should be close to 1 and the c_{final} must be close to 0 so that the cooling schedule is slow, because that the next control parameter is updated as $c_{k+1} = \beta \cdot c_k$. In fact, β and c_{final} is set to fix the steps before the termination of simulated annealing procedure, which will be discussed in Sect. 5. Third, the procedure initializes the optimum solution, or the current shortest vector, $\mathbf{v}''$ as the last basis vector $\mathbf{b}_n$, or its y-sparse representation $(0, \ldots, 0, 1)$. Fourth, we compute the lattice vector $\mathbf{v}$ that corresponds to the y-sparse representation of the initial vector $\mathbf{y}$.

After that, the procedure enters a **while**-loop which performs the cooling process iteratively and search the shortest vector. Entering the loop, it generates a new vector $\mathbf{y}'$ from our current vector $\mathbf{y}$ by calling the subroutine PERTURB() as in the next subsection, which choose a new vector from the neighbourhood of the current vector at random. Note that the perturbation process, or the generation mechanism, operates on the y-sparse representations of the lattice vectors. Then, we compute the lattice vector $\mathbf{v}'$ corresponding the newly-generated y-sparse representation $\mathbf{y}'$. After that, the procedure decides whether we accept the new vector $\mathbf{v}'$ as follows: if the new vector $\mathbf{v}'$ is shorter than the current vector $\mathbf{v}$, i.e., $\|\mathbf{v}'\| < \|\mathbf{v}\|$, we accept that new one by setting $accept$ as 1; however, if the new vector $\mathbf{v}'$ is longer, we first choose a random value r from $[0, 1)$, and, then, compare it to the exponential of the minus difference of the lengths of the two vector divided by the current control parameter c_k, i.e., $\exp(-\frac{\|\mathbf{v}'\| - \|\mathbf{v}\|}{c_k})$, and, finally, if r is smaller, we accept $\mathbf{v}'$, and we reject $\mathbf{v}'$ if r is not. Clearly, we accept the new vector $\mathbf{v}'$ at a 100 % probability if $\mathbf{v}'$ is shorter than $\mathbf{v}$, and the acceptance probability for the $\mathbf{v}'$ that is longer obeys the Boltzmann

Table 2. SIMULATEDANNEALING()

Input: B $= (\mathbf{b}_1, \mathbf{b}_2, \ldots, \mathbf{b}_n)$, **B***, the bound α, and the initial vector $\mathbf{y} = (y_1, \ldots, y_n)$;
Output: The shortest vector $\mathbf{v}'' = (v_1'', \ldots, v_n'') \in \mathcal{L}(\mathbf{B})$.

1. Initialize c_0, β, and c_{final}; //c_0 is high, β is close to 1, and c_{final} is close to 0
2. Let $k \leftarrow 0$;
3. Let $\mathbf{v}'' \leftarrow \mathbf{b}_n$, $\mathbf{y}'' \leftarrow (0, \ldots, 0, 1)$;
4. Compute $\mathbf{v} \sim \mathbf{y}$;
5. **While** $\|\mathbf{v}\| \leq \lambda_1$ **and** $c_k > c_{final}$ **do**
   (a) $\mathbf{y}' \leftarrow$ PERTURB($\mathbf{B}, \mathbf{B}^*, \alpha, \mathbf{y}$);
   (b) Compute $\mathbf{v}' \sim \mathbf{y}'$;
   (c) **If** $\|\mathbf{v}'\| < \|\mathbf{v}\|$ **then** $accept \leftarrow 1$
   (d) **else**
      (1) Choose $r \in [0, 1)$ randomly;
      (2) **If** $r < \exp(-\frac{\|\mathbf{v}'\| - \|\mathbf{v}\|}{c_k})$ **then** $accept \leftarrow 1$
      (3) **else** $accept \leftarrow 0$;
   (e) **If** $accept = 1$ **then**
      (1) Let $\mathbf{y} \leftarrow \mathbf{y}'$, $\mathbf{v} \leftarrow \mathbf{v}'$;
      (2) Let $c_{k+1} \leftarrow \beta \cdot c_k$, $k \leftarrow k + 1$;
      (3) **If** $\|\mathbf{v}\| < \|\mathbf{v}''\|$ **then** $\mathbf{v}'' \leftarrow \mathbf{v}$, $\mathbf{y}'' \leftarrow \mathbf{y}$;
6. **Return** $\mathbf{v}''$.

distribution under the current control parameter c_k. Finally, if the new vector is not accepted, the subroutine goes back to the start of the loop and generates new vectors and decide whether to accept again in the same way, and, if the new vector is accepted ($accept = 1$), it updates the current vector as the newly-generated one by $\mathbf{v}' \leftarrow \mathbf{v}$ and $\mathbf{y}' \leftarrow \mathbf{y}$, and updates the control parameter c_{k+1} as $\beta \cdot c_k$, and replaces k by $k + 1$. If the newly-generated vector $\mathbf{v}'$ is shorter than the current optimum solution, or the current shortest vector, $\mathbf{v}''$, the $\mathbf{v}''$ is updated as $\mathbf{v}'$ and $\mathbf{y}''$ is set as $\mathbf{y}'$.

The loop iterates itself until the shortest vector is found, i.e., $\|\mathbf{v}''\| < \lambda_1$, or the control parameter has been approaching 0, i.e., $c_k < c_{final}$. Finally, the subroutine returns the best vector $\mathbf{v}''$ ever found during the process of annealing.

3.4 Perturbation: The Generation Mechanism

In this subsection, we describe in detail the generation mechanism of the simulated annealing algorithm for the shortest vector problem. As shown in Table 3, the subroutine PERTURB() generates, on a given vector $\mathbf{y}$, a new vector $\mathbf{y}'$, which is feasible (in the hypersphere of short vectors as discussed in Sect. 2), from its neighbourhood randomly, by adding a small perturbation, on the space of y-sparse representations of lattice vectors.

The subroutine starts with setting the loop control variable p as 0, and enters a **while**-loop. In the loop, the subroutine chooses an index i from the second half of the elements in $\mathbf{y}$ randomly, since only the second half elements are considered as discussed in Sect. 2. After that, it generates a new vector $\mathbf{y}'$ by updating the ith element y_i in $\mathbf{y}$ as y_{i+1} or y_{i-1} at random. Equivalently, if we define the neighbourhood set of a vector $\mathbf{y}$ as the set of all vectors $\mathbf{y}'$ satisfying that the ℓ_1-distance between the new vector $\mathbf{y}'$ and the starting vector $\mathbf{y}$ is 1, i.e., $\|\mathbf{y}' - \mathbf{y}\|_1 = 1$, or

$$neighbour(\mathbf{y}) = \{\mathbf{y}' \in \mathbb{Z}^n | \|\mathbf{y}' - \mathbf{y}\|_1 = 1\},$$

or, likewise, under the notation of lattice vectors,

$$neighbour(\mathbf{v}) = \{\mathbf{v}' \in \mathcal{L}(\mathbf{B}) | \|\mathbf{y}' - \mathbf{y}\|_1 = 1, \mathbf{v} \sim \mathbf{y}, \mathbf{v}' \sim \mathbf{y}'\},$$

then the process of generation mechanism is choosing a vector from its neighbourhood set randomly, and the probability of each neighbour vector obeys the uniform distribution. However, if the newly-generated vector is the zero vector, then p is set as 1 to execute the loop one more time; likewise, if the ith element y_i in $\mathbf{y}$ is larger than the bound α_i or if the sum of all the squared norm $\|\mathbf{b}_i^*\|^2$ multiplied by the squared elements y_i^2 is larger than λ_1, the new vector $\mathbf{y}'$ is not feasible, that is, the new vector $\mathbf{y}'$ is beyond the hypersphere of the short vectors:

$$\sum_{i=1}^{n} y_i^2 \|\mathbf{b}_i\|^2 < 4\lambda_1^2, \tag{5}$$

and, therefore, p should also be set as 1 to choose another vector. The loop terminates while a feasible and nonzero vector $\mathbf{y}'$ is generated, and, after that, the newly-generated vector $\mathbf{y}'$ is returned.

4 Convergence Proof of the Simulated Annealing Algorithm

In this section, we presents the convergence proof of our simulated annealing algorithm for the shortest vector problem. In Subsect. 4.1, we build a mathematical model for our simulated annealing by means of a Markov chain, and in Subsect. 4.2, based on the Markov chain, we prove that our simulated annealing will always converge to the shortest vector under certain conditions.

4.1 Mathematical Model: An Inhomogeneous Markov Chain

As in Sect. 3, the simulated annealing algorithm for the shortest vector problem can be viewed as an algorithm that consecutively transform the current configuration (vector) into a new configuration from its neighbourhood set. This mechanism is best described mathematically as a Markov chain: generating a sequence of configurations, or trials, in which the new trial is only dependent on the previous one as in [17]. In the case of our simulated annealing algorithm for the shortest vector problem, it is clear that the newly-generated vector only

Table 3. PERTURB()

Input: B, B* $= (\mathbf{b}_1^*, \ldots, \mathbf{b}_n^*)$, $\alpha = (\alpha_1, \ldots, \alpha_n)$, and an integer vector $\mathbf{y} = (y_1, \ldots, y_n)$;
Output: Another *feasible* integer vector $\mathbf{y}' = (y_1', \ldots, y_n')$ such that $\mathbf{y}' \in neighbour(\mathbf{y})$.

---
1. Let $p \leftarrow 0$;
2. **While** $p = 0$ **do**
  //choose a new vector $\mathbf{y}' \in neighbour(\mathbf{y})$
  (a) Choose $i \in \{\lfloor \frac{n}{2} \rfloor, \ldots, n\}$ randomly;
  (b) Choose $sgn \in \{0, 1\}$;
  (c) **If** $sgn = 0$ **then**
    (1) $\mathbf{y}' \leftarrow (y_1, \ldots, y_{i-1}, y_i + 1, y_{i+1}, \ldots, y_n)$
  (d) **else**
    (1) $\mathbf{y}' \leftarrow (y_1, \ldots, y_{i-1}, y_i - 1, y_{i+1}, \ldots, y_n)$;
  (e) **If** $\mathbf{y}' = (0, 0, \ldots, 0)$ **then** $p \leftarrow 0$
  (f) **else if** $|y_i'| > \alpha_i$ **then** $p \leftarrow 0$
  (g) **else**
    (1) $sum \leftarrow \sum_{j=1}^n y_j^2 \cdot \|\mathbf{b}_j^*\|^2$;
    (2) **If** $sum > \lambda_1^2$ **then** $p \leftarrow 0$
    (3) **else** $p \leftarrow 1$;
4. **Return** $\mathbf{y}' = (y_1', \ldots, y_n')$.

---

depends on the current vector, and has nothing to do with the vectors before the current one.

We define the $S = \{\mathbf{v}_i\}_{1 \leq i \leq |S|}$ as the state set of all feasible vectors in hypersphere of short vectors as Eq. 5 in which we assume that the vectors are in a nondescending order in terms of their Euclidean norms (those with the same norm is in an arbitrary order), which is clearly finite. As far as the simulated annealing is concerned, the conditional probability $p_{ij}(k-1, k)$ in which $1 \leq i, j \leq |S|$ and $k \in \mathbb{Z}^n$ denotes the probability that the kth trial is a transition from the ith vector to the jth vector in S. Therefore, we can define the $\mathbf{q}^{(k)} = [p_1^{(k)}, p_2^{(k)}, \ldots, p_{|S|}^{(k)}]$ as the probabilities for all the vectors in S after k transitions, and the transition probabilities are $p_{ij}(k-1, k)$ and the $|S| \times |S|$-matrix $\mathbf{P} = \{p_{ij}(k-1, k)\}_{1 \leq i, j \leq |S|}$ constitutes the transition matrix for the simulated annealing algorithm. Thereby, we build a Markov chain $\{X_k\}_{k=0,1,\ldots}$ with its probability transition matrix $\mathbf{P}$.

The transition probabilities depend on the value of the control parameter c_k at the kth trial. Since the control parameter c_k is deceasing during the process of simulated annealing, the transition probabilities are not constant and, thence, the Markov chain is inhomogeneous. To be more precise, the transition probability transition matrix $\mathbf{P} = \{p_{ij}(c_k)\}_{1 \leq i, j \leq |S|, k=0,1,\ldots}$ at the kth trial can be defined as:

$$p_{ij}(c_k) = \begin{cases} g_{ij}(c_k) \cdot a_{ij}(c_k) & \text{if } j \neq i, \\ 1 - \sum_{\ell=1, \ell \neq i}^{|S|} g_{ij}(c_k) \cdot a_{ij}(c_k) & \text{if } j = i. \end{cases}$$

in which $g_{ij}(c_k)$ denotes the generation probability of generating $\mathbf{v}_j$ from $\mathbf{v}_j$ and $a_{ij}(c_k)$ the acceptance probability of accepting $\mathbf{v}_j$ from $\mathbf{v}_i$ ($\mathbf{v}_i, \mathbf{v}_j \in S$). Clearly, $\mathbf{P}$ is a stochastic matrix. Since the generation mechanism of our simulated annealing is to choose a new vector from the neighbourhood set uniform randomly and each vector has exactly $2n$ vectors in its neighbourhood, the generation probability $g_{ij}(c_k)$ is:

$$g_{ij}(c_k) = \begin{cases} \frac{1}{2n} & \text{if } \mathbf{v}_j \in neighbour(\mathbf{v}_i), \\ 0 & \text{otherwise.} \end{cases}$$

Therefore, the generation probability is a constant independent of c_k, i.e., $g_{ij}(c_k) = g_{ij}$. Similarly, since the acceptance probability obeys the Boltzmann distribution, the acceptance probability $a_{ij}(c_k)$ can be expressed as:

$$a_{ij}(c_k) = \begin{cases} \min\{1, \exp(-\frac{\|\mathbf{v}_j\| - \|\mathbf{v}_i\|}{c_k})\} & \text{if } \mathbf{v}_j \in neighbour(\mathbf{v}_i), \text{ nonzero, and feasible,} \\ 0 & \text{otherwise.} \end{cases}$$

Thus, an inhomogeneous Markov chain $\{X_k\}_{k=0,1,\ldots}$ is built up for the simulated annealing algorithm for the shortest vector problem, by which we will propose the convergence proof of the algorithm in the next subsection.

4.2 Convergence Proof

Theorem 1 (Convergence of the simulated annealing for SVP). *If $\{X_k\}_{k=0,1,\ldots}$ is the inhomogeneous Markov chain aforementioned in Subsect. 4.1 for the simulated annealing algorithm for the shortest vector with the control parameter c_k as described in Sect. 3, the Markov chain will converge to the shortest lattice vector, or:*

$$\lim_{k \to \infty} Pr\{X_k \in S_{opt}\} = 1,$$

under the three following conditions that

$$c_k > c_{k+1},$$

$$\lim_{k \to \infty} c_k = 0,$$

and

$$c_k > \frac{\Gamma}{\log_2 k},$$

in which $k = 0, 1, \ldots$, and the optimum set $S_{opt} = \{\mathbf{v} \in \mathcal{L}(\mathbf{B}) | \|\mathbf{v}\| = \lambda_1(\mathbf{B})\}$, and Γ is a constant, whatever the initial distribution is.

Proof. The theorem that inhomogeneous Markov chain converges to the global minima is a direct consequence of the proof in [24] by proving the strong ergodicity of the inhomogeneous Markov chain (see [29] for weakly and strongly ergodic inhomogeneous Markov chain), and an explicit expression for Γ can be found in [6, 11, 24]. □

Theorem 1 ensures the convergence of the simulated annealing algorithm for the shortest vector problem to the globally minimal solution, or the shortest vector, under the circumstances that the control parameter decreases at a rate as slow as $c_k > \frac{\Gamma}{\log_2 k}$ for some constant Γ. Therefore, the simulated annealing will guarantee to, at a probability 1, find the shortest vector if it cools down slowly enough and the algorithm runs long enough.

5 A Practical Simulated Annealing Algorithm: Worst-Case Time Complexity

Although the simulated annealing algorithm for the shortest vector problem guarantees to converge to the optimum solution, or the shortest lattice vector, with probability 1 if it satisfies the three conditions as aforementioned in Theorem 1, yet any practical implementation will only be an approximation algorithm for the reasons that the algorithm cannot runs for an infinite time to make probability converge to 1, and, therefore, the tiny probability of failure always existed for any practical simulated algorithm. In this section we devote to the implementation details for a practical simulated algorithm will find as short a lattice vector as possible within a fixed number of transitions (or steps), and, thereafter, estimate the time complexity of the approximation simulated annealing algorithm.

The implementation details refers to fixing the three parameters: the initial and final control parameters, c_0 and c_{final}, and the decreasing coefficient β for cooling scheduling. We follows the methodology in [19] as follows. The initial control parameter is chosen to be large enough for the initial distribution is close to uniform, or, i.e., (n denotes the rank of the lattice below)

$$\exp\left(-\frac{\max_i\{\|\mathbf{v}_i\| - \lambda_1\}}{c_0}\right) > \chi_0,$$

in which χ_0 is close to 1, and we have $c_0 > \frac{\max_i\{\|\mathbf{v}_i\| - \lambda_1\}}{-\ln(\chi_0)}$. In our simulated annealing algorithm, we set χ_0 as 0.9 and approximates $\max_i\{\|\mathbf{v}_i\| - \lambda_1\}$ as $n \times \|\mathbf{b}_n\|$.

The final control parameter c_{final} is chosen to make sure that the probability for the current configuration in the Markov chain to be ϵ more than λ_1 is less than a small real number θ, and, then, as in [19], we have

$$c_{final} \leq \frac{\epsilon}{\ln(|S| - 1) - \ln\theta},$$

in which S is the set of feasible lattice vectors in the hypersphere as described in Subsect. 2.3. In our algorithm, we set θ as 10^{-n} and ϵ as λ_1/n.

As for the decreasing coefficient for the control parameter β, it is necessary for the decrement to satisfy the third condition in Theorem 1: $c_k > \mathcal{O}((\ln k)^{-1})$, and, [19] shows that the decrement for the control

$$c_{k+1} = c_k \cdot \left(1 + \frac{\gamma \cdot c_k}{\max_i\{\|\mathbf{v}_i\| - \lambda_1\}}\right)^{-1}.$$

for a small real number δ. Still, we set $\max_i\{\|\mathbf{v}_i\| - \lambda_1\}$ as $n \times \|\mathbf{b}_n\|$ and δ as $1/n$, and, thereby, figure out β.

If we choose the parameters in this way, [1,19] and clearly proves that our algorithm will terminate within $\mathcal{O}(\ln(|S|))$ steps in terms of cooling the control parameter, or, namely, the worst-case time complexity for our simulated annealing. Moreover, Theorem 2 in [13] implies that there exists at most $poly(n) \cdot n^{\frac{n}{2e}+o(n)}$ feasible lattice vectors in the hypersphere, or, that is,

$$|S| = \mathcal{O}(n) \cdot n^{\frac{n}{2e}+o(n)}.$$

Therefore, we have

$$\mathcal{O}(\ln(|S|)) = \mathcal{O}(\ln n \cdot \frac{n}{2e} \cdot \ln n) = \mathcal{O}(n \cdot \ln^2 n),$$

which means that our simulated annealing algorithm will be of polynomial time complexity, thereby obtaining our worst-case time complexity.

One might argue that, though the algorithm terminates within polynomial time, there is no guarantees for proximity of the final vector to optimal solutions, or, the shortest vector. Therefore, we perform experiments in the next section to show the efficiency of the simulated annealing algorithm.

6   Experimental Results

In this section, we perform some experiments of our simulated annealing algorithm for the shortest vector problem. All algorithms are impelemnted using C++ with Victor Shoup's Number Theory Library (NTL) version 6.1.0 [30], and we perform our experiments on a workstation with 16 Intel Xeon 2.6 Ghz CPUs and 64 G RAM under a Red Hat Linux Server release 5.6. Experiments of our random algorithm are performed on the random lattices of dimension 20-100 from SVP challenge benchmarks [27], and the running times are compared to those of Kannan-Helfrich enumeration algorithm [14]. All the random bases are all generated by $seed = 0$, and are all preprocessed by a subroutine of at least $\frac{n}{6}$-**BKZ** reduction before experiments. The running times of simulated annealing algorithm is output by averaging the running times of 20-100 attempts for each dimensions (only the successful attempts are considered). Note that all the implementation details (parameters) follows the discussion in Sect. 5 that precedes.

Figure 1 shows the sequence of the newly-generated vectors by one successful attempt of our simulated annealing algorithm for the random lattices of dimension 40 (seed 0), and the algorithm generates in sequence 1355 lattice vectors and consumes 0.720881 seconds before it reaches the shortest vector of a Euclidean norm approximately 1702. As shown in the figure, the lengths (Euclidean norms) of the sequence of vectors takes on an quite chaotic look (ranged between 3500 and 1500), which is averaged at almost 2500 and shows no obvious trend of decreasing. Therefore, we can see that the control parameter should ensure that the algorithm should at any time escape from any local minimum, which

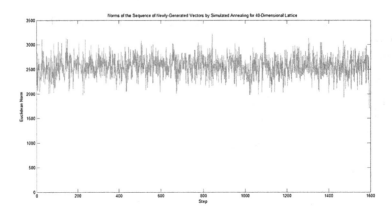

Fig. 1. Norms of the sequence of Newly-Generated vectors by simulated annealing (Dimensional 40)

require that the cooling schedule should be slow as discussed in Subsect. 4.2 (a decreasing coefficient close to 1 is necessary).

Figure 2 compares the running time of our simulated annealing algorithm for the shortest vector problem to the famous Kannan-Helfrich algorithm. As shown in the figure, our simulated annealing algorithm runs less than 1 s for lattices of dimension less than 40, and it runs still quite fast up to dimension 100. The running time comparison implies that our simulated annealing algorithm gains advantages over Kannan-Helfrich algorithm in two aspects: first, it enjoys an exponential speedup to Kannan's algorithm (as discussed above in Sect. 5 our simulated algorithm will terminates within a polynomial time complexity), and, second, it succeeds in finding the shortest vector of the lattice of dimension up to 100 while Kannan's algorithm found up to dimensional 72 before the running time becomes extremely long. Therefore, the simulated annealing algorithm outperforms the famous Kannan's algorithm and is an efficient and fast algorithm for the shortest vector problem.

7   Conclusion

In this paper, we propose a novel simulated annealing algorithm for the shortest vector problem through y-sparse representations of short lattice vectors. The algorithm is, thereafter, proved to converge to the shortest vector at a probability 1, under the conditions that the initial control parameter is high and that cooling scheduling of the control parameter is slow enough. Then, an approximation version of the algorithm as described before is proposed, which is of polynomial time complexity, and it is implemented to perform experiments on a variety of lattices in SVP challenges. The experimental results show that this new algorithm outperforms the famous Kannan's algorithm for SVP and output shortest vectors in lattices of higher dimensions. In conclusion, our newly-proposed simulated annealing algorithm for the shortest vector problem is an completely new and efficient SVP solver, which paves a new road for research into SVP algorithms.

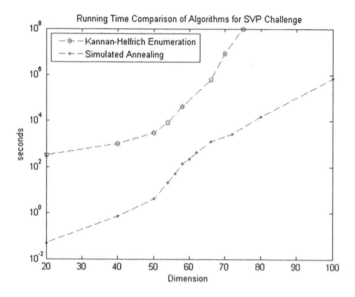

Fig. 2. Running time comparison of SVP algorithms

8   Future Work

In the future, we will continue to challenge some random lattices in SVP challenge with higher dimensions (more than 100) using our simulated annealing algorithm, and compare the results with more currently widely-used practical SVP algorithms: such as the enumeration with extreme pruning, and the genetic algorithm for SVP. Additionally, we will attempts to establish a theoretical relationship between quality of the final vector in simulated annealing (how close the final vector is to the shortest one in lattices) and the running time.

References

1. Aarts, E.H., Laarhoven, V.P.: Statistical cooling: a general approach to combinatorial optimization problems. Philips J. Res. **40**(4), 193–226 (1985)
2. Ajtai, M.: Generating hard instances of lattice problems (extended abstract). In: STOC, pp. 99–108 (1996)
3. Ajtai, M.: The shortest vector problem in ℓ_2 is np-hard for randomized reductions. In: Proceeding of the 30th Symposium on the Theory of Computing (STOC 1998), pp. 284–406 (1998)
4. Ajtai, M., Dwork, C.: A public-key cryptosystem with worst-case/average-case equivalence. In STOC, pp. 284–293 (1997)
5. Ajtai, M., Kumar, R., Sivaumar, D.: A sieve algorithm for the shortest lattice vector problem. In: Proceedings of the 33th annual ACM symposium on Theory of computing (STOC 2001) 33, pp. 601–610 (2001)
6. Anily, S., Federgruen, A.: Simulated annealing methods with general acceptance probabilities. J. Appl. Probab. **24**, 657–667 (1987)

7. Černý, V.: Thermodynamical approach to the traveling salesman problem: an efficient simulation algorithm. J. Optim. Theory Appl. **45**(1), 41–51 (1985)
8. Chen, Y., Nguyen, P.Q.: BKZ 2.0: Better lattice security estimates. In: Lee, D.H., Wang, X. (eds.) ASIACRYPT 2011. LNCS, vol. 7073, pp. 1–20. Springer, Heidelberg (2011)
9. Ding, D., Zhu, G., Wang, X.: A genetic algorithm for searching shortest lattice vector of svp challenge. Cryptology ePrint Archive, Report 2014/489 (2014). http://eprint.iacr.org/
10. Gama, N., Nguyen, P.Q., Regev, O.: Lattice enumeration using extreme pruning. In: Gilbert, H. (ed.) EUROCRYPT 2010. LNCS, vol. 6110, pp. 257–278. Springer, Heidelberg (2010)
11. Geman, S., Geman, D.: Stochastic relaxation, gibbs distributions, and the bayesian restoration of images. IEEE Trans. Pattern Anal. Mach. Intell. **6**, 721–741 (1984)
12. Goldreich, O., Goldwasser, S., Halevi, S.: Public-key cryptosystems from lattice reduction problems. In: Kaliski Jr, B.S. (ed.) CRYPTO 1997. LNCS, vol. 1294, pp. 112–131. Springer, Heidelberg (1997)
13. Hanrot, G., Stehlé, D.: Improved analysis of kannan's shortest lattice vector algorithm. In: Menezes, A. (ed.) CRYPTO 2007. LNCS, vol. 4622, pp. 170–186. Springer, Heidelberg (2007)
14. Kannan, R.: Improved algorithms for integer programming and related lattice problems. In: Proceedings of the 15th Symposium on the Theory of Computing (STOC 1983) 15, pp. 99–108 (1983)
15. Kannan, R.: Minkowski's convex body theorem and integer programming. Math. Oper. Res. **12**, 415–440 (1987)
16. Kirkpatrick, S., Gelatt, C.D., Vecchi, M.P., et al.: Optimization by simmulated annealing. Science **220**(4598), 671–680 (1983)
17. Lawler, G. F. Introduction to Stochastic Processes. CRC Press, Boca Raton (1995)
18. Lenstra, A.K., Lenstra, H.W., Lovász, L.: Factoring polynomials with rational coefficients. Mathematische Annalen **261**(4), 513–534 (1982)
19. Lundy, M., Mees, A.: Convergence of an annealing algorithm. Math. Prog. **34**(1), 111–124 (1986)
20. Micciancio, D., Goldwasser, S.: Complexity of Lattice Problems: A Cryptographic Perspective. The Springer International Series in Engineering and Computer Science, vol. 671. Kluwer Academic Publishers, Boston (2002)
21. Micciancio, D., Regev, O.: Worst-case to average-case reductions based on gaussian measure. In: Proceedings of the 45rd annual symposium on foundations of computer science - FOCS 2004 (Rome, Italy), October 2004, pp. 371–381. IEEE. Journal verion in SIAM Journal on Computing
22. Micciancio, D., Regev, O.: Worst-case to average-case reductions based on gaussian measure. SIAM J. Comput. **37**(1), 267–302 (2007). Preliminary version in FOCS 2004
23. Micciancio, D., Voulgaris, P.: A deterministic single exponential time algorithm for most lattice problems based on voronoi cell computations. In: Proceedings of the 42th annual ACM symposium on Theory of computing (STOC 2010) 42, pp. 351–358 (2010)
24. Mitra, D., Romeo, F., Sangiovanni-Vincentelli, A.: Convergence and finite-time behavior of simulated annealing. In: 24th IEEE Conference on Decision and Control, vol. 24, pp. 761–767. IEEE (1985)
25. Nguyen, P.Q., Vidick, T.: Sieve algorithms for the shortest vector problem are practical. J. Math. Crypt. **2**(2), 181–207 (2008)

26. Regev, O.: New lattice-based cryptographic constructions. J. ACM **51**(6), 899–942 (2004)
27. Schneider, M., Gamma, N.: Svp challenge (2010). http://www.latticechallenge.org/svp-challenge/
28. Schnorr, C.P.: A hierarchy of polynomial lattice basis reduction algorithms. Theor. Comput. Sci. **53**, 201–224 (1987)
29. Seneta, E.: Non-negative Matrices and Markov Chains, 2nd edn. Springer Publishers, New York (2006)
30. Shoup, V.: Number theory c++ library (ntl) vesion 6.0.0 (2010). http://www.shoup.net/ntl/
31. van Emde Boas, P.: Another np-complete partition problem and the complexity of computing short vectors in a lattice. Technical Report, Mathematisch Instituut, Universiteit van Amsterdam 81–04 (1981)
32. Wang, X., Liu, M., Tian, C., Bi, J.: Improved nguyen-vidick heuristic sieve algorithm for shortest vector problem. In: Proceedings of the 6th ACM Symposium on Information, Computer and Communications Security. ACM, pp. 1–9 (2011)

Rerandomizable Threshold Blind Signatures

Veronika Kuchta$^{(\boxtimes)}$ and Mark Manulis

Department of Computing, University of Surrey, Guildford, UK
v.kuchta@surrey.ac.uk, mark@manulis.eu

Abstract. This paper formalizes the concept of *threshold blind signatures* (TBS) that bridges together properties of the two well-known signature flavors, blind signatures and threshold signatures. Using TBS users can obtain signatures through interaction with t-out-of-n signers without disclosing the corresponding message to any of them. Our construction is the first TBS scheme that achieves security in the standard model and enjoys the property of being rerandomizable. The security of our construction holds according to most recent security definitions for blind signatures by Schröder and Unruh (PKC 2012) that are extended in this work to the threshold setting.

Rerandomizable TBS schemes enable constructions of distributed e-voting and e-cash systems. We highlight how TBS can be used to construct the first e-voting scheme that simultaneously achieves privacy, soundness, public verifiability in the presence of *distributed* registration authorities, following the general approach by Koenig, Dubuis, and Haenni (Electronic Voting 2010), where existence of TBS schemes was assumed but no construction given. As a second application, we discuss how TBS can be used to distribute the currency issuer role amongst multiple parties in a decentralized e-cash system proposed by Miers et al.(IEEE S&P 2013).

1 Introduction

Blind Signatures. Blind signatures, introduced by Chaum [21], allow users to obtain a signature on some message through interaction with the signer in a way that doesn't expose the message. This property, which is called *blindness* is the distinctive property of blind signatures, in addition to the *unforgeability* requirement, which guarantees that no more signatures can be produced in addition to those output through the interaction with the signer. Blind signatures are considered as an important building block for a variety of applications, including e-voting [9,10,31,45] and e-cash schemes [21], anonymous credential systems [15] and oblivious transfer [19]. Security properties and constructions of blind signatures have been explored in numerous subsequent works: Pointcheval and Stern [52] defined and proved the security requirements for blind signatures in the random oracle model. Juels et al. [41] defined a blind signature scheme which is secure under general complexity assumptions. Recently, Schröder and Unruh [54] showed that security definitions from [52] have some drawbacks and came up with an improved definition of honest-user unforgeability. A lot of

© Springer International Publishing Switzerland 2015
M. Yung et al. (Eds.): INTRUST 2014, LNCS 9473, pp. 70–89, 2015.
DOI: 10.1007/978-3-319-27998-5_5

work has been done on the constructions of blind signature schemes, both in the random oracle model, e.g. [1,4,7,11,53], and in the standard model, e.g. [3,9,17,32,39,42,48,51].

Threshold Signatures. Threshold signatures, introduced by Desmedt [26] distribute the ability to sign a message across t-out-of-n signers. This distribution process is typically carried out using secret sharing techniques and is therefore helpful for the distribution of trust in various cryptographic applications. In addition, threshold signatures can be used to achieve reliability and thus improve on the availability of services. Security properties and constructions of threshold signatures have been explored in [33,47,56]. Well-known constructions of threshold signatures in the random oracle model under the RSA assumptions have been proposed by Desmedt and Frankel [27] and Shoup [56]. Boldyreva [11] showed how to construct threshold signatures in the random oracle model in Gap Diffie-Hellman groups. More recently, Li et al. [46] distributed the signing process of the well-known Waters signature scheme in the standard model under the CDH assumption in bilinear groups.

Our Contribution: Threshold Blind Signature Schemes. In this work we formalize the concept of threshold blind signatures (TBS) and present an instantiation that enables the user to obtain a signature through interaction with a distributed set of n signers on some message of user's choice without revealing any information about the message. Each signer is in possession of a secret key share which is used in the signing process. The distribution of secret key shares in our scheme is performed by a trusted dealer, albeit alternative methods, e.g. [34], can also be applied. The signature generation process cannot be forged unless the adversary corrupts t signers. The blindness property ensures that even if all n potential signers are corrupted no information about the message is leaked. When defining these properties for TBS we adapt new security definitions from Schröder and Unruh [54], introduced originally for blind signatures, to the threshold setting. The requirements modeled for blind signatures in [54] are considered as being stronger than those given previously by Pointcheval and Stern [52]. In particular, they prevent an attack by which the adversary queries the signing oracle twice on the same message and then outputs a forgery on a different message.

Our TBS scheme is built based on the techniques underlying the blind signature scheme introduced by Okamoto [51] that deploys bilinear groups. Our TBS is more than an adaptation of the scheme from [51] to a threshold setting since we introduce further changes to the original construction to enhance its performance. In particular, by using non-interactive zero-knowledge (NIZK) proof techniques from [35,37] we can remove several rounds of interaction between the user and the signers, thus obtaining the same round-optimality as in case of (non-threshold) blind signatures in [29]. The NIZK proof from [35], which is based on the DLIN assumptions, gives us concurrent security for the overall TBS construction in the Common Reference String (CRS) model.

The standard assumptions and stronger definitions of security make our scheme superior to the existing TBS constructions from [43,57] that were proven secure in the random oracle model with respect to the (weaker) definitions from [52], which in turn makes them vulnerable to attacks against blind signatures identified in [54]. Our TBS construction enjoys the re-randomization property, which makes it especially attractive for a range of applications such as distributed e-voting and e-cash. We show how our TBS scheme can be used to realize e-voting in presence of distributed registration authorities and decentralized e-cash in presence of distributed currency issuers.

Applications of TBS. The use of blind signatures in e-voting schemes goes back to Chaum [21] and various e-voting schemes utilizing blind signatures have been introduced since then, e.g. [6,8,9,22,31,50]. The blindness property in most e-voting constructions is necessary to ensure privacy of the submitted votes, while the unforgeability property is used for authentication. The corresponding signature is typically issued by the registration authority, which is supposed to check the voter is eligible to participate in the election. The use of threshold blind signatures in this context is a helpful alternative for the case where the registration authority needs to be distributed across multiple not necessarily fully trustworthy entities. Such distributed approach for voter registration has been proposed by Koenig, Dubuis, and Haenni [44] assuming existence of threshold blind signatures, yet without offering concrete constructions of this primitive. As proven in [44], existence of a public registration board is necessary in order to prevent potential abuses. Public verifiability, originally defined in [40], is a property that guarantees the validity of the election outcome, preventing voting authorities from biasing the results. We show that our re-randomizable TBS construction can be used to obtain an e-voting scheme where the registration authority can be distributed across multiple parties and where the property of public verifiability holds simultaneously. In our construction we follow the template from [44]. Our scheme also achieves public verifiability as it was required in [44] because the voters send their votes together with signatures to a public board such that each voter can complain if he does not find his vote on the board or if he is generally suspicious about the content on the board. We provide an publicly verifiable e-voting scheme, which guarantees extended security in the signing process because of the threshold setting. Since the power of one signing authority is distributed amongst a number of signers, the signature on a vote will be accepted if and only if t out of n signers provide their signatures on the blinded vote to public board.

Our TBS scheme can be used to construct distributed e-cash. The concept of e-cash was introduced by Chaum [21] and later refined in [13,14,23,30,41,52]. A threshold approach was used by Camenisch et al. [18] in the design of endorsed e-cash schemes to provide fairness for the user. By utilizing threshold setting, the user creates n endorsements for one coin, of which any t can be used to reconstruct the coin. The e-cash scheme by Zhou [59] uses threshold cryptography to enable traceability of the issued e-coins. The secret sharing of the key and probabilistic encryption algorithm enable threshold management of private key and the scheme avoids the misuse of identity tracing and currency tracing in fair e-cash scheme. Miers et al. [49] recently described the common problem of many e-cash protocols that fundamentally rely on the issuer of e-coins being trusted and mentioned the distribution of his role amongst multiple issuers as a possible solution. We describe how our TBS scheme can offer such a standard-model solution for distributed e-cash schemes.

2   Building Blocks and Hardness Assumptions

In this section we recall several hardness assumptions and building blocks that will be used in our work.

Definition 1 (Bilinear Groups). *Let* $\mathcal{G}(1^\lambda)$, $\lambda \in \mathbb{N}$ *be an algorithm that on input a security parameter* 1^λ *outputs the description of two cyclic groups* $\mathbb{G}_1 = \langle g_1 \rangle$ *and* $\mathbb{G}_2 = \langle g_2 \rangle$ *of prime order* q *with* $|q| = 1^\lambda$, *where possibly* $\mathbb{G}_1 = \mathbb{G}_2$, *and an efficiently computable* $e : \mathbb{G}_1 \times \mathbb{G}_2 \to \mathbb{G}_T$ *with* $\mathbb{G}_T$ *being another cyclic group of order* q. *The group pair* $(\mathbb{G}_1, \mathbb{G}_2)$ *is called* bilinear *if* $e(g_1, g_2) \neq 1$ *and* $\forall u \in \mathbb{G}_1$, $v \in \mathbb{G}_2$, $\forall a, b \in \mathbb{Z}$: $e(u^a, v^b) = e(u, v)^{ab}$.

Definition 2 (DLin-Assumption). *Let* $\mathbb{G}$ *be a cyclic group of order* q. *The DLin assumption states that given a tuple* $(g, g^x, g^y, g^{xa}, g^{yb}, g^c)$ *for random* $a, b, x, y, c \in \mathbb{Z}_q^*$, *it is hard to decide whether* $c = a + b$. *When* $(g, u = g^x, v = g^y)$ *is fixed, a tuple* (u^a, v^b, g^{a+b}) *is called a linear tuple, whereas a tuple* (u^a, v^b, g^c) *for a random and independent* c *is called a random tuple. Adversaries advantage in solving the assumption is negligible.*

Definition 3 (CDH-Assumption). *Let* $\mathbb{G}, \mathbb{G}_T$ *be two groups of prime order* q. *Let* $e : \mathbb{G} \times \mathbb{G} \to \mathbb{G}_T$ *be a bilinear map and let* $\langle g \rangle = \mathbb{G}$ *be the generator of* $\mathbb{G}$. *Let* $\mathcal{A}_{CDH}$ *be an adversary taking as input the security parameter* λ. *Suppose that* $a, b \leftarrow \mathbb{Z}_q^*$ *are randomly chosen.* $\mathcal{A}_{CDH}$ *is to solve the following problem: Given* g, g^a, g^b *compute the* g^{ab}. *Let* ϵ *be the advantage of algorithm* $\mathcal{A}$ *in solving the CDH assumption if*

$$\left| Pr[\mathcal{A}(g, g^a, g^b) = g^{ab}] \right| \geq \epsilon(\lambda).$$

Non-Interactive Zero-Knowledge Proof [36]. A non-interactive proof system $(\mathcal{G}, \mathcal{K}, \mathcal{P}, \mathcal{V})$ for a relation R with setup consists of four PPT algorithms: a setup algorithm $\mathcal{G}$, a common reference string (CRS) generation algorithm $\mathcal{K}$, a prover $\mathcal{P}$ and a verifier $\mathcal{V}$. The setup algorithm outputs public parameters I and a commitment key ck. The CRS generation algorithm takes I as input and outputs a CRS ρ. The prover $\mathcal{P}$ takes as input (I, ρ, x, ω), where x is the statement and ω is the witness, and outputs a proof π. The verifier $\mathcal{V}$ takes as input (I, ρ, x, π) and outputs 1 if the proof is acceptable and 0 otherwise. $(\mathcal{G}, \mathcal{K}, \mathcal{P}, \mathcal{V})$ is non-interactive proof system for R if it has the following properties:

Completeness. A non-interactive proof is complete if an honest prover can convince an honest verifier whenever the statement belongs to the language and the prover holds a witness testifying to this fact. For all adversaries $\mathcal{A}$ we have: $Pr[(I, ck) \leftarrow \mathcal{G}(1^\lambda); \rho \leftarrow \mathcal{K}(I, ck); (x, \omega) \leftarrow \mathcal{A}(I, \rho); \pi \leftarrow \mathcal{P}(I, \rho, x, \omega) : \mathcal{V}(I, \rho, x, \pi) = 1$ if $(I, x, \omega) \in R] = 1$.

Soundness. A non-interactive proof is sound if it is impossible to prove a false statement. We say $(\mathcal{G}, \mathcal{K}, \mathcal{P}, \mathcal{V})$ is perfectly sound if for all adversaries $\mathcal{A}$ we have: $Pr[(I, ck) \leftarrow \mathcal{G}(1^\lambda); \rho \leftarrow \mathcal{K}(I, ck); (x, \pi) \leftarrow \mathcal{A}(I, \rho); \pi \leftarrow \mathcal{P}(I, \rho, x, \omega):$ $\mathcal{V}(I, \rho, x, \pi) = 0$ if $x \notin L] = 1$.

Knowledge Extraction. We say that $(\mathcal{G}, \mathcal{K}, \mathcal{P}, \mathcal{V})$ is a proof of knowledge for R if there exists a knowledge extractor $\mathcal{E} = (\mathcal{E}_1, \mathcal{E}_2)$ with the following properties: For all PPT adversaries $\mathcal{A}$ we have $Pr[(I, ck) \leftarrow \mathcal{G}(1^\lambda); \rho \leftarrow \mathcal{K}(I, ck):$ $\mathcal{A}(I, \rho) = 1] = Pr[I \leftarrow \mathcal{G}(1^\lambda); (\rho, \xi) \leftarrow \mathcal{E}_1(I) : \mathcal{A}(I, \rho) = 1]$. For all adversaries $\mathcal{A}$ holds $Pr[(I, ck) \leftarrow \mathcal{G}(1^\lambda); (\rho, \xi) \leftarrow \mathcal{E}_1(I, ck); (x, \pi) \leftarrow \mathcal{A}(I, \rho); \omega \leftarrow$ $\mathcal{E}_2(\rho, \xi, x, \pi) : \mathcal{V}(I, \rho, x, \pi) = 0$ or $(x, \omega) \in R] = 1$.

Zero-Knowledge. We say that $(\mathcal{G}, \mathcal{K}, \mathcal{P}, \mathcal{V})$ is a NIZK proof if there exists a PPT simulator $(\mathcal{S}_1, \mathcal{S}_2)$ such that for all PPT adversaries $\mathcal{A}$ we have $Pr[(I, ck) \leftarrow \mathcal{G}(1^\lambda); \rho \leftarrow \mathcal{K}(I, ck) : \mathcal{A}(I, \rho) = 1] \approx Pr[I \leftarrow \mathcal{G}(1^\lambda); (\rho, \tau) \leftarrow$ $\mathcal{S}_1(I) : \mathcal{A}(I, \rho) = 1]$, and for all adversaries $\mathcal{A}: Pr[(I, ck) \leftarrow \mathcal{G}(1^\lambda); (\rho, \tau) \leftarrow$ $\mathcal{S}_1(I, ck); (x, \omega) \leftarrow \mathcal{A}(I, \rho, \tau); \pi \leftarrow \mathcal{P}(I, \rho, x, \omega) : \mathcal{A}(\pi) = 1] = Pr[(I, ck) \leftarrow$ $\mathcal{G}(1^\lambda); (\rho, \tau) \leftarrow \mathcal{S}_2(I, ck); (x, \omega) \leftarrow \mathcal{A}(I, \rho, \tau); \pi \leftarrow \mathcal{P}(I, \rho, x, \omega) : \mathcal{A}(\pi) = 1]$, where $\mathcal{A}$ outputs $(I, x, \omega) \in R$.

3 Threshold Blind Signatures

A threshold blind signature scheme gives the user the ability to get a signature on a message without revealing its content and it distributes the secret key among a certain number of signers. We observe a $t-$out-of-n threshold blind signature scheme. It means that it is not possible to construct a valid blind signature on a message by contacting less than t-out-of-n servers. The threshold blind signature scheme is applicable to many constructions of cryptographic schemes because of its role in the decentralization the power of the signer.

Definition 4 (Threshold Blind Signature). *A t-out-of-n threshold blind signature scheme TBS in a Common Reference String model consists of the following four algorithms:*

TBParGen(1^λ): *A PPT algorithm takes as input the security parameter 1^λ and outputs public parameters I (possibly containing a common reference string crs_{TBS}).*

KGen(I): *On input public parameters I this algorithm outputs a secret share sk_i for each signer S_i, $i \in \{1, \ldots, n\}$ and a public key pk.*

TBSign($\cdot$): *This is a protocol between a user $\mathcal{U}$ and the signers S_i, $i \in \{1, \ldots, n\}$. The input of $\mathcal{U}$ is pk and a message m. The input of each server S_i is the secret share sk_i. The protocol results in a signature σ output by $\mathcal{U}$.*

TBVerify(pk, m, σ): *A deterministic algorithm which on input a public key pk, message m, a signature σ outputs 1 if the signature is valid and 0 otherwise.*

TBS Unforgeability. We recall the unforgeability definition for blind signatures by Schröder and Unruh [54] and adopt it to the threshold setting. This definition requires that $(m_i^*, \sigma_i^*) \neq (m_j, \sigma_j)$ for all i, j and $(m_i^*, \sigma_i^*) \neq (m_j^*, \sigma_j^*)$

for i, j with $i \neq j$, which in comparison to the earlier definition by Pointcheval and Stern [52] allows to tell which message is being signed in a given interaction. It is assumed that the adversary randomly chooses up to $(t-1)$ out of n servers. When an adversary corrupts a server, it is given the entire computation history of that server, and it gets control of the server for the running time of the system. An adversary against unforgeability of TBS has the target to generate $q_S + 1$ valid message/signature pairs after it has interacted at most q_S times with the honest signer.

Definition 5 (Unforgeability). *A threshold blind signature scheme $TBS = (\mathsf{TBParGen}, \mathsf{KGen}, \mathsf{TBSign}, \mathsf{TBVerify})$ is unforgeable if for all PPT adversaries $\mathcal{A}$ the probability that the following experiment $\mathbf{Unforge}_{\mathcal{A}}^{TBS}(\lambda)$ evaluates to 1 is negligible in the security parameter λ.*

1. $I \leftarrow \mathsf{TBParGen}(1^\lambda)$
2. $(sk_i, pk) \leftarrow \mathsf{KGen}(I)$ *for all* $i \in \{1, \ldots, n\}$
3. $\{i_1, \ldots, i_{n-t+1}\} \leftarrow \mathcal{A}(pk)$
4. $(\sigma_1^*, m_1^*, \ldots, (\sigma_{q_S+1}^*, m_{q_S+1}^*)) \leftarrow \mathcal{A}^{\mathcal{O}^{\mathsf{TBSign}}(\cdot)}(sk_1, \ldots, sk_{t-1})$.
5. *If* $\mathsf{TBVerify}(pk, m_i^*, \sigma_i^*) = 1$ *for all* $i \in [1, q_S + 1]$ *and* $(m_i^*, \sigma_i^*) \neq (m_j^*, \sigma_j^*)$ *for all* $j \in [1, q_{S+1}], j \neq i$ *then return 1, otherwise return 0.*

$\mathcal{O}^{\mathsf{TBSign}}(\cdot)$ *is an oracle that executes the* $\mathsf{TBSign}(sk_i, m)$ *protocol on behalf of all uncorrupted servers* $i_1, \ldots, i_{n-t+1}$. *The total number of invoked* TBSign *protocol sessions is denoted by* q_S.

TBS Blindness. The TBS blindness property prevents signers from linking generated signatures to corresponding sessions of the signing protocol. Therefore, it should be impossible for a malicious signer $\mathcal{A}$ to decide on the order in which two messages, m_0 and m_1, were signed in two protocol sessions with an honest user $\mathcal{U}$.

Definition 6 (Blindness). *A threshold blind signature scheme $TBS = (\mathsf{TBParGen}, \mathsf{KGen}, \mathsf{TBSign}, \mathsf{TBVerify})$ is called blind if for any PPT adversary $\mathcal{A}$ the probability that the following experiment $\mathbf{TBlind}_{\mathcal{A}}^{TBS}(\lambda)$ evaluates to 1 exceeds $1/2$ by at most a negligible amount in the security parameter λ.*

1. $I \leftarrow \mathsf{TBParGen}(1^\lambda)$
2. $(m_0, m_1, pk, st_{find}) \leftarrow \mathcal{A}(I, \mathbf{find})$
3. *Choose* $b \xleftarrow{r} \{0, 1\}$
4. *Execute* $\sigma_b \leftarrow \mathsf{TBSign}(pk, m_b)$ *and* $\sigma_{1-b} \leftarrow \mathsf{TBSign}(pk, m_{1-b})$ *sessions on behalf of user $\mathcal{U}$. If* $\sigma_b = \perp$, *or* $\sigma_{1-b} = \perp$ *then* $(\sigma_b, \sigma_{1-b}) \leftarrow (\perp, \perp)$.
5. $b^* \leftarrow \mathcal{A}(\mathbf{guess}, \sigma_0, \sigma_1)$.
6. *If* $b = b^*$, *then return 1, otherwise return 0.*

A Note on Key Generation. There exist several approaches for the distribution of keys amongst multiple signers. The approach by Shamir [55] applies secret sharing and distributes secret key shares to corresponding signers through a trusted dealer. The protocol by Feldman [28] minimizes this trust assumption on the dealer by requiring the latter to broadcasts information that can then be used by the signers to individually check the validity of their shares and detect incorrect shares at reconstruction time. The key generation protocol by Gennaro et al. [34] proceeds in a pure distributed fashion, where each signer defines its own share of the secret key and participates in a protocol with all remaining signers to setup the key. During the protocol parties can determine malicious signers those contributions will be dropped. The distributed key generation protocol by Abe and Fehr [2] for discrete logarithm-based keys achieves adaptive security in the non-erasure model and avoids the use of interactive zero knowledge proofs.

4 TBS Construction in the Standard Model

4.1 Our TBS Scheme

In this section we present our TBS scheme based on the techniques underlying the Okamoto's blind signature scheme [51] and the NIZK proof from [36]. We assume existence of a trusted dealer for the distribution of secret key shares.

Parameter Generation: The algorithm $\texttt{TBParGen}(1^\lambda)$ outputs the common reference string $CRS = (\mathbb{G}, \mathbb{G}_T, q, g, e, ck)$, where $ck = (u'_k, u_{k,j})$, $j = \{1, \ldots, n\}$, and $k = \{1, 2, 3\}$ is the commitment key. The perfect binding key consists of the following values $u_{1,j} = \left(u'_1\right)^{\xi_{1,j}}$, $u_{2,j} = \left(u'_2\right)^{\xi_{2,j}}$, $u_{3,j} = \left(u'_3\right)^{\xi_{1,j}+\xi_{2,j}+\zeta}$; $\xi_{1,j}, \xi_{2,j}, \zeta \xleftarrow{r} \mathbb{Z}_q^*$ and $u'_1 = g^\rho, u'_2 = g^\tau, u'_3 = g$. The corresponding extraction key is given by $xk = (ck, \rho, \tau, \zeta)$. During the generation process of perfectly hiding key, the algorithm outputs the following trapdoor key $tk_j = (ck, \xi_{1,j}, \xi_{2,j})$, $j = \{1, \ldots, \ell\}$.

Key Generation: The algorithm $\texttt{KGen}(I)$ picks $x \xleftarrow{r} \mathbb{Z}_q$, computes $g_1 = g^x$, It then picks a random polynomial $f \xleftarrow{r} \mathbb{Z}_q[Z]$ of degree $t - 1$, with $t \leq n$ being a threshold and $f(0) = x$. Let $f(z) = \sum_{i=1}^{t-1} a_i z_i$. The algorithm computes $x_i = f(i)$ for each server $i \in \{1, \ldots, n\}$. Let $\boldsymbol{vk} = (vk_1, \ldots, vk_n) = (g^{x_1}, \ldots, g^{x_n})$. The outputs consists of the public key $pk = (g_1, g_2, \boldsymbol{vk})$ and a separate secret share $sk_i = g_2^{x_i}$ for each S_i, $i \in \{1, \ldots, n\}$.

Signature generation: The $\texttt{TBSign}$ protocol on a ℓ-bit message $m = (\mu_1, \ldots, \mu_\ell)$ proceeds in two stages:

Stage 1: For all $i = \{1, \ldots, n\}$, user $\mathcal{U}$ chooses a random $r_i \xleftarrow{r} \mathbb{Z}_q^*$ and computes $X_i \leftarrow \left(u'_1 \prod_{j=1}^{\ell} u_{1,j}^{\mu_j}\right)^{r_i}$. $\mathcal{U}$ then prepares a NIZK proof for the well-formedness of X_i. This proof consists of two parts $\pi_i^{(1)}$ and $\pi_i^{(2)}$. It first part $\pi_i^{(1)}$ proves that all μ_j are bits using the NIZK proof from [37]. The user randomly selects $\alpha_{k,j} \xleftarrow{r} \mathbb{Z}_q^*$ for $k = 1, 2, 3$ and computes $A_{k,j} = \left(u'_k\right)^{\alpha_{k,j}} u_{k,j}^{\mu_j}$ for $j = \{1, \ldots, \ell\}, k = \{1, 2, 3\}$. U proves to each server S_i knowledge of α_j such that $A_{k,j} = \left(u'_k\right)^{\alpha_{k,j}}$ for $\mu_j = 0$ or $A_{k,j} = \left(u'_k\right)^{\alpha_{k,j}} u_{k,j}$ for $\mu_j = 1$

and $j = \{1, \ldots, \ell\}$, $k = \{1, 2, 3\}$. For each S_i the corresponding NIZK proof $\pi_i^{(1)} = (\bar{\pi}_1, \ldots, \bar{\pi}_\ell)$ consists of ℓ components $\bar{\pi}_j$, $j = \{1, \ldots, \ell\}$. Each of these proofs $\bar{\pi}_j = (\pi_{11}, \pi_{12}, \pi_{13}, \pi_{21}, \pi_{22}, \pi_{23})$ is computed as follows using a randomly chosen $t_j \xleftarrow{r} \mathbb{Z}_q^*$:

$$\pi_{11} = \left(u_{1,j}^{2\mu_j - 1}\left(u_1'\right)^{\alpha_{1,j}}\right)^{\alpha_{1,j}}$$

$$\pi_{12} = u_{2,j}^{(2\mu_j - 1)\alpha_{2,j}}\left(u_2'\right)^{\alpha_{1,j}\alpha_{2,j} - t_j}$$

$$\pi_{13} = u_{3,j}^{(2\mu_j - 1)\alpha_{1,j}}\left(u_3'\right)^{(\alpha_{1,j} + \alpha_{2,j})\alpha_{1,j} + t_j}$$

$$\pi_{21} = u_{1,j}^{(2\mu_j - 1)\alpha_{2,j}}\left(u_1'\right)^{\alpha_{1,j}\alpha_{2,j} + t_j}$$

$$\pi_{22} = \left(u_{2,j}^{2\mu_j - 1}\left(u_2'\right)^{\alpha_{2,j}}\right)^{\alpha_{2,j}}$$

$$\pi_{23} = u_{3,j}^{(2\mu_j - 1)\alpha_{2,j}}\left(u_3'\right)^{(\alpha_{1,j} + \alpha_{2,j})\alpha_{2,j} - t_j}$$

U sends the proofs π_i and the commitments $\{A_{k,j}\}_{k=\{1,2,3\}, j=\{1,\ldots,\ell\}}$ to the corresponding server S_i that checks the following verification equations:

$$e(u_1', \pi_{11}) = e(A_{1,j}, A_{1,j}u_{1,j}^{-1}),$$

$$e(u_2', \pi_{22}) = e(A_{2,j}, A_{2,j}u_{2,j}^{-1}),$$

$$e(u_3', \pi_{33}) = e(A_{3,j}, A_{3,j}u_{3,j}^{-1}),$$

$$e(u_1', \pi_{12})e(u_2', \pi_{21}) = e(A_{1,j}, A_{2,j}u_{2,j}^{-1})e(A_{2,j}, A_{1,j}u_{1,j}^{-1}),$$

$$e(u_1', \pi_{13})e(u_3', \pi_{31}) = e(A_{1,j}, A_{3,j}u_{3,j}^{-1})e(A_{3,j}, A_{1,j}u_{1,j}^{-1}),$$

$$e(u_3', \pi_{23})e(u_3', \pi_{32}) = e(A_{2,j}, A_{3,j}u_{3,j}^{-1})e(A_{3,j}, A_{2,j}u_{2,j}^{-1}),$$

for each $j = \{1, \ldots, \ell\}$ and $\pi_{33} = \pi_{1t}\pi_{2t}$, $t = \{1, 2, 3\}$. The server accepts $\pi_i^{(1)}$ if all verification equations hold.

In the second part $\pi_i^{(2)}$ user $\mathcal{U}$ proves to each server S_i the knowledge of $\{r_i, \beta_{k,i}, \delta_{i,j}\}_{k \in [3], j \in [\ell]}$ using the NIZK techniques from [35] and values $A_{k,j}$, $\alpha_{k,j}$ $k \in \{1, 2, 3\}; j \in \{1, \ldots, \ell\}$ that were used to compute $\pi_i^{(1)}$ by proving that $X_i = \left(\prod_{j=1}^\ell A_{1,j}\right)^{r_i}(u_1')^{\beta_{1,i}}$ and $X_i = (u_1')^{r_i}\prod_{j=1}^\ell u_{1,j}^{\delta_{i,j}}$, where $\beta_{1,i} = r_i - r_i \sum_{j=1}^\ell \alpha_{1,j}$ and $\delta_{i,j} = r_i\mu_j$, $j \in [\ell], i \in [n]$. This proof involves building commitments $B_{k,i} = \left(\prod_{j=1}^\ell A_{k,j}\right)^{r_i}(u_k')^{\beta_{k,i}}$ and $\hat{B}_{k,i} = (u_k')^{r_i}\prod_{j=1}^\ell u_{k,j}^{\delta_{i,j}}$, $k = \{1, 2, 3\}, i = \{1, \ldots, n\}, j = \{1, \ldots, \ell\}$. Note that $B_{1,i} = \hat{B}_{1,i} = X_i$. This effectively binds both parts of the proof to X_i. $\mathcal{U}$ splits $B_{k,i}$ and $\hat{B}_{k,i}$ into ℓ commitments such that $B_{k,i,j} = A_{k,j}^{r_i}\left(u_k'\right)^{\beta_{k,i}}$ and $\hat{B}_{k,i,j} = \left(u_{k,j}^{\delta_{i,j}}(u_k')^{r_i}\right)$.

The user makes then a NIZK proof for the Pedersen commitment for each of these components. We refer to Sect. 4.5 [35] for further details on the construction of $\pi_i^{(2)}$ proof that is used in this second part. Each $\pi_i^{(2)}$ consists of $6(\ell - 1) + 2$ components. Each server $S_i, i = \{1, \ldots, n\}$ verifies $\pi_i^{(2)}$ and proceeds if the proof is valid.

Stage 2: If S_i accepts the NIZK proof in Stage 1, it randomly chooses $d_i \xleftarrow{r} \mathbb{Z}_q^*$ and uses its secret key share $sk_i = g_2^{x_i}$ to compute $Y_{i1} \leftarrow sk_i X_i^{d_i}$ and $Y_{i2} \leftarrow g^{d_i}$. Finally, S_i sends its signature share $\sigma_i = (Y_{i1}, Y_{i2})$ to $\mathcal{U}$. For each received $\sigma_i = (\sigma_{i1}, \sigma_{i2})$, $\mathcal{U}$ checks the equation $e(Y_{i1}, g) = e(g_2, vk_i) \cdot e(X_i, Y_{i2})$ using the corresponding verification key $vk_i \in \boldsymbol{vk}$ and if successful chooses a random $s_i \xleftarrow{r} \mathbb{Z}_q^*$, and computes

$$\sigma_{i1} \leftarrow Y_{i1} \left(u' \prod_{j=1}^{\ell} u_j^{\mu_j} \right)^{s_i} \quad \text{and} \quad \sigma_{i2} \leftarrow Y_{i2}^{r_i} g^{s_i}.$$

Assume that $\mathcal{U}$ collected t shares σ_i from corresponding servers S_i, $i = 1, \ldots, t$. $\mathcal{U}$ first computes the Lagrange coefficients $\lambda_1, \ldots, \lambda_t \in \mathbb{Z}_q$ such that $x = f(0) = \sum_{i=1}^{t} \lambda_i f(i)$ and then $\sigma_1 = \prod_{i=1}^{t} (\sigma_{i1})^{\lambda_i}$ and $\sigma_2 = \prod_{i=1}^{t} (\sigma_{i2})^{\lambda_i}$. Finally, $\mathcal{U}$ outputs $\sigma = (\sigma_1, \sigma_2)$ as the resulting signature. (Note that σ has the same form as in the Okamoto's blind signature scheme from [51]).

Verification: The algorithm $\mathtt{TBVerify}(pk, m, \sigma)$ first parses pk as $(g_1, g_2, u', (u_1, \ldots, u_\ell))$, m as $(\mu_1, \ldots, \mu_\ell)$, and σ as (σ_1, σ_2) and outputs 1 if and only if $e(\sigma_1, g) = e(g_2, g_1) \cdot e\left(u' \prod_{j=1}^{\ell} u_j^{\mu_j}, \sigma_2\right)$.

4.2   Security Analysis

The unforgeability of our TBS scheme is proven in Theorem 1 through a direct reduction to the CDH assumption. Note that the blind signature scheme by Okamoto [51] those techniques we partially apply in TBS was proven to be unforgeable using a reduction to the original Waters signature scheme [58] that in turn holds under the CDH assumption.

Theorem 1 (Unforgeability). *Our TBS scheme is unforgeable in the common reference string model assuming the hardness of the CDH assumption from Definition 3 and the soundness property of the NIZK proof from [36].*

Proof To prove the above theorem we construct a simulator $\mathcal{C}$ which is given the CDH challenge (g, g^a, g^b) from Definition 3 and is internally using the unforgeability adversary $\mathcal{A}$ to compute g^{ab}. By ϵ we denote the success probability of $\mathcal{A}$ in forging the threshold blind signature. The interaction of $\mathcal{C}$ with $\mathcal{A}$ proceeds according to the following description.

Setup: To generate the public parameters the challenger $\mathcal{C}$ sets $l = 4q_S$ and chooses a random vector of length ℓ: $\boldsymbol{a} = (a_1, \ldots, a_\ell)$, where each is chosen uniformly and random in the interval between 0 and $l - 1$ and ℓ denotes the number of bits of a message m. Then it chooses a random $b' \xleftarrow{r} \mathbb{Z}_q$, and the vector $\boldsymbol{b} = (b_1, \ldots, b_\ell) \xleftarrow{r} \mathbb{Z}_q$. Next the challenger $\mathcal{C}$ sets the following public parameters $u' = g_1^{q-tl+a'} g^{b'}$ and $u_j = g_1^{a_j} g^{b_j}$, where t is the threshold number of the scheme. The public parameters $(g, g_1, g_2, u', \boldsymbol{u})$ are sent to the adversary $\mathcal{A}$. We assume for our scheme that the adversary corrupts $t-1$ servers $S_{i_1}, \ldots, S_{i_{t-1}}$. Let $\hat{S}$ be the set of indexes i_k of corrupted servers.

The challenger $\mathcal{C}$ generates secret shares sk_i of the private key sk for the corrupted servers in the following way: It picks $t - 1$ random integers $x_1, \ldots, x_{t-1} \in \mathbb{Z}_q$. Let $f \in \mathbb{Z}_q[Z]$ be the $t - 1$ polynomial, which satisfies $f(0) = x_i$ for $i = 1, \ldots, t - 1$. The challenger $\mathcal{C}$ gives the secret key shares $sk_i = g_2^{x_i}$ to the adversary $\mathcal{A}$.

The challenger $\mathcal{C}$ also generates the verification keys vk_i, which are useful to prove the correctness of secret shares. It sets $vk_i = g^{f(i)}$, such that the verification keys generate a vector $(vk_1, \ldots, vk_n) = (g^{f(1)}, \ldots, g^{f(n)})$ for the above defined polynomial f. It is easy for the challenger $\mathcal{C}$ to construct the verification keys for the corrupted servers from the set $\hat{S}$, because $f(i)$ equals to x_i, which are known to the challenger $\mathcal{C}$. Let $\check{S}$ denote the set of uncorrupted servers $(S_{i_t}, \ldots, S_{i_n})$. $\mathcal{C}$ has to compute the Lagrange coefficients $\lambda_{0,i}, \ldots, \lambda_{t-1,i} \in \mathbb{Z}_q$ such that $f(i) = \lambda_{0,i} f(0) + \sum_{k=1}^{t-1} \lambda_{k,i} f(k)$, where $\{i_1, \ldots, i_{t-1}\}$ are the indexes from the set $\hat{S}$ of corrupted servers and $\{i_t, \ldots, i_n\}$ are the indexes of uncorrupted servers. The Lagrange coefficients are then computed as follows:

$$\lambda_{k,i} = \prod_{k' \in \check{S} \setminus \{i\}} \frac{(k - k')}{(i - k')},$$

where $k \in \hat{S}$ is the index of a corrupted server and k' is the index of an uncorrupted server. It is easy to determine these Lagrange coefficients because they are independent from f. As a next step, $\mathcal{C}$ sets for $i \in \check{S}$ and $g_1 = g^x$:

$$vk_i = g_1^{\lambda_{0,i}} vk_1^{\lambda_{1,i}} \cdots vk_{t-1}^{\lambda_{t-1,i}} = g_1^{\lambda_{0,i}} g^{f(1)\lambda_{1,i}} \cdots g^{f(t-1)\lambda_{t-1,i}}$$
$$= g_1^{\lambda_{0,i}} g^{\sum_{k=1}^{t-1} f(k)\lambda_{k,i}} = g^{f(i)}.$$

Once $\mathcal{C}$ has computed all the verification keys vk_i, it outputs them to $\mathcal{A}$.

Signature Share Query: Once the adversary $\mathcal{A}$ has the verification keys it provides up to q_S signature share generation queries to the TBSign oracle according to the experiment in Definition 5. The oracle queries are processed by $\mathcal{C}$ that has to output a signature share $\sigma_i = (Y_{i1}, Y_{i2})$ on input $(X_i, A_{k,j}, B_{k,i}, \hat{B}_{k,i}, \pi_i^{(1)}, \pi_i^{(2)})$, for $i = \{1, \ldots, n\}, j = \{1, \ldots, \ell\}, k = \{1, 2, 3\}$. The proofs $\pi_i^{(1)}$ and $\pi_i^{(2)}$ ensure that $X_i = \left(u_1' \prod_{j=1}^{\ell} u_{1,j}^{\mu_j}\right)^{r_i}$ due to their soundness property as proven in [35, 37]. We note that the perfect binding property of the commitment scheme guarantees the soundness of the NIZK proof $(\pi_i^{(1)}, \pi_i^{(2)})$. Since our commitment scheme in Sect. 4.1 contains perfect binding keys it provides the existence of an extraction key which allows extraction of the values r_i

and $\mu_1, \ldots, \mu_\ell$. For more details on the proof we refer to [35,37]. The challenger $\mathcal{C}$ extracts $r_i, \mu_1, \ldots, \mu_\ell$ and prepares a signature on these values using the CDH challenge. $\mathcal{C}$ first defines two functions for $m = (\mu_1, \ldots, \mu_\ell)$:

$$F(m) = (q - tl) + a' + \sum_{i=1}^{\ell} a_i \mu_i \text{ and } G(m) = b' + \sum_{i=1}^{\ell} b_i \mu_i.$$

Additionally we define the following binary function:

$$K(m) = \begin{cases} 0, & \text{if } a' + \sum_{j=1}^{\ell} a_j \mu_j \equiv 0 \mod l \\ 1, & \text{otherwise.} \end{cases}$$

Upon the computation of $u' \prod_{j=1}^{\ell} u_j^{\mu_j} = g_1^{q-tl+a'} g^{b'} \prod_{j=1}^{\ell} g_1^{a_j \mu_j} g^{b_j \mu_j}$ and using $F(m)$ and $G(m)$ the challenger has to return signature shares $\sigma_i = (Y_{i1}, Y_{i2})$. These are computed by $\mathcal{C}$ using the Lagrange coefficients $\lambda_{0,i}, \lambda_{1,i}, \ldots, \lambda_{t-1,i} \in \mathbb{Z}_q$ such that $f(i) = \lambda_{0,i} f(0) + \sum_{k=1}^{t-1} \lambda_{k,i} f(k)$ using the technique from Boneh and Boyen [12]. $\mathcal{C}$ then picks $r_i' \in \mathbb{Z}_q$, and outputs the following signature tuple $\sigma_i = (Y_{i,1}, Y_{i,2})$, where $Y_{i,1} = g^{b\left(\lambda_{0,i} f(0) + \sum_{k=1}^{t-1} \lambda_{k,i} f(k)\right)} g^{aF(m)r_i'} g^{G(m)r_i'}$ and $Y_{i,2} = g^{r_i'}$, where $r_i' = r_i - \frac{b\lambda_{0,i}}{F(m)}$. Note that $f(0) = -\frac{G(m)}{F(m)}$. The signature σ_i satisfies the verification equation $e(Y_{i1}, g) = e(g_2, vk_i) e(u' \prod_{j=1}^{\ell} u_j^{\mu_j}, Y_{i2})$ since

$$e(g_2, vk_i) e(u' \prod_{j=1}^{\ell} u_j^{\mu_j}, Y_{i,2})$$

$$= e\left(g^b, g^{\lambda_{0,i} f(0) + \sum_{k=1}^{t-1} \lambda_{k,i} f(k)}\right) e\left(g^{aF(m)} g^{G(m)}, g^{r_i'}\right)$$

$$= e\left(g, g^{b\left(\lambda_{0,i} f(0) + \sum_{k=1}^{t-1} \lambda_{k,i} f(k)\right)}\right) e\left(g^{aF(m)r_i'} g^{G(m)r_i'}, g\right)$$

$$= e\left(g^{b\left(\lambda_{0,i} f(0) + \sum_{k=1}^{t-1} \lambda_{k,i} f(k)\right)} g^{aF(m)\left(r_i - \frac{b\lambda_{0,i}}{F(m)}\right)} g^{G(m)\left(r_i - \frac{b\lambda_{0,i}}{F(m)}\right)}, g\right)$$

$$= e(Y_{i,1}, g).$$

In order to complete the simulation without aborting, it is required that all signature queries on m have $K(m) \neq l$. In this case, if $F(m) \neq 0$ then $\mathcal{C}$ is able to simulate the signature on the requested m; otherwise, $\mathcal{C}$ will not be able to generate such signature and the simulation aborts.

Extraction: The execution of this step corresponds to the fourth step from the experiment in Definition 5, where $\mathcal{A}$ sets $\sigma_i^* = (\sigma_{i1}^*, \sigma_{i2}^*)$ for $i \in \{1, q_S\}$ as a valid signature share for a message $m^* = (\mu_1^*, \ldots, \mu_\ell^*)$, which was not queried before. As next, we define a function $Q(m^*, \boldsymbol{q}, A')$, where $A' = (a', a_1, \ldots, a_\ell)$ are the simulated values, and $\boldsymbol{q} = (q_1, \ldots, q_S)$ as

$$Q(m^*, \boldsymbol{q}, A') = \begin{cases} 0, & \text{if } \forall_{j=1}^{s} q_j : K_{q_j}(m_j) = 1, \text{ and } a' + \sum_{j=1}^{\ell} a_j \mu_j^* \equiv 0 \mod l \\ 1, & \text{otherwise.} \end{cases}$$

The function evaluates to 0 if all signature queries will not cause an abort for a given choice of values A' and the function $a' + \sum_{j=1}^{\ell} a_j \mu_j^*$ mod l which equals $F(m^*)$ mod l vanishes for the values $m^* = (\mu_1^*, \ldots, \mu_\ell^*)$. That means if $F(m^*) = 0$ then $\mathcal{C}$ can extract g^{ab} by computing

$$g^{ab} = \left(\frac{\sigma_{i1}}{\sigma_{i2}^{G(m)d_i} \, (g^b)^{\sum_{k=1}^{t-1} \lambda_{k,i} f(k)}} \right)^{-\frac{1}{d_i \lambda_{0,i}}}.$$

Therefore we can consider the probability over the simulation values $(\mu_1^*, \ldots, \mu_\ell^*)$ as If $F(m^*)$ is not 0, then we have $Q(m^*, \boldsymbol{q}, A') = 1$ and the extraction aborts with probability $Pr[Q(m^*, \boldsymbol{q}, A') = 1]$. $\mathcal{C}$ repeats the above showed steps q_S times. If all q_S rounds are completed, $\mathcal{A}$ outputs at least $q_S + 1$ valid signatures with different messages, where at least one valid message-signature pair is different from the q_S valid messages-signatures given from $\mathcal{C}$ algorithm.

Analysis: The probability of success of an adversary $\mathcal{A}$ can be compared with the probability that $\mathcal{C}$ aborts in the simulation, which happens either if $F(m_i) = 0$ for a signature query on m_i or if $F(m^*) \neq 0$. The probability for $F(m^*) \neq 0$ can be bounded using following lemma.

Lemma 1 ([38]). *Let* $X, Y_1, \ldots, Y_q \subseteq [l]$ *such that holds* $|X|, |Y_i| \geq d$ *and* $|(X \setminus Y_i) \cup (Y_i \setminus X)| \geq d$ *for some* $d \geq 1$ *and all* i. *Then, we have*

$$Pr\left[a(X) = 0 \wedge \forall i \in [q] : a(Y_i) \neq 0\right] \geq \left(1 - C \cdot q \cdot \frac{\sqrt{\ell}}{d \cdot \sqrt{w}}\right) \cdot \frac{D}{\sqrt{d \cdot w}}$$

for $a(X) = \sum_{i \in X} x_i$ *and for fixed constants* C, D *that do not depend on values* ℓ, w, d, q, X *and the* Y_i

To apply this lemma to our analysis we set $X := m^*$ and $Y_i = m_i$ for $i \in [1, \ldots, q_S]$, $a(X) = F(m^*)$ and $a(Y_i) = K_q,(m_j)$. ℓ denotes the bit-length of the message and w equals in our scheme to the length of the vector $\mathbf{a}$, such that $w = \ell$. It also holds that $|(X \setminus Y_i) \cup (Y_i \setminus X)| \geq$. We consider that *abort* denotes the event that the simulation fails. This happens either because $F(m_i) = 0$ for a signature query on m_i, or because $F(m^*) \neq 0$. The lemma above provides an upper bound of $1 - \Theta(1/q)$. We conclude that $Pr[a(X) = 0 \wedge \forall i \in [q] : a(Y_i) \neq 0]$ corresponds to $Pr[\overline{abort}]$. The proof in [38] for Lemma 1 showed that the upper bound can be estimated by $\frac{D\sqrt{\chi}}{4C} \frac{1}{q_S}$, where $\chi = d/\ell$. Since χ is a constant, then the probability $P[\overline{abort}]$ has a lower bound of $\Theta(1/q)$. That means that the lower bound of the probability $Pr[\overline{abort}]$ is $\epsilon(1/q)$. This completes the proof of unforgeability of our TBS scheme. ◻

Theorem 2 (Blindness). *Our TBS scheme is blind in the common reference string model assuming the hardness of the DLin assumption from Definition 2, the perfect hiding property and the zero-knowledge property of the NIZK proof from [35, 37].*

Proof. The full proof of blindness is given in Appendix B.

5   Applications

Having introduced a new TBS construction, we highlight now its application to distributed e-voting and to distributed e-cash systems.

Distributed Verifiable E-voting. We recall first the general concept of an e-voting scheme and highlight functionalities of its algorithms based on [20]: a *voter* V is a party that is authorized by a *voting authority* to submit votes. The *tallying authority* collects individual votes and tally the results of the election to obtain the outcome. A *public board* which can be considered as a broadcast channel makes its content public to all parties and each party can add information to the board but not remove or modify any of the published contents. This board is typically used for the purpose of universal verifiability [25] of the e-voting process.

Koenig et al. [44] presented a generic template for e-voting protocols with distributed voting authorities assuming existence of threshold blind signatures, yet without offering concrete constructions of the latter schemes. Their generic e-voting protocol was shown to satisfy the security properties from [45]. By using our rerandomizable threshold blind signature scheme we therefore enable the actual construction of such distributed e-voting scheme.

The resulting scheme proceeds as follows. Each voting authority A_i, $i = 1, \ldots, n$ is in possession of the secret share sk_i from our TBS construction, and the corresponding public key pk is assumed to be published on the public board. Each voter encrypts its vote using the public key of the tallying authority and executes the TBSign protocol with each of n voting authorities to obtain a threshold blind signature σ on the encrypted vote if at least t out of n authorities provide their shares. It combines all signature share to a common threshold blind signature σ. This signature and the encrypted vote are sent to public board. The tallying authority decrypts each vote and publishes its content on the board together with the corresponding proof of decryption. The votes can then be counted and verified publicly.

In general, an e-voting scheme is required to provide the following properties that we recall informally here. More formal definitions can be found in [8,31, 45] The first important property is *privacy*, which means that individual votes remain hidden. The *soundness* property prevents dishonest voters from biasing the voting process. Finally, the *public verifiability* property ensures that anyone can check that the votes has been counted to prevent potential falsifications of counting process.

We briefly discuss why our construction offers privacy, soundness and public verifiability. Privacy follows from the fact, that the e-voting scheme is based on our TBS scheme. This guarantees in case that all authorities collude against a voter, the voter's privacy remains preserved due to the blindness property of the TBS scheme. Soundness is satisfied because of the TBS unforgeability. Public verifiability is satisfied because the decrypted votes, corresponding ciphertext as well as proofs for correct decryption of votes are published on the public board.

Distributed E-cash. We recall the basic functionality of an e-cash scheme [23]. The parties involved are the banks, users and merchants. Any user can *withdraw* an e-coin from her account at the bank and then *spend* it at some merchant. The merchant can then *deposit* the received coin on its account in the bank. In a distributed e-cash the role of the bank is split amongst n currency issuers, who are involved in the process of coin generation. This distributed approach helps to mitigate the threat of dishonest banks as suggested by Miers et al. [49].

Our TBS scheme can be used as a building block for a distributed e-cash scheme as discussed in the following. Upon withdrawal the user requests a coin by choosing a random unique coin identifier r and by executing the TBSign protocol over a secure channel with n issuers, each in possession of a secret key share sk_i. After obtaining at least t valid signature shares on r the user can compute the blind signature σ which resembles the coin. The coin spending protocol is performed over a secure channel through which the user sends its coin (σ, r) to the merchant, who in turn can check the validity of the coin by executing the TBVerify algorithm. If the coin is valid, the merchant establishes a secure connection with the bank aiming to deposit it on its account. In order to avoid double-spending the bank must check that no coins with identifier r were previously spent using the coin database that is maintained by the bank. If the coin passes this check then the bank deposits it on the merchant's account.

A (distributed) e-cash scheme is supposed to fulfill the following three common properties [16, 18]. The *anonymity* property means that even if $t-1$ dishonest issuers conspire with malicious merchants, the coin withdrawal and spending phases performed by the user should remain unlinkable. The *balance* property prevents coalitions of malicious users and merchants from depositing more coins than were originally withdrawn. The *(ex)culpability* property implies that any dishonest user who is willing to spend one coin twice is caught and that no coalition of at most $t-1$ malicious issuers with merchants is able to accuse an honest user of double-spending.

We discuss briefly the security of the above approach. The anonymity property follows from the blindness of TBS signatures, which guarantees that spending of a coin (r, σ) cannot be linked to the corresponding withdrawal phase. The balance property follows from unforgeability of TBS signatures and the requirement on the bank to check that coin identifiers r do not repeat. If r does not repeat and more coins were deposited than issued then at least of those coins would resemble a TBS forgery. The (ex)culpability property does not rely on the security of the TBS scheme and follows from the authentication property of secure channels between the user and the issuers upon withdrawal and between the user and the merchant upon spending. More precisely, from the authentication requirement on such channels. In order to accuse an honest user of double spending malicious issuers and merchants would need to come up with a transcript of the spending protocol authenticated by the user that shows the attempt to spend the same coin (r, σ) twice. Similarly, in order to catch a dishonest user who double-spends a coin honest issuers and merchants would be able to present two transcripts of the spending protocol authenticated by this user.

6   Conclusion

We proposed the first standard-model construction of (re-randomizable) thresh-old blind signatures (TBS), where signatures can be obtained in a blind way through interaction with n signers of which t are required to provide their sig-nature shares. The stronger security notions for TBS schemes formalized in our work extend the definitions from [54] to the threshold setting. We further showed how our TBS construction can be used to realize a distributed e-voting protocol following the template from [44] that guarantees privacy, soundness and public verifiability in presence of distributed voting authorities. As a second applica-tion we discussed construction of a distributed e-cash scheme, which achieves the desirable properties of anonymity, balance, and (ex)culpability, and where n issuers are involved in the generation of coins, a measure suggested in [49] to address the trust problem in non-distributed e-cash scenarios.

A   Blind Signature Scheme by Okamoto [51]

Our construction is influenced by the techniques underlying the following blind signature scheme from [51].

BParGen(1^λ): Generate the public bilinear group parameters $I = (\mathbb{G}, \mathbb{G}_T, q, g, e)$.
KGen(I): Pick $x \xleftarrow{r} \mathbb{Z}_q^*$ and generators $g_2, u', u_1, \ldots, u_n \xleftarrow{r} \mathbb{G}$ and set $g_1 \leftarrow g^x$.
  Output $pk = (g, g_1, g_2, u', u_1, \ldots, u_n)$ and $sk = g_2^x$.
BSign($\cdot$): Let $m \in \{0, 1\}^n$ be a message and μ_i the i-th bit of m. User U selects
  $r \xleftarrow{r} \mathbb{Z}_p^*$ and computes $X \leftarrow \left(u' \prod_{i=1}^{n} u_i^{\mu_i} \right)^r$ and sends X to the signer S. U
  additionally provides to S that it knows $(r, \mu_1, \ldots, \mu_n)$ with $\mu_i \in \{0, 1\}$ for
  X using the following witness indistinguishable Σ protocol:
  U selects $\delta_1, \ldots, \delta_n \xleftarrow{r} \mathbb{Z}_p^*$, computes $M_i = u_i^{\mu_i}(u')^{\delta_i}$, $(i = 1, \ldots, n)$ and
    sends $(M_1, \ldots, M_n)$ to S.
  U proves to S that U knows δ_i such that $M_i = (u')^{\delta_i}$ for $\mu_i = 0$ or $M_i = u_i(u')^{\delta_i}$ for $\mu_i = 1$, where $i \in [1, n]$. This proof can be realized by a Σ
    protocol which was described in [5].
  U proves to S that U knows $(t, \beta, \gamma_1, \ldots, \gamma_n)$ such that $X = \left(\prod_{i=1}^{n} M_i \right)^t \cdot$
  $(u')^\beta$, and $X = (u')^t \prod_{i=1}^{n} u_i^{\gamma_i}$, where $\beta \leftarrow t - t(\sum_{i=1}^{n} \delta_i) \mod p$ and $\gamma_i \leftarrow t\mu_i$.
  If S accepts in the above protocol then it selects $d \xleftarrow{r} \mathbb{Z}_p^*$, computes $Y_1 \leftarrow g_2^x X^d$, $Y_2 \leftarrow g^d$, and sends (Y_1, Y_2) to U. U eventually selects $s \xleftarrow{r} \mathbb{Z}_p^*$ and
  computes a blind signature $\sigma = (\sigma_1, \sigma_2)$, where

$$\sigma_1 \leftarrow Y_1 \left(u' \prod_{i=1}^{n} u_i^{\mu_i} \right)^s \quad \text{and} \quad \sigma_2 \leftarrow Y_2^r g^s.$$

BVerify(pk, m, σ): Parse pk as $(g, g_1, g_2, u', u_1, \ldots, u_n)$ and σ as (σ_1, σ_2). If

$$e(\sigma_1, g) = e(g_1, g_2)e \left(\sigma_2, u' \prod_{i=1}^{n} u_i^{\mu_i} \right)^{-1} \quad \text{output 1; otherwise output 0.}$$

The unforgeability and blindness of the scheme were proven in [51] based on the unforgeability of the Waters scheme [58] and the security of "OR" proofs [24].

B   Proof of Theorem 2 (Blindness)

Proof. We assume that the proposed signature scheme is not blind. That means the existence of a dishonest signer $\mathcal{S}^*$, which can guess b correctly with a non-negligible advantage $1/2 + \epsilon$. We construct an algorithm $\mathcal{C}$ which can break the security of the DLIN assumption as follows. Given the public parameters $pp = (\mathbb{G}, \mathbb{G}_T, q, e, g)$, the DLIN problem instance $(g^a, g^b, g^c) = (u_1', u_2', u_3')$ the challenger $\mathcal{C}$ computes $(u_{1,j}, u_{2,j}, u_{3,j}) = \left(\left(u_1'\right)^{\xi_{1,j}}, \left(u_2'\right)^{\xi_{2,j}}, \left(u_3'\right)^{\xi_{3,j}} \right)$, with $\xi_{1,j}, \xi_{2,j}, \xi_{3,j} \in \mathbb{Z}_q^*$, and $\xi_{3,j} = \xi_{1,j} + \xi_{2,j}$, $j \in \{1, \ldots, \ell\}$. $\mathcal{C}$ gives $(pp, pk, u_1', u_2', u_3', u_{1,j}, u_{2,j}, u_{3,j})$ to $\mathcal{S}^*$ as CRS. $\mathcal{S}^*$ gives $\mathcal{C}$ a public key $pk = (g_1, g_2, \mathbf{vk})$ and two messages $m_0, m_1 \in \mathbb{Z}_q^*$. The challenger $\mathcal{C}$ checks if $pk \in \mathbb{G}$ and $m_0, m_1 \in \mathbb{Z}_q^*$. If it holds $\mathcal{C}$ picks a random bit $b \in \{0, 1\}$. $\mathcal{C}$ chooses $r_i \in \mathbb{Z}_q^*$ and computes $X_{i,0} = (u_1' \prod_{j=1}^{\ell} u_{1,j}^{\mu_{j,0}})^{r_i}$ and $X_{i,1} = (u_1' \prod_{j=1}^{\ell} u_{1,j}^{\mu_{j,1}})^{r_i}$ for $m_b = (\mu_{1,b}, \ldots, \mu_{\ell,b}), b \in \{0, 1\}$. $\mathcal{C}$ executes the both NIZK protocols from Sect. 4.1 to prove $\mathcal{S}^*$ that $\mathcal{C}$ knows $(r_i, \mu_{1,b}, \ldots, \mu_{\ell,b})$ for both messages $m_b = \{m_0, m_1\}$. From the proofs in [35,37] follows that for $u_{3,j} = \left(u_3'\right)^{\xi_{1,j} + \xi_{2,j}}$ the commitments are perfect hiding and the two parameter initializations are indistinguishable under the DLIN assumption. Therefore the commitments on the messages m_b and m_{1-b} leak no information about the message. The perfect hiding property of commitments allows to simulate NIZK proofs $(\pi_{i,0}^{(1)}, \pi_{i,0}^{(2)})$ and $(\pi_{i,1}^{(1)}, \pi_{i,1}^{(2)})$, that remain indistinguishable from real proofs as shown in Sect. 4.4, [35]. $\mathcal{C}$ outputs $X_{i,b} X_{i,1-b}$ and the simulated NIZK proofs $(\pi_{i,b}^{(1)}, \pi_{i,b}^{(2)})$ and $(\pi_{i,1-b}^{(1)}, \pi_{i,1-b}^{(2)})$, where $\pi_{i,b}^{(1)}$ is the first part of NIZK proof, which is built to the commitment $X_{i,b}$ and $\pi_{i,b}^{(2)}$ is the corresponding second part of NIZK proof to the commitment $X_{i,b}$. Analogously are defined the proofs $(\pi_{i,1-b}^{(1)}, \pi_{i,1-b}^{(2)})$. After completing the NIZK protocol the challenger $\mathcal{C}$ acts as a honest user and proceeds in the same manner as the real one. $\mathcal{C}$ sends his outputs to the dishonest signer $\mathcal{S}^*$. The challenger $\mathcal{C}$ executes the signing process first on behalf of $\mathcal{U}_b$ on input $(pk, X_{i,b}, \pi_{i,b}^{(1)}, \pi_{i,b}^{(2)})$ and then on behalf of $\mathcal{U}_{1-b}$ on input $pk, X_{i,1-b}, \pi_{i,1-b}^{(1)}, \pi_{i,1-b}^{(2)})$. Since the commitments and the proofs do not leak any information about the message, the output $\sigma_{i,b}$ of the signing protocol on behalf of $\mathcal{U}_b$ is indistinguishable from the output $\sigma_{i,1-b}$ of the protocol on behalf of $\mathcal{U}_{1-b}$. If $\mathcal{S}^*$ rejects to sign one of the inputs $(X_{i,b}, \pi_{i,b}^{(2)})$ or $(X_{i,1-b}, \pi_{i,1-b}^{(2)})$, then for the corresponding output holds $\sigma_b = \perp$ or $\sigma_{1-b} = \perp$. This means that the both resulting signatures are set to $\perp$, and $\mathcal{S}^*$, does not gain any advantage if he would try to hinder the game execution. Otherwise, after finishing the signing phase of the blind signature for $\mathcal{U}_b$ and $\mathcal{U}_{1-b}$, $\mathcal{C}$ checks the validity of the obtained signatures for $\mathcal{U}_0$, $\mathcal{U}_1$ by computing the follows $e(Y_{i,b,1}, g) = e(g_2, vk_i)e(X_{i,b}, Y_{i,b,2})$. If both of the signatures $\sigma_{i,b}, \sigma_{i,1-b}$ are valid, $\mathcal{C}$ gives them to $\mathcal{S}^*$. If only one of them is valid, $\mathcal{C}$ outputs $\perp$. $\mathcal{C}$ obtains then the output b' of $\mathcal{S}^*$. If $b = b'$, $\mathcal{C}$ outputs $\beta \leftarrow 0$, otherwise it outputs $\beta \leftarrow 1$.

Analysis: Observe that if $b = b'$ then $(u_{1,j}, u_{2,j}, u_{3,j})$ for $j = \{1, \ldots, \ell\}$ are DLIN tuples with $(u_{1,j}, u_{2,j}, u_{3,j}) = \left(\left(u_1' \right)^{\xi_{1,j}}, \left(u_2' \right)^{\xi_{2,j}}, \left(u_3' \right)^{\xi_{3,j}} \right)$, with $\xi_{3,j} = \xi_{1,j} + \xi_{2,j}$ and $(u_1', u_2', u_3') = (g^a, g^b, b^c)$. In this case the challenger outputs $b_{DLIN} = 1$ and σ_b, σ_{1-b} are perfectly simulated. Therefore $Pr[b_{DLIN} = 1|b = b'] = 1/2$ Whether the challenger $\mathcal{C}$ outputs $\perp$ or two valid signatures σ_0, σ_1 depends only the adversary's reply, i.e. whether its reply σ_i satisfies the verification process or not. Therefore it is completely independent from b, since the distribution of X_0 and X_1 are indistinguishable from each other. Hence $Pr[b_{DLIN} = 0|b \neq b'] = 1/2 + \epsilon$. Eventually it follows that the success probability in DLIN problem is $1/2(1/2) + 1/2(1/2 + \epsilon) = 1/2 + \epsilon/2$, which contradicts the DLIN assumption, for negligible ϵ. $\qquad\square$

References

1. Abe, M.: A secure three-move blind signature scheme for polynomially many signatures. In: Pfitzmann, B. (ed.) EUROCRYPT 2001. LNCS, vol. 2045, pp. 136–151. Springer, Heidelberg (2001)

2. Abe, M., Fehr, S.: Adaptively secure Feldman VSS and applications to universally-composable threshold cryptography. In: Franklin, M. (ed.) CRYPTO 2004. LNCS, vol. 3152, pp. 317–334. Springer, Heidelberg (2004)

3. Abe, M., Fuchsbauer, G., Groth, J., Haralambiev, K., Ohkubo, M.: Structure-preserving signatures and commitments to group elements. In: Rabin, T. (ed.) CRYPTO 2010. LNCS, vol. 6223, pp. 209–236. Springer, Heidelberg (2010)

4. Abe, M., Ohkubo, M.: A framework for universally composable non-committing blind signatures. In: Matsui, M. (ed.) ASIACRYPT 2009. LNCS, vol. 5912, pp. 435–450. Springer, Heidelberg (2009)

5. Abe, M., Okamoto, T.: Provably secure partially blind signatures. In: Bellare, M. (ed.) CRYPTO 2000. LNCS, vol. 1880, pp. 271–286. Springer, Heidelberg (2000)

6. Baudron, O., Fouque, P., Pointcheval, D., Stern, J., Poupard, G.: Practical multi-candidate election system. In: Proceedings of the Twentieth Annual ACM Symposium on Principles of Distributed Computing, PODC 2001, pp. 274–283. ACM (2001)

7. Bellare, M., Namprempre, C., Pointcheval, D., Semanko, M.: The power of RSA inversion oracles and the security of Chaum's RSA-based blind signature scheme. In: Syverson, P.F. (ed.) FC 2001. LNCS, vol. 2339, pp. 309–328. Springer, Heidelberg (2002)

8. Benaloh, J.C., Tuinstra, D.: Receipt-free secret-ballot elections (extended abstract). In: Proceedings of the 26th Annual ACM Symposium on Theory of Computing, pp. 544–553. ACM (1994)

9. Blazy, O., Fuchsbauer, G., Pointcheval, D., Vergnaud, D.: Signatures on randomizable ciphertexts. In: Catalano, D., Fazio, N., Gennaro, R., Nicolosi, A. (eds.) PKC 2011. LNCS, vol. 6571, pp. 403–422. Springer, Heidelberg (2011)

10. Blazy, O., Fuchsbauer, G., Pointcheval, D., Vergnaud, D.: Short blind signatures. J. Comput. Secur. **21**(5), 627–661 (2013)

11. Boldyreva, A.: Threshold signatures, multisignatures and blind signatures based on the gap-diffie-hellman-group signature scheme. In: Desmedt, Y.G. (ed.) PKC 2003. LNCS, vol. 2567, pp. 31–46. Springer, Heidelberg (2002)

12. Boneh, D., Boyen, X.: Short signatures without random oracles. In: Cachin, C., Camenisch, J.L. (eds.) EUROCRYPT 2004. LNCS, vol. 3027, pp. 56–73. Springer, Heidelberg (2004)

13. Brands, S.: Untraceable off-line cash in wallets with observers. In: Stinson, D.R. (ed.) CRYPTO 1993. LNCS, vol. 773, pp. 302–318. Springer, Heidelberg (1994)

14. Brands, S.A.: An efficient off-line electronic cash system based on the representation problem. Technical report, Amsterdam, The Netherlands (1993)

15. Camenisch, J., Groß, T.: Efficient attributes for anonymous credentials. In: Proceedings of the 2008 ACM Conference on Computer and Communications Security, CCS 2008, pp. 345–356. ACM (2008)

16. Camenisch, J.L., Hohenberger, S., Lysyanskaya, A.: Compact e-Cash. In: Cramer, R. (ed.) EUROCRYPT 2005. LNCS, vol. 3494, pp. 302–321. Springer, Heidelberg (2005)

17. Camenisch, J.L., Koprowski, M., Warinschi, B.: Efficient blind signatures without random oracles. In: Blundo, C., Cimato, S. (eds.) SCN 2004. LNCS, vol. 3352, pp. 134–148. Springer, Heidelberg (2005)

18. Camenisch, J., Lysyanskaya, A., Meyerovich, M.: Endorsed e-cash. In: 2007 IEEE Symposium on Security and Privacy (S&P 2007), pp. 101–115. IEEE Computer Society (2007)

19. Camenisch, J.L., Neven, G., Shelat, A.: Simulatable adaptive oblivious transfer. In: Naor, M. (ed.) EUROCRYPT 2007. LNCS, vol. 4515, pp. 573–590. Springer, Heidelberg (2007)

20. Cetinkaya, O., Cetinkaya, D.: Verification and validation issues in electronic voting. Electron. J. e-Government **5**, 117–126 (2007)

21. Chaum, D.: Blind signatures for untraceable payments. CRYPTO 1982, pp. 199–203. Springer, Heidelberg (1982)

22. Chaum, D.: Elections with unconditionally-secret ballots and disruption equivalent to breaking RSA. In: Günther, C.G. (ed.) EUROCRYPT 1988. LNCS, vol. 330, pp. 177–182. Springer, Heidelberg (1988)

23. Chaum, D., Fiat, A., Naor, M.: Untraceable electronic cash. In: Goldwasser, S. (ed.) CRYPTO 1988. LNCS, vol. 403, pp. 319–327. Springer, Heidelberg (1990)

24. Cramer, R., Damgård, I.B., Schoenmakers, B.: Proof of partial knowledge and simplified design of witness hiding protocols. In: Desmedt, Y.G. (ed.) CRYPTO 1994. LNCS, vol. 839, pp. 174–187. Springer, Heidelberg (1994)

25. Cramer, R., Gennaro, R., Schoenmakers, B.: A secure and optimally efficient multi-authority election scheme. In: Fumy, W. (ed.) EUROCRYPT 1997. LNCS, vol. 1233, pp. 103–118. Springer, Heidelberg (1997)

26. Desmedt, Y.G.: Society and group oriented cryptography: a new concept. In: Pomerance, C. (ed.) CRYPTO 1987. LNCS, vol. 293, pp. 120–127. Springer, Heidelberg (1988)

27. Desmedt, Y.G., Frankel, Y.: Shared generation of authenticators and signatures. In: Feigenbaum, J. (ed.) CRYPTO 1991. LNCS, vol. 576, pp. 457–469. Springer, Heidelberg (1992)

28. Feldman, P.: A practical scheme for non-interactive verifiable secret sharing. In: 28th Annual Symposium on Foundations of Computer Science, pp. 427–437. IEEE Computer Society (1987)

29. Fischlin, M.: Round-optimal composable blind signatures in the common reference string model. In: Dwork, C. (ed.) CRYPTO 2006. LNCS, vol. 4117, pp. 60–77. Springer, Heidelberg (2006)

30. Franklin, M., Yung, M.: Towards provably secure efficient electronic cash. Technical report TR CUSC-018-92, Columbia University, Department of Computer Science (1993). Also in: Lingas, A., Carlsson, S., Karlsson, R. (eds.): ICALP 1993. LNCS, vol. 700. Springer, Heidelberg (1993)

31. Fujioka, A., Okamoto, T., Ohta, K.: A practical secret voting scheme for large scale elections. In: Zheng, Y., Seberry, J. (eds.) AUSCRYPT 1992. LNCS, vol. 718, pp. 244–251. Springer, Heidelberg (1993)

32. Garg, S., Rao, V., Sahai, A., Schröder, D., Unruh, D.: Round optimal blind signatures. In: Rogaway, P. (ed.) CRYPTO 2011. LNCS, vol. 6841, pp. 630–648. Springer, Heidelberg (2011)

33. Gennaro, R., Jarecki, S., Krawczyk, H., Rabin, T.: Robust threshold DSS signatures. In: Maurer, U.M. (ed.) EUROCRYPT 1996. LNCS, vol. 1070, pp. 354–371. Springer, Heidelberg (1996)

34. Gennaro, R., Jarecki, S., Krawczyk, H., Rabin, T.: Secure distributed key generation for discrete-log based cryptosystems. In: Stern, J. (ed.) EUROCRYPT 1999. LNCS, vol. 1592, pp. 295–310. Springer, Heidelberg (1999)

35. Groth, J.: Simulation-sound NIZK proofs for a practical language and constant size group signatures. In: Lai, X., Chen, K. (eds.) ASIACRYPT 2006. LNCS, vol. 4284, pp. 444–459. Springer, Heidelberg (2006)

36. Groth, J.: Short pairing-based non-interactive zero-knowledge arguments. In: Abe, M. (ed.) ASIACRYPT 2010. LNCS, vol. 6477, pp. 321–340. Springer, Heidelberg (2010)

37. Groth, J., Ostrovsky, R., Sahai, A.: Non-interactive Zaps and New Techniques for NIZK. In: Dwork, C. (ed.) CRYPTO 2006. LNCS, vol. 4117, pp. 97–111. Springer, Heidelberg (2006)

38. Hofheinz, D., Jager, T., Knapp, E.: Waters Signatures with Optimal Security Reduction. In: Fischlin, M., Buchmann, J., Manulis, M. (eds.) PKC 2012. LNCS, vol. 7293, pp. 66–83. Springer, Heidelberg (2012)

39. Horvitz, O., Katz, J.: Universally-composable two-party computation in two rounds. In: Menezes, A. (ed.) CRYPTO 2007. LNCS, vol. 4622, pp. 111–129. Springer, Heidelberg (2007)

40. Juels, A., Catalano, D., Jakobsson, M.: Coercion-resistant electronic elections. In: Proceedings of the 2005 ACM Workshop on Privacy in the Electronic Society, WPES 2005, pp. 61–70. ACM (2005)

41. Juels, A., Luby, M., Ostrovsky, R.: Security of blind digital signatures. In: Kaliski Jr, B.S. (ed.) CRYPTO 1997. LNCS, vol. 1294, pp. 150–164. Springer, Heidelberg (1997)

42. Kiayias, A., Zhou, H.-S.: Equivocal blind signatures and adaptive UC- security. In: Canetti, R. (ed.) TCC 2008. LNCS, vol. 4948, pp. 340–355. Springer, Heidelberg (2008)

43. Kim, J.-H., Kim, K., Lee, C.S.: An efficient and provably secure threshold blind signature. In: Kim, K. (ed.) ICISC 2001. LNCS, vol. 2288, pp. 318–327. Springer, Heidelberg (2002)

44. Koenig, R.E., Dubuis, Haenni, R.: Why public registration boards are required in e-voting systems based on threshold blind signature protocols. In: Electronic Voting 2010, EVOTE 2010, 4th International Conference, Co-organized by Council of Europe, Gesellschaft für Informatik and E-Voting.CC, vol. 167 LNI, pp. 255–266. GI (2010)

45. Lee, B., Kim, K.: Receipt-free electronic voting scheme through collaborationf of voter and honest verifier. In: Proceeding of JW-ISC 2000, pp. 101–108 (2000)

46. Li, J., Yuen, T.H., Kim, K.: Practical threshold signatures without random oracles. In: Susilo, W., Liu, J.K., Mu, Y. (eds.) ProvSec 2007. LNCS, vol. 4784, pp. 198–207. Springer, Heidelberg (2007)

47. Lysyanskaya, A., Peikert, C.: Adaptive security in the threshold setting: from cryptosystems to signature schemes. In: Boyd, C. (ed.) ASIACRYPT 2001. LNCS, vol. 2248, pp. 331–350. Springer, Heidelberg (2001)

48. Meiklejohn, S., Shacham, H., Freeman, D.M.: Limitations on transformations from composite-order to prime-order groups: the case of round-optimal blind signatures. In: Abe, M. (ed.) ASIACRYPT 2010. LNCS, vol. 6477, pp. 519–538. Springer, Heidelberg (2010)

49. Miers, I., Garman, C., Green, M., Rubin, A.D. : Zerocoin: Anonymous distributed e-cash from bitcoin. In: 2013 IEEE Symposium on Security and Privacy, SP 2013, pp. 397–411. IEEE Computer Society (2013)

50. Okamoto, T.: An electronic voting scheme. In: Terashima, N., Altman, E. (eds.) Advanced IT Tools. IFIP, pp. 21–30. Springer, Heidelberg (1996)

51. Okamoto, T.: Efficient blind and partially blind signatures without random oracles. In: Halevi, S., Rabin, T. (eds.) TCC 2006. LNCS, vol. 3876, pp. 80–99. Springer, Heidelberg (2006)

52. Pointcheval, D., Stern, J.: Provably secure blind signature schemes. In: Kim, K., Matsumoto, T. (eds.) ASIACRYPT 1996. LNCS, vol. 1163, pp. 252–265. Springer, Heidelberg (1996)

53. Pointcheval, D., Stern, J.: New blind signatures equivalent to factorization (extended abstract). In: Proceedings of the 4th ACM Conference on Computer and Communications Security CCS 1997, pp. 92–99. ACM (1997)

54. Schröder, D., Unruh, D.: Security of blind signatures revisited. In: Fischlin, M., Buchmann, J., Manulis, M. (eds.) PKC 2012. LNCS, vol. 7293, pp. 662–679. Springer, Heidelberg (2012)

55. Shamir, A.: How to share a secret. Commun. ACM **22**(11), 612–613 (1979)

56. Shoup, V.: Practical threshold signatures. In: Preneel, B. (ed.) EUROCRYPT 2000. LNCS, vol. 1807, pp. 207–220. Springer, Heidelberg (2000)

57. Vo, D.L., Zhang, F., Kim, K.: A new threshold blind signature scheme from pairings (2003)

58. Waters, B.: Efficient identity-based encryption without random oracles. In: Cramer, R. (ed.) EUROCRYPT 2005. LNCS, vol. 3494, pp. 114–127. Springer, Heidelberg (2005)

59. Zhou, X.:Threshold cryptosystem based fair off-line e-cash. In: Proceedings on the 2nd International Symposium on Intelligent Information Technology, pp. 692–696 (2008)

Verifiable Computation of Large Polynomials

Jiaqi Hong[1,2], Haixia Xu[1]([✉]), and Peili Li[1,2]

[1] Institute of Informassurance and Communication Security Research Center,
CAS, Beijing, China
[2] Graduate University of the Chinese Academy of Sciences, Beijing, China
{hongjiaqi,xuhaixia,lipeili}@iie.ac.cn

Abstract. Due to the proliferation of powerful cloud service, verifiable computation, which makes a computationally weak client perform intensive computations possible through outsourcing tasks to a powerful server, is attracting increasing attention. The correctness of the returned result should be verified as the server may be not trusted.
In this paper, we present a verifiable computation protocol on large polynomials, which can be publicly verified by any parties in the network. Compared with verifiable computation protocol presented by Backes et al., which is on quadratic, multi-variable polynomials, our verifiable computation protocol is on high degree, multi-variable polynomials and publicly verifiable.

Keywords: Verifiable computation · Amortized · Pre-computation · Public verification

1 Introduction

Verifiable computation makes it possible for personal computers to perform intensive computations through outsourcing computation tasks to a powerful cloud. In the age of cloud, resources are becoming more centralized. Individuals lacking computational capacity need only to buy the corresponding service in the cloud instead of purchasing their own expensive equipments to perform computation tasks. In this way, not only individual performs its computations cheap, but also resources in the cloud shared by many individuals are made full use of. So this kind of service model is attracting increasing attention.

In this paper, we name those who want to outsource intensive tasks clients, and those who have powerful resources servers. As the server may be not honest, the returned result should be verified by client to avoid malicious behavior from dishonest server. The cost of verification must be cheaper than the cost of preforming the computation, otherwise this outsourcing task will make no sense. In many instances, other parties in the network except for the client want to use this computation result. It will be better if the returned result can be publicly verified by all parties. For example, a doctor asks a server to perform a computation on the data of his patient, nurses need the computation result for better nursing. If the nurses have the ability to verify the result, they can get access to the correct result even if the doctor is not online. Many cases like this.

M. Yung et al. (Eds.): INTRUST 2014, LNCS 9473, pp. 90–104, 2015.
DOI: 10.1007/978-3-319-27998-5_6

1.1 Related Work

Verifiable computation has a large body of prior works. There are two branches, one is on general functions, the other is on specific functions. Researches on general functions often used the method of knowledge proofs to verify the correctness of the returned result [10,11,17,18,25]. Until Gentry et al. constructed a fully homomorphic encryption over idea lattices [13], several verifiable computation protocols on general functions using the fully homomorphic encryption appeared [1,8,16,21]. A representative is Gennaro et al.' research work [16]. The authors combined fully homomorphic encryption with Yao's garbled circuit and used the range of the circuit to verify the result. Researches on specific functions utilized the special structure of the outsourcing function. So those researches often focused on polynomial and matrix computations [2,4,12,26]. Of course, there were some other researches on linear algebra [23] and exponential operations [19]. Our verifiable computation protocol is on large polynomials which have a large scale of variables and are in high degree. This kind of polynomials has an extensive use in important statistics. We will introduce two notions most relevant with our work in the following, one is amortized verifiable computation, the other is public verification.

Amortized Verifiable Computation. This notion was proposed by Gennaro et al. [16], it was widely used in many of the later works about verifiable computation [2,4,7,12,26]. The client performs a pre-computation for a specific function, then the returned result from server can be verified by client in a cheap cost. Although the pre-computation cost may be as expensive as the cost of performing the outsourcing function, this function can be performed several times on different inputs by server. After several computations, this expensive pre-computation cost can be amortized.

Benabbas et al. [4] followed this amortized notion and they proposed a novel method on verifiable polynomial computations. They utilized pseudorandom functions which have closed-form efficiency to generate a series of numbers as new polynomial coefficients according to specific polynomial structure. Clients use the reconstructed polynomial to verify the correctness of the returned result. As the reconstructed polynomial can be efficiently computed, this protocol is efficient on the amortized notion. The randomness of their pseudorandom functions are based on decisional Diffie-Hellman assumption.

Backes et al. [2] proposed another brand new method on verifiable quadratic polynomial computations. They combined homomorphic MAC with verifiable computation and this is also a representative of amortized verifiable computation. Clients preform a pre-computation on the multi-variable, quadratic polynomial first, then the returned result from the server can be verified by computing a quadratic polynomials on two variables. The verification is pretty efficient. Though their protocol is on quadratic polynomials, their work is irradiative. The security of their homomorphic MAC is based on decision linear assumption.

Public Verification. Recently, two works on verifiable computation can be publicly verified. Parno et al. [24] used the primitive of attribute-based encryption.

As we know, the ciphertext of an attribute-based encryption can be decrypted only if the attribute makes a function true. They used the result of an attribute under a one way function as public verification key. Any verifier performs this one-way function on the returned result to check the equality with the public verification key to verify the result. Their protocol was suitable for functions that can be expressed as poly-size Boolean Formulas. Another work is by Fiore et al. [12]. They followed the work of Benabbas et al. [4] and made the result publicly verifiable by combining it with a bilinear map. They used a pseudorandom function proposed by Lewko and Waters [22], and reduced the security of their verifiable computation protocol on co-computational Diffie-Hellman assumption.

In publicly verifiable computation protocols, the client performs an off-line pre-computation according to the function only, and then performs on-line pre-computations under inputs. The result of the on-line pre-computation should be public to allow a public verification. Expensive off-line pre-computation cost will be amortized to each on-line pre-computation if this function will be outsourced many time to server under different inputs.

1.2   Our Contribution

In this paper, we present a verifiable computation protocol on large polynomials. We call polynomials of a large scale of variables and in high degree large polynomials. This kind of polynomials has a significant use in statistics. We follow the idea of Backes et al. [2] and extend their protocol to a more generally applicable case. Their protocol is about verifiable computation on quadratic polynomials, while our protocol is on high degree polynomials and the result can be publicly verified. One challenge is that their protocol restricted in quadratic polynomial because their basic tool, homomorphic MAC, was constructed over a bilinear map. In a bilinear map setting, multiplication of exponents can be performed at most once. If we want to construct a verifiable computation protocol on high degree polynomials, the multilinear map is intuitive. Fortunately, Garg et al. [14] made a plausible lattices-based construction of multilinear map. Though this multilinear map is not efficient enough now, this cannot stop people from using it for new constructions [6,15,20,26]. This multilinear map makes sense in our verifiable computation protocol as the pre-computation is performed on integers first, and then encodes the result to a group element in multilinear map. Another challenge is that the randomness of their pseudorandom function used for constructing homomorphic MAC is based on decision linear assumption which will no longer hold in a multilinear map setting. So we construct a new pseudorandom function based on subgroup decisional assumption to build our verifiable computation protocol. What's more, this pseudorandom function has a better performance in reducing the pre-computation cost than the pseudorandom function used by Backes et al.. The last challenge is to realize public verification. We follow the idea of Fiore et al. [12] to make a publicly verifiable computation protocol. The security is based on co-computational Diffie-Hellman assumption.

Assume the outsourcing polynomial is of m variables and degree at most d in each monomial. The main features of our protocol are as follows:

- Our protocol is a publicly verifiable computation protocol on large polynomials.
- We follow the idea of amortized verifiable computation. The off-line pre-computation cost is $O((m+1)^d)$, same as the cost of performing the outsourcing polynomial computation. The on-line pre-computation cost is $O(d)$ in addition with a multilinear map operation. After several computations on different inputs, off-line pre-computation cost can be amortized.

2 Preliminaries

Notation. If S is a set, $x \overset{U}{\longleftarrow} S$ denotes uniformly choosing an element x from S. If $\mathcal{A}$ is an algorithm, $x \leftarrow \mathcal{A}(\cdot)$ denotes the process of running $\mathcal{A}$ on some appropriate input and assigning its output to x. Let $n \in \mathbb{N}$ be the security parameter, lastly we abbreviate $param$ for public parameter, PPT for probabilistic polynomial time and PRF for pseudorandom function.

2.1 Multilinear Maps

One of our basic tool is the multilinear map. Garg et al. [14] made a plausible lattices based construction, then Coron et al. [9] made another construction over integers. Their multilinear map is a graded encoding system in fact, here we review an intuitive definition of it. The groups in this paper are all cyclic groups with order $N = pq$, where p, q are both n-bit primes.

Definition 1 (Multilinear Map). *Let $\vec{G} = (\mathbb{G}_1, \ldots, \mathbb{G}_k)$ be a sequence of cyclic groups each of order N, and g_i be a canonical generator of $\mathbb{G}_i$. There exist a set of bilinear maps $\{e_{i,j} : \mathbb{G}_i \times \mathbb{G}_j \rightarrow \mathbb{G}_{i+j} | i, j \geq 1 \wedge i + j \leq k\}$, which satisfy the following operations:*

$$e_{i,j}(g_i^a, g_j^b) = g_{i+j}^{ab} : \forall a, b \in \mathbb{Z}_N.$$

when the context is obvious, we drop the subscripts i and j, such as, $e(g_i^a, g_j^b) = g_{i+j}^{ab}$.

Let $\mathcal{G}(1^n, k)$ denote a multilinear map generator with a security parameter n and a positive integer k which indicates the required encoding level as its inputs. The output of $\mathcal{G}(1^n, k)$ is a multilinear map $\Gamma_k = (N, \mathbb{G}_1, \ldots, \mathbb{G}_k, g_1, \ldots, g_k, e)$ as described before. In a multilinear map setting, multiplication of the exponents in high degree is possible without restriction of degree 2 as in a bilinear map setting.

2.2   Pseudorandom Function

Here, we review a definition of PRF. A PRF consists of two algorithm, KeyGen and $F_K(\cdot)$. Assume that the domain of the PRF is $\mathcal{X}$ and the range is $\mathcal{Y}$, KeyGen produces a secret key K while $F_K(\cdot)$ produces $y \in \mathcal{Y}$ according to K and an input $x \in \mathcal{X}$. A definition of PRF is as follows:

Definition 2 (PRF). F *is a pseudorandom function if for every* PPT *adversary* $\mathcal{A}$, *there exists a negligible function* $neg(\cdot)$ *such that for all* n:

$$|Pr[\mathcal{A}^{F_K(\cdot)}(1^n, param) = 1] - Pr[\mathcal{A}^{R(\cdot)}(1^n, param) = 1]| \le neg(n)$$

where $R : \mathcal{X} \to \mathcal{Y}$ *is a random function.*

2.3   Computational Assumptions

Let $\Gamma_k = (N, \mathbb{G}_1, \ldots, \mathbb{G}_k, g_1, \ldots, g_k, e) \leftarrow \mathcal{G}(1^n, k)$ be a k-linear map. We review the (k, l)-Multilinear Diffie-Hellman Inversion assumption suggested by Sahai et al. [20]:

Definition 3 ((k, l)-MDHI). *Given* Γ_k *and* $g_1, g_1^a, \ldots, g_1^{a^l} \in \mathbb{G}_1$, *where* $a \xleftarrow{U} \mathbb{Z}_N$, *the advantage of an adversary* $\mathcal{A}$ *in finding out* $g_k^{a^{kl+1}}$ *is*

$$\mathbf{Adv}_{\mathcal{A}}^{mdhi} = |Pr[\mathcal{A}(\Gamma_k, g_1^a, \ldots, g_1^{a^l}) = g_k^{a^{kl+1}}]|.$$

For any PPT *adversary* $\mathcal{A}$, *there exists a negligible function* $neg(\cdot)$ *such that for all* n, $\mathbf{Adv}_{\mathcal{A}}^{mdhi}(n) \le neg(n)$.

The subgroup decisional assumption was first suggested by Boneh et al. [3]. Given $\mathbb{G}_i$ with order $N = pq$ and $u \xleftarrow{U} \mathbb{G}_i$, it is hard to determine whether u belongs to subgroup $\mathbb{G}_i^q$ or not.

Definition 4 (SDA$_i$). *Given* $\mathbb{G}_i$ *and* $u \xleftarrow{U} \mathbb{G}_i$, *the advantage of an adversary* $\mathcal{A}$ *in determining whether* u *belongs to subgroup* $\mathbb{G}_i^q$ *or not is*

$$\mathbf{Adv}_{\mathcal{A}}^{sda_i} = |Pr[\mathcal{A}(\mathbb{G}_i, u) = 1] - Pr[\mathcal{A}(\mathbb{G}_i, u^p) = 1]|.$$

For any PPT *adversary* $\mathcal{A}$, *there exists a negligible function* $neg(\cdot)$ *such that for all* n, $\mathbf{Adv}_{\mathcal{A}}^{sda_i}(n) \le neg(n)$.

Zhang et al. [26] proved that subgroup decisional assumption holds for Γ_k if SDA_i holds for every $\mathbb{G}_i$, $i = 1, \ldots, k$.

The last one is the co-computational Diffie-Hellman assumption suggested by Boneh et al. [5].

Definition 5 (co-CDH Assumption). *Given* Γ_k *and* g_1^a, g_2^b, *where* $a, b \xleftarrow{U} \mathbb{Z}_N$, *the advantage of an adversary* $\mathcal{A}$ *in finding out* g_1^{ab} *is*

$$\mathbf{Adv}_{\mathcal{A}}^{cdh} = Pr[\mathcal{A}(\Gamma_k, g_1^a, g_2^b) = g_1^{ab}].$$

For any PPT *adversary* $\mathcal{A}$, *there exists a negligible function* $neg(\cdot)$ *such that for all* n, $\mathbf{Adv}_{\mathcal{A}}^{cdh}(n) \le neg(n)$.

2.4   Basic Model

Now we review a basic publicly verifiable computation model. The client performs an off-line pre-computation according to the outsourcing function only through the following **KeyGen** algorithm, and then performs an on-line pre-computation on specific inputs through the following **ProbGen** algorithm. The result of the on-line pre-computation should be public to allow a public verification. The server runs the **Compute** algorithm and returns a σ_y. Any third party can verify the returned computation result and output a value y or an error $\perp$.

Let $\mathcal{F}$ be a family of functions. A publicly verifiable computation protocol $\mathcal{VC}$ for $\mathcal{F}$ is as follows:

- **KeyGen**$(1^n, f) \rightarrow (SK, PK, EK)$. With a security parameter n and $f \in \mathcal{F}$, key generation algorithm produces secret key SK, public key PK, and evaluation key EK. Send EK to server. This is the off-line pre-computation on f.
- **ProbGen**$(PK, SK, x) \rightarrow (\sigma_x, VK_x)$. With an input x in the domain of f, the problem generation algorithm allows the client to produce an input encoding σ_x and a public verification key VK_x. This is the on-line pre-computation on specific input x.
- **Compute**$(PK, EK, f, \sigma_x) \rightarrow \sigma_y$. With PK, EK, f and σ_x, this algorithm allows server to perform a computation on f and return a σ_y to the verifier.
- **Verify**$(PK, VK_x, \sigma_y) \rightarrow y/\perp$. With PK, VK_x, and σ_y, this algorithm allows any party to verify the result and return a value y or an error $\perp$.

A verifiable computation protocol is secure if it holds the following properties: correctness and soundness. Simply, correctness is the value output by an honest server can be verified correctly.

Definition 6 (Correctness). *For any $f \in \mathcal{F}$, any $(SK, PK, EK) \leftarrow$* **KeyGen**$(1^n, f)$, *any $x \in Dom(f)$, if $(\sigma_x, VK_x) \leftarrow$* **ProbGen**$(PK, SK, x)$ *and $\sigma_y \leftarrow$* **Compute**(PK, EK, f, σ_x), *then the output of* **Verify**(PK, VK_x, σ_y) *is $f(x)$ with all but negligible probability.*

Soundness is any PPT adversary $\mathcal{A}$ cannot persuade a verifier to accept an incorrect computation result. Define the following experiment:

$\mathbf{Exp}_{\mathcal{A}}^{PubVer}[\mathcal{VC}, f, l, n]$:
$(SK, PK, EK) \leftarrow$ **KeyGen**$(1^n, f)$,
For $i = 1$ to l:
$\quad x_i \leftarrow \mathcal{A}(PK, EK, \sigma_{x,1}, VK_{x,1}, \ldots, \sigma_{x,i-1}, VK_{x,i-1})$,
$\quad (\sigma_{x,i}, VK_{x,i}) \leftarrow$ **ProbGen**(PK, SK, x_i);
$x^* \leftarrow \mathcal{A}(PK, EK, \sigma_{x,1}, VK_{x,1}, \ldots, \sigma_{x,l}, VK_{x,l})$,
$(\sigma_{x^*}, VK_{x^*}) \leftarrow$ **ProbGen**(PK, SK, x^*),
$\hat{\sigma}_y \leftarrow \mathcal{A}(PK, EK, \sigma_{x,1}, VK_{x,1}, \ldots, \sigma_{x,l}, VK_{x,l}, \sigma_{x^*}, VK_{x^*})$,
$\hat{y} \leftarrow$ **Verify**$(PK, VK_{x^*}, \hat{\sigma}_y)$,
If $\hat{y} \neq \perp$ and $\hat{y} \neq f(x^*)$, output 1, else output 0.

For any $n \in \mathbb{N}$, any function $f \in \mathcal{F}$, the advantage of an adversary $\mathcal{A}$ making at most $l = poly(n)$ queries in the above experiment against $\mathcal{VC}$ is

$$\mathbf{Adv}_{\mathcal{A}}^{\text{PubVer}}(\mathcal{VC}, f, l, n) = Pr[\mathbf{Exp}_{\mathcal{A}}^{\text{PubVer}}[\mathcal{VC}, f, l, n] = 1]$$

Definition 7 (Soundness). *A verifiable computation protocol $\mathcal{VC}$ is sound for $\mathcal{F}$, if for any $f \in \mathcal{F}$ and any PPT adversary $\mathcal{A}$ there exists a negligible function $neg(\cdot)$ such that for all n, $\mathbf{Adv}_{\mathcal{A}}^{\text{PubVer}}(\mathcal{VC}, f, l, n) \leq neg(n)$.*

3   Multi-labeled Program

The idea of our work is inspired by Backes et al.' multi-labeled verifiable computation protocol [2]. Briefly describing the conception of multi-labeled program and its corresponding verifiable computation protocol will help readers appreciate our work more easily.

In a multi-labeled program, a pair of labels $L = (\Delta, \tau)$ is used to identify a set of input message, where Δ is data set identifier and τ is input identifier. For an instance, if we want to record the weather condition per hour in a day, then we should keep track of temperature, humidity, sunlight and so on hourly. $\tau = (\tau_1, \tau_2, \cdots)$ labels temperature, humidity, sunlight etc. respectively, while Δ labels time. Regard the recordings in each hour as one data set. Different Δ_i labels different data sets, then τ can be reused to label inputs in different data sets. A pair $L = (\Delta, \tau)$ can uniquely identify a set of inputs while any single Δ or τ can not. Please refer to [2] for details.

The authors proposed a verifiable computation protocol on quadratic polynomials of m variables using multi-label. The verification cost is the cost of performing a quadratic polynomial on two variables. This verifiable computation protocol is efficient if m is large enough. We briefly review their protocols in the following:

Assume that the outsourcing function f is a quadratic polynomial of m variables. For every input x_i, $i = 1, \ldots, m$, client generates two pairs of pseudorandom values according to their labels such as: $(u_i, v_i) \leftarrow F_{K_1}(\tau_i)$, $(a, b) \leftarrow F_{K_2}(\Delta)$, where F is a PRF and K_1, K_2 are secret keys of F. Client chooses $\alpha \xleftarrow{U} \mathbb{Z}_N$ as its secret key and sets $y_0^{(i)} = x_i$, $Y_1^{(i)} = (g^{u_i a + v_i b - x_i})^{\frac{1}{\alpha}}$, $Y_2^{(i)} = 1 \in \mathbb{G}_1$ for $i = 1, \ldots, m$, sends m tuples (y_0, Y_1, Y_2) to server. Server computes σ_y according to the arithmetic circuit of f gate by gate:

- **Addition.** If the gate is an addition gate, assume values on two input wires are respectively $y_0^{(1)}$ and $y_0^{(2)}$. Compute (y_0, Y_1, Y_2) as follows:

$$y_0 = y_0^{(1)} + y_0^{(2)}, Y_1 = Y_1^{(1)} \cdot Y_1^{(2)},$$

$$Y_2 = Y_2^{(1)} \cdot Y_2^{(2)}.$$

– **Multiplication.** If the gate is a multiplication gate, assume values on two input wires are respectively $y_0^{(1)}$ and $y_0^{(2)}$. Compute (y_0, Y_1, Y_2) as follows:

$$y_0 = y_0^{(1)} \cdot y_0^{(2)}, Y_1 = (Y_1^{(1)})^{y_0^{(2)}} \cdot (Y_1^{(2)})^{y_0^{(1)}},$$

$$Y_2 = e(Y_1^{(1)}, Y_1^{(2)}).$$

– **Mulplication with constant.** If the gate is a multiplication gate, the value of one input wire is a constant c, the value of another input wire is $y_0^{(1)}$. Compute (y_0, Y_1, Y_2) as follows:

$$y_0 = c \cdot y_0^{(1)}, Y_1 = (Y_1^{(1)})^c,$$

$$Y_2 = (Y_2^{(1)})^c.$$

After finishing the computation, server sets $\sigma_y = (y_0, Y_1, Y_2)$ and returns it to client. The verification equation is:

$$W = e(g, g)^{y_0} \cdot e(Y_1, g)^{\alpha} \cdot Y_2^{\alpha^2}, \tag{1}$$

where W is computed by client in two steps. Firstly, the client performs a pre-computation on the outsourcing quadratic polynomial f to obtain a quadratic polynomial on two variables:

$$p(z_1, z_2) = f(\rho_1(z_1, z_2), \ldots, \rho_m(z_1, z_2))$$

where $\rho_i(z_1, z_2) = u_i z_1 + v_i z_2$, $(u_i, v_i) \leftarrow F_{K_1}(\tau_i)$. Then, when the client wants to outsource this polynomial computation on specific inputs, it generates $(a, b) \leftarrow F_{K_2}(\Delta)$ according to data set label Δ and computes $W = p(a, b)$. If Eq. (1) holds, the returned σ_y is honestly computed and y_0 is the correct computation result. Otherwise, client outputs $\bot$. This polynomial f can be outsourcing many times on different inputs and the verification cost is the cost of performing a quadratic polynomial computation on two variables. The correctness and soundness of this protocol have been proved by Backes et al. [2].

This verifiable computation protocol can deal with polynomials in degree at most 2 as it is in the setting of bilinear map. If we just extend it to high degree polynomial using multilinear map, the verification cost will be a two variables polynomial of the same high degree. Unfortunately, the decision linear assumption which the protocol reduces the randomness of its PRF on no longer holds in a multilinear map setting. We construct a variant of the PRF which has a better performance in reducing the on-line pre-computation cost while realizing public verification.

4 Our Protocol

In this section, we present a publicly verifiable computation protocol on large polynomials. Assume that the outsourcing polynomial f is of m variables and degree at most d. We follow the idea of multi-labeled program and use a pair of labels $L = (\Delta, \tau_i)$ to identify input x_i, for all $i = 1, \ldots, m$. In the following, we will introduce our PRF first, then give a detailed verifiable computation protocol built on our PRF.

4.1   PRF with Amortized Closed-Form Efficiency

The randomness of our PRF is based on the subgroup decisional assumption.

PRF:

- KeyGen(1^n): Let $\Gamma_k = (N, \mathbb{G}_1, \ldots, \mathbb{G}_k, g_1, \ldots, g_k, e) \leftarrow \mathcal{G}(1^n, k)$. Choose two secret keys k_1, k_2 for PRFs $F'_{k_{1,2}} : \{0,1\}^n \rightarrow \mathbb{Z}_N$. Output $K = \{p, q, k_1, k_2\}$ and public parameter $param = \Gamma_k$.
- $F_K(x)$: On input x, generate a pair of values (a, b) according to its label $L = (\Delta, \tau)$ such as: $a \leftarrow F'_{k_1}(\tau)$, and $b \leftarrow F'_{k_2}(\Delta)$, where $\Delta \in \{0,1\}^n$ and $\tau \in \{0,1\}^n$. Output $F_K(x) = g_1^{pab}$.

Theorem 1. *If F' is a pseudorandom function and the SDA assumption holds for Γ_k, then PRF is a pseudorandom function.*

Proof. The proof follows by a standard hybrid argument.

Game 0: this is the real game described above for PRF.

Game 1: this is Game 0 except that $F'_{k_1}(\tau)$ is replaced by a random function $\Phi_1 : \{0,1\}^n \rightarrow \mathbb{Z}_N$. It is easy to argue that Game 1 is indistinguishable with Game 0.

Game 2: this is Game 1 except that $F'_{k_2}(\Delta)$ is replaced by a random function $\Phi_2 : \{0,1\}^n \rightarrow \mathbb{Z}_N$. Similarly to the previous case, one can easily argue that Game 2 is indistinguishable with Game 1.

Game(3, j): let Q_Δ be the upper bound on the number of distinct Δ queried by adversary $\mathcal{A}$. If $S = \{\Delta_1, \ldots, \Delta_{Q_\Delta}\}$ is the ordered set of Δ queried by $\mathcal{A}$, then, for $0 \leq j \leq Q_\Delta$, we define the following partial sets of S: $S_{\leq j} = \{\Delta_i \in S : i \leq j\}$ and $S_{>j} = \{\Delta_i \in S : i > j\}$. Then we define Game $(3, j)$ same as Game 2 except that queries (Δ, τ) where $\Delta \in S_{\leq j}$ are answered with a random value R chosen uniformly in $\mathbb{G}_1$, whereas queries (Δ, τ) where $\Delta \in S_{>j}$ are answered with $R = g^{pab}$ where $a \leftarrow \Phi_1(\tau)$ and $b \leftarrow \Phi_2(\Delta)$.

As one can notice, Game $(3, 0)$ is the same as Game 2, while Game $(3, Q_\Delta)$ is the game where all queries are answered with freshly random values in $\mathbb{G}_1$, just like $\mathcal{A}$ is getting access to a truly random oracle from $\mathcal{X}$ to $\mathbb{G}_1$. If for every $1 \leq j \leq Q_\Delta$, Game $(3, j-1)$ is computationally indistinguishable from Game $(3, j)$ under the subgroup decisional assumption holds for Γ_k, the proof can be done. So we prove the following lemma:

Lemma 1. *If subgroup decisional assumption holds for Γ_k, then $|Pr[G_{3,j-1}] - Pr[G_{3,j}]|$ is negligible for $1 \leq j \leq Q_\Delta$.*

The key tool of our proof is the following lemma which shows the function $f_b(U) = U^{pb}$ is a weak PRF under the subgroup decisional assumption.

Lemma 2. *If the subgroup decisional assumption holds for Γ_k then function $f_b(U) = U^{pb}$, where $b \xleftarrow{U} \mathbb{Z}_N$, is a weak PRF.*

Proof. For a tuple (g_1, g_1^a, g_1^{pab}), we rename g_1^a as U and g_1^{pab} as V. Given such (U, V), challenger can create polynomially-many binary pairs (U_i, V_i) which have the same form, all V_i are random values in subgroup $\mathbb{G}_1^q$. If there exist a PPT adversary who can distinguish $f_b(U_i)$ with a random function, whose output is a random value in $\mathbb{G}_1$, in a non-negligible probability, then the challenger can solve subgroup decisional problem with the same probability.

Proof (Lemma 1). Now we show that any PPT adversary $\mathcal{A}$ who has non-negligible probability in distinguish Game $(3, j-1)$ with Game $(3, j)$ can build a PPT challenger $\mathcal{C}$ who distinguishes the weak PRF $f_b(U) = U^{pb}$ with a random function in the same probability.

$\mathcal{C}$ receives as input $param = \Gamma_k$ and gets access to an oracle which outputs a binary pair (U, V) on each query. Recall that if $\mathcal{O} = \mathcal{O}_f$, then $V = U^{pb}$ where b is the secret key of the weak PRF f. Otherwise, if $\mathcal{O} = \mathcal{O}_R$, then V is randomly chosen in $\mathbb{G}_1$. In both case, U is randomly chosen at every new query.

$\mathcal{C}$ runs the simulation for $\mathcal{A}$ as follows.

Assume that Q_τ is the upper bound on the number of distinct τ queried by $\mathcal{A}$. Let (Δ, τ) be query from $\mathcal{A}$, and assume that $(\Delta, \tau) = (\Delta_k, \tau_i)$ for $1 \leq k \leq Q_\Delta$ and $1 \leq i \leq Q_\tau$. $\mathcal{C}$ answers (Δ_k, τ_i) as follows.

- If $k \leq j-1$, then $\mathcal{C}$ chooses $R \xleftarrow{U} \mathbb{G}_1$ uniformly and returns R.
- If $k > j$, then $\mathcal{C}$ chooses $b_k \xleftarrow{U} \mathbb{Z}_N$ and queries the oracle $\mathcal{O}_f$. Return $R = f_{b_k}(U_i)$.
- If $k = j$, then $\mathcal{C}$ returns $R = V_i$

Basically, the simulator is implicitly setting $b_j = b$ where b is the secret key of the weak PRF f. Let $G_{3,j}$ be the event that Game $(3, j)$ outputs 1 which is run by adversary $\mathcal{A}$. Finally, $\mathcal{C}$ outputs the same bit b as $\mathcal{A}$ outputs b.

When $\mathcal{C}$ gets access to the weak PRF, where $V_i = f_b(U_i)$, then $\mathcal{C}$ is simulating Game $(3, j-1)$. On the other hand, when $\mathcal{C}$ gets access to a random function, where V_i is random and independent of U_i, then $\mathcal{C}$ simulates the view of Game $(3, j)$. That are $Pr[\mathcal{C}^{\mathcal{O}_f} = 1] = Pr[G_{3,j-1}]$ and $Pr[\mathcal{C}^{\mathcal{O}_R} = 1] = Pr[G_{3,j}]$. We have:

$$|Pr[\mathcal{C}^{\mathcal{O}_f} = 1] - Pr[\mathcal{C}^{\mathcal{O}_R} = 1]| = |Pr[G_{3,j-1}] - Pr[G_{3,j}]|$$

The simulation is perfect, and Lemma 1 has been proved.

The PRF helps to amortize the pre-computation cost. For a specific polynomial f, which is of m variables and in degree d, the client performs the pre-computation in two steps. In STEP 1, the client transforms this polynomial to a one variable, degree d polynomial ρ in a cost $O((m+1)^d)$, the same as the cost of performing the computation on f. In STEP 2, the client performs a computation on ρ with cost $O(d)$. Details as follows:

STEP 1.

This is off-line pre-computation. Generate $a_i \leftarrow F'_{k_1}(\tau_i)$ according to input identifier τ_i for $i = 1, \ldots, m$, where $F'_{k_1}(\cdot)$ is the pseudorandom function to produce an exponent a. Set $\rho_i(z) = pa_i \cdot z$ for $i = 1, \ldots, m$. Obviously, all $\rho_i(z)$

are degree-1 polynomial on variable z with no constant. Perform the computation of f on $\rho_1(z), \ldots, \rho_m(z)$ to get a new one variable, degree d polynomial $\rho(z)$:

$$\rho(z) = f(\rho_1(z), \ldots, \rho_m(z)).$$

It is worth noting that the above computation can be done off-line by client as it is only related to function. The input identifier can be reused many times for a specific polynomial f as long as data set identifier is different. The cost of this step is $O((m+1)^d)$.

STEP 2.

This is on-line pre-computation. Generate $b \leftarrow F'_{k_2}(\Delta)$ according to data set identifier, where $F'_{k_2}(\cdot)$ is the pseudorandom function to produce an exponent b. Perform the computation of $\rho(z)$ on b, the result is $\rho(b)$ and computation cost is $O(d)$.

When performing STEP 2, the input of polynomial f has been identified. STEP 2 can be performed many times on different inputs for a specific polynomial f, the cost of off-line pre-computation can be amortized if this polynomial f will be performed many times on different inputs. So, the cost of pre-computation will be low on average.

4.2   Construction

Our verifiable computation protocol on large polynomials utilizes the PRF above. Let f be the outsourcing polynomial, assume it is a polynomial of m variables and in degree d. Details as follows:

- **KeyGen**$(1^n, k, f) \rightarrow (SK, PK, EK)$. This is key generation algorithm run by client. Generate a k-linear map, $\Gamma_k = (N, \mathbb{G}_1, \ldots, \mathbb{G}_k, g_1, \ldots, g_k, e) \leftarrow \mathcal{G}(1^n, k)$, where $k = d + 2$. Choose $\alpha \xleftarrow{U} \mathbb{Z}_N$ uniformly. Choose secret keys of PRF as described before, $K = (k_1, k_2)$. Run STEP 1 to generate a one variable, degree d polynomial $\rho(z)$. Set $ek = (ek_0, ek_1, \ldots, ek_i, \ldots, ek_d)$ where $ek_i = g_d^{\alpha^i}{}_{-i+1}$.
  The secret key $SK = (k_1, k_2, p, q, \alpha)$, the public key $PK = \Gamma_k$. The evaluation key $EK = ek$, send it to server.
- **ProbGen**$(SK, PK, x) \rightarrow (\sigma_x, VK_x)$. This is problem generation algorithm run by client. Run STEP 2 to get the result $\rho(b)$ and set the public verification key as $VK_x = g_{d+2}^{\rho(b)}$.
  Run PRF to get $R_i = g_1^{p a_i b}$ for each input x_i, $i = 1, \ldots, m$. Set $\sigma_i = (y_0^{(i)}, Y_1^{(i)}, Y_2^{(i)})$, where $y_0^{(i)} = x_i \in \mathbb{Z}_N$, $Y_1^{(i)} = (R_i \cdot g_1^{-x_i})^{\frac{1}{\alpha}} \in \mathbb{G}_1$, $Y_2^{(i)} = 1 \in \mathbb{G}_1$. Set $\sigma_x = (\sigma_1, \ldots, \sigma_m)$, send it to server.
- **Compute**$(PK, EK, f, \sigma_x) \rightarrow \sigma_y$. Given the evaluation key EK, σ_x, PK and the outsourcing polynomial f, server computes a σ_y as follows. For our convenience to describe, we interpret $f(x)$ as $f(x) = \sum_{i=1}^{s} f_i p_i(x)$, where for each monomial $f_i p_i(x)$ we interpret it further as $f_i p_i(x) = f_i \prod_{j=1}^{d} x_{i_j}$, where $0 \leq i_1, \ldots, i_d \leq m$, x_0 denotes constant 1, while $x_1, \ldots, x_m$ denote the m

variables. Server computes $\sigma_y = (y_0, Y_1, Y_2)$ according to each monomial first and then adds the s triples (y_0, Y_1, Y_2) together, details as follows:
Initiate $y_0 = 0$, $Y_1 = 1 \in \mathbb{G}_{d+1}$, $Y_2 = 1 \in \mathbb{G}_{d+1}$.

For $i = 1, \ldots, s$:

If $i_1 = \ldots = i_d = 0$, then:
$\quad y_{0i} = f_i, Y_{1i} = ek_0, Y_{2i} = ek_0;$
Else, let $\bar{j}$ be such that $i_{\bar{j}} \geq 1$ and $i_{\bar{j}+1} = \cdots = i_d = 0:$

$$Y_{2i} = e(Y_1^{(i_1)}, Y_1^{(i_2)}, \ldots, Y_1^{(i_{\bar{j}})}, ek_{\bar{j}}),$$

$$Y_{1i} = e(Y_1^{(i_1)}, \ldots, Y_1^{(i_{\bar{j}-1})}, ek_{\bar{j}-1})^{v_0^{(i_{\bar{j}})}} \cdot e(Y_1^{(i_1)}, \ldots, Y_1^{(i_{\bar{j}-2})}, Y_1^{(i_{\bar{j}})}, ek_{\bar{j}-1})^{v_0^{(i_{\bar{j}-1})}}$$

$$\cdots e(Y_1^{(i_1)}, Y_1^{(i_3)} \ldots, Y_1^{(i_{\bar{j}})}, ek_{\bar{j}-1})^{v_0^{(i_2)}} \cdot e(Y_1^{(i_2)}, \ldots, Y_1^{(i_{\bar{j}})}, ek_{\bar{j}-1})^{v_0^{(i_1)}}$$

$$\cdot e(Y_1^{(i_1)}, \ldots, Y_1^{(i_{\bar{j}-2})}, ek_{\bar{j}-2})^{v_0^{(i_{\bar{j}-1})} v_0^{(i_{\bar{j}})}} \cdots e(Y_1^{(i_3)}, \ldots, Y_1^{(i_{\bar{j}})}, ek_{\bar{j}-2})^{v_0^{(i_1)} v_0^{(i_2)}}$$

$$\cdots$$

$$e(Y_1^{(i_1)}, ek_1)^{v_0^{(i_2)} \cdots v_0^{(i_{\bar{j}})}} \cdots e(Y_1^{(i_{\bar{j}})}, ek_1)^{v_0^{(i_1)} \cdots v_0^{(i_{\bar{j}-1})}},$$

$$y_{0i} = v_0^{(i_1)} \cdots v_0^{(i_{\bar{j}})},$$
set $y_{0i} = f_i y_{0i}, Y_{1i} = (Y_{1i})^{f_i},$ and $Y_{2i} = (Y_{2i})^{f_i};$

set $y_0 = y_0 + y_{0i}$, $Y_1 = Y_1 \cdot Y_{1i}$, and $Y_2 = Y_2 \cdot Y_{2i}$.
Server sets $\sigma_y = (y_0, Y_1, Y_2)$ and returns σ_y to verifier.
- **Verify**$(PK, VK_x, \sigma_y) \to y/\bot$. Any third party who wants to verify the result checks the following equation:

$$g_{d+2}^{y_0} \cdot e(Y_1, g_1) \cdot e(Y_2, g_1) = VK_x \tag{2}$$

If the equation holds, verifier outputs y_0 as the correct computation result. Otherwise, outputs an error symbol $\bot$.

First we show the correctness of the protocol briefly. Recall that $ek_i = g_{d-i+1}^{\alpha^i}$, if σ_y is honestly calculated by server, there is

$$g_{d+2}^{y_0} \cdot e(Y_1, g_1) \cdot e(Y_2, g_1) = g_{d+2}^{\rho(b)}, \tag{3}$$

Notice that $VK_x = g_{d+2}^{\rho(b)}$, then Eq. (2) holds. The honest result returned from server can be verified correctly.

Now we show the soundness of our protocol. If (k, l)-MDHI assumption holds in Γ_k, any PPT adversary can't get any secret keys from public key PK and evaluation key EK.

Theorem 2. *If co-CDH assumption holds in Γ_k, then any PPT adversary $\mathcal{A}$ making at most $l = poly(n)$ queries has advantage*

$$Adv_{\mathcal{A}}^{PubVer}(\mathcal{VC}, f, l, n) \leq neg(n),$$

where $neg(\cdot)$ is a negligible function.

Proof. The proof follows by a standard hybrid argument based on the following games:

Game 0: this is the real game same as $\mathbf{Exp}_{\mathcal{A}}^{PubVer}(\mathcal{VC}, f, l, n)$.

Game 1: this is Game 0 except for the following change in the evaluation of $\rho(b)$. For any x asked by the adversary during the game, instead of computing $\rho(b)$ using the STEP 1 and STEP 2, which is efficient in an amortized notion, an inefficient one step evaluation $\rho(b) = f(\rho_1(b), \ldots, \rho_m(b))$ is used. One can easily argue that Game 1 is indistinguishable with Game 0.

Game 2: this is Game 1 except that PRF is replaced by a truly random function $R : \{0,1\}^n \times \{0,1\}^n \to \mathbb{G}_1$. Let R be a set of m random values generated by this random function where R is a set of m numbers. One can easily argue that Game 2 is indistinguishable with Game 1 as the randomness of our PRF.

Now we show if there exists a PPT adversary $\mathcal{A}$ who can win in Game 2 with a non-negligible probability, then there is a challenger $\mathcal{C}$ who can solve the co-CDH problem with the same probability.

$\mathcal{C}$ takes as input a group description Γ_k, chooses $r \xleftarrow{U} \mathbb{Z}_N$. For a query $x = (x_1, \ldots, x_m)$ from $\mathcal{A}$, $\mathcal{C}$ chooses m random values $\beta_1, \ldots, \beta_m \in \mathbb{Z}_N$, sets $R^{(i)} = g_1^{\beta_i}$, for $i = 1, \ldots, m$. all $R^{(i)} = g_1^{\beta_i}$ are random values in $\mathbb{G}_1$. Set $\sigma_x = (\sigma_1, \ldots, \sigma_m)$ where $\sigma_i = (y_0^{(i)}, Y_1^{(i)}, Y_2^{(i)})$, $y_0^{(i)} = x_i$, $Y_1^{(i)} = (R^{(i)} \cdot g_1^{-x_i})^{\frac{1}{r}}$, $Y_2^{(i)} = 1 \in \mathbb{G}_1$. Set $ek = (ek_0, ek_1, \ldots, ek_i, \ldots, ek_d)$ where $ek_i = g_{d-i+1}^{r^i}$. $\mathcal{C}$ computes $VK_x = g_{d+2}^{f(\beta_1, \ldots, \beta_m)}$ and returns VK_x and σ_x to $\mathcal{A}$. The distribution of VK_x and σ_x are exactly the same as the one in Game 2.

Finally, let $\sigma_y^* = (y_0^*, Y_1^*, Y_2^*, W^*)$ be the output of $\mathcal{A}$ at the end of the game, such that for some x^* chosen by $\mathcal{A}$ it holds $\mathbf{Verify}(PK, VK_{x^*}, \sigma_{y^*}) = y^*$, $y^* \neq \perp$ and $y^* \neq f(x^*)$. By verification, this means that

$$g_{d+2}^{y_0^*} \cdot e(Y_1^* \cdot Y_2^*, g_1) = VK_x. \tag{4}$$

Let $\sigma_y = (y_0, Y_1, Y_2, W)$ be the correct output of the computation. Then, by correctness it also holds:

$$g_{d+2}^{y_0} \cdot e(Y_1 \cdot Y_2, g_1) = VK_x. \tag{5}$$

Dividing the verification Eq. (4) by (5),

$$g_{d+2}^{y_0^* - y_0} = e(Y_1/Y_1^* \cdot Y_2/Y_2^*, g_1). \tag{6}$$

That is, for a false y_0^*, $\mathcal{A}$ can find a Y_1^* and a Y_2^* to satisfy Eq. (6) in a non-negligible probability, then $\mathcal{B}$ solves the co-CDH problem with the same probability.

5   Conclusion

In this paper, we propose a delegated computation protocol on high degree polynomials over a large amount of variables which allows public verification. Assume that the delegated polynomial is of m variables and degree at most d. The off-line pre-computation cost is $O((m + 1)^d)$, same as the cost of performing the outsourcing polynomial computation. The on-line pre-computation cost is $O(d)$ in addition with a multilinear map operation. Using the notion of amortization, off-line pre-computation cost can be amortized if the client delegates the same function f several times on different inputs. This protocol is efficient in average.

Acknowledgment. This work is supported by the National Natural Science Foundation of China (No.61379140) and the National Basic Research Program of China (973 Program) (No. 2013CB338001). The authors wish to acknowledge the anonymous referees for helpful suggestions.

References

1. Barbosa, M., Farshim, P.: Delegatable homomorphic encryption with applications to secure outsourcing of computation. In: Dunkelman, O. (ed.) CT-RSA 2012. LNCS, vol. 7178, pp. 296–312. Springer, Heidelberg (2012)
2. Backes, M., Fiore, D., Reischuk., R. M.: Verifiable delegation of computation on outsourced data. In: CCS 2013, pp. 863–874. ACM press (2013). A full version is avaliable at http://eprint.iacr.org/2013/469 (2013)
3. Boneh, D., Goh, E.-J., Nissim, K.: Evaluating 2-DNF formulas on ciphertexts. In: Kilian, J. (ed.) TCC 2005. LNCS, vol. 3378, pp. 325–341. Springer, Heidelberg (2005)
4. Benabbas, S., Gennaro, R., Vahlis, Y.: Verifiable delegation of computation over large datasets. In: Rogaway, P. (ed.) CRYPTO 2011. LNCS, vol. 6841, pp. 111–131. Springer, Heidelberg (2011)
5. Boneh, D., Lynn, B., Shacham, H.: Short signatures from the weil pairing. In: Boyd, C. (ed.) ASIACRYPT 2001. LNCS, vol. 2248, pp. 514–532. Springer, Heidelberg (2001)
6. Catalano, Dario, Fiore, Dario, Gennaro, Rosario, Nizzardo, Luca: Generalizing homomorphic MACs for arithmetic circuits. In: Krawczyk, Hugo (ed.) PKC 2014. LNCS, vol. 8383, pp. 538–555. Springer, Heidelberg (2014)
7. Choi, S.G., Katz, J., Kumaresan, R., Cid, C.: Multi-client non-interactive verifiable computation. In: Sahai, A. (ed.) TCC 2013. LNCS, vol. 7785, pp. 499–518. Springer, Heidelberg (2013)
8. Chung, K.-M., Kalai, Y., Vadhan, S.: Improved delegation of computation using fully homomorphic encryption. In: Rabin, T. (ed.) CRYPTO 2010. LNCS, vol. 6223, pp. 483–501. Springer, Heidelberg (2010)
9. Coron, J.-S., Lepoint, T., Tibouchi, M.: Practical multilinear maps over the integers. In: Canetti, R., Garay, J.A. (eds.) CRYPTO 2013, Part I. LNCS, vol. 8042, pp. 476–493. Springer, Heidelberg (2013)
10. Cormode, G., Mitzenmacher, M., Thaler, J.: Practical Verified Computation with Streaming Interactive Proofs. In: ITCS 2012, pp. 90–112. ACM press, New York (2012)

11. Cormode, G., Thaler, J., Yi, K.: Verifying computations with streaming interactive proofs. Proc. VLDB Endowment **5**(1), 25–36 (2011)
12. Fiore, D., Gennaro, R.: Publicly Verification delegation of large polynomials and matrix computations, with applications. In: CCS 2012, pp. 501–512. ACM press, New York (2012)
13. Gentry, C.: A fully homomorphic encryption scheme. In: Stanford University (2009)
14. Garg, S., Gentry, C., Halevi, S.: Candidate multilinear maps from ideal lattices. In: Johansson, T., Nguyen, P.Q. (eds.) EUROCRYPT 2013. LNCS, vol. 7881, pp. 1–17. Springer, Heidelberg (2013)
15. Garg, S., Gentry, C., Halevi, S., Sahai, A., Waters, B.: Attribute-based encryption for circuits from multilinear maps. In: Canetti, R., Garay, J.A. (eds.) CRYPTO 2013, Part II. LNCS, vol. 8043, pp. 479–499. Springer, Heidelberg (2013)
16. Gennaro, R., Gentry, C., Parno, B.: Non-interactive verifiable computing: outsourcing computation to untrusted workers. In: Rabin, T. (ed.) CRYPTO 2010. LNCS, vol. 6223, pp. 465–482. Springer, Heidelberg (2010)
17. Goldwasser, S., Kalai, Y.T., Rothblum, G.N.: Delegating computation: interactive proofs for muggles. In STOC 2008, pp. 113–122. ACM press, New York (2008)
18. Goldwasser, S., Lin, H., Rubinstein, A.: Delegation of computation without rejection problem from designated verifier cs-proofs. In: IACR Cryptology ePrint Archive, avaliable at http://eprint.iacr.org/2011/456 (2011)
19. Hohenberger, S., Lysyanskaya, A.: How to securely outsource cryptographic computations. In: Kilian, J. (ed.) TCC 2005. LNCS, vol. 3378, pp. 264–282. Springer, Heidelberg (2005)
20. Hohenberger, S., Sahai, A., Waters, B.: Full domain hash from (leveled) multilinear maps and identity-based aggregate signatures. In: Canetti, R., Garay, J.A. (eds.) CRYPTO 2013, Part I. LNCS, vol. 8042, pp. 494–512. Springer, Heidelberg (2013)
21. López-Alt, A., Tromer, E., Vaikuntanathan, V.: On-the-fly multiparty computation on the cloud via multikey fully homomorphic encryption. In: STOC 2012, pp. 1219–1234. ACM press (2012)
22. Lewko, A.B., Waters, B.: Efficient pseudorandom functions from the dicisional linear assumption and weaker variants. In: CCS 2009, pp. 112–120. ACM press, New York (2009)
23. Mohassel, P.: Efficient and secure delegation of linear algebra. In: IACR Cryptology ePrint Archive, avaliable at http://eprint.iacr.org/2011/605, (2011)
24. Parno, B., Raykova, M., Vaikuntanathan, V.: How to delegate and verify in public: verifiable computation from attribute-based encryption. In: Cramer, R. (ed.) TCC 2012. LNCS, vol. 7194, pp. 422–439. Springer, Heidelberg (2012)
25. Rothblum, G.N., Vadhan, S., Wigderson, A.: Interactive proofs of proximity: delegating computation in sublinear time. In: STOC 2013, pp. 793–802. ACM press, New York (2013)
26. Zhang, L.F., Safavi-Naini, R.: Private outsourcing of polynomial evaluation and matrix multiplication using multilinear maps. In: Abdalla, M., Nita-Rotaru, C., Dahab, R. (eds.) CANS 2013. LNCS, vol. 8257, pp. 329–348. Springer, Heidelberg (2013)

A Characterization of Cybersecurity Posture from Network Telescope Data

Zhenxin Zhan[1], Maochao Xu[2], and Shouhuai Xu[1($\boxtimes$)]

[1] Department of Computer Science,
University of Texas at San Antonio, San Antonio, USA
zhenxin.zhan.dr@gmail.com, shxu@cs.utsa.edu
[2] Department of Mathematics, Illinois State University, Normal, USA
mxu2@ilstu.edu

Abstract. Data-driven understanding of cybersecurity posture is an important problem that has not been adequately explored. In this paper, we analyze some real data collected by CAIDA's network telescope during the month of March 2013. We propose to formalize the concept of cybersecurity posture from the perspectives of three kinds of time series: the number of victims (i.e., telescope IP addresses that are attacked), the number of attackers that are observed by the telescope, and the number of attacks that are observed by the telescope. Characterizing cybersecurity posture therefore becomes investigating the phenomena and statistical properties exhibited by these time series, and explaining their cybersecurity meanings. For example, we propose the concept of *sweep-time*, and show that sweep-time should be modeled by stochastic process, rather than random variable. We report that the number of attackers (and attacks) from a certain country dominates the total number of attackers (and attacks) that are observed by the telescope. We also show that substantially smaller network telescopes might not be as useful as a large telescope.

Keywords: Cybersecurity data analytics · Cybersecurity posture · Network telescope · Network blackhole · Darknet · Cyber attack sweep-time · Time series data

1 Introduction

Network telescope [26] (aka blackhole [5,10], darknet [3], or network sink [38], possibly with some variations) is a useful instrument for monitoring unused, routeable IP address space. Since there are no legitimate services associated to these unused IP addresses, traffic targeting them is often caused by attacks. This allows researchers to use telescope-collected data (together with other kinds of data) to study, for example, worm propagation [4,23,25,30], denial-of-service (DOS) attacks [17,24], and stealth botnet scan [12]. Despite that telescope data can contain unsolicited — but not necessarily malicious — traffic that can be

© Springer International Publishing Switzerland 2015
M. Yung et al. (Eds.): INTRUST 2014, LNCS 9473, pp. 105–126, 2015.
DOI: 10.1007/978-3-319-27998-5_7

caused by misconfigurations or by Internet background radiation [15,28,36], analyzing telescope data can lead to better understanding of *cybersecurity posture*, an important problem that has yet to be investigated.

Our Contributions. In this paper, we empirically characterize cybersecurity posture based on a dataset collected by CAIDA's /8 network telescope (i.e., 2^{24} IP addresses) during the month of March 2013. We make the following contributions. **First**, we propose to characterize cybersecurity posture by considering three time series: the number of victims, the number of attackers, and the number of attacks. To the best of our knowledge, this is the first formal definition of cybersecurity posture. **Second**, we define the notion of *sweep-time*, namely the time it takes for most telescope IP addresses to be attacked at least once. We find that sweep-time cannot be described by a probabilistic distribution, despite that a proper subset of the large sweep-times follows the power-law distribution. We show that an appropriate stochastic process can instead describe the sweep-time. This means that when incorporating sweep-time in theoretical cybersecurity models, it cannot always be treated as a random variable and may need to be treated as a stochastic process. **Third**, we find that the total number of attackers that are observed by the network telescope is dominated by the number of attackers from a certain country X.[1] Moreover, we observe that both the number of attackers from country X and the total number of attackers exhibit a strong periodicity. Although we cannot precisely pin down the root cause of this *dominance and periodicity* phenomenon, it does suggest that thoroughly examining the traffic between country X and the rest of the Internet may significantly improve cybersecurity. **Fourth**, we investigate whether or not substantially smaller network telescopes would give approximately the same statistics that would be offered by a single, large network telescope. This question is interesting on its own and, if answered affirmatively, could lead to more cost-effective operation of network telescopes. Unfortunately, our analysis shows that substantially smaller telescopes might not be as useful a single, large telescope (of 2^{24} IP addresses).

Related Work. One approach to understanding cybersecurity posture is to analyze network telescope data. Studies based on telescope data can be classified into two categories. The first category analyzes telescope data *alone*, and the present study falls into this category. These studies include the characterization of Internet background radiation [28,36], the characterization of scan activities [1], and the characterization of backscatter for estimating global DOS activities [17,24]. However, we analyze cybersecurity posture, especially with regard to attacks that are likely caused by malicious worm, virus and bot activities. This explains why we exclude the backscatter data (which is filtered as noise in the present paper). The second category of studies analyzes telescope data together with other kinds of relevant data. These studies include the use of tele-

---

[1] We were fortunate to see the real, rather than anonymized, attacker IP addresses, which allowed us to aggregate the attackers based on their country code. Our study was approved by IRB.

scope data and network-based intrusion detection and firewall logs to analyze Internet intrusion activities [39], the use of out-of-band information to help analyze worm propagation [4,23,25], and the use of active interactions with remote IP addresses to filter misconfiguration-caused traffic [28]. There are also studies that are somewhat related to ours, including the identification of one-way traffic from data where two-way traffic is well understood [1,7,15,20,33].

The other approach to understanding cybersecurity posture is to analyze data collected by honeynet-like systems (e.g., [5,6,21,29,40]). Unlike network telescopes, these systems can interact with remote computers and therefore allow for richer analysis, including the automated generation of attack signatures [18, 37].

To the best of our knowledge, we are the first to formally define *cybersecurity posture* via three time series: the number of victims, the number of attackers, and the number of attacks.

The rest of the paper is organized as follows. Section 2 describes the data and defines cybersecurity posture. Section 3 briefly reviews some statistical preliminaries. Section 4 defines and analyzes the sweep-time. Section 5 investigates the dominance and periodicity phenomenon exhibited by the number of attackers. Section 6 investigates whether substantially smaller network telescopes would be sufficient or not. Section 7 discusses the limitations of the present study. Section 8 concludes the paper.

2   Representation of Data and Definition of Cybersecurity Posture

Data Description. The data we analyze was collected between 3/1/2013 and 3/31/2013 by CAIDA's network telescope, which is a passive monitoring system based on a globally routeable but unused /8 network (i.e., 1/256 of the entire Internet IP v4 address space) [31]. Since a network telescope passively collects unsolicited traffic, the collected traffic would contain *malicious traffic* that reaches the telescope (e.g., automated malware spreading), but may also contain *non-malicious traffic* — such as Internet background radiation (e.g., backscatter caused by the use of spoofed source IP addresses that happen to belong to the telescope) and misconfiguration-caused traffic (e.g. mistyping an IP address by a remote computer). This means that pre-processing the raw data is necessary. At a high level, we will analyze data D_1 and D_2, which are sets of *flows* [8] and are obtained by applying the pre-processing procedures described below.

Data D_1. Based on CAIDA's standard pre-processing [32], the collected IP packets are organized based on eight fields: source IP address, destination IP address, source port number, destination port number, protocol, TTL (time-to-live), TCP flags, and IP length. The flows are reassembled from the IP packets and then classified into three classes: *backscatter*, *ICMP request* and *other*. At a high level, backscatter traffic is identified via TCP SYN+ACK, TCP RST, while ICMP request is identified via ICMP type 0/3/4/5/11/12/14/16/18. (A similar classification method is used in [36].) Since (i) backscatter-based analysis of DOS attacks has been conducted elsewhere (e.g., [17,24]), and (ii) ICMP has been mainly used to launch DOS attacks (e.g., *ping flooding* and *smurf or fraggle* attacks [19,24,35]), we disregard the traffic corresponding to *backscatter* and *ICMP request*. Since we are more interested in analyzing cybersecurity posture corresponding to attacks that are launched through the TCP/UDP protocols, we focus on the TCP/UDP traffic in the *other* category mentioned above. We call the resulting data D_1, in which each TCP/UDP flow is treated as a distinct attack.

Data D_2. Although (i) D_1 already excludes the traffic corresponding to *backscatter* and *ICMP request*, and (ii) D_1 only consists of TCP/UDP flows in the *other* category mentioned above, D_1 may still contain flows that are caused by misconfigurations. Eliminating misconfiguration-caused flows in network telescope data is a hard problem because network telescope is passive (i.e., not interacting with remote computers [16]). Indeed, existing studies on recognizing misconfiguration-caused traffic had to use payload information (e.g., [22]), which is however beyond the reach of network telescope data. Note that recognizing misconfiguration-caused traffic is even harder than recognizing one-way traffic already (because misconfiguration can cause both one-way *and* two-way traffic), and that solving the latter problem already requires using extra information (such as two-way traffic [1,7,15,20,33]). These observations suggest that we use some heuristics to filter probable misconfiguration-caused flows from D_1. Our examination shows that, for example, 50 % (81 %) attackers launched 1 attack (≤ 9 attacks, correspondingly) against the telescope during the month. We propose to extract D_2 by filtering from D_1 the flows that correspond to remote IP addresses that initiate fewer than 10 flows/attacks during the month. This heuristic method filters possibly many, if not most, misconfiguration-caused flows in D_1. Even though the ground truth (i.e., which TCP/UDP flows correspond to malicious attacks) is not known, D_2 might be closer to the ground truth than D_1.

Data Representation. In order to analyze the TCP/UDP flow data D_1 and D_2, we represent the flows through time series at some *time resolution r*. We consider two time resolutions (because a higher resolution leads to more accurate statistics): hour, denoted by "H," and minute, denoted by "m." For a given time resolution of interest, the total time interval $[0, T]$ is divided into short periods $[i, i + 1)$ according to time resolution $r \in \{H, m\}$, where $i = 0, 1, \ldots, T - 1$, and $T = 744$ h (or $T = 4,464$ min) in this case. We organize the flows into time series from three perspectives:

- the number of *victims* (i.e., network telescope IP addresses that are "hit" by remote attacking IP addresses contained in D_1 or D_2) per time unit at time resolution r,
- the number of *attackers* (i.e., the remote attacking IP addresses contained in D_1 or D_2) per time unit at time resolution r, and

- the number of *attacks* per time unit at time resolution r (i.e., TCP/UDP flows initiated from remote attacking IP addresses in D_1 or D_2 are treated as attacks).

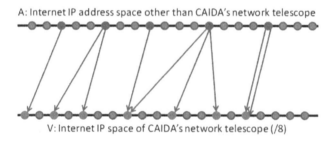

A: Internet IP address space other than CAIDA's network telescope

V: Internet IP space of CAIDA's network telescope (/8)

Fig. 1. Illustration of the attacker-victim relation during time interval $[i, i + 1)$ at time resolution $r \in \{H, m\}$ in D_1: each dot represents an IP address, a red-colored dot represents an attacking IP address (i.e., attacker), a pink-colored dot represents a victim, each arrow represents an attack (i.e., TCP/UDP flow), the number of attackers is $|A(r; i, i + 1)| = 5$, the number of victims is $|V(r; i, i + 1)| = 7$, and the number of attacks is $y(r; i, i + 1) = 9$. The same holds for data D_2 (Color figure online).

As illustrated in Fig. 1, let V be CAIDA's fixed set of telescope IP addresses, and A be the rest of IP addresses in cyberspace, where $|A| = 2^{32} - |V|$. The major notations are (highlighted and) defined as follows:

- V, A: the set of CAIDA's network telescope IP addresses and the set of the rest IP v4 addresses, respectively.
- $r \in \{H, m\}$: time resolution (H: per hour; m: per minute).
- $V(r; i, i + 1) \subseteq V$ and $V'(r; i, i + 1) \subseteq V$: the sets of *victims* attacked at least once during time interval $[i, i + 1)$ at time resolution r in D_1 and D_2, respectively.
- $V(r; i, j) = \bigcup_{\ell=i}^{j-1} V(r; \ell, \ell + 1)$ and $V'(r; i, j) = \bigcup_{\ell=i}^{j-1} V'(r; \ell, \ell + 1)$: the cumulative set of victims that are attacked at some point during time interval $[i, j)$ at time resolution r in data D_1 and D_2, respectively.
- $V(r; 0, T)$ and $V'(r; 0, T)$: the sets of victims that are attacked at least once during time interval $[0, T)$ in D_1 and D_2, respectively. Note that these sets are actually independent of time resolution r, but we keep r for notational consistence.
- $A(r; i, i + 1) \subseteq A$ and $A'(r; i, i + 1) \subseteq A$: the sets of *attackers* that launched attacks against some $v \in V$ during time interval $[i, i + 1)$ at time resolution r in D_1 and D_2, respectively.
- $y(r; i, i+1)$ and $y'(r; i, i+1)$: the numbers of *attacks* that are launched against victims belonging to $V(r; i, i+1)$ and $V(r; i, i+1)$ during time interval $[i, i+1)$, respectively.

Cybersecurity Posture. We define cybersecurity posture as:

Definition 1 (cybersecurity posture). *For a given time resolution r and network telescope of IP address space V, the* cybersecurity posture *as reflected by telescope data D_1 is described by the phenomena and (statistical) properties exhibited by the following three time series:*

- *the number of victims $|V(r; i, i + 1)|$,*
- *the number of attackers $|A(r; i, i + 1)|$, and*
- *the number of attacks $y(r; i, i + 1)$,*

where $i = 0, 1, \ldots$. Similarly, we can define cybersecurity posture corresponding to D_2.

Based on the above definition of cybersecurity posture, the main research task is to characterize the phenomena and statistical properties of the three time series (e.g., how can we predict them?) and the similarity between them. As a first step, we characterize, by using data D_1 as an example, the *number* of victims $|V(r; i, i + 1)|$ rather than the *set* of victims $V(r; i, i + 1)$, and the *number* of attackers $|A(r; i, i+1)|$ rather than the *set* of attackers $A(r; i, i+1)$. We leave the characterization of the *sets* of victims and attackers to future study. Moreover, we characterize the *number* of attacks $y(r; i, i+1)$ rather than the specific classes of attacks, because telescope data does not provide rich enough information to recognize specific attacks. In Sect. 7, we will discuss limitations of the present study, including the ones that are imposed by the heuristic data pre-processing method for obtaining D_1 and D_2.

3   Statistical Preliminaries

We briefly review some statistical concepts and models dealing with time series data, while referring their formal descriptions and technical details to [11, 13, 27, 34].

Brief Review of Some Statistical Concepts. Time series can be described by statistical models, such as the AutoRegressive Integrated Moving Average (ARIMA) model and the Generalized AutoRegressive Conditional Heteroskedasticity (GARCH) model that will be used in the paper. The ARIMA model is perhaps the most popular class of time series models in the literature. It includes many specific models, such as random walk, seasonal trends, stationary, and non-stationary models [11]. ARIMA models cannot accommodate high volatilities of time series data, which however can be accommodated by GARCH models [13]. GARCH models also can capture many phenomena, such as dynamic dependence in variance, skewness, and heavy-tails [34].

In order to find accurate models for describing time series data, we need to do model selection. There are many model selection criteria, among which the Akaike's Information Criterion (AIC) is widely used. This criterion is based on appropriately balancing between goodness of fit and model complexity. It is defined in such a way that the smaller the AIC value, the better the model [27].

Measuring the Difference (or Distance) Between Two Time Series. We need to measure the difference (or distance) between two time series: $Z_1, Z_2, \ldots$ and $Z'_1, Z'_2, \ldots$, where $Z_i \geq 0$ and $Z'_i \geq 0$ for $i = 1, 2, \ldots$. This difference measure characterizes:

1. the *fitting* error, where the Z_i time series may represent the observed values and the Z'_i time series may represent the fitted values;
2. the *prediction* error, where the Z_i time series may correspond to the observed values and the Z'_i time series may correspond to the predicted values;
3. the *approximation* error, where the Z_i time series may describe the observed values and the Z'_i time series may describe the values that may be inferred (i.e., estimated or approximated) from other data sources (e.g., we may want to know whether or not the statistics derived from the data collected by a large network telescope can be inferred from the data collected by a much smaller network telescope).

For conciseness, we use the standard and popular measure known as Percent Mean Absolute Deviation (PMAD) [2]. Specifically, suppose $Z_t, Z_{t+1}, \ldots, Z_{t+\ell}$ are given data, and $Z'_t, Z'_{t+1}, \ldots, Z'_{t+\ell}$ are the fitted (or predicted, or approximated) data. The overall fitting (or prediction, or approximation) error (or the PMAD value) is defined as $\frac{\sum_{j=t}^{t+\ell} |Z_j - Z'_j|}{\sum_{j=t}^{t+\ell} Z_j}$. The closer to 0 the PMAD value, the better the fitting (or prediction, or approximation). We note that our analysis is not bound to the PMAD measure, and it is straightforward to adapt our analysis to incorporate other measures of interest.

Measuring the Shape Similarity Between Two Time Series. Two time series may be very different from a measure such as the PMAD mentioned above, but may be similar to each other in their shape (perhaps after some appropriate re-alignments). Therefore, we may need to measure such shape similarity between two time series. Dynamic Time Warping (DTW) is a method for this purpose. Intuitively, DTW aims to align two time series that may have the same shape and, as a result, the similarity between two time series can be captured by the notion of *warping path* (aka *warping function*). The closer the warping path to the diagonal, the more similar the two time series. We use the DTW algorithm in the R software package, which implements the algorithm described in [14].

4   Characteristics of Sweep-Time

The Notion of Sweep-Time. Figure 2 describes the times series of $|V(H; i, i+1)|$ in D_1 and $|V'(H; i, i+1)|$ in D_2. Using D_1 as example, we make the following observations (similar observations can be made for D_2). First, there is a significant volatility at the 632nd hour, during which the number of victims is as low as $4,377,079 \approx 2^{22}$. Careful examination shows that the total number of attackers during the 632nd hour is very small, which would be the cause. Second, most telescope IP addresses are attacked within a single hour. For example, $15,998,907$, or 96% of $|V(H; 1, 733)|$, telescope IP addresses are attacked at least

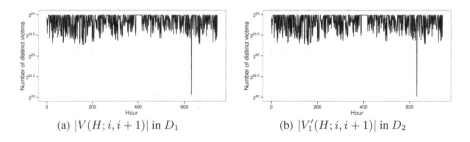

(a) $|V(H; i, i+1)|$ in D_1 (b) $|V_1'(H; i, i+1)|$ in D_2

Fig. 2. Time series of number of victims: $|V(H; i, i+1)|$ in D_1 and $|V_1'(H; i, i+1)|$ in D_2.

once during the first hour. Third, no victims other than $V(H; 0, 703)$ are attacked during the time interval $[704, 744)$.

Since Fig. 2 shows that there are a large number of victims per hour, we ask the following question: How long does it take for most telescope IP addresses to be attacked at least once? That is: **How long does it take for $\tau \times |V(H; 0, T)|$ victims to be attacked at least once, where $0 < \tau < 1$?** This suggests us to define the following notion of *sweep-time*, which is relative to the observation start time.

Definition 2 (sweep-time). *With respect to D_1, the* sweep-time *starting at the ith time unit of time resolution r, denoted by I_i, is defined as:*

$$\left| \bigcup_{\ell=i}^{I_i-1} V(r; \ell, \ell+1) \right| < \tau \times |V(H; 0, T)| \leq \left| \bigcup_{\ell=i}^{I_i} V(r; \ell, \ell+1) \right|.$$

Corresponding to data D_2, we can define sweep-time I_i' as:

$$\left| \bigcup_{\ell=i}^{I_i'-1} V'(r; \ell, \ell+1) \right| < \tau \times |V'(H; 0, T)| \leq \left| \bigcup_{\ell=i}^{I_i'} V'(r; \ell, \ell+1) \right|.$$

By taking into consideration the observation starting time i, we naturally obtain two time series of sweep-time: $I_0, I_1, \ldots$ for D_1 and $I_0', I_1', \ldots$ for D_2. We want to characterize these two time series of sweep-time.

Characterizing Sweep-Time. Since *per-hour* time resolution gives a coarse estimation of sweep-time, we use *per-minute* time resolution for better estimation of it. Figure 3 plots the time series of sweep-time $I_0, I_{10}, I_{20}, \ldots$, namely a sample of $I_0, I_1, \ldots, I_{10}, \ldots, I_{20}, \ldots$ because it is too time-consuming to consider the latter entirely (time resolution: minute).

Figure 3 suggests that the sweep-time time series exhibit similar shape. Accordingly, we use the DTW method to characterize their similarity. Recall that DTW aims to align two time series via the notion of *warping path*, such that the closer the warping path to the diagonal, the more similar the two time series.

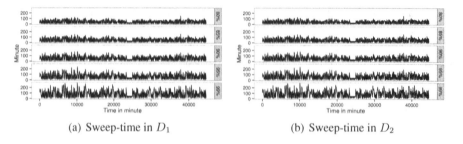

(a) Sweep-time in D_1 (b) Sweep-time in D_2

Fig. 3. Time series plots of sweep-time (y-axis) with respect to $\tau \in \{80\%, 85\%, 90\%, 95\%, 99\%\}$, where the x-axis represents the observation starting time that is sampled at every 10 minutes. In other words, the plotted points are the sample $(0, I_0), (10, I_{10}), (20, I_{20}), \ldots$ rather than $(0, I_0), (1, I_1), \ldots$.

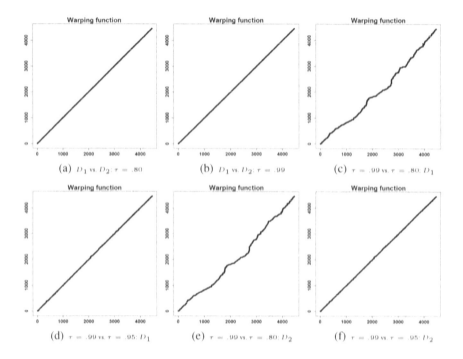

Fig. 4. DTW-based similarities of sweep-time time series with different threshold τ.

Figure 4 confirms the above observation, by presenting some examples of the warping paths (the others are omitted due to space limitation). Specifically, Figs. 4(a) and (b) show that the sweep-times with respect to $\tau = .80$ and $\tau = .99$ are almost the same in D_1 and D_2, respectively. Figures 4(c) and (d) show that for two thresholds, say τ_1 and τ_2, the smaller the $|\tau_1 - \tau_2|$, the more similar the two respective time series of sweep-time in D_1. Figures 4(e) and (f) demonstrate the same phenomenon for D_2.

The above discussion suggests that the notion of sweep-time is not sensitive to spatial threshold τ. This leads us to ask: **What is the distribution of sweep-time?** However, this question makes sense only when the time series is stationary. By using an augmented Dickey-Fuller test [34], we conclude that the sample of the sweep-time series, namely $I_0, I_{10}, I_{20}, \ldots$, is *not* stationary, which means that time series $I_0, I_1, I_2 \ldots$ is not stationary (otherwise, the sample should be stationary). Therefore, we cannot use a single distribution to characterize the sweep-time; Instead, we have to characterize the sweep-time as a stochastic process. In order to identify good time series models that can fit the sweep-time, we need to identify some statistical properties that are exhibited by sweep-time. In particular, we need to know if the sweep-time is heavy-tailed, meaning that the sweep-times greater or equal to $x_{\min}$ exhibit the power-law distribution, where $x_{\min}$ is called *cut-off* parameter.

Table 1. Power-law test statistics of the sweep-time with respect to *spatial threshold* $\tau \in \{80\%, 85\%, 90\%, 95\%, 99\%\}$, where α is the fitted power-law exponent, $x_{\min}$ is the cut-off parameter, $KS \in [0.04, 0.06]$ is the Kolmogorov-Smirnov statistic [9] for comparing the fitted power-law distribution and the data (meaning that the fitting is good) as indicated by that the p-values are $\gg 0.05$, and "$\# \geq x_{\min}$" represents the number of sweep-times that are greater than or equal to $x_{\min}$ (i.e., the number of sweep-times that are used for fitting).

τ	α	$x_{\min}$	KS	p-value	$\# \geq x_{\min}$	τ	α	$x_{\min}$	KS	p-value	$\# \geq x_{\min}$
Dataset D_1 with time resolution 1-min						Dataset D_2 with time-resolution 1-min					
80%	7.89	78	.05	.14	475	80%	8.46	82	.05	.19	391
85%	8.46	94	.04	.52	385	85%	8.37	95	.04	.36	379
90%	8.89	118	.06	.42	244	90%	9.24	120	.05	.39	237
95%	9.52	148	.05	.68	193	95%	12.82	170	.04	.99	72
99%	13.67	215	.04	.98	131	99%	15.23	224	.04	.99	94

Table 1 summarizes the power-law test statistics of the sweep-time with cut-off parameter $x_{\min}$. We observe that for both D_1 and D_2, all the α values (i.e., the fitted power-law exponents) are very large. For spatial threshold $\tau = 80\%$ in D_1, we have $x_{\min} = 78$ minutes, meaning that the number of sweep-times that are greater than or equal to $x_{\min}$ is 475 (or 10.6% out of 4,462). As spatial threshold τ increases, $x_{\min}$ increases and the number of sweep-times greater than or equal to $x_{\min}$ decreases. We also observe that for the same τ, D_1 and D_2 have similar $x_{\min}$ values, which means that the filtered attack traffic in D_1 does not affect the power-law property of the data.

The above analysis suggests that in order to fit the sweep-time, we should use a model that can accommodate the power-law property. Therefore, we use the ARMA+GARCH model, where ARMA accommodates the stable sweep-times smaller than $x_{\min}$, and GARCH, with skewed student t-distribution, accommodates the power-law distributed sweep-times (which are greater than or equal to x_{min}). Consider spatial threshold $\tau = .99$ as an example. Figures 5(a) and (b) plot the observed data and the fitting model for sweep-time I_t (observation starting time t):

$$I_t - \mu_t = \phi_1(I_{t-1} - \mu_{t-1}) + \phi_2(I_{t-2} - \mu_{t-2}) + \epsilon_t,$$

where $\mu_t = \mu + \xi\sigma_t$ is the dynamic mean composed of a constant term μ and standard deviation σ_t of the error term, $\sigma_t^2 = \omega + \alpha_1\epsilon_{t-1} + \beta_1\sigma_{t-1}^2$, $\mu_t = E[I_t]$, ϵ_t is the error term at time t, and $\sigma_t^2 = E(y_t - \mu_t)^2$ is the variance modeled via the standard GARCH(1,1) process. The fitting errors (PMAD values) are .121 and .119 for D_1 and D_2, respectively.

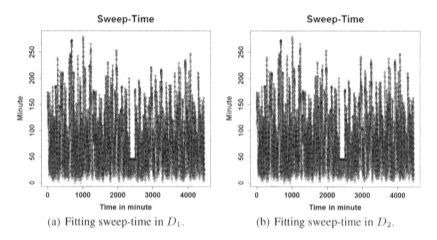

(a) Fitting sweep-time in D_1.   (b) Fitting sweep-time in D_2.

Fig. 5. Fitting sweep-time with spatial threshold $\tau = .99$ (time resolution: minute), where black-colored dots are observed sweep-time values, red-colored dots are fitted values (Color figure online).

It was known that malware can infect almost all susceptible computers within a very short period of time (e.g., the Slammer worm [23]), meaning that the sweep-time with respect to *a specific observation starting time* is very small. This is, in a sense, re-affirmed by our study. However, for continuous attacks that are based on a bag of attacking tools, sweep-time should be better modeled with respect to *any* (rather than a specific) observation starting time. We are the first to show that the sweep-time *cannot* be modeled by a random variable (which would make the model in question easier to analyze though). This leads to the following insight, which could guide future development of advanced cybersecurity models.

Insight 1. *When one needs to model the sweep-time (i.e., the time it takes for each IP address of a τ-portion of a large network space to be attacked at least once), it should be modeled by a stochastic process rather than a random variable.*

5   A Phenomenon Exhibited by Attacking IP Addresses

For each attacker IP address, we can use the WHOIS service to retrieve its country code. Figure 6 plots the origins of attackers that contribute to most of the attackers (per country code). Note that the category "others" in D_1 include 6,894,900 attacker IP addresses (or 1.7 % of the total number of attackers) whose country codes cannot be retrieved from the WHOIS service. The category "others" in D_2 include 10,740 attacker IP addresses (or 0.01 % of the total number of attackers) whose country codes cannot be retrieved from the WHOIS service. This means that many attacker IP addresses whose country codes cannot be retrieved from the WHOIS service are filtered. Moreover, the attacker IP addresses with no country code do not have a significant impact on the result. We observe that country X contributes 30 % of the attackers in D_1 and 76 % of the attackers in D_2. Country Y contributes 28 % of the attackers in D_1 and 3 % of the attackers in D_2. This is caused by the fact that 50 % attackers from country X and 98 % attackers from country Y launch fewer than 10 attacks during the month, and therefore do not appear in D_2. This prompts us to study the relationship between two time series: the total number of attackers and the number of attackers from country X.

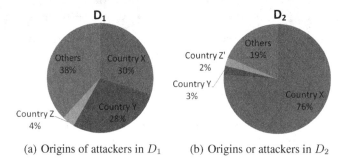

(a) Origins of attackers in D_1 (b) Origins or attackers in D_2

Fig. 6. During the month, three countries, which are anonymized as X, Y and Z, contribute to most of the attackers in D_1; whereas countries X, Y and Z' ($Z' \neq Z$) contribute to most of the attackers in D_2.

The Dominance and Periodicity Phenomenon. Figure 7 compares the times series of the total number of attackers observed by the telescope and the time series of the number of attackers from country X in D_1 and D_2, respectively. For D_1, Fig. 7(a) shows that the total number of attackers during time interval $[455, 630]$, namely the 176 hours between the 455th hour (on March 19, 2013) and the 630th hour (on March 27, 2013), is substantially greater than its counterpart during the other hours. This is caused by the substantial increase in the number of attackers from country Y, despite that we do not know the root cause behind the substantial increase of attackers in country Y. For D_2, Fig. 7(b) does not exhibit the same kind of substantial increase during the interval $[455, 630]$, meaning that many of the "emerging" attackers from country Y are filtered (because they launched fewer than 10 attacks during the month).

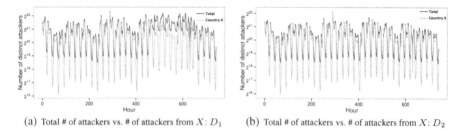

(a) Total # of attackers vs. # of attackers from X: D_1 (b) Total # of attackers vs. # of attackers from X: D_2

Fig. 7. The dominance and periodicity phenomenon exhibited by two time series: the total number of attackers versus the number of attackers from country X (time resolution: hour).

Figure 7 further suggests a surprising consistency between the two time series. Specifically, when the number of attackers from country X is large (small), the total number of attackers is large (small). For D_1, this is confirmed by Figs. 8(a) and (b), which clearly show that the same periodicity is exhibited by the total number of attackers and by the number of attackers from country X. For D_2, this is confirmed by Figs. 8(c) and (d), which clearly show that the same periodicity is exhibited by the total number of attackers and by the number of attackers from country X. We observe that the wave bases are periodic with a period of 24 hours. After looking into the time zone of country X, we find that the wave bases (i.e., that least number of attackers) correspond to the hour between 12:00 noon and 1 pm local time. One may speculate that this is caused by computers possibly being put into the hibernate mode (during lunch time). This may not be true because during the night hours, more computers would be put into the hibernate mode (or even powered off) and therefore even fewer attackers would be observed. However, this is not shown by the data. One perhaps more plausible explanation is that the attacking computers may be coordinated or controlled (for example) by botnets.

While we defer the detailed characterization of the phenomenon to Appendix A, we summarize the phenomenon as:

Phenomenon 1 (The dominance and periodicity phenomenon exhibited by the number of attackers). *The time series of the total number of attackers and the time series of the number of attackers from a particular country X exhibit the same periodicity. Moreover, the total number of attackers is dominated by the number of attackers from country X.*

6   Inferring Global Cyber Security Posture from Smaller Monitors

In this section, we explore whether it is possible to use small network telescopes to approximate bigger telescopes, from the perspectives of estimating/inferring the number of victims, attackers and attacks. Answering this question is interesting on its own, and could lead to more cost-effective operations of network telescopes.

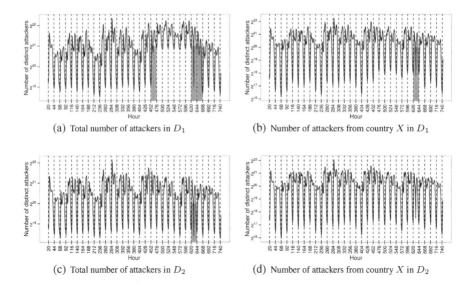

(a) Total number of attackers in D_1 (b) Number of attackers from country X in D_1

(c) Total number of attackers in D_2 (d) Number of attackers from country X in D_2

Fig. 8. Elaboration of the dominance and periodicity phenomenon (time resolution: hour).

Methodology. We divide the /8 network telescope into B equal-size blocks of IP addresses, where each block is called a *small telescope*. We want to know whether we can infer the number of victims (or attackers, or attacks) that are observed by the /8 telescope during time interval $[t, t + 1)$ at time resolution H (i.e., per hour), denoted by $Y(H; t, t + 1)$, from the number of victims (or attackers, or attacks) observed by b small telescopes, where $b << B$, during the same time interval, denoted by $Y_1(H; t, t+1), \ldots, Y_b(H; t, t+1)$. In other words, we want to know whether the following equation would hold:

$$Y(H; t, t + 1) = c + \sum_{i=1}^{b} \phi_i Y_i(H; t, t + 1),$$

where c is some constant and ϕ_i's are coefficients. Naturally, we can use the same PMAD measure to evaluate the estimation/inference error.

Whenever feasible, we want to consider all possible combinations of b small telescopes. For $B = 16$, there are $\binom{16}{b}$ combinations; for $B = 256$, there are $\binom{256}{b}$ combinations. For $B = 256$ and $b \geq 4$, the number of combinations becomes prohibitive. This suggests that we first cluster the $B = 256$ blocks into b groups based on the DTW measure, and then sample one block from each of the b groups. In a sense, this corresponds to the *best-case* scenario sampling because we need the prior information about the groups or clusters. If the sample statistics cannot approximate the statistics derived from the data collected by the /8 telescope, we can conclude that small telescopes are not as useful as the large telescope.

Characterizing Inference Errors of Small Telescopes. From the perspective of inferring the number of victims from small telescopes, Table 2 summarizes

Table 2. PMAD-based measurement of the inference error when using b (out of the B) small telescopes to approximate the number of victims that are observed by the larger /8 telescope, where "SD" stands for standard deviation.

	Min	Mean	Median	Max	SD		Min	Mean	Median	Max	SD
D_1 with $B = 16$: PMAD values						D_1 with $B = 256$: PMAD values					
$b = 1$	.0372	.0472	.0464	.0682	.0083	$b = 1$	.0563	.0689	.0678	.1131	.0074
$b = 2$	.0230	.0322	.0302	.0586	.0067	$b = 2$	.0414	.0535	.0521	.1129	.0057
$b = 3$	.0167	.0232	.0239	.0301	.0033	$b = 3$	.0330	.0447	.0436	.0982	.0049
D_2 with $B = 16$: PMAD values						D_2 with $B = 256$: PMAD values					
$b = 1$	.0384	.0484	.0478	.0702	.0085	$b = 1$	.0799	.1136	.1155	.1155	.0071
$b = 2$	.0235	.0340	.0311	.0702	.0079	$b = 2$	.0731	.1119	.1155	.1155	.0096
$b = 3$	.0167	.0267	.0251	.0702	.0066	$b = 3$	.0755	.0787	.0787	.0839	.0021

the inference errors, in terms of the min, mean, median and max PMAD values of all the considered combinations of sample blocks, as well as the standard deviation of the PMAD values. For D_1 and $B = 16$, we observe that a single small telescope (out of the 16 telescopes of size 2^{20} IP addresses) would give good approximation of the number of victims that would be obtained based on the larger /8 network telescope. This is because the maximum PMAD errors is .0682, namely 6.82 % approximation error. For D_1 and $B = 256$, the mean approximation error is 6.89 % for $b = 1$ (i.e., using one small telescope), and 5.35 % for $b = 2$ (i.e., using two small telescopes) and 4.47 % for $b = 3$ (using three small telescopes). For D_2 and $B = 16$, we observe a similar phenomenon as in the case of D_1 and $B = 16$. However, for D_2 and $B = 256$, the mean approximation errors is significantly larger than in the case of D_1 and $B = 256$, namely 11.36 %, 11.19 % and 7.87 % for $b = 1, 2, 3$, respectively. Therefore, we can conclude that from the perspective of the number of victims, a single telescope of size 2^{20} IP addresses would give approximately the same result as the telescope of size 2^{24}, and 3 randomly selected small telescope of size 2^{16} would give approximately the same result as the telescope of size 2^{24}. That is, the small telescope could be used instead.

Due to space limitation, we defer to Appendix B the characterizations on inferring the number of attackers from small telescopes and on inferring the number of attacks from small telescopes. Based on these characterizations, we draw the following insight:

Insight 2. *For estimating the number of victims, substantially small telescopes could be used instead. However, for estimating the number of attackers and the number of attacks, substantially small telescopes might not be sufficient.*

The above discrepancy between the number of victims and the numbers of attackers/attacks is possibly caused by the following: The victims are somewhat "uniformly" attacked, but the attackers and attacks are far from "uniformly" distributed. Moreover, a *single* attacker that scans the large telescope's IP address space will make it easy to estimate the number of victims from small telescopes.

7   Limitations of the Study

The present study has several limitations, which are inherent to the data but not to the methodology we use. First, our analysis treats each remote IP address as a unique attacker. This is not accurate when the remote networks using Network Address Translation (NAT), because remote attackers from the same network can be "aggregated" into a single attacker. If many networks in country X indeed use NAT, then the actual number of attackers from country X is indeed larger, although the number of attacks from country X is not affected by NAT.

Second, the characteristics presented in the paper inherently depend on the nature of network telescope. For example, D_1 and D_2 still may contain some misconfiguration-caused, non-malicious traffic. Due to the lack of interactions between network telescope and remote computers (an inherent limitation of network telescopes), it is hard to know the ground truth [16]. Therefore, better filtering methods are needed so as to make the data approximate the ground truth as closely as possible.

Third, it is possible that some attackers are aware of the network telescope and therefore can instruct their attacks to bypass it. As a consequence, the data may not faithfully reflect the cybersecurity posture.

Fourth, the data collected by the network telescope does not contain rich enough information that would allow us to conduct deeper analysis, such as analyzing the global characteristics of specific attacks. Moreover, the data is a "coarse" sample of the ground-truth cybersecurity posture because the first flow from a remote attacker may be a scan/probe activity, or a first attack attempt against a specific port.

8   Conclusion

We have studied the cybersecurity posture based on the data collected by CAIDA's network telescope. We have found that the *sweep-time* should be characterized as a stochastic process rather than a random variable. We also have found that the *total* number of attackers (and attacks) that are observed by the network telescope is largely determined by the number of attackers from a single country. There are many interesting problems for future research, such as: How can we (more) accurately predict the time series? To what extent they are predictable?

Acknowledgement. We thank CAIDA for sharing with us the data analyzed in the paper. This work was supported in part by ARO Grant #W911NF-13-1-0141 and NSF Grant #1111925.

A   Characterization of the Dominance and Periodicity Phenomenon Exhibited by Attackers

Now we quantify the similarity between the two time series via Dynamic Time Warping (DTW), fitted model, and prediction accuracy.

Similarity Based on DTW. Figure 9(a) plots the warping path between the total number attackers in D_1 and the total number of attackers in D_2. The two time series are very similar to each other, except for the time interval $[452, 668]$ as suggested by Figs. 8(a) and (c). Figure 9(b) plots the warping path between the two time series plotted in Fig. 7(a), namely the total number of attackers in D_1 and the total number of attackers from country X in D_1. It shows that the two time series are very similar to each other except during the time interval $[455, 630]$, as suggested by Figs. 8(a) and (b). Figure 9(c) plots the warping path between the two time series plotted in Figs. 8(b) and (d). It shows that the two time series are almost identical to each other, and that the filtering of rarely seen attackers/attacks does not manipulate the periodic structure of the time series of the number of attackers from country X. Figure 9(d) plots the warping path between the two time series plotted in Figs. 8(c) and (d), which indeed are almost identical to each other.

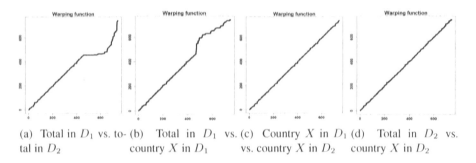

(a) Total in D_1 vs. to-   (b) Total in D_1 vs.   (c) Country X in D_1 (d) Total in D_2 vs.
tal in D_2 country X in D_1 vs. country X in D_2 country X in D_2

Fig. 9. DTW statistics between the times series of the total number of attackers and the time series of the number of attackers for country X.

Similarity Based on Fitted Models. Since both the time series exhibit periodicity, we use the multiplicative seasonal ARIMA model to fit the two time series in D_1 and D_2, respectively. The model parameters are: nonseasonal orders (p, d, q), and seasonal orders (P, D, Q), and seasonal period $s = 24$ based on the above discussion of periodicity. For model selection, the parameter sets are:

- $(p, d, q) \in [0, 5] \times \{0, 1\} \times [0, 5]$;
- $(P, D, Q) \in [0, 5] \times \{0, 1\} \times [0, 5]$.

According to the AIC criterion (briefly reviewed in Sect. 3), the two time series in both D_1 and D_2 prefer to the following model:

$$W_t = \phi_1 W_{t-1} + e_t + \theta_1 e_{t-1} + \Phi_1 W_{t-24} + \Phi_2 W_{t-48} +$$
$$\Theta_1 e_{t-24} + \Theta_2 e_{t-48} + \Theta_3 e_{t-96},$$

where $W_t = |A(r; t, t+1)| - |A(r; t-24, t-23)|$. Table 3 summarizes the fitting results. We observe that the two fitted models in D_1 are similar to each other in terms of coefficients, and that the two fitted models in D_2 are almost identical to each other.

Table 3. Coefficients in the fitted models of the total number of attackers and the number of attackers from country X.

	ϕ_1	θ_1	Φ_1	Φ_2	Θ_1	Θ_2	Θ_3
Fitted model of total number of attackers in D_1: PMAD=.08							
Coefficients	.91	.38	1.22	-.98	-2.15	2.11	-.86
Fitted model of number of attackers from country X in D_1: PMAD=.06							
Coefficients	0.82	.39	1.22	-.99	-2.19	2.16	-.91
Fitted model of total number of attackers in D_2: PMAD=.08							
Coefficients	.79	.4	1.21	-.99	-2.18	2.16	-.9
Fitted model of number of attackers from country X in D_2: PMAD=.07							
Coefficients	.79	.4	1.21	-.99	-2.19	2.16	-.9

Similarity Based on Prediction Accuracy. Table 4 summarizes the PMAD values for 1, 4, 7 and 10 hours ahead-of-time prediction of the number of attackers during the last 96 hours in both D_1 and D_2. For D_1, we observe that 1-h ahead-of-time predictions for the number of attackers from country X and the total number of attackers are reasonably accurate (with PMAD value .093 and .092, or 9.3 % and 9.2 % prediction error, respectively); whereas the predictions for 4, 7 and 10 hours ahead-of-time are not useful. For D_2, we observe similar prediction results, namely that 1-h ahead-of-time predictions lead to 7.5 % prediction error for the total number of attackers and 9.5 % prediction error for the number of attackers from country X.

Table 4. PMAD values for $h = 1, 4, 7, 10$ hours ahead-of-time predictions on the total number of attackers and on the number of attackers from country X, as observed by the telescope.

	$h=1$	$h=4$	$h=7$	$h=10$		$h=1$	$h=4$	$h=7$	$h=10$
D_1: PMAD values					D_2: PMAD values				
Total	.092	.244	.333	.404	Total	.075	.156	.180	.177
Country X	.093	.208	.240	.245	Country X	.095	.203	.230	.224

B   Further Characterizations on the Inference Errors of Small Telescopes

Inferring the Number of Attackers From Small Telescopes. Similarly, we would like to infer the number of attackers based on small telescopes. Table 5 summarizes the inference errors in terms of the min, mean, median and max PMAD values of all the considered combinations of sample blocks, as well as the standard deviation of the PMAD values. For D_1 and $B = 16$, we observe that 3 small telescopes (out of the 16 telescopes of size 2^{20} IP addresses) would give good approximation of the number of attackers that would be obtained based on the network telescope of size 2^{24}. This is because the maximum PMAD value is 7.34%. For D_2 and $B = 16$, we observe that using 4 small telescopes of size 2^{20} does not lead to good approximation. For $B = 256$, neither D_1 nor D_2 leads to obtain good enough approximation. These suggest that using significantly small telescopes may not lead to robust results.

Table 5. PMAD-based measurement of the inference error when using b (our of the B) small telescopes to approximate the number of attackers that are observed by the larger /8 telescope, where "SD" stands for standard deviation.

	Min	Mean	Median	Max	SD		Min	Mean	Median	Max	SD
D_1 with $B = 16$: PMAD values						D_2 with $B = 16$: PMAD values					
$b = 1$	.0663	.0921	.0916	.1329	.0205	$b = 1$	.1689	.1863	.1857	.2221	.0119
$b = 2$	.0626	.0797	.0752	.1194	.0138	$b = 2$	.1476	.1784	.1798	.2221	.0089
$b = 3$	.0624	.0700	.0713	.0734	.0033	$b = 3$	.1395	.1730	.1755	.2221	.0098
$b = 4$	.0593	.0685	.0693	.0734	.0036	$b = 4$	.1355	.1664	.1676	.1874	.0103
D_1 with $B = 256$: PMAD values						D_2 with $B = 256$: PMAD values					
$b = 1$	.1273	.2499	.3037	.4303	.0983	$b = 1$	.2967	.4346	.4387	.4396	.0172
$b = 2$	.1121	.1929	.1467	.4303	.0846	$b = 2$	.1712	.4307	.4387	.4396	.0240
$b = 3$	.1251	.1510	.1447	.3287	.0232	$b = 3$	.1713	.2369	.2352	.3730	.0422
$b = 4$	.1250	.1491	.1445	.2935	.0188	$b = 4$	.1664	.2261	.2329	.2542	.0231
$b = 5$	.1236	.1474	.1435	.2762	.0168	$b = 5$	.1618	.1658	.1664	.1674	.0015

Inferring the Number of Attacks From Small Telescopes. From the perspective of inferring the number of attacks, Table 6 summarizes the inference errors as in the above. For D_1 and $B = 16$, we observe that 3 small telescopes (out of the 16 telescopes of size 2^{20} IP addresses) would give good approximation of the number of attacks that would be obtained based on the larger network telescope of size 2^{24}. This is because the maximum PMAD errors is .0761, namely 7.61% approximation error. For D_1 and $B = 256$, the mean approximation error is 9.58% for $b = 5$ (i.e., using 5 small telescopes instead), which is marginally acceptable. For D_2 and $B = 16$, we observe that using 4 small telescopes of size 2^{20} can lead to worst-case approximation error 8.27%. For D_2 and $B = 256$, we observe that using 5 small telescopes of size 2^{16} does not lead to good approximation. That is, substantially small telescope may not be as useful as the large telescope.

Table 6. PMAD-based measurement of the inference error when using b (our of the B) small telescopes to approximate the number of attacks observed by the /8 telescope, where "SD" stands for standard deviation.

	Min	Mean	Median	Max	SD		Min	Mean	Median	Max	SD
D_1 with $B = 16$: PMAD values						D_2 with $B = 16$: PMAD values					
$b = 1$	.0755	.0890	.0883	.1106	.0095	$b = 1$	.0765	.0901	.0895	.1130	.0096
$b = 2$	.0436	.0629	.0589	.0953	.0138	$b = 2$	.0444	.0655	.0659	.1130	.0150
$b = 3$	.0363	.0488	.0435	.0761	.0125	$b = 3$	.0343	.0518	.0461	.1130	.0138
$b = 4$	.0272	.0392	.0369	.0739	.0100	$b = 4$	.0267	.0395	.0372	.0827	.0094
D_1 with $B = 256$: PMAD values						D_2 with $B = 256$: PMAD values					
$b = 1$	.1166	.1322	.1311	.1917	.0098	$b = 1$	.1344	.1909	.1938	.1938	.0117
$b = 2$	.0853	.1069	.1056	.1917	.0092	$b = 2$	.1197	.1881	.1938	.1938	.0160
$b = 3$	.0699	.0916	.0897	.1917	.0086	$b = 3$	.1323	.1409	.1387	.1602	.0079
$b = 4$	.0835	.0984	.0970	.1700	.0069	$b = 4$	.1310	.1383	.1366	.1520	.0069
$b = 5$	.0818	.0958	.0945	.1582	.0064	$b = 5$	.1256	.1312	.1310	.1419	.0048

References

1. Allman, M., Paxson, V., Terrell, J.: A brief history of scanning. In: Proceedings of ACM IMC 2007, pp. 77–82 (2007)
2. Armstrong, J.S.: Principles of Forecasting: A Handbook for Researchers and Practitioners, vol. 30. Springer, New York (2001)
3. Bailey, M., Cooke, E., Jahanian, F., Myrick, A., Sinha, S.: Practical darknet measurement. In: Proceedings of 2006 Annual Conference on Information Sciences and Systems, pp. 1496–1501 (2006)
4. Bailey, M., Cooke, E., Jahanian, F., Watson, D.: The blaster worm: then and now. IEEE Secur. Priv. **3**(4), 26–31 (2005)
5. Bailey, M., Cooke, E., Jahanian, F., Nazario, J., Watson, D., et al.: The internet motion sensor-a distributed blackhole monitoring system. In: Proceedings of NDSS 2005 (2005)
6. Barford, P., Chen, Y., Goyal, A., Li, Z., Paxson, V., Yegneswaran, V.: Employing honeynets for network situational awareness. In: Jajodia, S., Liu, P., Swarup, V., Wang, C. (eds.) Cyber Situational Awareness. Advances in Information Security, vol. 46, pp. 71–102. Springer, New York (2010)
7. Brownlee, N.: One-way traffic monitoring with iatmon. In: Proceedings of PAM 2012, pp. 179–188 (2012)
8. Claffy, K., Braun, H., Polyzos, G.: A parameterizable methodology for internet traffic flow profiling. IEEE J. Sel. Areas Commun. **13**(8), 1481–1494 (1995)
9. Clauset, A., Shalizi, C.R., Newman, M.E.J.: Power-law distributions in empirical data. SIAM Rev. **51**(4), 661–703 (2009)
10. Cooke, E., Bailey, M., Mao, Z.M., Watson, D., Jahanian, F., McPherson, D.: Toward understanding distributed blackhole placement. In: Proceedings of ACM Worm 2004, pp. 54–64 (2004)

11. Cryer, J., Chan, K.: Time Series Analysis With Applications in R. Springer, New York (2008)
12. Dainotti, A., King, A., Claffy, K., Papale, F., Pescapè, A.: Analysis of a "/0" stealth scan from a botnet. In: Proceedings of ACM IMC 2012, pp. 1–14 (2012)
13. Engle, R.F.: Autoregressive conditional heteroscedasticity with estimates of the variance of united kingdom inflation. Econometrica: J. Econometric Soc. 50(4), 987–1007 (1982)
14. Giorgino, T.: Computing and visualizing dynamic time warping alignments in R: the dtw package. J. Stat. Softw. 31(7), 1–24 (2009)
15. Glatz, E., Dimitropoulos, X.: Classifying internet one-way traffic. In: Proceedings of ACM IMC 2012, pp. 37–50 (2012)
16. Gringoli, F., Salgarelli, L., Dusi, M., Cascarano, N., Risso, F., Claffy, K.: Gt: picking up the truth from the ground for internet traffic. SIGCOMM Comput. Commun. Rev. 39(5), 12–18 (2009)
17. Hussain, A., Heidemann, J., Papadopoulos, C.: A framework for classifying denial of service attacks. In: Proceedings of ACM SIGCOMM 2003, pp. 99–110 (2003)
18. Kreibich, C., Crowcroft, J.: Honeycomb: creating intrusion detection signatures using honeypots. SIGCOMM Comput. Commun. Rev. 34(1), 51–56 (2004)
19. Lau, F., Rubin, S.H., Smith, M.H., Trajkovic, L.: Distributed denial of service attacks. In: Proceedings of 2000 IEEE International Conference on Systems, Man, and Cybernetics, vol. 3, pp. 2275–2280 (2000)
20. Lee, D.J., Brownlee, N.: Passive measurement of one-way and two-way flow lifetimes. SIGCOMM Comput. Commun. Rev. 37(3), 17–28 (2007)
21. Li, Z., Goyal, A., Chen, Y., Paxson, V.: Towards situational awareness of large-scale botnet probing events. IEEE Trans. Inf. Forensics Secur. 6(1), 175–188 (2011)
22. Li, Z., Goyal, A., Chen, Y., Kuzmanovic, A.: Measurement and diagnosis of address misconfigured p2p traffic. IEEE Netw. 25(3), 22–28 (2011)
23. Moore, D., Paxson, V., Savage, S., Shannon, C., Staniford, S., Weaver, N.: Inside the slammer worm. IEEE Secur. Priv. 1(4), 33–39 (2003)
24. Moore, D., Shannon, C., Brown, D., Voelker, G., Savage, S.: Inferring internet denial-of-service activity. ACM Trans. Comput. Syst. 24(2), 115–139 (2006)
25. Moore, D., Shannon, C., Brown, J.: Code-red: a case study on the spread and victims of an Internet worm. In: Proceedings of ACM IMW 2002, pp. 273–284 (2002)
26. Moore, D., Shannon, C., Voelker, G.M., Savage, S.: Network telescopes, Technical report. Department of Computer Science and Engineering, University of California, San Diego (2004)
27. Neter, J., Kutner, M.H., Nachtsheim, C.J., Wasserman, W.: Applied linear statistical models, vol. 4. Irwin, Chicago (1996)
28. Pang, R., Yegneswaran, V., Barford, P., Paxson, V., Peterson, L.: Characteristics of internet background radiation. In: Proceedings of ACM IMC 2004, pp. 27–40 (2004)
29. Provos, N.: A virtual honeypot framework. In: Proceedings of USENIX Security Symposium, pp. 1–14 (2004)
30. Shannon, C., Moore, D.: The spread of the witty worm. IEEE Secur. Priv. 2(4), 46–50 (2004)
31. CAIDA UCSD Network Telescope. http://www.caida.org/
32. CAIDA UCSD Network Telescope. http://www.caida.org/tools/measurement/corsaro/docs/plugins.html
33. Treurniet, J.: A network activity classification schema and its application to scan detection. IEEE/ACM Trans. Netw. 19(5), 1396–1404 (2011)

34. Tsay, R.S.: Analysis of Financial Time Series. Wiley, New york (2010)
35. Weiler, N.: Honeypots for distributed denial-of-service attacks. In: Proceedings of IEEE Workshop on Enabling Technologies: Infrastructure for Collaborative Enterprises (WET-ICE 2002), pp. 109–114 (2002)
36. Wustrow, E., Karir, M., Bailey, M., Jahanian, F., Huston, G.: Internet background radiation revisited. In: Proceedings of ACM IMC 2010, pp. 62–74 (2010)
37. Yegneswaran, V., Giffin, J., Barford, P., Jha, S.: An architecture for generating semantic aware signatures. In: Proceedings of Usenix Security Symposium (2005)
38. Yegneswaran, V., Barford, P., Plonka, D.: On the design and use of internet sinks for network abuse monitoring. In: Jonsson, E., Valdes, A., Almgren, M. (eds.) RAID 2004. LNCS, vol. 3224, pp. 146–165. Springer, Heidelberg (2004)
39. Yegneswaran, V., Barford, P., Ullrich, J.: Internet intrusions: global characteristics and prevalence. In: Proceedings of ACM SIGMETRICS 2003, pp. 138–147 (2003)
40. Zhan, Z., Xu, M., Xu, S.: Characterizing honeypot-captured cyber attacks: statistical framework and case study. IEEE Trans. Inf. Forensics Secur. **8**(11), 1775–1789 (2013)

Key-Exposure Protection in Public Auditing with User Revocation in Cloud Storage

Hua Guo[1]([⊠]), Fangchao Ma[2], Zhoujun Li[1], and Chunhe Xia[2]

[1] State Key Laboratory of Software Development Environment,
Beihang University, Beijing 100191, China
hguo@buaa.edu.cn
[2] Beijing Key Laboratory of Network Technology, School of Computer Science
and Engineering, Beihang University, Beijing, China

Abstract. With the development of cloud data storage, more and more data owners are choosing to store their data in the Cloud and share them as a group. To protect integrity of sharing data, data are signed before they are stored on the cloud. When a user is revoked from the group, the revoked user's signature can be converted to the existing group member's signature by the cloud to preserve the revocation's efficiency. Accordingly, the public auditing should be done by the third party auditor using the existing group member's public key. As a basic secure requirement, the cloud sever should not know the existing group member's private key even if he obtains the revoked user's private key. In this paper, we propose a new public auditing protocol in which a public verifier is always able to audit the integrity of shared data even if some part of shared data has been re-signed by the cloud. By integrating the proxy re-signature with random masking technique, the new public auditing protocol satisfies the basic secure requirement. In addition, we prove the security of the new protocol, and finally compare it with other existing public auditing protocols and show that the new mechanism provides a good key-exposure protection for the existed public auditing protocol for shared data without losing the communication and computation efficiency.

Keywords: Public auditing · Shared data · Cloud storage · User revocation · Key-exposure protection

1 Introduction

Cloud data storage allows data owners to move data from their local computing systems to the cloud. A cloud data storage service consists of three different participants, namely the cloud server, the third party auditor (TPA) and users. The cloud server has ample storage space and provides data storage and sharing services for users. The TPA is able to provide data auditing service based on requests from users, without downloading the entire file. Cloud user stores large amount of data or files on a cloud server. In a group, there are two types of

© Springer International Publishing Switzerland 2015
M. Yung et al. (Eds.): INTRUST 2014, LNCS 9473, pp. 127–136, 2015.
DOI: 10.1007/978-3-319-27998-5_8

users: an original user who is the creator of the shared data, and a number of group users who can access and modify the data. Shared data is further divided into a number of blocks.

Instead of the initial investment of expensive infrastructure setup, large equipments and daily maintenance cost, the data owners only need to pay the space they actually use, e.g., cost-per-gigabyte-stored model. Additionally, data owners can rely on the Cloud to access data at will. More and more data owners are choosing to store their data in the Cloud. After an original group user creates shared data and stores them on the cloud server, every group user is able to access and modify shared data so that he can share the latest version of the shared data with the rest of the group. To protect sharing data integrity, all of the data, including the data created by the original group user and the data modified by the different group users, should be signed before they are stored on the cloud server. Thus different data blocks are signed by different users due to data modifications performed by different users. Since each data block is signed by a group user, a public verifier, such as a third party auditor (TPA), can check data integrity in the cloud without downloading the entire data, referred to as public auditing.

Public auditing allows data integrity to be publicly checked without completely downloading the data. Ateniese et al. [1] are the first to propose the model of Provable Data Possession (PDP), which allows a verifier to publicly check the correctness of a clients data stored at an untrusted storage using RSA-based homomorphic authenticators. Later, Ateniese et al. [2] presented a dynamic version of the prior PDP scheme to support dynamic operations based on hash function and symmetric key encryption. However, it looses two important properties, i.e., the publicly verifiability and fully dynamic data operations. Subsequently, Erway et al. [3] introduced Dynamic Provable Data Possession by using authenticated dictionaries, which are based on rank information. Juels et al. [4] proposed a POR model to ensure both data possession and retrievability. Unfortunately, this mechanism prevents efficient extension for updating data. Shacham and Waters [5] designed an improved PDP scheme based on BLS signatures, which is not publicly verifiable and only provides a user with a limited number of verification requests. In 2012, Wang et al. [6] presented data integrity checking approaches to achieve public auditability, storage correctness, privacy-preserving, batch auditing, lightweight, dynamic data support and error location and recovery. To preserve users confidential data from the TPA, Wang et al. [7] propose a public mechanism using random maskings, which also supports batch auditing. In Zhu et al. [8] public auditing mechanism, the fragment structure is exploited to reduce the storage of signatures, and index hash tables are used to provide dynamic operations for users. Meanwhile, to preserve the identity of the signer on each block from the users and the TPA, Wang et.al. [9] proposed a mechanism for public auditing shared data in the cloud for a group of users by using ring signature-based homomorphic authenticators. The auditing mechanism in [10] is designed to preserve identity privacy for a large number of users. However, it fails to support public auditing.

When a revoked user leaves the group, for security reason, the signatures generated by this revoked user should be re-signed by an existing user in the group since they are no longer valid to the group. The most efficient way is allowing the cloud to convert the revoked user's signatures to the existing group member's signatures. Accordingly, the public auditing process should be done by the third party auditor using the existing group member's public key. There are two basic secure requirements: (1) the cloud, who is not in the same trusted domain with each user, is only able to convert a signature of the revoked user into a signature of an existing user(say Alice) on the same block, but it cannot sign arbitrary blocks on behalf of either the revoked user or an existing user; (2) the cloud sever should not know the existing group member's private key even if he obtains the revoked user's private key.

Previous works pay a lot of attentions on auditing the integrity of personal data and preserve identity privacy from the TPA. Very recently, Wang et al. [11] presented a public auditing mechanism with efficient user revocation in an untrusted cloud by utilizing proxy re-signatures [12]. In Wang et al. scheme, the first secure requirement is satisfied, which the second one is lost. More precisely, if a revoked user (Say B) leaks his private key to the cloud, the private key of the valid group member (say A) would be computed by the cloud. As a direct result, the untrusted cloud server would repudiate the deletion messages since he can sign a message using A's private key which would bring in a dispute. One straightforward method to deal with this threat is that user A updates his private key after the re-sign private key is distributed to the cloud. However, this would bring in a huge amount of communication and computation resource since A has to download all of his signed message and re-sign them and re-upload them to the cloud sever, which takes away the benefits which Wang et al. scheme brings in. Therefore, how to preserve the private key's security of the group member even if the private key of the revoked user is compromised is the problem we are going to tackle in this paper.

In this paper, we construct a new public auditing mechanism by integrating the proxy re-signature with random masking technique, to guarantee that the compromising of the revoked user's private key would not affect the security of the private keys of the users in the group. Thus the new scheme is more secure than the exist public auditing scheme, i.e., after the auditing protocol's execution, the cloud could not learn any knowledge about the private keys of the users in the group even if the private key of the revoked user is compromised. We prove the security and compare the efficiency of our proposed schemes with the state-of-the-art.

The rest of this paper is arranged as follows. In Sect. 2, we introduce several cryptographic primitives. In Sect. 3, detailed design and security analysis of the new mechanism are presented. Section 4 analyzes the efficiency of the new mechanism. Finally Sect. 5 give the concluding remark of the whole paper.

2   Preliminaries

In this section, we introduce the background knowledge that will be used for the new scheme. We give the basic definition and properties of bilinear pairings and the computational problems.

We first revisit the "admissible bilinear map" [13] and the Computational Diffie-Hellman problem, which play central roles in our scheme.

The admissible bilinear map $\hat{e}$ is defined over two groups of the same prime order p denoted by $\mathcal{G}_1$ and $\mathcal{G}_2$ in which the Computational Diffie-Hellman problem is hard. More formally, we have the following definition:

Definition 1. *(Bilinear Map) Let $\mathcal{G}_1$ and $\mathcal{G}_2$ are two multiplicative cyclic groups of the same order p. Let g denote a generator of $\mathcal{G}_1$. An admissible pairing is a bilinear map $e : \mathcal{G}_1 \times \mathcal{G}_1 \to \mathcal{G}_2$ which has the following properties:*

- *Bilinear: given $u, v \in \mathcal{G}_1$ and $a, b \in Z$, we have $e(u^a, v^b) = e(u, v)^{ab}$.*
- *Non-degenerate: $e(g, g) \neq 1$.*
- *Computable: e is efficiently computable.*

Throughout this paper, we will simply use the term "Bilinear map" to refer to the admissible bilinear map defined above.

We now revisit the Computational Diffie-Hellman (CDH) problem and the Discrete Logarithm (DL) Problem.

Definition 2. *(Computational Diffie-Hellman Problem) For $a, b \in Z_p$, given $g, g^a, g^b \in \mathcal{G}_1$ as input, output $g^{ab} \in \mathcal{G}_1$.*

The CDH assumption holds in $\mathcal{G}_1$ if it is computationally infeasible to solve the CDH problem in $\mathcal{G}_1$.

Definition 3. *(Discrete Logarithm (DL) Problem.) For $a \in Z_p$, given $g, g^a \in \mathcal{G}_1$ as input, output a.*

The DL assumption holds in $\mathcal{G}_1$ if it is computationally infeasible to solve the DL problem in $\mathcal{G}_1$.

3   Construction of the New Public Auditing Mechanism

In this section, we will show how to construct the public auditing mechanism for shared data. As in Wang et al. [11] scheme, we also designate the cloud as the proxy to translate signatures for users in the group, and mandate the revoked user's signatures to be translated to the original user.

3.1   Scheme Details

The new public auditing mechanism consists of six algorithms: KeyGen, ReKey, Sign, ReSign, ProofGen, ProofVerify.

- KeyGen. This algorithm is run by every user in the group to generates his/her public key and private key.
- ReKey. This algorithm is run among the revoked user, the existing group user and the cloud server. After the execution, the algorithm helps the cloud to output a re-signing key.

- Sign. This algorithm is run by all of the group user, i.e., the original user computes signatures on the shared data blocks he creates; the group user computes the signature on the modified share data block.
- ReSign. This algorithm is run by the cloud server. After a user revoked from the group, the cloud uses the re-signing key to re-sign the blocks which were previous signed by the revoked user.
- ProofGen. This algorithm is run by the cloud to generate a proof of possession of shared data.
- ProofVerify. This algorithm is run by a public verifier to check the correctness of a proof.

Let $\mathcal{G}_1$ and $\mathcal{G}_2$ be two groups of order p, g be a generator of $\mathcal{G}_1$, $e : \mathcal{G}_1 \times \mathcal{G}_1 \to \mathcal{G}_2$ be a bilinear map, w be a random element of $\mathcal{G}_1$. The global parameters are $(e, p, \mathcal{G}_1, \mathcal{G}_2, g, w, H, H)$, where H is a hash function with $H : \{0,1\}^* \to \mathcal{G}_1$ and H is a hash function with $H : \{0,1\}^* \to Z_q$. The total number of blocks in shared data is n, and shared data is described as $M = (m_1, \ldots, m_n)$. The total number of users in the group is d.

- **KeyGen.** For user u_i, he randomly generates $x_i \in Z_p$, and outputs his public key $pk_i = g^{x_i}$ and private key $sk_i = x_i$. Without loss of generality, user u_1 is assumed to be the original user, who is the creator of shared data. The original user also creates a public user list (UL), which contains ids of all the users in the group and is signed by the original user.
- **ReKey.** Assume that private and authenticated channels exist between each pair of entities, and there is no collusion. The cloud generates a re-signing key k_{ij} as follows:
  - The cloud generates a random $r \in Z_p$ and sends it to the revoked user u_i;
  - User u_i sends r/x_i to user u_j , where $sk_i = x_i$;
  - User u_j randomly selects $r_j \in Z_p$ and sends $(\frac{r(r_j + \lambda x_j)}{x_i}, g^{r_j})$ to the cloud, where $sk_j = x_j$ and $\lambda = h(g^{r_j})$;
  - The cloud recovers $k_{i \to j} = \frac{r_j + \lambda x_j}{x_i} \in Z_p$.
- **Sign.** Given private key $sk_i = x_i$, block $m_k \in Z_p$ in shared data M and its block identifier id_k, where $k \in [1, n]$, user u_i outputs the signature on block m_k as $\sigma_k = (H(id_k)w^{m_k})^{x_i} \in \mathcal{G}_1$.
- **ReSign.** When user u_i is revoked from the group, the cloud is able to convert signatures of user u_i into signatures of user u_j on the same block. More specifically, given re- signing key $k_{i \to j}$, public key pk_i, signature σ_k, block m_k and block identifier id_k, the cloud first checks that $e(\sigma_k, g) =?e(H(id_k)w^{m_k}, pk_i)$. If the verification result is 0, the cloud outputs ; otherwise, it outputs

$$\sigma_k = \sigma_{k_{i \to j}} = (H(id_k)w^{m_k})^{x_i(\frac{\lambda x_j + r_j}{x_i})} = (H(id_k)w^{m_k})^{\lambda x_j + r_j}.$$

After the re-signing, the original user removes user u_is id from UL and signs the new UL.
- **ProofGen.** To audit the integrity of shared data, the TPA generates an auditing message as follows:

- Randomly picks a c-element subset L of set $[1, n]$ to locate the c selected random blocks that will be checked in this auditing task.
- Generates a random $y_l \in Z_q$, for $l \in L$ and q is a much smaller prime than p.
- Outputs an auditing message $(l, y_l)_{l \in L}$, and sends it to the cloud.

After receiving an auditing message, the cloud generates a proof of possession of shared data M. More concretely,

- The cloud divides set L into d subset $L_1, \cdots, L_d$, where L_i is the subset of selected blocks signed by user u_j. And the number of elements in subset L_i is c_i. Clearly, we have $c = \sum_{i=1}^{d} c_i$, $L = L_1 \bigcup \cdots \bigcup L_d$ and $L_i \bigcap L_j = \emptyset$, for $i \neq j$.
- For each set $L_i (i \neq 1)$, the cloud computes $\alpha_i = \sum_{l \in L_i} y_l m_l \in Z_p$, $\beta_i = \Pi_{l \in L_i} \sigma_l^{y_l} \in G_1$.
  For L_1, the cloud computes $\alpha_{11} = \sum_{l \in L_{11}} y_l m_l$, $\beta_{11} = \Pi_{l \in L_{11}} \sigma_l^{y_l}$, and $\alpha_{12} = \sum_{l \in L_{12}} y_l m_l$, $\beta_{12} = \Pi_{l \in L_{12}} \sigma_l^{y_l}$, separately.
- Finally, the cloud outputs an auditing proof $(\alpha, \beta, \gamma, id_{l \mid l \in L})$, where $\alpha = (\alpha_1, \cdots, \alpha_d)$, $\beta = (\beta_1, \cdots, \beta_d)$, $\gamma = g^{r_1}$, $\alpha_1 = (\alpha_{11}, \alpha_{12})$, $\beta_1 = (\beta_{11}, \beta_{12})$;

- **ProofVerify.** With an auditing proof $(\alpha, \beta, \gamma, id_{l \mid l \in L})$, an auditing message $(l, y_l)_{l \in L}$, and all the existing users public keys $(pk_1, \cdots, pk_d)$, the TPA checks the correctness of this auditing proof as
  - For $l \in L_i (i \neq 1)$, the cloud checks if the equation

$$e(\prod_{i=2}^{d} \beta_i, g) = \prod_{i=2}^{d} e(\prod_{l \in L_i} H(id_l)^{y_l} \dot{w}^{\alpha_i}, pk_i)$$

holds or not.
  - For $l \in L_{11}$, the cloud checks if the equation

$$e(\beta_{11}, g) = e(\prod_{l \in L_{11}} H(id_l)^{y_l} \cdot w^{\alpha_{11}}, pk_1)$$

holds or not.
  - For $l \in L_{12}$, the cloud checks if the equation

$$e(\beta_{12}, g) = e(\prod_{l \in L_{12}} H(id_l)^{y_l} \cdot w^{\alpha_{12}}, pk_1)^{\lambda} \cdot e(\prod_{l \in L_{12}} H(id_l)^{y_l} \cdot w^{\alpha_{12}}, \gamma)$$

holds or not.

If the result is 1, the verifier believes that the integrity of all the blocks in shared data M is correct. Otherwise, the verifier outputs 0.

Remark. Similar to Wang et al. [11] scheme, since we utilize the bilinear pairings, our public auditing protocol also supports batch auditing, i.e., the TPA can perform multiple auditing tasks simultaneously.

3.2 Security Analysis of the Public Auditing Mechanism

In this subsection, we will give a security analysis of our public auditing protocol, including the storage correctness and the key-privacy-preserving.

Storage Correctness Guarantee. We need to prove that the cloud sever can not generate valid response toward TPA without faithfully storing the data.

Theorem 1. *Given shared data M and its signatures, a verifier is able to correctly check the integrity of shared data M.*

Proof: The correctness of our mechanism can be verified by proving the following equations. Based on the properties of bilinear maps, the correctness of the equations are presented as the following:

– For $l \in L_i(i \neq 1)$, the cloud checks the equation

$$
e(\prod_{i=2}^{d} \beta_i, g) = \prod_{i=2}^{d} e(\prod_{l \in L_i} \sigma_l^{y_l}, g)
$$

$$
= \prod_{i=2}^{d} e(\prod_{l \in L_i} (H(id_l)w^{m_l})^{x_i y_l}, g)
$$

$$
= \prod_{i=2}^{d} e(\prod_{l \in L_i} H(id_l)^{y_l} \cdot \prod_{l \in L_i} w^{m_l y_l}, g^{x_i})
$$

$$
= \prod_{i=2}^{d} e(\prod_{l \in L_i} H(id_l)^{y_l} \cdot w^{\alpha_i}, pk_i)
$$

– For $l \in L_{11}$, the cloud checks the equation

$$
e(\beta_{11}, g) = e(\sigma_l^{y_l}, g)
$$

$$
= e(\prod_{l \in L_{11}} (H(id_l)w^{m_l})^{x_i y_l}, g)
$$

$$
= e(\prod_{l \in L_{11}} H(id_l)^{y_l} \cdot \prod_{l \in L_{11}} w^{m_l y_l}, g)
$$

$$
= e(\prod_{l \in L_{11}} H(id_l)^{y_l} \cdot w^{\alpha_{11}}, pk_1).
$$

– For $l \in L_{12}$, the cloud checks the equation

$$
e(\beta_{12}, g) = e((\sigma_l')^{y_l}, g)
$$

$$
= e(\prod_{l \in L_{12}} (H(id_l)w^{m_l})^{(r_1 + \lambda x_i)y_l}, g)
$$

$$
= e(\prod_{l \in L_{12}} H(id_l)^{y_l} \cdot \prod_{l \in L_{12}} w^{m_l y_l}, g^{r_1 + \lambda x_i})
$$

$$
= e(\prod_{l \in L_{12}} H(id_l)^{y_l} \cdot w^{\alpha_{12}}, pk_1)^{\lambda} \cdot e(\prod_{l \in L_{12}} H(id_l)^{y_l} \cdot w^{\alpha_{12}}, \gamma)
$$

Theorem 2. *For the cloud, it is computational infeasible to generate a forgery of an auditing proof under our mechanism.*

Proof: The difference between our scheme and Wang et al. scheme [11] is the way of the re-sign key's generation, i.e., random masking technology is used to generate the re-key is our scheme. Accordingly, the verification of some signatures from the original data creator U_1 is a little different from Wang et al. scheme. However, we find that these difference does not affect the proof process when we adapt Wang et al. strategy to our scheme. Thus the proof of our scheme is similar to that of Wang et al. scheme.

Key-Privacy Guarantee: We want to make sure that TPA can not derive the existing group user's private key, even when he colludes with the revoked user. This is equivalent to prove the following Theorem.

Theorem 3. *For the cloud sever, it is computational infeasible to compute the target signer's private key, even if with the help of the revoked user's private key under our mechanism.*

Proof: According to the description of the public auditing protocol, we know that if the cloud sever colludes with the revoked user, he has the knowledge of this revoked user's private key a. We need to prove that the cloud sever cannot obtain the existing group user's private key to whom the signatures from the revoked user are translated. In our scheme, this existing group user is assumed to be the shared data's creator U_1. The hardness of this problem lies in the hardness of the DL problem (given g and g^k, it is hard to compute k). More specifically, when the cloud sever makes "re-key" query to U_1, the challenger returns "$(\frac{k+h(g^k)b}{a}, g^k)$" to the cloud sever. Note that with the value of g^k, due to the hardness of discrete-log assumption, the value k is still hidden against the cloud sever. Thus, privacy of U_1's private key b is guaranteed from k.

4   Efficiency Analysis and Comparison

In this section, we discuss the communication and computation cost of our mechanism by comparing the new scheme with Wang et al. [11] scheme. Suppose d is the number of existing users in the group, c is the number of selected blocks, $|n|$ is the size of an element of set $[1, n]$, $|q|$ is the size of an element of Z_q, $|p|$ is the size of an element of $\mathcal{G}_1$ or Z_p, $|id|$ is the size of a block identifier. Additional, suppose $Exp_{\mathcal{G}_1}$ denotes one exponentiation in $\mathcal{G}_1$, $Mul_{\mathcal{G}_1}$ denotes one multiplication in $\mathcal{G}_1$, Pair denotes one pairing operation on $e : \mathcal{G}_1 \times \mathcal{G}_1 \rightarrow \mathcal{G}_2$, and $Hash_{\mathcal{G}_1}$ denotes one hashing operation in $\mathcal{G}_1$.

From Table 1, we can find that the communication cost of our scheme is a little high than that of Wang et al. scheme. More precisely, in our scheme, the size of an auditing message $(l, y_l)_{l \in L}$ is $c \cdot (|n| + |q|)$ bits and the size of an auditing proof $\{\alpha, \beta, \gamma, id_{l l \in L}\}$ is $3d \cdot |p| + c \cdot (|id|)$ bits. Therefore, the total communication cost of an auditing task is $3d \cdot |p| + c \cdot (|id| + |n| + |q|)$ bits. The total communication cost of an auditing task is $2d \cdot |p| + c \cdot (|id| + |n| + |q|)$ bits,

Table 1. The comparison of the proposed scheme and Wang et al. scheme.

Schemes	Scheme 1 [11]	Ours
Computation (Exp+Mul+Pair)	$(c+d)Exp_{\mathcal{G}_1} + (c+2d)Mul_{\mathcal{G}_1}$ $+dMul_{\mathcal{G}_2} + (d+1)Pair + cHash_{\mathcal{G}_1}$	$(c+d)Exp_{\mathcal{G}_1} + (c+2d)Mul_{\mathcal{G}_1}$ $+dMul_{\mathcal{G}_2} + (d+2)Pair + cHash_{\mathcal{G}_1}$ $+(c_{12})Exp_{\mathcal{G}_2}$
Communication	$2d \cdot |p| + c \cdot (|id| + |n| + |q|)$	$3d \cdot |p| + c \cdot (|id| + |n| + |q|)$
Allow PKC^a	no	yes

^{a}PKC means private-key-compromising

therefore the number of the communication cost our scheme is more than that of Wang et al. scheme is $d \cdot |p|$. Therefore the communication cost of our scheme is almost the same as that of Wang et al. scheme.

In terms of the communication cost, as shown in ReSign of our mechanism, the cloud first verifies the correctness of the original signature on a block, and then computes a new signature on the same block with a re-signing key. The computation cost of re-signing a block in the cloud is $2Exp_{\mathcal{G}_1} + Mul_{\mathcal{G}_1} + 2Pair + Hash_{\mathcal{G}_1}$. The cloud can further reduce the computation cost of the re-signing on a block to $Exp_{\mathcal{G}_1}$ by directly re-signing it without verification. The public auditing performed by the TPA ensures that the re-signed blocks are correct. Thus, the computation cost of an auditing task in our mechanism is $(c+d)Exp_{\mathcal{G}_1}+(c+2d)Mul_{\mathcal{G}_1}+(d+2)Pair+dMul_{\mathcal{G}_2}+cHash_{\mathcal{G}_1}+(c_{12})Exp_{\mathcal{G}_2}$. The total communication cost of an auditing task is $(c+d)Exp_{\mathcal{G}_1} + (c+2d)Mul_{\mathcal{G}_1} + (d+1)Pair + dMul_{\mathcal{G}_2} + cHash_{\mathcal{G}_1}$, therefore the number of the computation cost our scheme is more than that of Wang et al. scheme is $1Pair + (c_{12})Exp_{\mathcal{G}_2}$. Since the computation capability of the TPA is strong enough so that we conclude that the computation cost of our scheme is almost the same as that of Wang et al. scheme.

We finally check the security when the revoked uses's private key is compromised to the cloud. From the scheme, we can find that when the revoked uses's private key is compromised to the cloud, the private key of the existed group user in our scheme is secure, while the private key of the existed group user in Wang et al. scheme is easy to be computed by the cloud.

5   Conclusion

We have presented a key-exposure protection public auditing mechanism for shared data in the cloud with efficient user revocation. The main advantage of this scheme, i.e., even if the revoked user's private key is compromised, the private keys of the users in the group are still kept secret from the cloud server, is obtained by combining the random masking technology with an enhanced proxy re-signature scheme which is proposed in this paper. In addition, a public verifier is always able to audit the integrity of shared data even if some part of shared data has been re-signed by the cloud. We analyzed the efficiency and the security of the new public auditing scheme, and showed that our mechanism provides a good key-exposure protection for a public auditing protocol for shared data, at the same time keeps the high communication and computation efficiency.

Acknowledgements. This work was supported by the National Natural Science Foundation of China (grant number 61300172), the Research Fund for the Doctoral Program of Higher Education (grant number 20121102120017) and the Fund of the State Key Laboratory of Software Development Environment (grant number SKLSDE-2014ZX-14), and the Fundamental Research Funds for the Central Universities grant number YWF-14-JSJXY-008).

References

1. Ateniese, G., Burns, R., Curtmola, R. et al.: Provable data possession at untrusted stores. In: The Proceedings of ACM CCS 2007, pp. 598–610 (2007)
2. Ateniese, G., Pietro, R.D., Mancini, L.V., Tsudik, G.: Scalable and efficient provable data possession. In: The Proceedings of ICST SecureComm 2008 (2008)
3. Erway, C., Kupcu, A., Papamanthou, C., Tamassia, R.: Dynamic provable data possession. In: The Proceedings of ACM CCS 2009, pp. 213–222 (2009)
4. Juels, A., Burton, J., Kaliski, S.: Proofs of retrievability for large files. In: The Proceedings of ACM CCS 2007, pp. 584–597 (2007)
5. Shacham, H., Waters, B.: Compact proofs of retrievability. In: Pieprzyk, J. (ed.) ASIACRYPT 2008. LNCS, vol. 5350, pp. 90–107. Springer, Heidelberg (2008)
6. Wang, C., Wang, Q., Ren, K., Cao, N., Lou, W.: Toward secure and dependable storage services in cloud computing. IEEE Trans. Serv. Comput. 5(2), 220–232 (2012)
7. Wang, C., Chow, S.S., Wang, Q., Ren, K., Lou, W.: Privacy-preserving public auditing for secure cloud storage. IEEE Transa. Comput. **62**(2), 275–362 (2013)
8. Zhu, Y., Wang, H., Hu, Z. et al.: Dynamic audit services for integrity verification of outsourced storage in clouds. In: The Proceedings of ACM SAC 2011, pp. 1550–1557 (2011)
9. Wang, B., Li, B., Li, H.: Oruta: privacy-preserving public auditing for shared data in the cloud. In: The Proceedings of IEEE Cloud 2012, pp. 95–302 (2012)
10. Wang, B., Li, B., Li, H.: Knox: privacy-preserving auditing for shared data with large groups in the cloud. In: Bao, F., Samarati, P., Zhou, J. (eds.) ACNS 2012. LNCS, vol. 7341, pp. 507–525. Springer, Heidelberg (2012)
11. Wang, B., Li, B., Li, H.: Public auditing for shared data with efficient user revocation in the cloud. In: The Proceedings of INFOCOM 2013, pp. 2904–2912 (2013)
12. Ateniese, G., Hohenberger, S.: Proxy re-signatures: new definitions, algorithms and applications. In: The Proceedings of ACM CCS 2005, pp. 310–319 (2005)
13. Boneh, D., Franklin, M.: Identity-based encryption from the weil pairing. In: Kilian, J. (ed.) CRYPTO 2001. LNCS, vol. 2139, pp. 213–229. Springer, Heidelberg (2001)

Software Behavior Model Measuring Approach of Combining Structural Analysis and Language Set

JingFeng Xue, Yan Zhang, ChangZhen Hu,
HongYu Ren, and ZhiQiang Li[✉]

School of Software, Beijing Institute of Technology, Beijing 100081, China
lizq@bit.edu.cn

Abstract. Structural analysis represented by FSMDiff algorithm is the main measuring approach for existing software behavior model which is based on finite state automata. This method just focus on the data structure of finite state automata as figure characteristics, however, as software behavior model, it is more important for finite state automaton to reflect the characteristics of software behavior. So we need to find out a method to distinguish the importance in the finite state automata between different state nodes. This paper shows how the output of the FSMDiff algorithm can provide a quantified expression of structural difference between two models. According to this, we also introduce the language-set analysis, which uses the depth-first traversal algorithm to solve the language set of finite state automata. Above all, we propose a new strategy of assigning weights for the local elements of software behavior model, which can fusion assigning weight results and structural analysis for evaluation of software behavioral models. Experiment results demonstrate the effectiveness and feasibility of software behavioral model measuring approach of combining structural analysis and language set, and laid the foundation for constructing evaluation system of software behavior model inference technology.

Keywords: Software behavior model · Finite state automata · Structural analysis · Language-set · FSMDiff algorithm

1 Introduction

The detection approach based on software behavior is a dynamic approach towards software running process. It forms the node sequence by tracking the state nodes which are produced during the software execution to describe the software behavior and detect behavior through establishing the knowledge base or building models [1]. Generally, software behavior inference technology for a specific type of software has generality, so software behavior model inference technology has local generality. How to evaluate the performance and accuracy of software behavior model inference technology is a topic which is worth studying. To measure the performance and accuracy of software behavior model inference technology, we usually compare it with the target software or the standard behavior model of program [2]. However, there is very little software or program will be able to make the standard behavior models

© Springer International Publishing Switzerland 2015
M. Yung et al. (Eds.): INTRUST 2014, LNCS 9473, pp. 137–150, 2015.
DOI: 10.1007/978-3-319-27998-5_9

during the development stage, the software behavior model in the process of development and testing is not required for developer and tester. So, to evaluate the software behavior model inference technology, we need to establish the test set first, the test set should contain the typical software and program set, and provide the standard behavior models of these software and program [3]. On the basic of this we establish a series of evaluation standard which is used to test the performance of software behavior model inference technology. The main purpose of this paper is to provide an automatic approach to compare target behavior model and standard behavior model, and make it the basic evaluation approach of evaluating the software behavior model inference technology.

2 The Research Object and Predefine

The basic method of evaluating the software behavior model inference technology is to compare the software behavior model which is built by this inference technology with standard software behavior model. In order to make the content of research more pertinence, this paper will revolve around software behavior model which is based on finite state automata to research and test. The detail of the finite state automata is shown in [4].

The language set of finite state automata can be seen as the union of path set from the start node to each final node. From the view of software behavior model, the content of this set is all behavior of the software that is represented by the software behavior model can execute. Each language of finite state automata is corresponds to a series of identifiable operation of software from start to finish [5]. The definition of the language of finite state automata is as in [6].

We use *Precision* and *Recall* as evaluation parameters in this paper. The measurement method of *Precision* and *Recall* was originally proposed by Van Rijsbergen in order to verify the accuracy of the information retrieval technology [7]. The detail of *Precision* and *Recall* is as follows:

Precision and *Recall*: given the set REL of related elements in standard object by comparing and the set RET of related elements in the target object by comparing, then, *Precision* and *Recall* are calculated as follows:

$$Precision = \frac{|REL \cap RET|}{|RET|}, \quad Recall = \frac{|REL \cap RET|}{|REL|}$$

Precision and *Recall* are generally used for comparing two similar objects, and let the two objects that to be compared become standard and target object separately, this pair of parameters will show the difference between target object and standard object. Generally speaking, if we see the target objects as a result, *Precision* shows that the accuracy of the result, *Recall* shows that the coverage of the correct parts of the result relative to the theoretically all correct result, that is the integrity.

3 Structure Comparison Algorithm of Software Behavior Model

3.1 The FSMDiff Algorithm

The FSMDiff algorithm identifies a selection of initial "landmark", and uses these to infer the difference between two finite state automatons. The "landmarks" are referred to as "key pair". The details of the algorithm are as follows [7]:

(1) Input of the algorithm are two finite state automatons named FSA_A and FSA_B, the first threshold t that is used to identify landmark pairs, the second one m, and the attenuation value k.

(2) The intermediate data includes the set of key-pairs $KPairs$, the structure to record the similarity score $PairsToScores$ and to store the union of the surrounding state pair of the key pair set $NPairs$.

(3) The function $computeScores()$ is used to compute the similarity score of all the state pair combination of two FSAs, and save the result in the structure to record the similarity score $PairsToScores$.

(4) The function $identifyLandmarks()$ selects key pair by two thresholds in $PairsToScores$ and add to the set.

(5) After step (2) to step (4), if the key pair set is empty, initial state of the two FSAs will be paired the key pair and added to the key pair set.

(6) Compute the union of the surrounding state pair set of all the key pair in the key pair set, and use the union and the key pair set to make a subtraction, if the result is not empty, then add a pair of key pair to the key pair set once a time in descending order based on the structure to record the similarity score $PairsToS-cores$, repeat this process until the result is an empty set, that is the union we get is the subset of the key pair set, now the content of the key pair set is the similar part of the two FSAs.

(7) Finally, add the states which in the target FSA but not in the key pair set to the $Added$ set, and add the states which in the standard FSA but not in the key pair set to the $Removed$ set. The $Added$ set and $Removed$ set are the concrete manifestation of the structural difference of the two FSAs, and the algorithm return $Added$ set and $Removed$ set as the output.

3.2 Computing *Precision* and *Recall*

FSMDiff returns the structural difference between two FSAs. For certain tasks, it is necessary to quantify the result which is in the form of set. We use *Precision* and *Recall* parameters for quantified expression. Combining the output of FSMDiff algorithm, assume that in the input of the algorithm, FSA_A is the standard FSA and FSA_B is the target FSA.

Let $REL = \Delta_A$ and $RET = \Delta_B$, where REL is the transition set of FSA_A and RET is the transition set of FSA_B; Or $REL = Q_A$ and $RET = Q_B$, where REL is the state set of FSA_A and RET is the state set of FSA_B. Then, the intersection of REL and RET is computed with the help of FSMDiff, and can be defined as follows [7]:

$$\text{REL} \cap \text{RET} = \{\delta \in (\Delta_A \cup \Delta_B) | \delta \ni (\text{Added} \cup \text{Removed})\}$$

$$\text{REL} \cap \text{RET} = \{q \in (Q_A \cup Q_B) | q \ni (\text{Added} \cup \text{Removed})\}$$

The value of |RET| and |REL| are the basic attributes of FSA, that is the number of states or transitions, and it is easy to solve.

3.3 Thinking of Structured Analysis Results

The structural analysis just focus on the data structure of finite state automata as figure characteristics, however the finite state automata as a software behavior model is not just a figure. As software behavior model, it is more important for finite state automata to reflect the characteristics of software behavior. Each node of software behavior model represents the different stages and states of a software execution whose significance and importance are different.

It is necessary to distinguish the two FSAs in the view of figure, but this is not perfect. Therefore, it is needed to find a method or strategy to distinguish the different meaning and importance of different states or transitions in their own finite state automata. For this purpose, this paper introduces the language set of finite state automata to distinguish the different meaning and importance of different states or transitions in their own finite state automata by solving the language set of finite state automata.

4 To Solve the Language Set of Software Behavior Model

4.1 The Introduction of Language Set

Due to *Precision* shows the accuracy of the compared result and *Recall* shows the integrity of the result, this pair of parameters got above is accurate for structural analysis result, but it is hard to really show the differences of function between software behavior models.

Therefore, it is necessary to assign weights for the local elements of FSA, we can certainly know the frequency of a state or transition appears in the language set by the language set of FSA, and then make sure the importance of the state node or transition. In this paper, the strategy we choose is based on solving the language set of the finite state automata, and count the frequency of various local elements appear in the language set, then calculate the score of the local elements of the finite state automata.

4.2 Assign Weights for the Local Elements of FSA

In this paper, we use depth-first traversal to solve the language set of FSA, and then solve the weights of each state and transition of FSA using the set. As shown in Fig. 1, we will introduce how to use the language set of FSA to assign weights for the state nodes and transitions respectively and analyze the results.

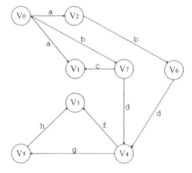

Fig. 1. The example of FSA.

4.2.1 Assign Weights for State Nodes

Assume that Fig. 1 corresponds to a FSA, and its initial node is V_0, termination nodes are V_1 and V_3, the steps of calculating the weight of each nodes of the FSA are as follows:

(1) Using the depth-first traversal to search the path set from V_0 to V_1 and from V_0 to V_3 respectively:

$$\text{PATH}(V_0, V_1) = \{(V_0, V_1), (V_0, V_7, V_1)\};$$
$$\text{PATH}(V_0, V_3) = \{(V_0, V_7, V_4, V_3), (V_0, V_7, V_4, V_5, V_3),$$
$$(V_0, V_2, V_6, V_4, V_3), (V_0, V_2, V_6, V_4, V_5, V_3)\}.$$

(2) Count the times of each state node appears in the language set:

$$T(V_0) = 6; T(V_1) = 2; T(V_2) = 2; T(V_3) = 4;$$
$$T(V_4) = 4; T(V_5) = 2; T(V_6) = 2; T(V_7) = 3.$$

(3) The sum of all the states appear in the language set is 25 times, thus, the weight of each state is as follows:

$$W(V_0) = 6/25; W(V_1) = 2/25; W(V_2) = 2/25;$$
$$W(V_3) = 4/25; W(V_4) = 4/25; W(V_5) = 2/25;$$
$$W(V_6) = 2/25; W(V_7) = 3/25.$$

4.2.2 Assign Weights for Transition

In a similar way, assume that Fig. 1 corresponds to a FSA, and its initial node is V_0, termination nodes are V_1 and V_3, the steps of calculating the weight of each transition of the FSA are as follows:

(1) First of all, define the structure of transition of the FSA as follows: (V_S, x, V_E), where V_S is the source state of the transition, V_E is the target state of the transition, x is the transition condition. Change the form of language set into transition sequence, the result is as follows:

$$PATH(V_0, V_1) = \{\{(V_0, a, V_1)\}, \{(V_0, b, V_7), (V_7, c, V_1)\}\};$$
$$PATH(V_0, V_3) = \{\{(V_0, b, V_7), (V_7, d, V_4), (V_4, f, V_3)\},$$
$$\{(V_0, b, V_7), (V_7, d, V_4), (V_4, g, V_5), (V_5, h, V_3)\},$$
$$\{(V_0, a, V_2), (V_2, b, V_6), (V_6, d, V_4), (V_4, f, V_3)\},$$
$$\{(V_0, a, V_2), (V_2, b, V_6), (V_6, d, V_4), (V_4, g, V_5), (V_5, h, V_3)\}\}.$$

(2) Count the times of each transition appears:

$$T(V_0, a, V_1) = 1; T(V_0, a, V_2) = 2; T(V_0, b, V_7) = 3;$$
$$T(V_2, b, V_6) = 2; T(V_4, g, V_5) = 2; T(V_4, f, V_3) = 2;$$
$$T(V_5, h, V_3) = 2; T(V_6, d, V_4) = 2;$$
$$T(V_7, c, V_1) = 1; T(V_7, d, V_4) = 2.$$

(3) The sum of all the transitions appear in the language set is 19 times, the weight of each transition is as follows:

$$T(V_0, a, V_1) = 1/19; T(V_0, a, V_2) = 2/19; T(V_0, b, V_7) = 3/19;$$
$$T(V_2, b, V_6) = 2/19; T(V_4, g, V_5) = 2/19; T(V_4, f, V_3) = 2/19;$$
$$T(V_5, h, V_3) = 2/19; T(V_6, d, V_4) = 2/19;$$
$$T(V_7, c, V_1) = 1/19; T(V_7, d, V_4) = 2/19.$$

4.2.3 Discussion of the Ways to Assign Weights

Above, we use the language set of FSA to assign weights for the state nodes and transitions respectively, the two ways are both assign weights for the local minimum elements of the FSA, it can be combined with the compared result of the topological structure of the FSA that is the output result (*Added, Removed*) of FSMDiff algorithm.

5 Measuring Approach of Combining Structural Analysis and Language Set

In this section, we put forward a weighted topological analysis approach which combining the structural analysis result of FSA and the analysis result of language set and apply it to the comparison of software behavior models.

5.1 Predefine to the Analysis Result of Language Set

Now give the formulation of weighting state nodes and transitions. Assume that:

$$\sum S = \{Q_{s1}, Q_{s2}, Q_{s3}, \ldots\}; \sum T = \{Q_{t1}, Q_{t2}, Q_{t3}, \ldots\}$$
$$\Delta T = \{P_{t1}, P_{t2}, P_{t3}, \ldots\}; \Delta S = \{P_{s1}, P_{s2}, P_{s3}, \ldots\}$$

Besides, the results based on the language set analysis are as follows:

$$REL_{Lang} = \{l_1, l_2, l_3, \ldots\}; RET_{Lang} = \{t_1, t_2, t_3, \ldots\}$$

Among them, $\sum T$ is the state set of the target FSA, and $\sum S$ is the state set of the standard FSA, ΔT is the transition set of the target FSA, and ΔS is the transition set of the standard FSA, $l_1, l_2, l_3, \ldots$ and $t_1, t_2, t_3, \ldots$ are the language of the standard finite state automata RET and the target finite state automata REL which get from the depth-first traversal algorithm, and it is consist of the state sequence or the transition sequence of the FSA.

Let the weights of the state nodes $W_{Qt1}, W_{Qt2}, W_{Qt3}, \ldots$ and $W_{Qs1}, W_{Qs2}, W_{Qs3}, \ldots$ express the weights of state nodes $Q_{t1}, Q_{t2}, Q_{t3}, \ldots$ and $Q_{s1}, Q_{s2}, Q_{s3}, \ldots$ in $\sum T$ and $\sum S$ respectively, its calculation method is to count the times of each state appears in the language set, for example, marking the state nodes of the FSA,

$$W_{Qt_x} = \frac{|\{l_i | Qt_x \in l_i\}|}{\sum |l_i|}, \quad i = 1, 2, \ldots, |REL_{Lang}|$$

Where $|l_i|$ expresses the number of state nodes included in the state l_i, then the expression means the radio of the number of states which contains Q_{tx} in the set REL_{Lang} and the sum of all the states appear in the language set.

Similarly, the calculation formula of the transition weights of the standard FSA is as follow:

$$W_{Pt_x} = \frac{|\{l_i | Pt_x \in l_i\}|}{\sum |l_i|}, \quad i = 1, 2, \ldots, |REL_{Lang}|$$

5.2 Predefine to the Result of Structural Analysis

Obtaining the weighting formula of the state nodes and transitions of the FSA, in order to calculate the distinction parameters *Precision* and *Recall* of the software behavior model, for example, the method of assignment based on states, let,

$$RET_{Top} = \sum T = \{Q_{t1}, Q_{t2}, Q_{t3}, \ldots\}$$
$$REL_{Top} = \sum S = \{Q_{s1}, Q_{s2}, Q_{s3}, \ldots\}$$
$$REL_{Top} \cap RET_{Top} = \{Q_1, Q_2, Q_3, \ldots\}$$

Where $REL_{Top} \cap RET_{Top}$ is the key pair set, and each element shows a pair of key pair which equivalent to each other. Then, the analysis results of FSMDiff algorithm are as follows:

$$Added = RET_{Top} - REL_{Top} \cap RET_{Top}$$
$$Removed = REL_{Top} - REL_{Top} \cap RET_{Top}$$

Besides, the Q_i of $REL_{Top} \cap RET_{Top}$, its weight is W_{LQi} in REL_{Lang} while its weight is W_{TQi} in RET_{Lang}.

5.3   The Extended *Precision* and *Recall*

To build the evaluation method of extended *Precision* and *Recall* is mainly combining the weights of different states and transitions based on the structural analysis of *Precision* and *Recall*.

Based on the method of state weight, the definition of evaluation parameters extended *Precision* and *Recall* are as follows:

$$Precision(Q) = \frac{\sum WT_{Qi}}{\sum W_{Qt_j}}, \quad (i = 1, 2, \ldots, |REL_{Top} \cap RET_{Top}|; \quad j = 1, 2, \ldots, |RET_{Top}|)$$

Where,

$$Q_i \in REL_{Top} \cap RET_{Top}, \quad Qt_j \in RET_{Top}$$

$$Recall(Q) = \frac{\sum WL_{Qi}}{\sum W_{Ql_j}}, \quad (i = 1, 2, \ldots, |REL_{Top} \cap RET_{Top}|; \quad j = 1, 2, \ldots, |REL_{Top}|)$$

Where,

$$Q_i \in REL_{Top} \cap RET_{Top}, \quad Ql_j \in REL_{Top}$$

Based on the method of transition weight, the definition of evaluation parameters extended *Precision* and *Recall* are as follows:

$$Precision(P) = \frac{\sum WT_{Pi}}{\sum W_{Pt_j}}, \quad (i = 1, 2, \ldots, |REL_{Top} \cap RET_{Top}|; \quad j = 1, 2, \ldots, |RET_{Top}|)$$

Where,

$$P_i \in REL_{Top} \cap RET_{Top}, \quad Pt_j \in RET_{Top}$$

$$Recall(P) = \frac{\sum WL_{Pi}}{\sum W_{Pl_j}}, \quad (i = 1, 2, \ldots, |REL_{Top} \cap RET_{Top}|; \quad j = 1, 2, \ldots, |REL_{Top}|)$$

Where,

$$P_i \in REL_{Top} \cap RET_{Top}, \quad Pl_j \in REL_{Top}$$

6   Experiments and Data Analysis

To verify the effectiveness of the software behavior model measuring approach of combining structural analysis and language set, this paper has carried on the experimental analysis. The experimental scheme is applying different software behavior model inference technology to the same software or program, which has a standard FSA and process tracking log file provided by the developers. The experiment will get two types of score, which are obtained by topology analysis method and analysis method of combining topology analysis and language set respectively. The performance of different evaluation methods are analyzed according to the difference of the two scores.

6.1   Experiment Object

As shown in Fig. 2, in the experiment, the software program of building software behavior model is a CVS client model. The model was built by Lo et al. [9] through a tracking results sample of the program. The model describes clearly the software behavior implementation of the CVS client, including software initialization, connect to the server, log on, search、delete、update of the version file and other basic operations, logout, disconnecting and other acts. In this paper, the software behavior model is named as FSA_S.

The evaluation objects of the experiment are Markov inference technique and EDSM inference technique. Markov inference technique was proposed by Cook and Wolf [10] to conclude the event-based software process model. And EDSM was originated from an inference algorithm in the field of grammar inference [11].

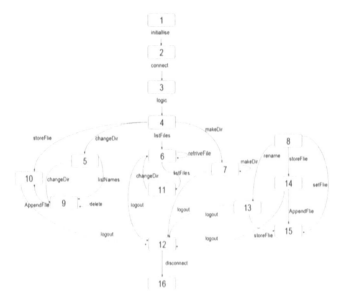

Fig. 2. Standard software behavior model.

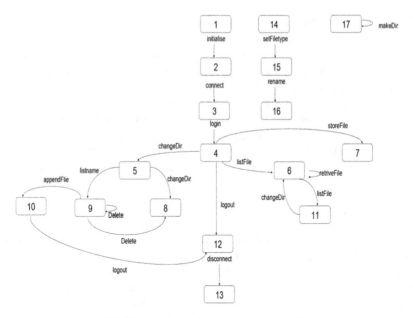

Fig. 3. Markov software behavior model.

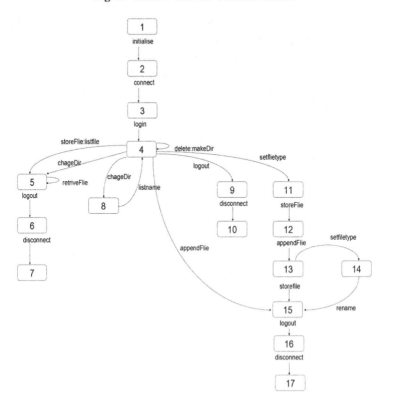

Fig. 4. EDSM software behavior model.

Neil et al. [7] use the Markov and EDSM inference techniques respectively to build the software behavior model of CVS procedure tracking samples published by Lo et al. [9].

The behavior model built by Markov inference technique is shown in Fig. 3, and it is named as FSA_{Markov}. The software behavior model built by EDSM inference technique is shown in Fig. 4, and it is named as FSA_{EDSM}.

6.2   Experiment Content

In order to compare the Markov and EDSM inference techniques, we need to compare and analysis the software behavior models built by the two inference techniques with the standard software behavior model respectively. The experiment objects are divided into two groups, the first group are FSA_{Markov} and FSA_S, the second one are FSA_{EDSM} and FSA_S. Carry out the following content respectively aim at the two groups of experiment objects:

(1) Solve the structural difference set of the target automata compared with the standard automata using FSMDiff algorithm and calculate the values of *Precision* and *Recall*.
(2) Calculate the language set of target automata and standard automata respectively using depth-first traversal, and calculate the weight of each state and transition in its own language set.
(3) Combining the structural difference set of target automata and standard automata and calculating result of local elements weight, calculate the value of before and after extended *Precision* and *Recall* based on state and transition respectively.

6.3   Experiment Data and Analysis

The experiment result is shown in Table 1, *Top_Precision* and *Top_Recall* express *Precision* and *Recall* of before extended, *Ex_Precision* and *Ex_Recall* express *Precision* and *Recall* of after extended. Q shows the result based on state, P shows the result based on transition.

Table 1. Comparison results of the two group software behavior models

	Top_Precision	Top_Recall	Ex_Precision	Ex_Recall
Markov: Q	0.65	0.69	0.72	0.55
Markov: P	0.76	0.59	0.78	0.54
EDSM: Q	0.65	0.69	0.52	0.49
EDSM: P	0.50	0.41	0.47	0.38

From the experiment result and data integration, we can get three types of comparison:

(1) Compare the data obtained by parameters calculation method before and after using extended *Precision* and *Recall*. Such as the two groups of data: $Ex\_Precision_{Markov}(Q)$, $Ex\_Recall_{Markov}(Q)$ and $Top\_Precision_{Markov}(Q)$, $Top\_Recall_{Markov}(Q)$, they are both the quantified expression of structural differences in the form of state node between Markov model and standard model, the difference is the first one refers to the language set of the software behavior model and the second one is only based on the results of the structural analysis, from Fig. 3 we can find that as target automata, Markov model misses the connectivity features relative to the standard automata. Break software into three FSAs to describe which could be described as one model, makes the function on the right side of Fig. 2 break away from the overall function and exist alone. The function of state nodes of Markov model relative to the standard model will be completed by the increased state nodes. Only in the views of structure based on state, the number of missing nodes and increased nodes is small, therefore, using the calculation formula of before extended *Precision* and *Recall*, we get the score of (0.65, 0.69). However, in the view of function, the increased state nodes of Markov model the same as matching nodes complete some functions of the standard model, we can consider that the function set of the model is close to the proper subset of the standard model, that is without considering the integrity of result, the accuracy of inference technique to build the model is very high. But the missing part of Markov model relative to the standard model is able to form a functional area, in this area the logic is complex and can complete many complex functions. Missing this area reduces seriously the integrity of Markov inference technique result. Combing the score of language set analysis after extended change from (0.65, 0.69) to (0.72, 0.55) is correspond to above.

(2) Compare the data obtained by the compared method based on different local elements. Such as the two groups of data: $Ex\_Precision_{Markov}(P)$, $Ex\_Recall_{Markov}(P)$ and $Ex\_Precision_{Markov}(Q)$, $Ex\_Recall_{Markov}(Q)$, they are both the quantified result of the structural differences of Markov model and the standard model using the extended *Precision* and *Recall*. The difference is, the structural analysis of first one is based on the transition and the second one is based on state node. For a finite state automata, with the growth of its size and complexity, the number of transition increases much faster than the number of state. As the Markov model is decomposed into three small FSAs, the ratio of the number of transition and the number of state is 1.24(21:17), if only compute the biggest one, the ratio is 1.38(18:13); EDSM model is mainly linear structure, the logic is simple, the ratio is 1.29(22:17). The larger scale of the compared elements, it will easier to reduce the contingency and error, we can find in Table 1 that pure structural analysis results based on state node is much lower than other similar results, 0.65 in Markov model compared to (0.72, 0.76, 0.78), 0.69 compared to (0.55, 0.59, 0.54); 0.65 in EDSM model compared to (0.52, 0.50, 0.47), 0.69 compared to (0.49, 0.41, 0.38). Based on this, when choosing the local elements for structure comparison, we should choose the larger scale one first.

(3) Compare the data of differences of software behavior model built by different inference techniques. Through the analysis of the above two kinds of comparison, we can see that the *Precision* and *Recall* calculation method of after extended is

better than the one of before extended, and based on the larger scale of local elements is a better comparison method. Therefore, in order to compare the Markov inference technique and EDSM inference technique, we choose $Ex\_Precision_{Markov}(P)$, $Ex\_Recall_{Markov}(P)$ and $Ex\_Precision_{EDSM}(P)$, $Ex\_Recall_{EDSM}(P)$ as the main basis of evaluation, the data is obtained by analyzing the transitions of the larger model, and using the calculation method of the extended *Precision* and *Recall* at the same time. The score of Markov model is (0.78, 0.54), compared with the score of EDSM model (0.47, 0.38), the two values are both higher, so, from the experiment result, we can think that Markov model is more "like" the standard model than the EDSM model, Markov inference technique is better for the software behavior model built by the CVS client.

How to understand the result? Is the performance of EDSM inference technique is lower than the performance of Markov inference technique? The experiment object is essentially the tracking samples of CVS client software. Firstly, the single sample has contingency, we cannot get the conclusion from the experiment result by one time. Secondly, the form of software is various, and the feature points of software behavior are also quite different. Intention of this paper and the final goal of the experiment are not to compare Markov inference technique and EDSM inference technique, but to verify the feasibility of software behavior model measuring approach based on finite state automata proposed in this paper. Based on the measuring approach, typical software behavior model library can be established, by providing a large number of standard software behavior model samples and tracking file samples to verify and compare the performance of various kinds of software behavior model inference technique.

7   Conclusion

Software behavior model which is based on finite state automata is an important model of software security detection. In this paper, we propose a software behavior model measuring approach of combining structural analysis and language set. By comparing the target software behavior model with the standard software behavior model, we can get the quantified criteria of the similarity of this model and the standard model. Meanwhile, based on the quantified result, apply this compared method to compare the model which is built by other inference techniques with the same standard model, we can evaluate the performance of other inference techniques further. To highlight different local elements of software behavior model based on finite state automata have different importance, based on the structural analysis of the finite state automata, we also analyzed the value of traversal algorithm in the software behavior model, and applied this algorithm to the strategy of obtaining the state weights and transition weights of software behavior model. In this paper, we broke through the pattern of analyzing and comparing software behavior model from a single view, on the basis of the structural analysis, merged into the analysis of language set, combined the advantage of the two methods to analyze and compare the software behavior model comprehensively, and experimented with the comparison of the software behavior model based on finite state automata. The results show that software behavior model measuring approach of combining structural analysis and language set is effective and feasible.

Acknowledgment. This work was supported by the Key Project of National Defense Basic Research Program of China (Grant No. B1120132031) and the Ph.D. Programs Foundation of Ministry of Education of China (Grant No. 20131101120043).

References

1. Wang, X.Z., Sun, L.C., Lu, Y.L.: Intrusion detection approach towards software behavior trustworthiness. J. Univ. Sci. Technol. China **41**(7), 626–635 (2011)
2. Peng, G.J., Tao, F., Zhang, H.G.: Research on theory model of software dynamic trustiness based on behavior integrity. In: Proceedings of the 2009 International Conference on Multimedia Information Networking and Security, pp. 130–134. Washington, DC, USA (2009)
3. Godefroid, E., Levin, M.Y., Molnar, D.: Automated whitebox fuzz testing. In: Proceedings of the 16th Annual Network & Distributed System Security Symposium, pp. 1–10. San Diego, USA (2008)
4. Borger, E.: Abstract state machines and high-level system design and analysis. Theor. Comput. Sci. **336**(2), 205–207 (2005)
5. Quante, J., Koschke, R.: Dynamic protocol recovery. In: Proceedings of the 14th International Working Conference on Reverse Engineering, pp. 219–228. Vancouver, Canada (2007)
6. Hopcroft, J., Motwani, R., Ullman, J.: Introduction to automata theory, languages and computation, 3rd edn. Addison-Wesley, New Jersey (2007)
7. Walkinshaw, N., Bogdanov, K.: Comparing software behavior models. Technical report: CS-08-16, The University of Sheffield, Sheffield, UK (2008)
8. Walkinshaw, N., Bogdanov, K., Johnson, K.: Evaluation and comparison of inferred regular grammars. In: Clark, A., Coste, F., Miclet, L. (eds.) ICGI 2008. LNCS (LNAI), vol. 5278, pp. 252–265. Springer, Heidelberg (2008)
9. Lo, D., Khoo, S.: QUARK: empirical assessment of automaton-based specification miners. In: Proceedings of the 13th Working Conference on Reverse Engineering, pp. 51–60. Benevento, Italy (2006)
10. Frossi, A., Maggi, F., Rizzo, G.L., Zanero, S.: Selecting and improving system call models for anomaly detection. In: Flegel, U., Bruschi, D. (eds.) DIMVA 2009. LNCS, vol. 5587, pp. 206–223. Springer, Heidelberg (2009)
11. Lang, K.J., Pearlmutter, B.A., Price, R.A.: In: Honavar, V.G., Slutzki, G. (eds.) ICGI 1998. LNCS (LNAI), vol. 1433, p. 1. Springer, Heidelberg (1998)

On Cache Timing Attacks Considering Multi-core Aspects in Virtualized Embedded Systems

Michael Weiß[1]([✉]), Benjamin Weggenmann[1], Moritz August[2], and Georg Sigl[2]

[1] Fraunhofer Institut AISEC, Garching, Germany
{michael.weiss,benjamin.weggenmann}@aisec.fraunhofer.de
[2] Technische Universität München, Munich, Germany
{moritz.august,sigl}@tum.de

Abstract. Virtualization has become one of the most important security enhancing techniques for embedded systems during the last years, both for mobile devices and cyber-physical system (CPS). One of the major security threats in this context is posed by side channel attacks. In this work, Bernstein's time-driven cache-based attack against AES is revisited in a virtualization scenario based on an actual CPS using the PikeOS microkernel virtualization framework. The attack is conducted in the context of the implemented virtualization scenario using different scheduler configurations. We provide experimental results which show that using dedicated cores for crypto routines will have a high impact on the vulnerability of such systems. We also compare the results to previous work in that field and our visualization directly shows the differences between cache architectures of the ARM Cortex-A8 and Cortex-A9. Further, a non-invasive countermeasure against timing attacks based on the scheduler of PikeOS is devised, which in fact increases the system's security against cache timing attacks.

Keywords: Cyber-physical system (CPS) · Virtualization · Trusted execution environment · Microkernel · AES · Cache timing · Embedded systems

1 Introduction

Former single-core real-time embedded systems used in the avionics and automotive industry are evolving to integrated ARM-based multi-core virtualized platforms, nowadays denoted as cyber-physical systems (CPSs). To save weight and costs of airplanes and vehicles, such systems run several user controlled applications beside security and safety critical applications side by side on the same physical system. Consider in-flight entertainment systems which provide users with the ability to connect their own untrusted devices, e.g., smart phones

Parts of this contribution were supported by the German Federal Ministry of Education and Research in the project *SIBASE* through grant number 01IS13020.

M. Yung et al. (Eds.): INTRUST 2014, LNCS 9473, pp. 151–167, 2015.
DOI: 10.1007/978-3-319-27998-5_10

and tablets. However, those systems also provide flight information which needs a connection to safety critical systems. That is why for instance the ARNIC-653 standard [3] demands for strict isolation and real-time constraints by statically configured partitions. A widely used real-time operating system framework in the avionics industry which provides partition separation according to ARNIC-653 is PikeOS [10]. It elaborates a microkernel and a user-space abstraction layer for this purpose.

However, none of those real-time operating systems have been examined for vulnerabilities to cache timing side channels circumventing the partition isolation considering influences of multi-core. Previous work mainly focuses on x86 systems in shared cloud scenarios.

Cache-based side channel attacks make use of a simple model to correlate the execution time of an algorithm with the state of the cache used by the CPU in charge. It is assumed that the execution time is lower if the data needed by the algorithm is already stored in a cache line (*cache-hit*). On the other hand, if the required data is not present in the cache and hence has to be loaded from the main memory (*cache-miss*), this will result in a longer execution time. This model is simple, but reasonable and only relies on the cache architecture of the CPU. In [24], we provide a suitable attack scenario, however only on a single-core system using an academic real-time framework, focusing on mobile phone devices. We use this as base for our research on how actual CPSs running in the cockpits and cabins of airplanes are vulnerable to cache-based timing side channel attacks.

Our main contributions are:

1. We adapted the virtualization-based attack scenario from [24] to a multi-core embedded system using the microkernel-based operating system framework PikeOS which has high relevance in avionic and also automotive industry.
2. We propose the discrete-time countermeasure which is based only on real-time configuration of the PikeOS scheduler as a drop-in update to existing systems and compare it to related approaches.
3. We elaborate different multi-core scheduler configurations and evaluate their vulnerability against time-driven cache attacks.
4. By comparing the attack results between single- and multi-core configurations, we are able to show that dedicated cores for crypto services leak the most information about the key.
5. Further, we compare our Cortex-A9-based setup against the Cortex-A8-based setup of [24], which leads to interesting patterns of key space reduction directly showing differences of the underlying cache architecture.

The rest of the paper is structured as follows: We provide background on cache based side-channels and related work in Sect. 2. In Sect. 3, the system architecture and attack scenario including the attacker model is described. We provide some more detailed background knowledge about the PikeOS scheduler in Sect. 4, before we describe the *discrete-time* countermeasure in Sect. 5. Experimental results of the attack performance under different scheduler configurations are evaluated in Sect. 6. Finally, the work is concluded in Sect. 7.

2   Background and Related Work

Cache-based attacks can be divided into three different categories, each having a different attacker model. *Time-driven* attacks [2,5,6,15,16] make use of the cache model in a very general way as they only require timing data of entire runs of a cryptographic algorithm, e.g., an encryption using AES. This corresponds to an attacker who has only very limited or coarse information about the cache. *Trace-driven* attacks [1,7] additionally require detailed information about the cache activity during single runs of the encryption, in particular the sequence of cache hits and misses caused by the memory accesses performed by the encryption algorithm. A trace can for instance be captured by profiling the power consumption while the encryption routine is running. This translates to an attacker, who has gained a substantial level of knowledge about the runtime cache behavior which in case of a power profile also requires physical access to the device. Finally, *access-driven* attacks [9,16] assume to have knowledge about the cache-sets accessed by the algorithm. The underlying assumption is therefore that the attacker can control the cache runtime behavior. In the *Prime+Probe* attack [16], for example, those areas of the cache that also hold the lookup tables of the attacked algorithm are filled by a spy process with own data before the encryption is triggered (*Prime*). After the encryption, the spy process measures the access time to its own data to see which parts have been evicted from the cache by the encryption algorithm (*Probe*). Now the attacker can deduce which parts of the lookup tables were accessed by the encryption and from this infer some or all bits of the secret key. As can be seen from the above explanations, time-driven attacks are the most widely applicable class of attacks since they do not require a strong attacker with fine grained access to the cache.

In [5], Bernstein proposes a cache-based timing attack to recover the secret key of an AES encryption on a remote server. Bernstein's paper contained no thorough analysis of the attack and no explanation why the attack is successful. Neve et al. fill this gap in [15] by presenting a full analysis of Bernstein's attack methodology and explaining the correlation model. They argue that Bernstein's original technique cannot be used easily as a real remote-only attack where timings need to be measured by the attacker. Moreover, they improve Bernstein's attack by also considering second round information and thus lowering the number of required samples. To get accurate timings, Bernstein avoided the noisy network channel between the attacked server and the attacker by measuring the encryption time directly on the server, which is a rather unrealistic scenario since the server needs to be modified. In virtualization environments, however, the noise is negligible since local communication channels with only a small and almost constant timing overhead are used, as shown in [24].

Ristenpart et al. [17] consider side-channel leakage in virtualization environments on the example of the Amazon EC2 cloud service. They show that there is cross virtual machine (VM) side-channel leakage. They used the access-driven *Prime+Probe* technique from [16] for analyzing the timing side-channel. However, Ristenpart et al. are not able to extract a secret encryption key from one VM. In [24], we considered a virtualization-based system where the trusted environment runs an AES server. Under the assumption that the untrusted environment could be hijacked by an attacker, we showed that a man-in-the-middle attack via an adapted version of the cache-timing attack by Bernstein [5] is generally able to significantly reduce the key space, thus making brute-force attacks feasible. The impact of noise under realistic workloads is examined by Spreitzer and Plos [19], who evaluate time-driven attacks on conventional mobile devices (ARM Cortex-A8 and A9). Unlike our approach, they consider noise induced by the Android operating system and applications running simultaneously on the device. However, they do so using a slightly unrealistic attacker model where the attacker captures timings in the very same process where the AES encryption routine is implemented and called, which likely reduces the effects of the OS and concurrent processes.

There are several ways to defend against time-driven cache timing attacks: One option is to switch to hardware-based implementations as provided by some processor manufacturers, e.g. Intel with its AES-NI instruction set [8], thus entirely avoiding cache-based attacks against the algorithm. If no hardware support is available, it is possible to change the implementation of the algorithm itself and get rid of the table lookups. While earlier software-based suggestions [13,14] were generally slow compared to table-based implementations, Kasper et al. [11] present an efficient constant-time implementation based on bit-slicing that is suitable for stream and packet encryption.

Kim et al. [12] present a novel countermeasure against cache-based side channel attacks in a virtualization environment called STEALTHMEM. This countermeasure works at hypervisor level by assigning dedicated cache lines to each CPU in a group of CPUs with shared L3 cache. These so-called stealth cache lines are never evicted; therefore, sensitive data, such as S-boxes in AES, can be stored in these cache lines without introducing cache or timing side channels for an attack. Stefan et al. [20] propose instruction-based scheduling to prevent cache-based timing attacks on a single CPU. Instead of having a fixed amount of time, a process has a fixed amount of instructions it can execute before the next process is scheduled. The authors examine a simple timing attack and show that this attack is prevented by the proposed scheduler with negligible increase in the size of binaries and execution time. These countermeasures require considerable changes to the hardware, the hypervisor, or the cryptographic algorithms, whereas neither of which is necessary for our approach. Lately, Varadarajan et al. [21] have proposed a similar approach to our *discrete-time* scheduler scheme for cloud systems which they call soft-isolation. In contrast to our approach for real-time based schedulers, their approach relies on a feature of the Xen hypervisor scheduler called minimum run time (MRT) guarantee.

3   Attack Scenario and System Architecture

We assume a Trusted Execution Environment (TEE) which separates two compartments, a trusted environment which provides crypto services and an untrusted environment which runs user applications. The secret keys used for encryption have a high security value and thus are only accessible in the trusted environment. A viable usage scenario is, e.g., to establish a VPN tunnel. The network protocol stacks of a rich operating system kernel are used in the untrusted environment while the payload is encrypted by a driver using an encryption service inside the trusted environment. Hence, the secret session keys cannot be compromised by an attacker in the rich OS.

For this work we adapted the virtualization based security architecture of [24], which is a realization of a TEE to a microkernel system, to PikeOS. PikeOS distinguishes between resource and time partitions. A resource partition in PikeOS denotes a separate address space protected by the microkernel, while time partitions are used to assign computation time to threads. The microkernel itself only implements the basic mechanism for IPC, scheduling, and separation of address spaces in privileged processor mode. Device drivers, higher level abstraction for inter-partition communication as well as virtual memory management are implemented in the user-space abstraction layer, called PikeOS System Software (PSSW). Native device drivers for secure devices can be implemented in their own partition also in user-space. Figure 1 illustrates this architecture including the attack scenario. In our scenario, the architecture comprises a rich environment which runs the untrusted user applications in one partition as well as a trusted environment that hosts the security and safety relevant trusted applications each in their own partition. Both environments are allowed to communicate with each other using protocol messages transmitted via the virtualization layer, which in our case is the PikeOS microkernel and its user-space abstraction layer PSSW. To exchange data between the trusted and untrusted applications, shared memory is used. The user applications may use the trusted applications via special device drivers integrated into the rich OS kernel.

The concrete attack scenario now assumes that an AES encryption server runs in the trusted environment. To launch an encryption, a user application

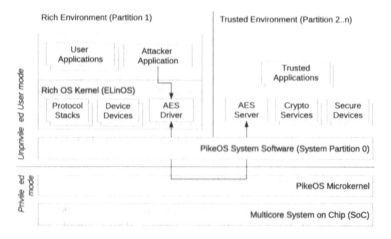

Fig. 1. Adapted virtualization based system security architecture

simply stores the plaintext in shared memory and calls the AES server through IPC. The ciphertext is then written back to the shared memory. In this scenario, an attacker has compromised the rich OS and wants to determine the key used by the AES server. As he has full access to the rich OS in the untrusted environment, he is able to launch as many encryptions as he likes with chosen plaintexts. This he could do either by hijacking running processes or deploying own code that directly uses the kernel of the rich OS. The attacker is therefore able to launch a time-driven attack as it was discussed above.

4   Scheduling in PikeOS

PikeOS features a special scheduler that uses a combination of *time-driven and priority-based scheduling* to account for the different needs of the applications. To allow for deterministic real-time responsiveness, the scheduler uses a time-driven approach. Every real-time application is statically assigned to a time slot of a defined length. The length of these time slots can vary between applications but has to stay within a certain relation to the length of the other time slots. Every application is periodically scheduled for the length of the slot it is assigned to. As every partition gets assigned a defined amount of CPU time at defined points in time, they are able to schedule real-time processes themselves. This, so far, is a standard approach for scheduling real-time applications. To also support non real-time applications, a straightforward extension of this approach is to just create a new time slot and assign all applications without timing constraints to it. Within this slot, a standard round robin scheduling scheme can be applied. However, this approach is inefficient since it wastes a lot of CPU time. The PikeOS scheduler refines this approach to a more efficient strategy. It might occur that the processes of a real-time application finish before its time slot end or that it does not have any processes to run at all. As it would harm the temporal determinism, the scheduler cannot simply switch to the next application in this situation. Rather than wasting this time, the PikeOS scheduler uses this excess CPU time to schedule applications with no real-time constraints. For this purpose, it leverages priority-based scheduling. All real-time applications are assigned the same mid-level priority number while low priority numbers are assigned to the other applications. Now, the scheduler continues to schedule the real-time applications periodically but uses the excess time to schedule the low-priority non real-time applications in a round robin fashion. In this way, no computing time is wasted and the overall amount of time needed to execute all applications decreases drastically when compared to a standard RTOS scheduler.

5   Discrete-Time Countermeasure

One main pitfall of novel countermeasures is that some of them require changes to already established systems that are too substantial to be easily implemented, hence making these countermeasures practically irrelevant. The *discrete-time* countermeasure that is presented in the following therefore aims at making cache based time-driven attacks infeasible for attackers while demanding as few changes and inducing as little overhead as possible. Assume the rich OS and the trusted environment are implemented as partitions in PikeOS and are hence handled by the scheduler. Now assume the attacker has compromised the rich OS and is able to launch the timing attack against the AES server that runs in a trusted partition. In order for the attacker to successfully carry out the attack, two conditions must be fulfilled:

1. He must be able to retrieve enough samples from the AES server, in the order of several hundred millions.
2. The samples must leak enough information for the correct hypothesis on the key to yield a higher correlation on average than all wrong hypotheses.

The *discrete-time* countermeasure aims at these two points. It works straight-forward in that both applications, the rich OS and the AES server in the trusted domain, are treated as real-time applications such that each is assigned an own time slot. Note, that it is not necessary for either of the two applications to have any real-time time constraints in order for the scheduler to be configured as described above. Using this configuration of the scheduler the time measured by the attack for one encryption t_{enc} is now given by Eq. 1.

$$t_{enc} = n \cdot t_{OS} + m \cdot t_{serv} \tag{1}$$

with t_{OS} being the length of the time slot of the rich OS and t_{serv} being the length of the time slot of the AES server. The two variables n and m represent the number of executions of the two time slots. Note that we ignore negligible timing quantities that are independent of the AES server, such as the remaining time in the slot of the rich OS after the encryption was requested and the time passing in the first slot of the rich OS after the encryption is done before the attacker's process is scheduled. As it can be easily verified the time is always a multiple of the two time slot lengths which gives rise to the countermeasure's name. This has two major effects on the attack. Firstly, as the scheduling for these two applications is strictly time-driven, the rich OS will be scheduled a number of times while still waiting for the encryption to finish and hence being idle. This will increase the time needed by an encryption in a way that, given carefully chosen values for t_{OS} and t_{serv}, a single encryption as it is needed for benign purposes can still be done without noticeable delay. However, a number of encryptions as needed for an attack will take a significantly larger amount of time. This already will make an attack time-wise more difficult. Secondly, as the information that can be gained by one sample is now very coarse-grained, there is only a very small correlation left between the timing information and occurring cache-misses or hits. This will make it very hard for the attacker to distinguish the correct key hypothesis from false ones and will increase the number of necessary samples. Therefore, the discrete-time countermeasure is a strong shield against the kind of attacks considered here. Furthermore, the countermeasure requires no change of any kind in the code and also causes arguably only little timing overhead. It is also straightforward to implement, can be extended to multiple applications and is most likely also applicable to other RTOS schedulers working in a similar manner as the PikeOS scheduler. Although not in the focus of this paper, access-driven attacks can be prevented similarly by a simple configuration in the scheduler to flush the cache when switching partitions.

6   Evaluation

To practically analyze the scenario presented in Sect. 3, we elaborated the following testbed. The untrusted runtime is implemented using the para-virtualized Linux distribution ELinOS (version 5.2) including the necessary code for the attacker to conduct the timing attack. The AES server in the trusted runtime is implemented as an application based on the native PikeOS API (version 3.3). Obviously, both applications have their own partition. To enable the communication between the two partitions, two unidirectional queuing ports and a shared memory page were set up. The rich OS and the AES server use these ports to communicate via a simple handshake protocol and use the shared page as buffer for plain- and ciphertexts. Queuing ports are unidirectional communication channels defined in the ARNIC-653 [3] standard that can be set up between two partitions statically at compile-time and then initialized at run-time by the applications.

As hardware platform, we chose the Freescale i.MX6 SabreLite board which comprises a Quad-Core ARM Cortex-A9 CPU with 1.2 GHz. The cache architecture consist of a 32 KB I- and D-Cache (L1) per Core and a 1 MB shared L2 cache. The L1 cache is 4-way associative and has a cache line size of 32 byte. For precise timing measurements, the ARM CCNT register was utilized as stated in [19, 24].

To analyze the success rate of Bernstein's timing attack, the effect of a broad range of parameters was examined. For the comparison between different values for these parameters, two criteria were used.

1. The number of different candidates for each key byte
2. The average position of the correct candidates in the ordered output lists

The first one directly gives information about how much the key space could be reduced by the attack. To quantitatively measure the effectiveness of the attack, this is therefore the best parameter. In the best case only one candidate, namely the correct one, remains for each byte and the key is hence revealed completely. But even only a significant reduction of the number of candidate bytes is already valuable to the attacker as he then can launch a brute-force attack in the reduced key space with the remaining possible values. However, this score does not use all information of the output of the attack. As the list of possible candidates for each key byte is ordered, it is interesting to know at which positions in these lists the correct values can be found. This is a measure for the ability of the attack to separate the correct hypotheses from the other remaining ones. In the best case, the correct value for each key byte always has the highest correlation and

Table 1. Summary of results for different scheduler configurations

No.	Utilized cores	Scheduler configuration	Average position	Remaining key-space
1	1	1 Core shared (Single)	3.625	$\approx 2^{80.2}$
2	4	4 Cores shared (Quad)	4.25	$\approx 2^{81.7}$
3	4	4 Cores Server, 1 core shared rich OS (Server)	4.0	$\approx 2^{82.8}$
4	2	1 Core dedicated each	4.0	2^{72}
5	4	2 Cores dedicated each	4.375	$\approx 2^{72.7}$

is therefore at first position in the list. That information is also of high interest to an attacker as he can use this information to significantly speed up his brute-force attack. Since he knows the correlation of all remaining possible byte values, he can order the possible keys by the correlation and then test for candidates with higher correlation first. This will usually require much less than the average $\frac{n}{2}$ guesses, n being the number of key candidates. Another approach to reduce brute-force complexity could be to use recently proposed key-rank estimation procedures [22,23] as shown by Spreitzer et al. [18].

For all the experiments summarized in Table 1, normal priority-based scheduling was used and the profiling and attack phase were done on the same device. This might not always be possible in a real-world setting, but was done to have an optimal setting for the evaluation. If not stated otherwise the attacked key was

$$0x21\ 53\ fc\ 73\ d4\ f3\ 4a\ 98\ 17\ 33\ bb\ 3f\ 18\ 92\ 00\ 8b$$

and both profiling and attack phase were conducted with 512 million samples to have approximately 2 million samples for each possible key candidate.

6.1 Identifying and Tuning of Attack Parameters

To reduce the noise in the measurements, Bernstein disregards all measurements above a certain threshold. In the original code, this threshold was set to a value fitting the timing behavior of his implementation. This value was therefore changed in this implementation. To evaluate the effect of this clipping, two different thresholds were investigated both with 512 million samples for attack and profiling phase. The threshold that was initially set to about 30,000 clock cycles higher than the average of the timing samples was compared to the threshold 20,000 above average. The results are displayed in Fig. 2a and Table 2a. The results clearly show that the lower threshold leads to a significant lower reduction of the key space. This implies that the timings lying in the interval between the two thresholds indeed contained information about the key. This also complies to the findings in [18], which show that the minimal timing attack of [4] does not leak any information on ARM.

One parameter that comes to mind very quickly when thinking about analyzing a side channel attack is the number of samples. One would assume that an increasing number of samples automatically results in a higher success rate

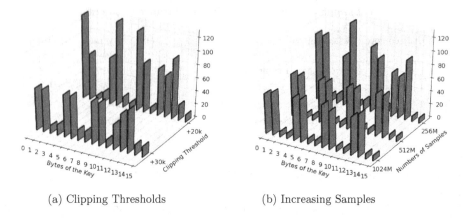

(a) Clipping Thresholds   (b) Increasing Samples

Fig. 2. Histograms describing the numbers of possible candidates for all bytes of the key and for varying clipping thresholds and samples

as the noise gets averaged out more and more, leaving only the relevant information behind. To verify this assumption, we conducted the attack with 256, 512 and 1024 million samples. The results are displayed in Fig. 2b. Table 2b shows the average position of the correct key byte candidates. As expected, increasing the number of samples does in fact also increase the success rate of the attack. However, the increase of the success rate shows a logarithmic behavior. This behavior is derived directly from the cache architecture. As only the upper k bits of a data word are used to index the cache lines, the timing behavior is independent of the lower bits. In the best case, the attack could therefore only reveal the upper k bits of each key byte. This explains the observed boundary of the reduction of the key space. It furthermore explains why the remaining number of possible values per byte is in almost all cases a power of 2. A similar behavior was described by Neve et al. [15]. This limitation only applies for aligned T-tables. In the case of disaligned T-tables, which is not the case in our setup, even more information might leak.

6.2   Single Core vs. Quad Core

The PikeOS scheduler allows the use of a CPU mask to specifically select the cores that shall run a partition. As each core has its own L1 cache but all cores share the L2 cache, it is interesting to examine how the success rate of the attack changes when only one or all cores are used. To do this, three different configurations were regarded. For the first one both partitions were run by a single shared core (configuration 1) while for the second one both partitions were run on all four cores (configuration 2). The third configuration involved the AES server running on all four cores (configuration 3) while ELinOS was assigned only one core. Note that the AES server application itself does not implement any concurrent block computation. The results are depicted in Fig. 3a and Table 1.

It can be seen that configuration 1 gave the best results for both criteria, and scenario 2 yielded the worst. This is understandable since in the first scenario,

Table 2. The average position of the correct key byte candidates for the different clipping thresholds and numbers of samples

(a)

Clipping Threshold	Average Position
+30k	3.625
+20k	4.3125

Measurements conducted with 512M samples in configuration 1, see Table 1

(b)

Number of Samples	Average Position
256M	5.5625
512M	4.0
1024M	3.875

Measurements conducted in configuration 4, see Table 1

the T-tables are stored in a single L1 cache and the L2 cache, whereas in scenario 2 the T-tables are most likely scattered over the four L1 caches and the L2 cache. This decreases the signal to noise ratio with high certainty and thus lowers the success rate. Additionally, when both the rich OS and the AES server use the same core, their cache usage will interfere which also reduces the quality of the timing samples. This effect is visible in the difference between scenarios 2 and 3. Although the AES server uses four cores in scenario 3 as well, it only interferes with the other application in one of them which leads to an overall better success rate of the attack.

Using a dedicated core for the AES server is expected to be not advisable because it reduces the noise. Therefore, it was investigated how the success rate of the attack is affected when the two partitions have one or two cores for their own in comparison to configuration 1 where both partitions share only one core. The results are shown in Fig. 3b. The use of dedicated cores leads to a significantly better success rate in terms of the total number of remaining key candidates. The setup with one dedicated core also shows a slight decrease in the average position of the correct key byte values. As it can be seen, assigning one core to each partition (configuration 4) thereby results in a slightly better attack result than using two dedicated cores (configuration 5). This can be explained by the already discussed effect of using multiple L1 caches. However, the slight increase of the average position compared to the scenarios where 4 cores are utilized seems to be caused by measurement inaccuracies. In summary, when using an ordinary priority based scheduling scheme on a multi-core system without any countermeasures, it is not recommended to use a dedicated core for the cryptographic algorithm as this would reduce the noise significantly.

6.3 Comparison to Fiasco Setup

In [24], we presented results for Bernstein's attack carried out in a very similar virtualization setting. In contrast to the hardware presented above, in [24] a platform based on a Cortex-A8 with 720 MHz was used. To implement the virtualization scenario, the Fiasco.OC microkernel together with L4Re was utilized. Note that we use the same key as in [24] to provide comparable results. In [24], we were able to reduce the byte value space of almost all bytes to 16 possibilities

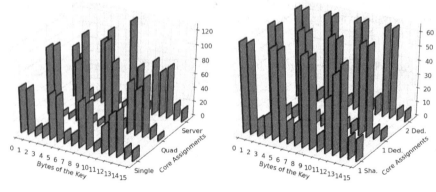

(a) Shared Scheduler Configurations   (b) Dedicated Scheduler Configurations

Fig. 3. Histograms describing the numbers of possible candidates for all bytes of the key and for varying scheduler configurations.

for the OpenSSL implementation of AES. For byte 3, no reduction was possible which seems to be an error of measurement, while bytes 7, 11 and 15 could only be reduced to 32 respectively 24 possible values. This result was achieved with 2 million samples for each byte value, translating to the overall number of 512 million samples that was also used in this work. Data about the position of the correct key byte values in the output lists was not provided.

The best result achieved in terms of the reduction of the key space in this work draws a very different picture. For one dedicated core for the ELinOS and the AES partition respectively, the highest reduction found was a reduction down to 8 possible values for the bytes 2,3,6,7,10,11,14,15. For the remaining bytes a reduction was possible only down to 64 different values. This pattern is interesting in itself as every consecutive 2-byte tuple seems to be highly correlated in the reduction capability. However, it is also very different from the result stated above. For this implementation, the maximally achieved reduction is twice as high as for the implementation in [24]. Nevertheless, only half of the bytes could be reduced that far while for the implementation in [24], nearly all byte spaces could be reduced to the respective minimum. Then again, in the implementation of this work all bytes could be reduced to at least 64 different values. This was not the case for the implementation using the Fiasco.OC kernel. Both implementations have in common that there seems to be a limit for the reduction of the key space that depends on the implementation. This was already mentioned above and is also stated in [24]. The two results are compared in Fig. 4. The difference of the reduction pattern reflects the different cache architectures in terms of the cache line size. On the Cortex-A8 with a 64 Byte cache line size every fourth key byte is harder to reduce, while on the Cortex-A9 with 32 Byte cache line size every first 2 bytes are harder to reduce. Both pattern repeat every 4 bytes, this is due to both caches are 4-way associative.

The total number of possible keys was reduced to 2^{72} for the worst PikeOS setup and to roughly 2^{70} for the Fiasco setup showing a slight advantage for

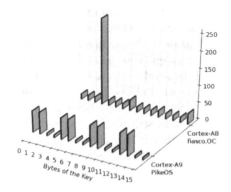

Fig. 4. Results on Cortex-A9 in this work compared to Cortex-A8 in [24]

the PikeOS setup. However, compared to the setup utilizing only one core with a reduction of key space to $\approx 2^{80.2}$, the PikeOS setup is about three orders of magnitude harder to attack.

6.4   Evaluation of the Countermeasure

To evaluate the effectiveness of the discrete-time countermeasure, a range of different scheduler configurations was tested. The ELinOS and the AES server partition were assigned one time slot each and the length of these slots was then varied. It was quickly found that the length of both slots would have to be in a certain relation in order to ensure that the rich OS and the AES server work correctly. One configuration that led to a behavior of the system indistinguishable from the behavior with simple priority-based scheduling was found to be to set the slot length to 5 ticks for both partitions. The default duration of one tick was set to 1 ms. Using this configuration, the delay of single AES encryptions increases significantly by roughly about 70 % while in contrast the encryption of a whole buffer with the size of a memory page may be conducted with only a small overhead of less than 30 %, see Table 3. This configuration was therefore chosen for the attack. Both partitions were assigned one dedicated core and the rest of the setup remained unchanged from previous experiments.

After running the profiling phase for one day we were able to retrieve ≈ 34 million samples. To capture the whole amount of 512 million samples for both

Table 3. Performance comparison of the countermeasure

Scheduling scheme	Average clock cycles per AES-block	
	one block (16 Byte)	one page (4 KByte)
Priority-based	$\approx 125,000$	≈ 1770
Discrete-time	$\approx 210,000$	≈ 2286

phases, this means a total run-time of about one month for the above config-uration of the scheduler. Remember that due to the different timing behavior induced by the countermeasure an even higher number of samples is needed in order to recover the key as good as possible. Therefore, it is reasonable to assume that for the attack to produce a useful output at least twice the num-ber of samples and hence, with the overhead caused by our countermeasure, even more than twice the time is needed. Even if the attacker would do the profiling phase off-line, he would still need to be able to access the system for about one month. It is very unlikely that such a computational intensive attack would remain unnoticed for the entire time frame. Furthermore, depending on the actual use of the AES server, a rescheduling of the key might occur during that time, too. It can be seen from this that the proposed countermeasure indeed protects a device very well while simultaneously requiring almost no effort to be set in place. Also, the user experience does not change with the countermeasure which might be an important factor for the mobile device market. The different run-times of one encryption for priority-based scheduling and the countermea-sure are shown in Table 3. For a more thorough evaluation of the discrete-time countermeasure, additional experiments need to be conducted.

Comparison to Other Countermeasures. In [12,20], two novel countermea-sures against cache-based attacks are introduced. Since these countermeasures target the same class of attacks as the discrete-time countermeasure, it is inter-esting to compare their approaches with ours. As the focus of this work was put on time-driven attacks, the comparison will focus on this aspect as well.

The STEALTHMEM countermeasure [12] tries to prevent both active and passive time- and access-driven attacks in virtualization environments. To that end, it uses dedicated cache lines in the shared cache for each CPU. Depending on the variant of STEALTHMEM used, this either reduces the total available amount of memory and shared cache, or it takes extra time to ensure that the stealth cache lines are not evicted from the cache. Both variants imply a small penalty in performance of about 5.9 % and 7.2 % respectively, and AES encryptions of 50,000 bytes are about 5 % slower with the first variant. Unlike the STEALTHMEM approach, the discrete-time countermeasure has no impact on the available cache and system memory. However, due to the larger time slots in our countermeasure, the overall performance degrades by about 30 % for AES encryptions on 4 KB of data, as explained above. To use STEALTHMEM, the hypervisor is extended with a special driver offering an API to the VMs that manages access to the dedicated cache lines. For Windows Server 2008 R2 with Hyper-V, this amounts to 5,000 lines of C code to be added to the hypervisor and 500 lines of C code added to the Windows boot loader modules. Furthermore, the implementations of cryptographic algorithms have to be modified to make use of the stealth cache lines via the provided API. For using our discrete-time countermeasure on the contrary, only a reconfiguration of the scheduler is needed. Neither the system nor the implementation of the cryptographic algorithm has to be changed. Also note that the required modification of the algorithm presents a potential pitfall. If not done correctly, some leakage remains and therefore breaks the countermeasure. Moreover, the amount of available cache lines that can be reserved for a core is limited so it has to be made sure that all relevant lookup tables fit inside to prevent information leakage.

The instruction-based scheduling scheme suggested in [20] aims at preventing cache-based attacks that exploit certain scheduling-induced race conditions between processes that arise due to the dependency of the execution time on the cache content. Both methods are similar in that they use a fixed value as their criterion for the scheduling. As the name implies, instruction-based scheduling uses a specified number of executed instructions as scheduling criterion. This prevents only those attacks that try to exploit the mentioned race conditions – but only when the processes are run on a single core. Furthermore, it is not sufficient to prevent time-driven attacks such as Bernstein's, since an attacker can still measure the total execution time which still depends on the cache. The discrete-time countermeasure on the other hand prevents this kind of race conditions even with multiple cores, and masks the overall execution time of an AES encryption. Also, instruction-based scheduling is a novel approach and hence not widely supported by current micro kernels. Therefore, extra effort has to be done to integrate it into existing systems or implement a new one which supports instruction-based scheduling. This is not the case for the discrete-time method, where our countermeasure can be readily configured. With respect to the overhead, both methods are fairly similar as they do not need any adaption of the applications and only induce a small time overhead.

7   Conclusion

In this work, we stated an attack scenario using a time-driven cache attack against embedded devices used in cyber-physical systems (CPSs) on the example of PikeOS, a microkernel-based operating system framework compliant to the ARNIC standard. We evaluated the attack with different scheduler configurations showing that dedicated cores for the crypto routine provide the most timing leakage. Further, we compared the results to a similar setup [24] for virtualization based Trusted Execution Environments (TEEs). We showed that using a shared core similar to the microkernel configuration in [24], the PikeOS setup of this work is about three orders of magnitude less vulnerable in reduction of key space, but at least almost one order of magnitude in the worst configuration using dedicated cores. Furthermore, we provided the scheduler based *discrete-time* countermeasure against time-driven cache attacks. Compared to other novel countermeasures, it does not depend on any hardware, software architecture or algorithm changes. Thus, our approach can be used as a drop-in configuration update for running CPSs, or other embedded platforms using a configurable scheduler.

References

1. Acıiçmez, O., Koç, Ç.K.: Trace-driven cache attacks on AES (short paper). In: Ning, P., Qing, S., Li, N. (eds.) ICICS 2006. LNCS, vol. 4307, pp. 112–121. Springer, Heidelberg (2006)
2. Acıiçmez, O., Schindler, W., Koç, Ç.K.: Cache based remote timing attack on the AES. In: Abe, M. (ed.) CT-RSA 2007. LNCS, vol. 4377, pp. 271–286. Springer, Heidelberg (2006)
3. Aeronautical Radio, Inc.: Avionics application software standard interface, ARNIC Specification p. 653 (1997)
4. Aly, H., ElGayyar, M.: Attacking AES using bernstein's attack on modern processors. In: Youssef, A., Nitaj, A., Hassanien, A.E. (eds.) AFRICACRYPT 2013. LNCS, vol. 7918, pp. 127–139. Springer, Heidelberg (2013)
5. Bernstein, D.J.: Cache-timing attacks on AES. Technical report (2005)
6. Bonneau, J., Mironov, I.: Cache-collision timing attacks against AES. In: Goubin, L., Matsui, M. (eds.) CHES 2006. LNCS, vol. 4249, pp. 201–215. Springer, Heidelberg (2006)
7. Gallais, J.-F., Kizhvatov, I., Tunstall, M.: Improved trace-driven cache-collision attacks against embedded AES implementations. In: Chung, Y., Yung, M. (eds.) WISA 2010. LNCS, vol. 6513, pp. 243–257. Springer, Heidelberg (2011)
8. Gueron, S.: Intel® advanced encryption standard (aes) instructions set. Technical report (2008)
9. Gullasch, D., Bangerter, E., Krenn, S.: Cache games - bringing access-based cache attacks on AES to practice. In: IEEE Symposium on Security and Privacy - S&P 2011. IEEE Computer Society (2011)
10. Kaiser, R., Wagner, S.: Evolution of the pikeos microkernel. In: First International Workshop on Microkernels for Embedded Systems, p. 50 (2007)
11. Käsper, E., Schwabe, P.: Faster and timing-attack resistant AES-GCM. In: Clavier, C., Gaj, K. (eds.) CHES 2009. LNCS, vol. 5747, pp. 1–17. Springer, Heidelberg (2009)
12. Kim, T., Peinado, M., Mainar-Ruiz, G.: Stealthmem: system-level protection against cache-based side channel attacks in the cloud. In: Presented as Part of the 21st USENIX Security Symposium (USENIX Security 2012), pp. 189–204, Bellevue, WA, USENIX (2012)
13. Könighofer, R.: A fast and cache-timing resistant implementation of the AES. In: Malkin, T. (ed.) CT-RSA 2008. LNCS, vol. 4964, pp. 187–202. Springer, Heidelberg (2008)
14. Matsui, M., Nakajima, J.: On the power of bitslice implementation on intel core2 processor. In: Paillier, P., Verbauwhede, I. (eds.) CHES 2007. LNCS, vol. 4727, pp. 121–134. Springer, Heidelberg (2007)
15. Neve, M., Seifert, J.-P., Wang, Z.: A refined look at bernstein's aes side-channel analysis. In: ASIACCS, p. 369 (2006)
16. Osvik, D.A., Shamir, A., Tromer, E.: Cache attacks and countermeasures: the case of AES. In: Pointcheval, D. (ed.) CT-RSA 2006. LNCS, vol. 3860, pp. 1–20. Springer, Heidelberg (2006)
17. Ristenpart, T., Tromer, E., Shacham, H., Savage, S.: Hey, you, get off of my cloud: exploring information leakage in third-party compute clouds. In: Proceedings of the 16th ACM Conference on Computer and Communications Security, CCS 2009, New York, NY, USA, pp. 199–212. ACM (2009)

18. Spreitzer, R., Gérard, B.: Towards more practical time-driven cache attacks. In: Naccache, D., Sauveron, D. (eds.) WISTP 2014. LNCS, vol. 8501, pp. 24–39. Springer, Heidelberg (2014)
19. Spreitzer, R., Plos, T.: On the applicability of time-driven cache attacks on mobile devices. In: Lopez, J., Huang, X., Sandhu, R. (eds.) NSS 2013. LNCS, vol. 7873, pp. 656–662. Springer, Heidelberg (2013)
20. Stefan, D., Buiras, P., Yang, E.Z., Levy, A., Terei, D., Russo, A., Mazières, D.: Eliminating cache-based timing attacks with instruction-based scheduling. In: Crampton, J., Jajodia, S., Mayes, K. (eds.) ESORICS 2013. LNCS, vol. 8134, pp. 718–735. Springer, Heidelberg (2013)
21. Varadarajan, V., Ristenpart, T., Swift, M.: Scheduler-based defenses against cross-vm side-channels. In: 23rd USENIX Security Symposium (USENIX Security 2014), San Diego, CA, pp. 687–702. USENIX Association, August 2014
22. Veyrat-Charvillon, N., Gérard, B., Renauld, M., Standaert, F.-X.: An optimal key enumeration algorithm and its application to side-channel attacks. In: Knudsen, L.R., Wu, H. (eds.) SAC 2012. LNCS, vol. 7707, pp. 390–406. Springer, Heidelberg (2013)
23. Veyrat-Charvillon, N., Gérard, B., Standaert, F.-X.: Security evaluations beyond computing power. In: Johansson, T., Nguyen, P.Q. (eds.) EUROCRYPT 2013. LNCS, vol. 7881, pp. 126–141. Springer, Heidelberg (2013)
24. Weiß, M., Heinz, B., Stumpf, F.: A cache timing attack on AES in virtualization environments. In: Keromytis, A.D. (ed.) FC 2012. LNCS, vol. 7397, pp. 314–328. Springer, Heidelberg (2012)

How to Choose Interesting Points for Template Attacks More Effectively?

Guangjun Fan[1]([⊠]), Yongbin Zhou[2], Hailong Zhang[2], and Dengguo Feng[1]

[1] State Key Laboratory of Computer Science, Institute of Software,
Chinese Academy of Sciences, Beijing, China
guangjunfan@163.com, feng@tca.iscas.ac.cn
[2] State Key Laboratory of Information Security, Institute of Information
Engineering, Chinese Academy of Sciences, Beijing, China
{zhouyongbin,zhanghailong}@iie.ac.cn

Abstract. Template attacks are widely accepted to be the most powerful side-channel attacks from an information theoretic point of view. For template attacks to be practical, one needs to choose some special samples as the interesting points in actual power traces. Up to now, many different approaches were introduced for choosing interesting points for template attacks. However, it is *unknown* that whether or not the previous approaches of choosing interesting points will lead to the best classification performance of template attacks. In this work, we give a negative answer to this important question by introducing a practical new approach which has completely different basic principle compared with all the previous approaches. Our new approach chooses the point whose distribution of samples approximates to a normal distribution as the interesting point. Evaluation results exhibit that template attacks based on the interesting points chosen by our new approach can achieve obvious better classification performance compared with template attacks based on the interesting points chosen by the previous approaches. Therefore, our new approach of choosing interesting points should be used in practice to better understand the practical threats of template attacks.

Keywords: Side-channel attacks · Power analysis attacks · Template attacks · Interesting points

1 Introduction

Side-channel attacks are one of the most important threats against modern cryptographic implementations. The basic idea of these attacks is to determine the key of a cryptographic device by exploiting its power consumption [11], its electromagnetic radiation [19], its execution time [18], and many more [20]. Traditional security notions (such as chosen-ciphertext security for public-key encryption schemes) do not provide any security guarantee against such attacks, and many implementations of provably secure cryptosystems were broken by side-channel attacks.

© Springer International Publishing Switzerland 2015
M. Yung et al. (Eds.): INTRUST 2014, LNCS 9473, pp. 168–183, 2015.
DOI: 10.1007/978-3-319-27998-5_11

Power analysis attacks have received such a large amount of attention because they are very powerful and can be conducted relatively easily. Therefore, let us focus exclusively on power analysis attacks. As an important method of power analysis attacks, template attacks were firstly proposed by Chari et al. in 2002 [1] and belong to the category of profiled side-channel attacks. Under the assumption that one (an actual attacker or an evaluator) has a reference device identical or similar to the target device, and thus be well capable of characterizing power leakages of the target device, template attacks are widely accepted to be the strongest side-channel attacks from an information theoretic point of view [1]. We note that, template attacks are also important tools to evaluate the physical security of a cryptographic device.

Template attacks consist of two stages. The first stage is the profiling stage and the second stage is the extraction stage. In the profiling stage, one captures some actual power traces from a reference device identical or similar to the target device and builds templates for each key-dependent operation with the actual power traces. In the extraction stage, one can exploit a small number of actual power traces measured from the target device and the templates obtained from the profiling stage to classify the correct (sub)key.

1.1 Motivations

Note that for real-world implementation of cryptography devices, a side-channel leakage trace (i.e. an actual power trace for the case of power analysis attacks) usually contains multiple samples corresponding to the target intermediate values. The reason is that the key-dependent operations usually take more than one instruction cycles. In addition, according to Nyquist-Shannon sampling theorem, the acquisition rate of the signal acquisition device is always set to be several times faster than the working frequency of the target cryptographic device.

For template attacks to be practical, it is paramount that not all the samples of an actual power trace are part of the templates. To reduce the number of samples and the size of the templates, one needs to choose some special points as the interesting points in actual power traces. Main previous approaches of choosing interesting points for template attacks can be divided into two kinds.

Approaches belong to the first kind try to choose the points which contain the most information about the characterized key-dependent operations as the interesting points with different principles. Classical template attacks [1] generally use the approaches belong to the first kind to choose interesting points. Moreover, many papers [2,3,5,10,12] suggested an accepted guideline for choosing interesting points for the approaches in the first kind. The accepted guideline is that one should only choose one point as the interesting point per clock cycle since more points in the same clock cycle do not provide more information. Disobeying this accepted guideline leads to poorer classification performance of template attacks even if a higher number of interesting points is chosen due to some numerical obstacles when one computes the inverse of the covariance matrices C_i (Please see Sect. 2.2 for more details.). Up to now, many different approaches of choosing interesting points which belong to the first kind were introduced. These approaches are *Correlation Power Analysis based approach* (Chap. 6 in [11]) (CPA), *Sum Of Squared pairwise T-differences based approach* [10] (SOST), *Difference Of Means based approach* [1] (DOM), *Sum Of Squared Differences based approach* [10] (SOSD), *Variance based approach* [15] (VAR), *Signal-to-Noise Ratios based approach* (pp. 73 in [11]) (SNR), *Mutual Information Analysis based approach* [16] (MIA), and *Kolmogorov-Smirnov Analysis based approach* [17] (KSA). One uses these approaches to choose the points which contain the most information about the characterized key-dependent operations as the interesting points by computing the signal-strength estimate $SSE(t)$ for each point P_t. For example, when one uses Correlation Power Analysis based approach to choose interesting points for template attacks, the signal-strength estimate $SSE(t)$ is measured by the coefficient of correlation between the actual power consumptions and the hypothetical power consumptions of a point P_t. For these approaches, in each clock cycle, the point with the strongest signal-strength estimate $SSE(t)$ is chosen as the interesting point.

Approaches belong to the second kind based on the principal components or Fisher's linear discriminant. *Principal Component Analysis based approach* [3] (PCA) and *Fisher's Linear Discriminant Analysis based approach* [9] (LDA) belong to the second kind. We note that, PCA-based template attacks is inefficient due to its high computational requirements [2] and may not improve the classification performance [7]. Therefore, PCA-based template attacks are not considered to be an approach which can be widely used to choose interesting points for template attacks. Moreover, LDA-based template attacks depends on the rare condition of equal covariances [4] (Please see Sect. 2.2 for more details.), which does not hold for most cryptographic devices. Therefore, it is not a better choice compared with PCA-based template attacks in most settings [4]. Due to these reasons, we ignore PCA-based template attacks as well as LDA-based template attacks and only consider the approaches of choosing interesting points for classical template attacks which are the *most* widely used profiled side-channel attacks in this paper.

However, up to now, it is still *unknown* that whether or not using the above approaches of choosing interesting points will lead to the best classification performance of template attacks. In other words, whether or not there exists other approaches which based on different basic principles will lead to better classification performance of template attacks is still *unclear*. If the answer to this question is negative, we can demonstrate that one can further improve the classification performance of template attacks by using the more advanced approach to choose interesting points rather than by designing some kind of improvements about the mathematical structures of the attacks. In this paper, we try to answer this important question.

1.2   Contributions

In this paper, we firstly present a new approach of choosing interesting points for template attacks which has completely different basic principle compared with all the previous approaches. The theoretical correctness of our new approach is supported by an important mathematical property of the multivariate Gaussian distribution and the Pearson's chi-squared test for goodness of fit [23].

Furthermore, we experimentally verified that template attacks based on the interesting points chosen by our new approach can achieve obvious better classification performance compared with template attacks based on the interesting points chosen by the previous approaches. This gives a negative answer to the question that whether or not using the previous approaches of choosing interesting points will lead to the best classification performance of template attacks.

Moreover, the computational price of our new approach is low and practical. Therefore, our new approach of choosing interesting points for template attacks can be used in practice to better understand the practical threats of template attacks.

1.3 Related Work

Template attacks were firstly introduced in [1]. Answers to some basic and practical issues of template attacks were provided in [2], such as how to choose interesting points in an efficient way and how to preprocess noisy data. Efficient methods were proposed in [4] to avoid several possible numerical obstacles when implementing template attacks.

Hanley et al. [12] presented a variant of template attacks that can be applied to block ciphers when the plaintext and ciphertext used are unknown. In [8], template attacks were used to attack a masking protected implementation of a block cipher. Recently, a simple pre-processing technique of template attacks, normalizing the sample values using the means and variances was evaluated for various sizes of test data [7].

Gierlichs et al. [10] made a systematic comparison of template attacks and stochastic model based attacks [22]. How to best evaluate the profiling stage and the extraction stage of profiled side-channel attacks by using the information-theoretic and the security metric was shown in [21].

1.4 Organization of This Paper

The rest of this paper is organized as follows. In Sect. 2, we briefly introduce basic mathematical concepts and review template attacks. In Sect. 3, we introduce our new approach of choosing interesting points for template attacks. In Sect. 4, we experimentally verify the effectiveness of the new approach in improving the classification performance of template attacks. In Sect. 5, we conclude the whole paper.

2 Preliminaries

In this section, we first introduce some basic mathematical concepts which are used in this paper, then briefly review template attacks.

2.1   Basic Mathematical Concepts

We first introduce the Gamma function and the chi-squared distribution. Then, we briefly introduce the concept of the goodness of fit of a statistical model and the Pearson's chi-squared test for goodness of fit.

Definition 1. *The Gamma function is defined as follows:*

$$\Gamma(x) = \int_0^\infty e^{-t}t^{x-1}dt,$$

where $x > 0$.

Definition 2. *The probability density function of the chi-squared distribution with k degrees of freedom (denoted by χ_k^2) is*

$$f(x; k) = \begin{cases} \frac{1}{\Gamma(\frac{k}{2})2^{k/2}}e^{-x/2}x^{(k-2)/2}, & x > 0; \\ 0, & x \leq 0, \end{cases}$$

where $\Gamma(\cdot)$ denotes the Gamma function.

The *goodness of fit of a statistical model* describes how well it fits a set of observations (samples). Measures of goodness of fit typically summarize the discrepancy between the observed values and the values expected under the statistical model in question. Such measures of goodness of fit can be used in statistical hypothesis testing. The *Pearson's chi-squared test for goodness of fit* [23] is used to assess the goodness of fit establishes whether or not an observed frequency distribution differs from a theoretical distribution. In the following, we will briefly introduce the Pearson's chi-squared test for goodness of fit.

Assume that, there is a population X with the following theoretical distribution:

$$H_0 : \Pr[X = a_i] = f_i \quad (i = 1, \ldots, k),$$

where a_i, f_i $(i = 1, \ldots, k)$ are known and $a_1, \ldots, a_k$ are pairwise different, $f_i > 0$ $(i = 1, \ldots, k)$.

One obtains n samples (denoted by $X_1, X_2, \ldots, X_n$) from the population X and uses the Pearson's chi-squared test for goodness of fit to test whether or not the hypothesis H_0 holds. We use the symbol ω_i to denote the number of samples in $\{X_1, X_2, \ldots, X_n\}$ which equal to a_i. If the number n is large enough, it will has that $\omega_i/n \approx f_i$, namely $\omega_i \approx nf_i$. The value nf_i can be viewed as the theoretical value (TV for short) of the category "a_i". The value ω_i can be viewed as the empirical value (EV for short) of the category "a_i". Table 1 shows the theoretical value and the empirical value of the category "a_i".

Clearly, when the discrepancy of the last two lines of Table 1 is smaller, the hypothesis H_0 increasingly seems to be true. It is well known that the Pearson's goodness of fit χ^2 statistic (denoted by Z) is used to measure this kind of discrepancy and is shown as follows:

$$Z = \sum_{i=1}^k (nf_i - \omega_i)^2/(nf_i). \tag{1}$$

Table 1. The theoretical value and the empirical value of each category

Category	a_1	a_2	$\cdots$	a_i	$\cdots$	a_k
TV	nf_1	nf_2	$\cdots$	nf_i	$\cdots$	nf_k
EV	ω_1	ω_2	$\cdots$	ω_i	$\cdots$	ω_k

The statistic Z can be exploited to test whether or not the hypothesis H_0 holds. For example, after choosing a constant Con under a given level, when $Z \leq Con$, one should accept the hypothesis H_0. When $Z > Con$, one should reject the hypothesis H_0. Now, let's consider a more general case. The following lemma about Z was given out by Pearson at 1900 [23] and the proof of Lemma 1 is beyond the scope of this paper.

Lemma 1. *If the hypothesis H_0 holds, when $n \to \infty$, the distribution of Z will approach to the chi-squared distribution with $k - 1$ degrees of freedom, namely χ^2_{k-1}.*

Assume that one computes a specific value of Z (denoted by Z_0) by a group of specific data. Let

$$L(Z_0) = \Pr[Z \geq Z_0 | H_0] \approx 1 - K_{k-1}(Z_0), \tag{2}$$

where the symbol $K_{k-1}(\cdot)$ denotes the distribution function of χ^2_{k-1}. *Clearly, when the probability $L(Z_0)$ is higher, the hypothesis H_0 increasingly seems to be true.* Therefore, the probability $L(Z_0)$ can be used as a tool to test the hypothesis H_0.

If the theoretical distribution of the population X is continuous, the Pearson's chi-squared test for goodness of fit is also valid. In this case, assume that, one want to test the following hypothesis:

H_1 : The distribution function of the population X is $F(x)$.

The distribution function $F(x)$ is continuous. To test the hypothesis H_1, one should set

$$-\infty = a_0 < a_1 < a_2 < \cdots < a_{k-1} < a_k = \infty,$$

and let $I_1 = (a_0, a_1], \cdots, I_i = (a_{i-1}, a_i], \cdots, I_k = (a_{k-1}, a_k)$. Moreover, one obtains n samples (denoted by $X_1, X_2, \ldots, X_n$) from the population X. Let ω_i denotes the cardinality of the set $\{X_j | X_j \in I_i, \; j \in \{1, 2, \ldots, n\}\}$ and

$$f_i = \Pr[x \in I_i, x \leftarrow X] = F(a_i) - F(a_{i-1}) \quad (i = 1, \ldots, k).$$

Then, one can also similarly compute $L(Z_0)$ (by equation (2)) to test the hypothesis H_1.

2.2   Template Attacks

Template attacks consist of two stages. The first stage is the profiling stage and the second stage is the extraction stage. We will introduce the two stages in the following.

The Profiling Stage. Assume that there exist K different (sub)keys $key_i, i = 0, 1, \ldots, K - 1$ which need to be classified. Also, there exist K different key-dependent operations $O_i, i = 0, 1, \ldots, K - 1$. Usually, one will built K templates, one for each key-dependent operation O_i. One can exploit some methods to choose N interesting points $(P_0, P_1, \ldots, P_{N-1})$. Each template is composed of a mean vector and a covariance matrix. Specifically, the mean vector is used to estimate the data-dependent portion of side-channel leakages. It is the average signal vector $\mathbf{M}_i = (M_i[P_0], \ldots, M_i[P_{N-1}])$ for each one of the key-dependent operations. The covariance matrix is used to estimate the probability density of the noises at different interesting points. It is assumed that noises at different interesting points approximately follow the multivariate normal distribution. A N dimensional noise vector $\mathbf{n}_i(\mathbf{S})$ is extracted from each actual power trace $\mathbf{S} = (S[P_0], \ldots, S[P_{N-1}])$ representing the template's key dependency O_i as $\mathbf{n}_i(\mathbf{S}) = (S[P_0] - M_i[P_0], \ldots, S[P_{N-1}] - M_i[P_{N-1}])$. One computes the $(N \times N)$ covariance matrix $\mathbf{C}_i$ from these noise vectors. The probability density of the noises occurring under key-dependent operation O_i is given by the N dimensional multivariate Gaussian distribution $p_i(\cdot)$, where the probability of observing a noise vector $\mathbf{n}_i(\mathbf{S})$ is:

$$p_i(\mathbf{n}_i(\mathbf{S})) = \frac{1}{\sqrt{(2\pi)^N |\mathbf{C}_i|}} exp\Big(-\frac{1}{2}\mathbf{n}_i(\mathbf{S})\mathbf{C}_i^{-1}\mathbf{n}_i(\mathbf{S})^T \Big) \quad \mathbf{n}_i(\mathbf{S}) \in \mathbb{R}^N. \quad (3)$$

In equation (3), the symbol $|\mathbf{C}_i|$ denotes the determinant of $\mathbf{C}_i$ and the symbol $\mathbf{C}_i^{-1}$ denotes its inverse. We know that the matrix $\mathbf{C}_i$ is the estimation of the true covariance $\mathbf{\Sigma}_i$. The condition of equal covariances [4] means that the leakages from different key-dependent operations have the same true covariance $\mathbf{\Sigma} = \mathbf{\Sigma}_0 = \mathbf{\Sigma}_1 = \cdots = \mathbf{\Sigma}_{K-1}$. In most settings, the condition of equal covariances does not hold. Therefore, in this paper, we only consider the device in which the condition of equal covariances does not hold.

The Extraction Stage. Assume that one obtains t actual power traces (denoted by $\mathbf{S}_1, \mathbf{S}_2, \ldots, \mathbf{S}_t$) from the target device in the extraction stage. When the actual power traces are statistically independent, one will apply maximum likelihood approach on the product of conditional probabilities (pp. 156 in [11]), i.e.

$$key_{ck} := argmax_{key_i} \Big\{ \prod_{j=1}^{t} \Pr[\mathbf{S}_j | key_i], i = 0, 1, \ldots, K - 1 \Big\},$$

where $\Pr[\mathbf{S}_j | key_i] = p_{f(\mathbf{S}_j, key_i)}(n_{f(\mathbf{S}_j, key_i)}(\mathbf{S}_j))$. The key_{ck} is considered to be the correct (sub)key. The output of the function $f(\mathbf{S}_j, key_i)$ is the index of a key-dependent operation. For example, when the output of the first S-box in the first round of AES-128 is chosen as the target intermediate value, one builds templates for each output of the S-box. In this case, $f(\mathbf{S}_j, key_i) = Sbox(m_j \oplus key_i)$, where m_j is the input plaintext corresponding to the actual power trace $\mathbf{S}_j$.

3   Our New Approach to Choose Interesting Points for Template Attacks

Now, we begin to introduce our new approach of choosing interesting points for template attacks. Firstly, we show the following Lemma whose proof is in Appendix A.

Lemma 2. *The marginal distribution of multivariate Gaussian distribution is a normal distribution.*

The main idea of our new approach is as follows. In template attacks, it is assumed that the distribution of the noises of multiple interesting points follows the multivariate Gaussian distribution. Moreover, by Lemma 2, we know that the marginal distribution of the multivariate Gaussian distribution is a normal distribution. Therefore, in classical template attacks, if the distribution of samples of each interesting point increasingly to approximate a normal distribution, the multivariate Gaussian distribution statistical model will increasingly to be suitable to be exploited to build the templates for template attacks. Otherwise, if the points whose distributions of samples are not similar to normal distributions are chosen as the interesting points, the multivariate Gaussian distribution will not be suitable to be exploited to build the templates and the classification performance of template attacks will be poor. Therefore, for each clock cycle, our new approach chooses the point whose distribution of samples is more approximate to a normal distribution than other points in the same clock cycle as the interesting point.

The Pearson's chi-squared test for goodness of fit can be used as a tool to assess whether or not the distribution of samples of each point approximates to a normal distribution. Specifically speaking, assume that, for a fixed point P_t, one obtains n samples $(X_1, X_2, \ldots, X_n)$ for a fixed operation on fixed data and computes:

$$\hat{\mu} = \frac{1}{n} \cdot \sum_{i=1}^{n} X_i, \quad s^2 = \frac{1}{n-1} \cdot \sum_{i=1}^{n} (X_i - \hat{\mu})^2. \tag{4}$$

Note that, in template attacks, one can operate the reference device as many times as possible and samples a large number of actual power traces in the profiling stage. Therefore, the value of n can be large enough. When the value of n is large enough, one can assume that the theoretical distribution of samples of the point P_t is the normal distribution $\mathcal{N}(\hat{\mu}, s^2)$ and to test whether this hypothesis holds by exploiting the Pearson's chi-squared test for goodness of fit as follows. The distribution function of the normal distribution $\mathcal{N}(\hat{\mu}, s^2)$ is denoted by $F(x; \hat{\mu}, s^2)$. Let $a_0 = -\infty, a_1 = \hat{\mu} - 2s, a_2 = \hat{\mu} - 1.5s, \ldots, a_9 = \hat{\mu} + 2s, a_{10} = +\infty$ and $I_1 = (-\infty, \hat{\mu} - 2s], I_2 = (\hat{\mu} - 2s, \hat{\mu} - 1.5s], \ldots, I_{10} = (\hat{\mu} + 2s, +\infty)$. Then, one computes $Z_0 = \sum_{i=1}^{10} (nf_i - \omega_i)^2 / (nf_i)$, where $f_i = F(a_i; \hat{\mu}, s^2) - F(a_{i-1}; \hat{\mu}, s^2)$ and $\omega_i = |\{X_j | X_j \in I_i, j \in \{1, 2, \ldots, n\}\}|$. After obtaining the statistic Z_0, one computes the value $L(Z_0)$ by using equation (2).

When the value of n is large enough, if the n samples $(X_1, X_2, \ldots, X_n)$ fit the normal distribution $\mathcal{N}(\hat{\mu}, s^2)$ well, the value $L(Z_0)$ will be high. Otherwise, the value $L(Z_0)$ will be low. Therefore, one can choose the interesting points based on the value $L(Z_0)$. For points in the same clock cycle, one computes the value $L(Z_0)$ of each point with the same actual power traces and chooses a point whose value $L(Z_0)$ is the highest one as the interesting point.

4   Experimental Evaluations

In this section, we will verify and compare the classification performance of template attacks based on the interesting points chosen by our new approach and the classification performance of template attacks based on the interesting points chosen by the previous approaches. Specifically speaking, our experiments are divided into two groups. In the first group, we tried to choose the interesting points by using different approaches. In the second group, we computed the classification performances of template attacks based on the interesting points chosen by different approaches.

For the implementation of a cryptographic algorithm with countermeasures, one usually first tries his best to use some methods to delete the countermeasures from actual power traces. If the countermeasures can be deleted, then one tries to recover the correct (sub)key using classical attack methods against unprotected implementation. For example, if one has actual power traces with random delays [14], he may first use the method proposed in [13] to remove the random delays from actual power traces and then uses classical attack methods to recover the correct (sub)key. The methods of deleting countermeasures from actual power traces are beyond the scope of this paper. Moreover, considering actual power traces without any countermeasures shows the upper bound of the physical security of the target cryptographic device. Therefore, we take unprotected AES-128 implementation as example.

The 1st S-box outputs of the 1st round of an unprotected AES-128 software implementation are chosen as the target intermediate values. The unprotected AES-128 software implementation is on an typical 8-bit microcontroller STC89C58RD+ whose operating frequency is 11 MHz. The actual power traces are sampled with an Agilent DSA90404A digital oscilloscope and a differential probe by measurement over a 20 Ω resistor in the ground line of the 8-bit microcontroller. The sampling rate was set to be 50 MS/s. The average number of actual power traces during the sampling process was 10 times. For our device, the condition of equal covariances does not hold. This means that the differences between different covariance matrixes $\mathbf{C}_i$ are very evident (can easily be observed from visual inspection).

In order to choose interesting points and to test the classification performance of template attacks, we generated three sets of actual power traces which are respectively denoted by Set A, Set B, and Set C. The actual power traces in Set A were used in the profiling stage. The actual power traces in Set B were used in the extraction stage. The actual power traces in Set C were used to choose interesting points. The Set A captured 20,000 actual power traces which were generated with a fixed main key and random plaintext inputs. The Set B captured 100,000 actual power traces which were generated with another fixed main key and random plaintext inputs. The Set C captured 110,000 actual power traces which were generated with a fixed main key and random plaintext inputs. Note that, we used the same device to generate the three sets of actual power traces, which provides a good setting for the focuses of our research.

4.1   Group 1

In all experiments, we chose 4 continual clock cycles about the target intermediate value (Note that, in our unprotected AES-128 software implementation, the target intermediate value only continued for 4 clock cycles.). In each clock cycle, there are 4 points. Therefore, there are 16 points (denoted by $P_0, P_1, \ldots, P_{15}$) totally[1]. Beside our new approach (denoted by CST), we also implemented all the other approaches of choosing interesting points for template attacks including CPA, SOST, DOM, SOSD, VAR, SNR, MIA, and KSA. All the approaches (CSF, CPA, SOST, DOM, SOSD, VAR, SNR, MIA, and KSA) used 110,000 actual power traces in Set C to choose interesting points. The leakage function of our device approximates the typical Hamming-Weight Model (pp. 40–41 in [11]). Therefore, we adopted this model for CPA, MIA, and KSA.

In order to get more accurate results, we conducted our new approach of choosing interesting points as follows. Due to the leakage function of our device approximates the typical Hamming-Weight Model, we chose 9 different values (denoted by $V_0, V_1, \ldots, V_8$) about the target intermediate value. The hamming weight of the 9 different values respectively are $0, 1, \ldots, 8$ (i.e. $HW(V_i) = i$, $i = 0, 1, \ldots, 8$). For each V_i ($i = 0, 1, \ldots, 8$), we selected 400 actual power traces in which the target intermediate value equals to V_i from Set C. Therefore, for each value V_i ($i = 0, 1, \ldots, 8$), there are 400 samples for each one of the 16 points ($P_0, P_1, \ldots, P_{15}$) and we computed the empirical mean value $\hat{\mu}$ and the empirical variance s^2 of the 400 samples for each one of the 16 points by equation (4). Then, for each V_i ($i = 0, 1, \ldots, 8$), we tried to assess the goodness of fit establishes whether or not the actual distribution of samples of the point P_i ($i \in \{0, 1, \ldots, 15\}$) differs from its assumed theoretical distribution $\mathcal{N}(\hat{\mu}, s^2)$ by computing the value $L(Z_0)$ with the 400 samples like that in Sect. 3. For the value V_i ($i = 0, 1, \ldots, 8$) and the point P_j ($j = 0, 1, \ldots, 15$), we computed the value $L(Z_0)$ and rewrote it by $L_{(i,j)}(Z_0)$. Then, we computed the value $L_j(Z_0)$ ($j = 0, 1, \ldots, 15$) for each one of the 16 points as follows:

$$L_j(Z_0) = \frac{1}{9} \cdot \sum_{i=0}^{8} L_{(i,j)}(Z_0), \ (j = 0, 1, \ldots, 15)$$

---

[1] The points $P_0, \ldots, P_3$ are in the first clock cycle. The points $P_4, \ldots, P_7$ are in the second clock cycle. The points $P_8, \ldots, P_{11}$ are in the third clock cycle. The points $P_{12}, \ldots, P_{15}$ are in the fourth clock cycle.

Table 2. The interesting points chosen by different approaches

Clock Cycle	1	2	3	4
CST	P_2	P_4	P_{11}	P_{12}
CPA	P_1	P_5	P_8	P_{12}
SOST	P_1	P_5	P_8	P_{12}
DOM	P_3	P_7	P_{10}	P_{12}
SOSD	P_3	P_7	P_{10}	P_{12}
VAR	P_3	P_7	P_{10}	P_{12}
SNR	P_3	P_7	P_{10}	P_{12}
MIA	P_1	P_5	P_8	P_{15}
KSA	P_1	P_5	P_8	P_{15}

and chose the interesting points based on the values $L_0(Z_0), \ldots, L_{15}(Z_0)$. In one clock cycle, the point with the highest $L_j(Z_0)$ is chosen as the interesting point.

In Table 2, we show the interesting points chosen by different approaches using the 110,000 actual power traces in Set C. From Table 2, we find that our approach chooses different interesting points in the first three clock cycles compared with other approaches.

4.2   Group 2

For simplicity, let n_p and n_e respectively denote the number of actual power traces used in the profiling stage and in the extraction stage. In this paper, we use the typical metric *success rate* [6] as the metric about the classification performance of template attacks.

In order to show the success rates of template attacks based on the interesting points chosen by different approaches under different attack scenarios, we conducted 4 groups of experiments. In these groups of experiments, the numbers of actual power traces used in the profiling stage are different. This implies that the level of accuracy of the templates in these groups of experiments are different. The higher number of actual power traces used in the profiling stage, the more accurate templates will be built. Moreover, in each groups of experiments, we still considered the cases that one can possess different numbers of actual power traces which can be used in the extraction stage.

Specifically speaking, in the 4 groups of experiments, we respectively chose 5,000, 10,000, 15,000, and 20,000 different actual power traces from Set A to build the 256 templates based on the interesting points chosen by different approaches in the profiling stage. Template attacks based on the interesting points chosen by approach A is denoted by the symbol "A-TA". We tested the success rates of template attacks based on the interesting points chosen by different approaches when one uses n_e actual power traces in the extraction stage as follows. We repeated the 9 attacks (CSF-TA, CPA-TA, SOST-TA, DOM-TA, SOSD-TA,

SNR-TA, VAR-TA, MIA-TA, and KSA-TA) 1,000 times. For each time, we chose n_e actual power traces from Set B uniformly at random and the 9 attacks were conducted with the same n_e actual power traces. We respectively recorded how many times the 9 attacks can successfully recover the correct subkey of the 1st S-box.

From Table 2, we find that the CPA approach and the SOST approach provide the same result of choosing interesting points. The DOM approach, the SOSD approach, the VAR approach, and the SNR approach provide the same result of choosing interesting points. Moreover, the MIA approach and the KSA approach provide the same result of choosing interesting points. The approaches which provide the same result of choosing interesting points will lead to the same classification performance of template attacks. Therefore, in order to show the success rates more clearly, we only show the success rates of CSF-TA, CPA-TA, DOM-TA, and MIA-TA in Fig. 1. The success rates of template attacks based on the interesting points chosen by different approaches when n_p equals to 5,000 and n_e equals to 4, 8, 12, 16, and 20 are shown in Table 3.

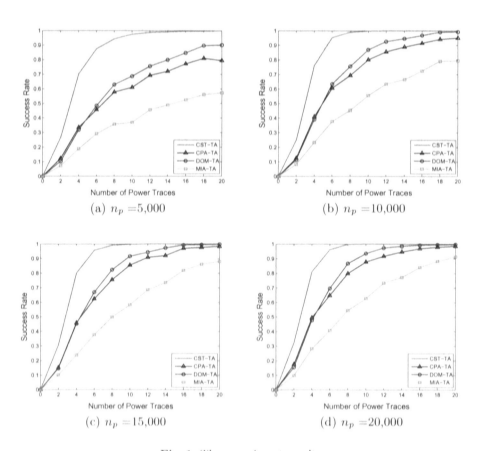

Fig. 1. The experiment results

Table 3. The success rates of template attacks when $n_p = 5,000$

n_e	4	8	12	16	20
CST-TA	0.70	0.94	0.99	1.00	1.00
CPA-TA	0.34	0.58	0.69	0.77	0.79
SOST-TA	0.34	0.58	0.69	0.77	0.79
DOM-TA	0.32	0.63	0.76	0.85	0.90
SOSD-TA	0.32	0.63	0.76	0.85	0.90
VAR-TA	0.32	0.63	0.76	0.85	0.90
SNR-TA	0.32	0.63	0.76	0.85	0.90
MIA-TA	0.19	0.36	0.46	0.53	0.57
KSA-TA	0.19	0.36	0.46	0.53	0.57

From Fig. 1 and Table 3, in all the attack scenarios, we find that template attacks based on the interesting points chosen by our new approach will achieve obvious higher success rates compared with template attacks based on the interesting points chosen by the previous approaches. For example, when $n_p = 5,000$ and $n_e = 4$, the success rate of CST-TA equals to 0.70, while the success rate of DOM-TA equals to 0.32. What's more, when $n_p = 5,000$, CST-TA only needs 7 actual power traces in the extraction stage to achieve success rate higher than 0.9, while DOM-TA needs 20 actual power traces in the extraction stage to achieve success rate higher than 0.9 under the same attack scenario. Therefore, we believe that using our new approach to choose the interesting points can effectively improve the classification performance of template attacks.

5 Conclusion

In this paper, we give a negative answer to the question that whether or not using the previous approaches of choosing interesting points will lead to the best classification performance of template attacks by introduction a new approach with completely different basic principle. Our new approach is based on the important mathematical property of the multivariate Gaussian distribution and exploits the Pearson's chi-squared test for goodness of fit.

Experiments verified that template attacks based on the interesting points chosen by our new approach will achieve obvious better classification performance compared with template attacks based on the interesting points chosen by the previous approaches. Moreover, the computational price of our new approach is low and practical. Therefore, our new approach of choosing interesting points can be used in practice to better understand the practical threats of template attacks. In the future, it is necessary to further verify our new approach in other devices such as ASIC and FPGA.

Acknowledgments. This work was supported by the National Basic Research Program of China (No. 2013CB338003), the National Natural Science Foundation of China

(Nos. 61472416, 61272478), and the National Key Scientific and Technological Project (No. 2014ZX01032401-001).

Appendix A: The Proof of Lemma 2

Proof: For simplicity, we only consider the case when $N = 2$. For the case $N > 2$, this Lemma holds similarly.

Let (ξ, η) denote a 2 dimensional random vector. The continuous distribution function and the probability density function of the 2 dimensional random vector respectively are $F(x, y)$ and $p(x, y)$. Then, the marginal distribution functions are as follows:

$$F_1(x) = \int_{-\infty}^{x} \int_{-\infty}^{\infty} p(u, y) du dy, \quad F_2(y) = \int_{-\infty}^{\infty} \int_{-\infty}^{y} p(x, u) dx du.$$

The marginal density functions are as follows:

$$p_1(x) = \int_{-\infty}^{\infty} p(x, y) dy, \quad p_2(y) = \int_{-\infty}^{\infty} p(x, y) dx.$$

For 2 dimensional multivariate Gaussian distribution, it has that

$$p(x, y) = \frac{1}{2\pi |\mathbf{C}|} exp\left\{ -\frac{1}{2}(x - a, y - b) \cdot \mathbf{C}^{-1} \cdot (x - a, y - b)^T \right\},$$

where

$$\mathbf{C} = \begin{pmatrix} \sigma_1^2 & r\sigma_1\sigma_2 \\ r\sigma_1\sigma_2 & \sigma_2^2 \end{pmatrix}$$

and the values $a, b, \sigma_1, \sigma_2, r$ are constant, $\sigma_1 > 0, \sigma_2 > 0, |r| < 1$. The probability density function $p(x, y)$ can be rewritten as follows

$$p(x, y) = \frac{1}{2\pi\sigma_1\sigma_2\sqrt{1 - r^2}} exp\left\{ -\frac{1}{2(1 - r^2)} \cdot \left[\frac{(x - a)^2}{\sigma_1^2} \right. \right.$$
$$\left. \left. -\frac{2r(x - a)(y - b)}{\sigma_1\sigma_2} + \frac{(y - b)^2}{\sigma_2^2} \right] \right\}.$$

Let

$$\frac{x - a}{\sigma_1} = u, \quad \frac{y - b}{\sigma_2} = v$$

and it has that

$$p_1(x) = \int_{-\infty}^{\infty} p(x, y) dy$$

$$= \frac{1}{2\pi\sigma_1\sqrt{1 - r^2}} \int_{-\infty}^{\infty} exp\left\{ -\frac{1}{2(1 - r^2)} \cdot [u^2 - 2ruv + v^2] \right\} dv$$

$$= \frac{1}{\sqrt{2\pi}\sigma_1} e^{-u^2/2} \int_{-\infty}^{\infty} \frac{1}{\sqrt{2\pi(1 - r^2)}} \cdot exp\left\{ -\frac{r^2u^2 - 2ruv + v^2}{2(1 - r^2)} \right\} dv$$

$$= \frac{1}{\sqrt{2\pi}\sigma_1} e^{-u^2/2} \int_{-\infty}^{\infty} \frac{1}{\sqrt{2\pi(1-r^2)}} e^{-(v-ru)^2/2(1-r^2)} dv$$

$$= \frac{1}{\sqrt{2\pi}\sigma_1} e^{-u^2/2} = \frac{1}{\sqrt{2\pi}\sigma_1} e^{-(x-a)^2/2\sigma_1^2}.$$

Therefore, $p_1(x)$ is the probability density function of the normal distribution $\mathcal{N}(a, \sigma_1^2)$. Similarly, we can prove that

$$p_2(y) = \frac{1}{\sqrt{2\pi}\sigma_2} e^{-(x-b)^2/2\sigma_2^2}.$$

In this way, Lemma 2 is proven. □

References

1. Chari, S., Rao, J.R., Rohatgi, P.: Template attacks. In: Kaliski Jr., B.S., Koç, Ç.K., Paar, C. (eds.) CHES 2002. LNCS, vol. 2523, pp. 13–28. Springer, Heidelberg (2003)
2. Rechberger, C., Oswald, E.: Practical template attacks. In: Lim, C.H., Yung, M. (eds.) WISA 2004. LNCS, vol. 3325, pp. 440–456. Springer, Heidelberg (2005)
3. Archambeau, C., Peeters, E., Standaert, F.-X., Quisquater, J.-J.: Template attacks in principal subspaces. In: Goubin, L., Matsui, M. (eds.) CHES 2006. LNCS, vol. 4249, pp. 1–14. Springer, Heidelberg (2006)
4. Choudary, O., Kuhn, M.G.: Efficient template attacks. In: Francillon, A., Rohatgi, P. (eds.) CARDIS 2013. LNCS, vol. 8419, pp. 253–270. Springer, Heidelberg (2014)
5. Bär, M., Drexler, H., Pulkus, J.: Improved template attacks. In: COSADE2010 (2010)
6. Standaert, F.-X., Malkin, T.G., Yung, M.: A unified framework for the analysis of side-channel key recovery attacks. In: Joux, A. (ed.) EUROCRYPT 2009. LNCS, vol. 5479, pp. 443–461. Springer, Heidelberg (2009)
7. Montminy, D.P., Baldwin, R.O., Temple, M.A., Laspe, E.D.: Improving cross-device attacks using zero-mean unit-variance mormalization. J. Cryptographic Eng. 3(2), 99–110 (2013)
8. Oswald, E., Mangard, S.: Template attacks on masking—resistance is futile. In: Abe, M. (ed.) CT-RSA 2007. LNCS, vol. 4377, pp. 243–256. Springer, Heidelberg (2006)
9. Standaert, F.-X., Archambeau, C.: Using subspace-based template attacks to compare and combine power and electromagnetic information leakages. In: Oswald, E., Rohatgi, P. (eds.) CHES 2008. LNCS, vol. 5154, pp. 411–425. Springer, Heidelberg (2008)
10. Gierlichs, B., Lemke-Rust, K., Paar, C.: Templates vs. stochastic methods. In: Goubin, L., Matsui, M. (eds.) CHES 2006. LNCS, vol. 4249, pp. 15–29. Springer, Heidelberg (2006)
11. Mangard, S., Oswald, E., Popp, T.: Power Analysis Attacks: Revealing the Secrets of Smart Cards. Springer, Berlin (2007)
12. Hanley, N., Tunstall, M., Marnane, W.P.: Unknown plaintext template attacks. In: Youm, H.Y., Yung, M. (eds.) WISA 2009. LNCS, vol. 5932, pp. 148–162. Springer, Heidelberg (2009)

13. Durvaux, F., Renauld, M., Standaert, F.-X., van Oldeneel tot Oldenzeel, L., Veyrat-Charvillon, N.: Efficient removal of random delays from embedded software implementations using hidden markov models. In: Mangard, S. (ed.) CARDIS 2012. LNCS, vol. 7771, pp. 123–140. Springer, Heidelberg (2013)
14. Coron, J.-S., Kizhvatov, I.: Analysis and improvement of the random delay countermeasure of CHES 2009. In: Mangard, S., Standaert, F.-X. (eds.) CHES 2010. LNCS, vol. 6225, pp. 95–109. Springer, Heidelberg (2010)
15. Mather, L., Oswald, E., Bandenburg, J., Wójcik, M.: Does my device leak information? An *a priori* statistical power analysis of leakage detection tests. In: Sako, K., Sarkar, P. (eds.) ASIACRYPT 2013, Part I. LNCS, vol. 8269, pp. 486–505. Springer, Heidelberg (2013)
16. Gierlichs, B., Batina, L., Tuyls, P., Preneel, B.: Mutual information analysis. In: Oswald, E., Rohatgi, P. (eds.) CHES 2008. LNCS, vol. 5154, pp. 426–442. Springer, Heidelberg (2008)
17. Whitnall, C., Oswald, E., Mather, L.: An exploration of the Kolmogorov-Smirnov test as a competitor to mutual information analysis. In: Prouff, E. (ed.) CARDIS 2011. LNCS, vol. 7079, pp. 234–251. Springer, Heidelberg (2011)
18. Kocher, P.C.: Timing attacks on implementations of Diffie-Hellman, RSA, DSS, and other systems. In: Koblitz, N. (ed.) CRYPTO 1996. LNCS, vol. 1109, pp. 104–113. Springer, Heidelberg (1996)
19. Gandolfi, K., Mourtel, C., Olivier, F.: Electromagnetic analysis: concrete results. In: Koç, Ç.K., Naccache, D., Paar, C. (eds.) CHES 2001. LNCS, vol. 2162, pp. 251–261. Springer, Heidelberg (2001)
20. European Network of Excellence (ECRYPT). The side channel cryptanalysis lounge. http://www.crypto.ruhr-uni-bochum.de/ensclounge.html
21. Standaert, F.-X., Koeune, F., Schindler, W.: How to compare profiled side-channel attacks? In: Abdalla, M., Pointcheval, D., Fouque, P.-A., Vergnaud, D. (eds.) ACNS 2009. LNCS, vol. 5536, pp. 485–498. Springer, Heidelberg (2009)
22. Schindler, W., Lemke, K., Paar, C.: A stochastic model for differential side channel cryptanalysis. In: Rao, J.R., Sunar, B. (eds.) CHES 2005. LNCS, vol. 3659, pp. 30–46. Springer, Heidelberg (2005)
23. Pearson, K.: On the criterion that a given system of deviations from the probable in the case of a correlated system of variables is such that it can be reasonably supposed to have arisen from random sampling. Philos. Mag. Ser. 5 **50**(302), 157–175 (1900)

NeuronVisor: Defining a Fine-Grained Cloud Root-of-Trust

Anbang Ruan$^{(\boxtimes)}$ and Andrew Martin

Department of Computer Science, University of Oxford, Oxford, UK
{anbang.ruan,andrew.martin}@cs.ox.ac.uk

Abstract. Security issues have become a significant barrier to the adoption of cloud computing services. Most existing security enhancements lack a well defined Root-of-Trust (RoT). Models for Trusted Clouds have been proposed, which establish RoT inside the cloud and vouch for the trustworthiness of the cloud services. However, these are often impractical due to cloud's dynamics and complexity. In this paper, we present the *NeuronVisor*, an abstract *Cloud Root-of-Trust (cRoT)* framework. NeuronVisor enforces decentralized attestations to capture trust dependency among interacting software components inside the cloud, and determines a single cRoT for each cloud application. This cRoT hides the cloud's internal by presenting a uniform interface for attesting to the trustworthiness of the entire cloud application and all its dependent services inside the cloud (the *Cloud TCB*). Our simulations show that, for more than 98 % times, one interrogation to the dynamically formed cRoT is able to identify the properties of more than 90 % of the nodes hosting a cloud application and its cloud TCB. Meanwhile, NeuronVisor achieves higher fault detection rate than the prevalent centralized cloud attestation scheme (CEN). It still achieves the same fault detection rate with CEN even when 90 % of the NeuronVisors are constantly tampered with and maliciously collaborating with each other.

1 Introduction

The Trusted Cloud concept has been proposed to integrate cloud systems with the Trusted Computing infrastructure [15,17]. Cloud nodes are equipped with built-in TPMs. Attestation services are implemented to gather the TPM-generated trust evidence from each node to report their genuine behaviors. With Trusted Clouds, customers are expecting to verify the fulfillment of their SLAs (Service Level Agreements) by attesting to the nodes hosting the declared services. However, the complexity and dynamics in cloud inhibit the effective implementation of these systems [13].

Currently, these schemes assume homogeneous property distributions among the cloud nodes [5]. For example, whether the entire cloud management nodes enforce ubiquitous non-discriminative VM scheduling policies, or whether all the compute nodes have the capabilities to enforce strong VM-isolation. Besides, they require customers to rely on a central delegate service to perform the attestations to the target services [15,17]. Customers then attest to this delegate to

© Springer International Publishing Switzerland 2015
M. Yung et al. (Eds.): INTRUST 2014, LNCS 9473, pp. 184–200, 2015.
DOI: 10.1007/978-3-319-27998-5_12

make sure that the attestation services are genuine and hence assume that their assets are secure, and the services they have purchased are trustworthy.

However, the cloud infrastructures have evolved to implement diverse supporting services for satisfying different applications' needs [8,21]. It has become common that cloud nodes implement different services. When customers are allowed to choose among these various supporting services from the cloud providers, the attestation delegate has to distinguish different attestation criteria for each of them. The effective and efficient implementation of this delegate will be a challenging task. On the other hand, customers' attestations to the delegate only prove its existence rather than effectiveness. In other words, without additional evaluations to the detailed configurations of this delegate, customers cannot be assured that their attestation requirements have been genuinely fulfilled. But this evaluation is especially complex, given the gigantic-scale and multi-tenancy of a cloud service provider.

From customers' perspective, a desirable cloud attestation scheme is to *directly* attest to the properties of their cloud applications and the cloud supporting services that their applications depend on. This requires the cloud providers to genuinely provide the customers with the information of all these related services. In short, this requires effective *cloud Trusted Computing Base (cTCB)* identification and attestation for applications. In our previous work, we proposed the RepCloud framework [13] to define the cTCB for a cloud application. RepCloud implements a decentralized attestation (DA) scheme, which enforces attestations among cloud nodes based on their the internal interaction patterns. With DA, the trust dependency is among the nodes are clearly determined, which facilitates the identification of a cloud application's cTCB. However, RepCloud lacks of a clearly defined RoT inside to cloud to effectively implement the cTCB attestations.

We observe that the difficulties in achieving effective and practical cloud attestations are generally caused by the problem of *insufficient abstraction of the Root-of-Trust model in cloud.* As a cloud hides its underlying hardware infrastructure and exposes only a uniform view of virtual resource to a cloud application, the RoT for the application should also have a uniform abstraction. A logical *cloud Root-of-Trust (cRoT)* for an application is thus desirable, which manages all the hardware RoT (TPMs) for the application and its cTCB (e.g. the supporting cloud services). The cRoT should scale with the application, and collect the trust information maintained by each TPM. It exports the aggregated trust information in a uniform interface to achieve cloud application attestations. With this abstraction, attestations to a cloud application are performed in the similar way as attestations to an application hosted on a single server: with one interaction with a cRoT, the properties of the entire cloud application and its cTCB are fetched and examined.

In this paper, we propose the NeuronVisor framework to approach the first step for implementing this cloud Root-of-Trust abstraction. NeuronVisor implements two functions: (1) it dynamically identifies the set of hardware Roots-of-Trust (i.e. TPMs) which maintain the trust evidence for each component of a cloud application and its cloud TCB; (2) it shares the trust evidence among these RoTs, so that querying any of the RoTs will aggregate all these relevant evidence. NeuronVisor therefore builds a foundation for constructing a *Cloud Chain-of-Trust*, which will organize and export the trust evidence of an entire cloud application and its cloud TCB in a uniform way to facilitate effective cloud attestations.

This paper is organized as follows: Sect. 2 presents the conceptual Neuron-Visor model. Section 3 illustrates the Neuron Web model for implementing this model for achieving the cloud Root-of-Trust semantics. In Sect. 4, the trust implications of the cRoT are evaluated with simulations, along with the design and implementation of the NeuronVisor prototype. However, due to the length limit of this paper, we leave the details to a longer version paper. Section 5 discusses the related work, and Sect. 6 concludes the paper and presents the future work.

2   NeuronVisor Framework

A NeuronVisor (or Neuron for short) is a software Root-of-Trust (i.e. TPM) management layer, deployed on each cloud node. A Neuron possesses only two properties: (1) whether it can genuinely attest to the properties of the upper-layer service components; (2) whether it can attest to other Neuron's properties. Each Neuron enforces autonomous *Decentralized Attestations* [13] to examine and disseminating these two properties of its interacting peers.

Neuron Structure. Figure 1 depicts the logical structure of the NeuronVisor layer (or *Neuron* for short). The Neuron Kernel (or kernel for short) is the core component of a Neuron. It stores the trust evaluations of its interacting Neurons. The kernel is maintained by the Attestation Module and the Trust Propagation Module. Attestation Module attests to target Neurons when new communications are initiated among their upper-layer service components, e.g. customers' Virtual Machines, or management service modules. It updates the kernel with the attestation information. Trust Propagation Module exchanges the updated kernel matrix with peer Neurons by adapting the Decentralized Attestation [13] protocols. A Neuron interacts with upper layer services through its Network Monitor and vTPM Manager. The Network Monitor intercepts the network traffics, and queries the integrity of the target Neuron from its kernel. Attestations to the Neuron will be initiated when necessary. The vTPM Manager exports necessary kernel matrix to the vTPMs to implement cloud attestations.

Two forms of attestations are implemented by each Neuron. Firstly, it implements the attestations to the properties of the upper layer services, e.g. Virtual Machines or cloud management service components. This is implemented by the vTPM Management Module. Secondly, it adapts the *Decentralized Attestation* [13] (DA) to attest to only the integrity of the peer NeuronVisor layer on all the interacting nodes, based their hosted services' communication patterns. It is achieved by the cooperation of the Attestation Module and the Network Monitor Module. By aggregating and disseminating this integrity information with the Trust Propagation Module, Neurons on the frequently communicating nodes

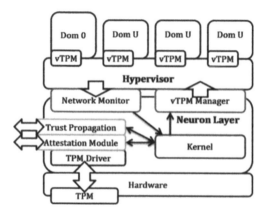

Fig. 1. Neuron structure.

form the *Neuron Web*, where the integrity of each Neuron is regularly examined. By sharing the properties of the upper layer services among the attested Neurons, this web forms a dynamic and scalable software layer to attest to all the hosted interacting services with a uniformed interface. By attesting to and querying any Neuron on this web, the properties of all dependent services are gathered. This dynamically formed Neuron Web thus achieves the cloud Root-of-Trust abstraction for attesting to a cloud application. In this paper, we mainly focus on this second form of attestation: how the decentralized attestations are implemented in the NeuronVisor framework to dynamically identify relevant Neurons, and effective share information among them.

Neuron Connections. Figure 2 depicts a simplified Neuron connection topology. With NeuronVisor, a node can communicate with the others only when its Neuron is *connected* to theirs. An established *connection* means the Neuron has successfully attested to the integrity of the target Neuron. The connected Neuron is referred to as the attesting Neuron's *neighbor*. These Neuron connections link semantic depending Neurons and form the Neuron Web.

An important property of a connection is its *strength*. It represents the *attestation relationship* from the Neuron to its neighbor. It is an integer indicating, in the view of the attesting Neuron, how the neighbor is attested to. The higher the strength, the harder it is for the connected Neuron to change its properties without being detected. Therefore, strength acts as an integrated evaluation based on past attestations to represent a Neuron's trustworthiness. Moreover, as suggested in [13], in an environment where nodes frequently attest to each other, an attestation ticket can be effectively reused for better reflecting the trust dependency and reducing redundant attestations. A Neuron thus also determines the strength value for its connection to a neighbor by analyzing the attestation relationship from other Neurons with the target neighbor. This relationship is gathered by the Trust Propagation module, and is maintained in the *Kernel*.

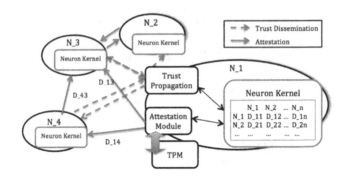

Fig. 2. Simplified neuron connection topology.

Neuron Kernel. As a Neuron only concerns the trustworthiness of its neighbors, the kernel of each Neuron maintains a *partial attestation graph* to record only the information of attestations performed to and from its neighbors. This is achieved by adapting the Decentralized Attestation (DA).

When a node initiates a new communication to another, its Neuron first attests to the target node's Neuron. This attestation establishes a new Neuron connection by adding an entry inside the kernel recording the attestation information. The Neuron then *fetches* all the attestation information from the neighbor's kernel, and *aggregates* it into its own kernel. Afterwards, the Neuron *propagates* back the updated data to all corresponding neighbors, who also depend on this data. These Neurons perform the aggregate-propagate procedure, until their kernels are stable, i.e. a new aggregation session only makes negligible changes to the kernel. This iterative information exchange thus allows the partial graph to contain all the attestation relationship of the neighbors [10,13]. It approaches a more accurate strength value assignment and achieves fast and accurate trust information dissemination.

For example, in Fig. 2, after Neuron N_1 attested to N_3, it fetches N_3's kernel. It updates related entries into its own kernel, and propagate these updates to N_2 and N_4, since they have also attested to N_4, which implies trust dependency. These two Neurons aggregate the updated information with their own kernel, and propagate the changed entries back. This may also contain the information that N_1 depends on, e.g. attestations towards N_3.

Connection Strength. The *attestation relationship* between two Neurons is modeled as an integer from their past attestation patterns maintained in the Neuron Kernel. This value represents a *Direct Trust* (D) towards the attested Neuron. For example, in Fig. 2, the solid arrows represent the direst trust relationship from attesting Neurons to the targets. *Transitive Trust* is deduced by connecting the direct trust from the attesting Neuron to a middle Neuron and from the middle one to the target. The strength value for the connection to the neighbor is in turn modeled from both the direct and the transitive trust through all possible middle Neurons. For example, in Fig. 2, the strength value for the connection from N_1 to N_3 is calculated from the direct trust D_13 and the transitive trust, which is calculated from D_14 and D_43. It represents how the neighbor is attested to both *directly* and *indirectly* by the attesting Neuron. This strength calculation helps NeuronVisor to reduce redundant attestations, while preventing malicious collaborative attacks [9] (Sect. 4).

The strength value decreases constantly, since it is calculated from the attestations, whose credibility degrades over time. When it reaches a *threshold*, the Neuron determines whether its upper layer services still have communications to their counterparts hosted on its neighbors. If not, it tears down the connection by removing the related entries from the Kernel. Otherwise, it attests to the neighbor and updates the strength. The new attestation information is also propagated to other neighbors, which will result in a new series of information exchanges. This strength degradation and repeated attestations thus allow the connection strength to better reflect the trust dependency and communication dynamics among interacting nodes. When a neighbor failed the attestation, it is identified as "unhealthy" and the connection is torn down. This failed node is reported, and will be examined or re-initiated. This information is also propagated among other neighbors, who will also reexamine the unhealthy target, and teal down the connection when necessary.

Neuron Web. Connected Neurons form a *Neuron web*. By choosing a *Center Neuron* and specifying the strength criteria, a *partial* Neuron web is identified, where all the connections have the strength satisfying the criteria. In NeuronVisor, the connection strength reflects the attestation relationship, which in turn indicates the communication patterns of the upper layer services. Therefore, by setting the criteria according to the services' communication patterns, the generated partial web will cover all the Neurons hosting these services. As only trusted Neurons are bound on a web, trust information can be shared freely among the Neurons without requiring interrogating the TPMs every time. As each Neuron possess the capabilities of attesting to the properties of the upper layer services, this web thus acts as the Cloud Root-of-Trust (cRoT) for all these services, which includes the connected application components and all participating cloud service components.

3   Neuron Web Model

In this section we define the mathematical model of the Neuron Web. We first present the basic notations, and then illustrate the procedures for the maintaining the Neuron Kernel. From the kernel, *Transitive Trust* is deduced to determine the connection *strength*.

3.1   Direct Trust

Neuron Kernel collects the results of the attestations performed by the Neuron and its neighbors. These attestations represent *Direct Trust* relationship among Neurons. Direct trust is modeled from the past attestation tickets. Two parts of information from tickets are usually used for modeling trust [13]: the *measurement* values, which are calculated and recorded by the Trusted Computing facilities [2,3] to represent the genuine properties of the Neuron, and a timestamp, t, which records the time the measurements are collected. However, since only the integrity of a Neuron is concerned, the measurements are replaced by a binary value: "healthy" or "unhealthy". It is implemented by examining whether these measurement values are located in a known-good values list [2]. Moreover, as an "unhealthy" Neuron will be isolated as soon as it is discovered, only the timestamp is necessary to represent this evaluation. Therefore, in NeuronVisor, a Neuron keeps an *Attestation History* (*AH*) for each healthy neighbor to record all timestamps for the *recent* attestations it performed to the neighbor.

In practice, repeatedly attestations have a minimum interval, which is determined by the time needed to fulfill a complete attestation [18]. We denote this interval as a *step* (τ). Thus the distance between two timestamps t_2 and t_1, can be expressed as the number of steps in between. It is denoted as $\Delta(t_2, t_1)$:

$$\Delta(t_2, t_1) = \lfloor \frac{(t_2 - t_1)}{\tau} \rfloor \tag{1}$$

With this definition, a *recent* ticket is defined as the one generated κ steps away from the current time, where κ is an integer constant chosen for a cloud implementation to suit its needs for balancing performance and security requirements.

We now model the direct trust from Neuron N_a to N_b at time t: $D_{a,b}(t)$. $D_{a,b}(t)$ is calculated by combining the timestamps maintained in the attestation history, which records the attestation tickets towards the neighbor (N_b). It is an integer interpreted as a *bitmap vector* with the length of κ. Each bit represents a timestamp one step away from its higher adjacent bit, and the highest bit indicates the time t. A bit is set to 1 when an attestation is performed at the step it stands for. Thus the direct trust, calculated as below, reflects all the recent successful attestations up to time t. $AH_j(t)$ denotes the attestation history for Neuron N_j at time t. As a step is defined as minimum attestation interval, different timestamps t in AH cannot indicate a same bit index. We can thus safely use *summation* instead of *bitwise OR* ("|") for setting the corresponding bits.

$$D_{i,j}(t) = \sum_{t_n \in AH_j(t)} 2^{\kappa - \Delta(t, t_n)} \tag{2}$$

This definition allows two evaluation values be compared. The larger one indicates the more recent an attestation is performed, and hence indicates a higher trust credibility. This property is used for modeling *Transitive Trust* described later. On the other hand, the bit pattern represents the past attestation pattern. It is further used to model a *Combined Trust* semantics, which will be briefly discussed in Sect. 3.4. In this paper, we focus on modeling the *Transitive Trust*.

Whenever an attestation is performed, the new evaluation value is calculated by shifting the original one rightwards $\Delta(t_{new}, t_{original})$ bits, and setting the highest bit to "1":

$$D_{i,j}(t_{new}) = 2^{\kappa}|\frac{D_{i,j}(t_{original})}{2^{\Delta t_{new}, t_{origial}}} \tag{3}$$

As each Neuron maintains a partial attestation graph, the attestation information regarding a pair of neighbors may be incomplete. Thus Neurons exchange their gathered information to approach a more accurate attestation relationship among Neurons. When two different version for a same attestation relationship are maintained by two Neurons at different time, e.g. $D_{i,j}^a(t_1)$ $D_{i,j}^b(t_2)$, they are first adjusted to a common time, and then merged together with the bitwise OR operation. In the rest of the paper, we use the superscript to denote that the data structure is maintained by a certain Neuron. Each data structure also contains a parameter t to indicate that its value is calculated for time t. We omit these two notations when the context is clear.

$$D_{i,j}(t_{new}) = \frac{D_{i,j}^a(t_1)}{2^{\Delta t_{new}, t_1}}|\frac{D_{i,j}^b(t_2)}{2^{\Delta t_{new}, t_2}} \tag{4}$$

3.2 Neuron Kernel

The kernel of a Neuron N_i is defined as a matrix $K_i(t)$ maintaining the direct trust it gathers at time t.

$$K_i(t) = \{D_{a,b}(t)\} \tag{5}$$

We define $D_{a,b} = 0$, when no attestation is performed from N_a to N_b. We also define $D_{a,a}$ equals the maximum direct trust value $2^{\kappa+1} - 1$, as a Neuron always trust itself. The set of N_i's neighbors (Nbr_i) is hence defined as:

$$\text{Nbr}_i = \{N_k \mid 0 < D_{i,k} < 2^{\kappa+1} - 1\} \tag{6}$$

Neuron Kernel is maintained as a *Global Trusted Matrix* in the Decentralized Attestation scheme [13]. Specifically, three steps are used to maintain the kernel: (1) the Neuron first *gathers* the direct trust for its neighbors by performing remote attestations to them; (2) it then *aggregates* entries in the kernel of the neighbors; and finally (3) it *disseminates* the updated kernel back to corresponding neighbors.

However, in NeuronVisor, only the *Direct Trust* values are disseminated, instead of the entire measurement values used in RepCloud [13]. This greatly increases the trust dissemination efficiency. For trust gathering, whenever a Neuron (N_i) attests to another (N_j), it updates the entry $D_{i,j}(t)$ in its kernel using Eq. (3). Other entries are also refreshed to adapt to the new time t. The trust aggregation is then performed by fetching the kernel of N_j: K^j. Entries in the retrieved K^j are merged with the corresponding ones in K^i by using Eq. 4. Finally, the updated entries are sent to the set of Neurons who also have dependency on this information. This set is determined by the I_b^a in [13].

3.3   Neuron Connections

The Connection Strength is calculated by aggregating the *Transitive Trust*. Transitive trust has been discussed in P2P systems [10,19,20]. Trust towards a "stranger" can be determined by consulting a "friend", whose trustworthiness is known and who knows the trustworthiness of the stranger. Similarly, the integrity of a Neuron can be assumed when it is attested to by a neighbor. Transitive Trust reuses trust information and reduces redundant attestations.

We first define a *transitive attestation path*, which contains a sequence of Neurons, with the former one attested to the following one. In NeuronVisor, only one-hop transitive trust is considered [12]. Thus each path contains three Neurons. The transitive trust, $T_{i,k,j}$, thus denotes the trust implication towards Neuron N_j, regarding a path containing $\{N_i, N_k, N_j\}$. As an attestation reflects the trustworthiness *up to* the time it is performed, the transitive trust should only represent the trustworthiness *up to* the time when all the Neuron on the attestation path are regarded as equally trustworthy. This means the value of this evaluation equals the smallest direct trust value along the path.

$$T_{i,k,j} = \min(D_{i,k}, D_{k,j}) \tag{7}$$

As $D_{i,i}$ is defined to equal the maximum value, transitive trust calculation also incorporates the direct trust: $T_{i,i,j} = D_{i,j}$.

We now define the *Connection Strength* $S_{i,j}$ from Neuron N_i to N_j. It is the maximum transitive trust value for all possible transitive paths from N_i to N_j. It represents the most recent time when N_j was iteratively attested to by N_i.

$$C_{i,j} = \max(\{T_{i,k,j} \mid D_{i,k} \neq 0 \wedge D_{k,j} \neq 0\}) \tag{8}$$

When the related entries in the kernel change, this trust value is updated. Every time when a node interacts with another one, this value is first adjusted to reflect the current time (by shifting rightwards Δ bits). It is then compared to the *threshold* value Φ. Only when it is larger, will the communication been enforced. Otherwise, the Neuron triggers a new attestation to the target, which will result in a new round of trust dissemination.

3.4   Discussions and Extensions

Connections Strength Interpretation. Neuron Connections help forming a cRoT (the Neuron Web). As discussed in Sect. 2, a partial Neuron Web is actually a dynamically formed centralized attestation domain, with the Center Neuron of the web as the attestation delegate. The strength values of the connections indicate how often these Neurons are attested to, directly or indirectly, by the center. Therefore, after choosing a VM as the cloud attestation target, customers have actually chosen an attestation domain, with the VM's underlying NeuronVisor as the center. After examining the integrity of this center Neuron, customers are able to infer the integrity of the other Neurons from the returned strength value matrix. As Neuron Connections are formed according to upper layer's communication relationship, this attestation domain preserves the application's dependency, which helps determining the cloud TCB.

Centralized Communications. Attestations among Neurons are based on decentralized communication patterns. However, in a cloud implementation, centralized communications still exist. For example, in OpenStack, the central Scheduler node talks to every Compute node regularly. However, the decentralized attestations are still enforced, because of the trust dissemination and transitive trust aggregation. When the Scheduler attests to a Compute, it fetches the Neuron Kernels of the Compute, which contains attestation information for other Computes. Therefore, the Scheduler only attests to the Compute nodes that have not been attested to recently. Decentralized attestations thus distribute the attestation responsibilities from a central delegate to all the cloud nodes. This prevents the single-point-of-failure. Meanwhile it reduces the complexity for managing the centralized attestation delegate.

Combined Trust. Transitive Trust represents the trust evaluation from a *local* view of a given Neuron. Aggregating all the local views, a *Combined Trust* can be deduced to represent the trust evaluation of each node in a global perspective [13]. Reputation systems have been proposed to deduce a global reputation value for each node from analyzing the mutual trust evaluations among interconnecting nodes [10,19,20]. Most systems model the global combined reputation of a node by using three criteria: (a) the past interaction patterns the target one has with its neighbors; (b) the number of nodes that have interacted with the target; (c) the relationship of these nodes. With NeuronVisor, the *Direct Trust* models the past interaction patterns, while in the Neuron Kernel, criteria b and c can also be deduced. Therefore, NeuronVisor builds a foundation for implementing the *Combined Trust*, which will help better representing the trust dynamics inside the cloud. We leave further investigation in this direction to our future work.

Cloud Attestations. As discussed in Sect. 2, a partial Neuron web is identified by choosing a Neuron as a center and setting a reference connection strength value. This web thus binds the recently communicated cloud nodes together. Moreover, as only genuine Neurons are bound with the Web, trust information can be safely shared among the Web. Therefore, when a center Neuron receives a cloud attestation request, it searches its Neuron kernel and locates the neighbors with satisfying connection strength. The trust information maintained by its neighbors is then queried. The center Neuron hence aggregates all these trust information, and return it as an integrated attestation ticket to the attester. We leave the design and implementation of a cloud attestation system based on the Neuron Web to our future work.

4   Evaluations

In this section, we first discuss the threats to our NeuronVisor framework, and how it defenses against them. We then evaluate NeuronVisor with simulations.

Due to the length limit of this paper, we will present the detailed evaluations and prototype implementation in a longer-version paper.

4.1   Security Analysis

Neuron Kernels are maintained in a peer-to-peer manner. Therefore, well-known attacks [9] against this decentralized trust management scheme need to be concerned. In NeuronVisor, a Neuron Kernel is updated with *trust aggregation* and *dissemination*. The kernel of a target Neuron is only merged after the target is attested to. Hence only kernel from a genuine Neuron is aggregated. On the other hand, a Neuron can only disseminate attestation results generated by itself. This prevents the *Self-Promotion* attack, which is performed by a tampered Neuron in order to improve its own connection strength value in the view of others, hence preventing it from being attested to. Moreover, reporting a target Neuron as "unhealthy" (the negative evaluation) will result in it being attested to by a management authority. When false report is discovered, the reporter is attested to. This helps to identify the *Slandering* attacks, with which the malicious Neuron's goal is to ruin the reputation of a target Neuron. Finally in NeuronVisor, Neuron identity is represented by the TPM identify, which can only be created by a Trust Third Party. Thus Sybil attacks are prevented.

Collusive attacks need further examinations. As we assume that the Neurons have identical implementations, as long as attackers have successful exploited one Neuron, it is not hard for them to take control of more Neurons with the same techniques. When a large number of tampered Neurons exist, they disseminate false trust information to promote the connection strength of each other. This may result in other Neuron to regard them as also "healthy". To guard against this attack, NeuronVisor uses the Transitive Trust to calculate the strength value. Thus only when the dissemination source Neuron has higher credibility, will its reported trust information be trusted. Moreover, during cloud attestation, as the center Neuron is attested to by customers, the collusive Neurons will ultimately identified. In this case, the larger the collusive group is, the higher chance it will be discovered. As a result, the damage of this attack is well controlled. The effects of the attacks to NeuronVisor are also examined with simulations next.

4.2   Simulations

We evaluated NeuronVisor with simulations. We modified the RepCloud simulator to implement the NeuronVisor protocols. We simulated a cloud deployed with 50 computing nodes, with each capable of hosting 16 VMs simultaneously. Cloud applications with size ranging from 4 to 10 VMs are deployed randomly to the computing nodes. Communications are enforced randomly among the VMs from the same application.

We use the *Tampered Interaction (TI)* counts as our evaluation criteria. TI occurs when a node interacts with a tampered target before the attack has been identified. For example, when a node is attacked at time t_a, which is in between two consecutive attestations at t_1 and t_2 respectively $(t_1 < t_a < t_2)$, all the interactions with the node during the time period between t_a to t_2 are regarded as tampered. We evaluate NeuronVisor (NT) by comparing its TI counts with a centralized attestation scheme's (CEN) with the same simulation configuration. In CEN, a central node attests to every node inside a cluster repeatedly with a predefined *interval*.

Figure 3 illustrates the Tampered Interaction counts of NeuronVisor (NT) simulations with different Φ values. The TI counts for centralized attestation scheme (CEN) with equivalent *interval* value is also presented. When Φ is low, NT achieves less TI counts with equivalent settings. This is because NT performs attestations according to communication needs. Thus attestations are distributed to better match communication patterns, while CEN uniformly attests to each node regardless their interaction semantics.

To evaluate cloud attestation, we simulate an attestation agent, which attests to every simulated cloud application in each simulation cycle. The agent chooses a random VM from the application, and calculates the reference strength value from the application's interaction factor. It then determines the set of Neurons from the selected VM's Neuron Kernel using the reference strength value. This set thus compose the cRoT for the application. The agent then examine whether the Neurons in the cRoT cover the actual Neurons hosting the entire application. We evaluate this scenario by increasing the average application size (AAS). As shown in Fig. 4, for more than 98 % of times, more than 90 % Neurons are identified by a single cloud attestation. When AAS increases, the coverage percentage decreases. This is because, regarding our simulation, the larger the application, the less likely a VM will communicate to every other VMs in a simulation cycle. In this case, customers initiate more attestation requests according to its application's internal communication patterns.

NeuronVisor effectively defends against both targeted and collaborative attacks. Targeted attacks simply disable the attacked Neuron. Disabled Neurons do not perform attestations to the others, and do not disseminate trust. All the upper layer communications are enforced without evaluating the trust-

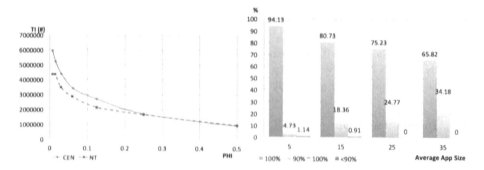

Fig. 3. Tampered interaction counts with different connection threshold.

Fig. 4. cRoT coverage with cloud attestation.

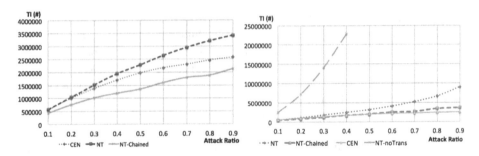

Fig. 5. Independent attacks.   **Fig. 6.** Malicious collusive attacks.

worthiness of the target node. In the simulation, we assume a tampered Neuron is fixed as soon as it is attested to. As shown in Fig. 5, the differences on the TI counts between NT and CEN increase as the attack ratio increases. This is because the more Neurons are tampered, the fewer nodes are attested to. When 90 % of Neurons are tampered, around 40 % more interactions are tampered than the CEN scheme. But NeuronVisor is still able to identify all tampered nodes, as the simulation run through the whole simulation. On the other hand, any single attack to the central delegate will disable the entire CEN.

We further proposed NT-chained. When discovering a tampered node (N_a), a Neuron performs additional attestations to all the node's neighbor, as these neighbors' were less likely to be attested to, since N_a were tampered and were unable to perform attestations. Hence they are more likely to be tampered as well, especially under high attack ratios. As depicted in the figure, NT-chained achieves even better TI detection performance than the CEN scheme.

Figure 6 presents how NeuronVisor defenses Collaborative Attacks. NT generally incurred a higher TI counts difference to the CEN scheme. When 90 % Neurons were tampered and maliciously collaborated with each other, more than one time of tampered interaction counts incurred. However, NeuronVisor still discovered all tampered nodes. For the NT-noTrans scheme, with which the transitive trust calculation were disabled, the TI counts raised fast. All the Neurons were tampered before the end of a simulation when the attack ratio is above 0.4. On the other hand, the NT-Chained enhancement still exhibited good TI detection rate, though its TI counts are slightly higher than the CEN counterparts when the attack ratios are high. As it is not common for a real cloud system to regularly has very high attack ratio (e.g. > 50 %), we believe that the NT-Chained enhancement is suitable for most scenarios.

4.3   Implementation

We implemented a proof-of-concept NeuronVisor in the XenServer-OpenStack architecture [?] (Fig. 7). Each node is deployed with a Xen hypervisor [6]. OpenStack facilities, e.g. Compute, Scheduler, etc. [1], are deployed in separated DomUs. Customer Virtual Machines are also deployed as DomUs.

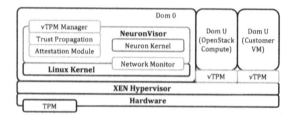

Fig. 7. NeuronVisor implementation with XenServer and OpenStack.

Referring to Fig. 1, NeuronVisor is running in a layer lower than Dom0, so that its TCB is reduced. This can be implemented by adopting the nested virtualization structure [21], or enforcing the DRTM [11] to protect the NeuronVisor application running inside Dom0. For simplicity, we currently implement NeuronVisor as a user-space application in Dom0. Therefore, the entire Dom0 serves as the NeuronVisor's TCB. Decentralized Attestations enforced by each Neuron thus verify the integrity of this entire chain-of-trust [2, 14], including the bootloader, Xen hypervisor, and the entire software stack loaded inside Dom0. Each Neuron is deployed with an expected measurement list, which records all the software components that are allowed to be loaded inside every Neuron's Dom0. The decentralized attestations thus examine whether each target Neuron's measurement list is included in this white-list, and deduce a binary attestation result: whether the Neuron is *healthy* or *unhealthy*.

The Network Monitor in our current prototype is implemented as the xtables extensions [4] to the Linux kernel in Dom0. It intercepts the network traffic sending from all the DomUs, and maintains a *Neuron Connection Strength Vector* data structure in the kernel space. It writes the IP address of the communication target through a kernel device. The Attestation Module running in the user space monitors this device and performs attestations described above to the NeuronVisor on the communication target node. This attestation then triggers the trust disseminations, and updates the Neuron Kernel. The resulting updated connection strength vector is written to the kernel device and stored in the Network Monitor module.

5   Related Work

Trusted Virtual Datacenter (TVDc) [7] provides strong isolation between workloads by enforcing a Mandatory Access Control (MAC) policy throughout a datacenter. It also provides integrity guarantees to each workload by leveraging a hardware root of trust in each platform to determine the identity and integrity of every piece of software running on a platform. Trusted Cloud Computing Platform (TCCP) [15] enables users to attest to the IaaS provider and determine whether or not the service is secure before they launch their virtual machines. Cloud verifier (CV) [17] generates integrity proofs for customers to verify the integrity and access control enforcement abilities of the cloud platform that protect the integrity of customer's application VMs in IaaS cloud. These trusted cloud systems share the similar centralized structure. Customers attest to the properties of the entire cloud altogether. As discussed in Sect. 1 and [13], implementation and management complexity limit their scalability.

Santos et al. proposes that the barriers for widely cloud adaptation are resulted from the insufficient capabilities of the current TCG model to represent trust semantics inside the cloud [16]. They proposed a new abstraction to let data be sealed and unsealed only by nodes whose configurations match a predefined policy. However, the genuinely enforcements of this architecture still needs to be attested to, which is still implemented by the centralized delegated scheme.

Abbadi [5] also identifies that cloud dynamics prohibit the practical trusted cloud implementation. He suggests that the cloud infrastructure is actually homogeneous. With this assumption, he proposes the *combined chain-of-trust*, which attests to a cluster of nodes who have the exactly same configuration together. These nodes exhibit a single nodes' properties, and hence the upper layer services are bound to one single combined COT. However, in this paper we proposed that when considering the supporting services as the TCB of the cloud application [8,21], the homogeneous assumptions are broken. In this case the detailed TCB properties are required for cloud application attestations.

6   Conclusion

In this paper we proposed the NeuronVisor framework, which defines a logical cloud Root-of-Trust (cRoT) abstraction for a cloud application. Our simulations showed that, with moderate overheads, NeuronVisor manages the trust dependency inside the cloud, while achieving higher fault detection rate than the centralized attestation schemes. Besides, NeuronVisor is robust against classic attacks on reputation systems, as it combines the trusted computing technology for evaluating trust. NeuronVisor builds an important foundation for implementing effective cloud attestations. Based on the cRoT abstraction, the properties of the entire cloud application and its cloud TCB can be obtained, with a very few interactions with the hardware RoT. This significantly reduces the design and implementation complexity of the attestation delegate. In our future work, we will design a property-based cloud attestation system and implement the NeuronVisor with a minimized TCB.

References

1. Openstack. http://www.openstack.org
2. Trusted computing group. http://www.trustedcomputinggroup.org
3. Trusted computing group. http://www.trustedcomputinggroup.org/resources/tpm_main_specification
4. Xtables-addons. http://xtables-addons.sourceforge.net

5. Abbadi, I.M.: Clouds trust anchors. In: Proceedings of the 2012 IEEE 11th International Conference on Trust, Security and Privacy in Computing and Communications (Washington, DC, USA, 2012), TRUSTCOM 2012, pp. 127–136. IEEE Computer Society (2012)

6. Barham, P., Dragovic, B., Fraser, K., Hand, S., Harris, T., Ho, A., Neugebauer, R., Pratt, I., Warfield, A.: Xen and the art of virtualization. In: Proceedings of the nineteenth ACM symposium on Operating systems principles (New York, NY, USA, 2003), SOSP 2003. ACM (2003)

7. Berger, S., Cáceres, R., Pendarakis, D., Sailer, R., Valdez, E., Perez, R., Schildhauer, W., Srinivasan, D.: Tvdc: managing security in the trusted virtual datacenter. SIGOPS Oper. Syst. Rev. **42**, 40–47 (2008)

8. Butt, S., Lagar-Cavilla, H.A., Srivastava, A., Ganapathy, V.: Self-service cloud computing. In: Proceedings of the 2012 ACM conference on Computer and communications security (New York, NY, USA, 2012), CCS 2012. ACM (2012)

9. Hoffman, K., Zage, D., Nita-Rotaru, C.: A survey of attack and defense techniques for reputation systems. ACM Comput. Surv. **42**(1), 1 (2009)

10. Kamvar, S.D., Schlosser, M.T., Garcia-Molina, H.: The eigentrust algorithm for reputation management in p2p networks. In: Proceedings of the 12th international conference on World Wide Web (New York, NY, USA, 2003), WWW 2003. ACM (2003)

11. McCune, J.M., Li, Y., Qu, N., Zhou, Z., Datta, A., Gligor, V., Perrig, A.: Trustvisor: Efficient tcb reduction and attestation. In: Proceedings of the 2010 IEEE Symposium on Security and Privacy (Washington, DC, USA, 2010), SP 2010. IEEE Computer Society (2010)

12. Piatek, M., Isdal, T., Krishnamurthy, A., Anderson, T.: One hop reputations for peer to peer file sharing workloads. In: Proceedings of the 5th USENIX Symposium on Networked Systems Design and Implementation (Berkeley, CA, USA, 2008), NSDI 2008. USENIX Association (2008)

13. Ruan, A., Martin, A.: Repcloud: achieving fine-grained cloud tcb attestation with reputation systems. In: Proceedings of the sixth ACM workshop on Scalable trusted computing (New York, NY, USA, 2011), STC 2011. ACM (2011)

14. Sailer, R., Zhang, X., Jaeger, T., van Doorn, L.: Design and implementation of a tcg-based integrity measurement architecture. In: Proceedings of the 13th conference on USENIX Security Symposium - Volume 13 (Berkeley, CA, USA, 2004), SSYM 2004. USENIX Association (2004)

15. Santos, N., Gummadi, K.P., Rodrigues, R. Towards trusted cloud computing. In Proceedings of the 2009 conference on Hot topics in cloud computing (Berkeley, CA, USA, 2009), HotCloud. USENIX Association (2009)

16. Santos, N., Rodrigues, R., Gummadi, K.P., Saroiu, S.: Policy-sealed data: a new abstraction for building trusted cloud services. In: Proceedings of the 21st USENIX conference on Security symposium (Berkeley, CA, USA, 2012), Security 2012. USENIX Association (2012)

17. Schiffman, J., Moyer, T., Vijayakumar, H., Jaeger, T., McDaniel, P.: Seeding clouds with trust anchors. In: Proceedings of the 2010 ACM workshop on Cloud computing security workshop (New York, NY, USA, 2010), CCSW 2010. ACM (2010)

18. Stumpf, F., Fuchs, A., Katzenbeisser, S., Eckert, C.: Improving the scalability of platform attestation. In: Proceedings of the 3rd ACM workshop on Scalable trusted computing (New York, NY, USA, 2008), STC 2008, ACM (2008)

19. Walsh, K., Sirer, E.G.: Experience with an object reputation system for peer-to-peer filesharing. In: Proceedings of the 3rd conference on Networked Systems Design & Implementation - Volume 3 (Berkeley, CA, USA, 2006), NSDI 2006. USENIX Association (2006)
20. Xiong, L., Liu, L.: Peertrust: Supporting reputation-based trust for peer-to-peer electronic communities, vol. 16, IEEE Educational Activities Department
21. Zhang, F., Chen, J., Chen, H., Zang, B.: Cloudvisor: retrofitting protection of virtual machines in multi-tenant cloud with nested virtualization. In: Proceedings of the Twenty-Third ACM Symposium on Operating Systems Principles (New York, NY, USA, 2011), SOSP 2011. ACM (2011)

A Privacy-Aware Access Model
on Anonymized Data

Xuezhen Huang[⊠], Jiqiang Liu, and Zhen Han

School of Computer and Information Technology,
Beijing Jiaotong University, Beijing 100044, China
pltree@163.com

Abstract. With development of information technology and communication, corporations and individuals will collect some digital information to support information-based decisions. However, under some conditions, if all original data are released, some privacy will be disclosed, which will threaten data security and data privacy. Therefore, data owners will take some security measures. Role-based access control may authorize related original data accessed by users according to their roles. Privacy-preserving technology release processed data to avoid privacy disclosure. Nevertheless, existing privacy-preserving technologies lack continuity and are quite inefficient. This paper establishes an access model about on anonymized data and combines with the foregoing two security measures. On the premise that data security and data privacy are ensured, there is more flexibility and diversity and work efficiency is improved as well.

Keywords: Privacy · Data security · Access control · Anonymity

1 Introduction

Nowadays, information technology and communication technology develop rapidly. Electronic information is widely applied. To carry out data analysis so as to support information-based decisions, the government, social institutions, companies and individuals will collect a mass of electronic information. Much information, such as health information and census information, involves individuals' privacy or sensitive information. It will threaten individuals' security and privacy. Illegal disclosure of lots of data will even threaten social security and stability. Therefore, data access must be controlled strictly.

At the beginning of the 1990s, Ferraiolo et al. from National Institute of Standards and Technology (NIST) put forward role-based access control (RBAC) [3]. In 2007, Ni et al. proposed privacy-aware role-based access control (P-RBAC) [10] to support implementation of privacy policies. However, in model of data release, it

Project was partially supported by Research Fund for the Doctoral Program of Higher Education of China (No. 20120009110007), Program for Innovative Research Team in University of Ministry of Education of China (No. IRT201206) and Program for New Century Excellent Talents in University (NCET-11-0565).

© Springer International Publishing Switzerland 2015
M. Yung et al. (Eds.): INTRUST 2014, LNCS 9473, pp. 201–212, 2015.
DOI: 10.1007/978-3-319-27998-5_13

will not be enough if we only have P-RBAC. The accuracy of the data that P-RBAC policy allows to visit is 1 or 0, i.e., all or none. Since its policy lacks flexibility, it will result in excessive data protection, which makes data fail to be used sufficiently, or lead to disclosure of data privacy and security.

On the premise that individual privacy is not disclosed, people pay attention to privacy-preserving data publishing (PPDP) to keep usability of data as much as possible. The original data which will be released, such as health data and statistical data, are tabular data generally. They contain three kinds of attributes: identifier which is the attribute that identifies single individuals only, such as ID card No., name and cell phone number; quasi-identifier attribute (QA) which can be used together to potentially identify one person, such as gender, birthday and age; and sensitive attribute (SA) which describes an individual privacy, such as disease and salary. When data are published, data owners may carry out anonymization processing for original data selectively and then published anonymous data to prevent sensitive data from being disclosed. For anonymous data, identifiers contained in original ones has been removed and then the risk that sensitive data may be disclosed is reduced via privacy-preserving technologies like generalization or perturbation. The anonymity is the trade-off between privacy security and usability [1, 2, 5, 7–9, 13–15].

Existing anonymity models usually design a reasonable anonymity model based on privacy requirement of receivers. Then they release data satisfying anonymity models. However, there may be several receivers applying for data for different requirements in practical environment. Thus, attribute sets and privacy policies are different. Some receivers only need to know statistical information with high generalization. For instance, hospitals need send data with different attribute sets to administrators of medicine companies and insurance companies. In addition, a certain department of a medicine company can only access data with high generalization, i.e., accuracy is lower than that of data which corporate administrators can access. This aims at preventing harm or loss for individuals and the company involved in such data, if individuals of a department leave the company with valuable data. Hence, they are authorized to access information they need to know only. For this, information need be processed many times for several receivers. The existing methods have low efficiency.

Considering advantages and disadvantages of RBAC, P-RBAC and PPDP, this paper establishes a model for Anonymized data Privacy-aware Role-based Access Control (AP-RBAC), which combines advantages of the foregoing two approaches. It uses anonymization modes to store data satisfying different privacy policies. Based on this, it enforces P-RBAC, which makes legal users access anonymized data that can satisfy privacy policies of their own roles. AP-RBAC model keeps advantages of access control and implements fine-grained access control to satisfy data security and privacy policy. At the same time, it introduces technical advantages of privacy protection and builds privacy-preserving model to generate anonymized data. This enables accuracy of accessed data to be more flexible and more controllable. For different privacy policies, several anonymized data need be generated. Therefore, this paper proposes three anonymity modes, including Hierarchy Anonymity Mode (HAM), Node Storage Mode (NSM) and Hybrid Mode (HM), to implement role access for different privacy policies rapidly and effectively.

2   Related Work

In order to protect data security and data privacy, current research mainly implements control in two aspects. On the one hand, it simply controls whether users have ability to access some original data. RBAC [3] involves the given ability to access information according to users' posts or roles, which is an effective access control method satisfying data confidentiality and integrity. P-RBAC [10] extends RBAC and adds privacy policy based on RBAC to satisfy privacy requirements of different roles. However, the difference in different roles' access to data also lies in the situation that they access data with different accuracies. For example, posts with higher titles can access data with high accuracy and a post of data center can even have authority to access original data.

On the other hand, it adopts anonymization. It establishes a privacy model for one privacy requirement, implements privacy algorithm for original data and issues data with lower accuracy. In the anonymization method, anonymity algorithm inputs original data that have not been processed and outputs anonymized data which satisfy privacy models.

Three representative privacy-preserving models will be introduced in the following. k-anonymity [13] requires that each record in the released data cannot be distinguished from other $k-1$ records on QA at least. For k pieces of records that cannot distinguish from one another on QA, we call them an equivalence class. k-anonymity requires each equivalence class should contain k pieces of records. l-diversity [9] requires each equivalence class in the released data should have l different SA values. Thus, we have $2 \leq l \leq k$ generally. Compared with k-anonymity, this model can resist attack of background knowledge and consistency more strongly. Reference [9] puts forward two other anonymity models entropy l-diversity and recursive (c, l)-diversity to restrain SA for choice in practical application. Based on constraint of k-anonymity principle, t-closeness [7] restrains distribution of SA within an equivalence class and requires the distribution distance between an equivalence class and the whole table cannot exceed t. Many other models are extended ones based on the three models [1,2,8,14,15].

Nevertheless, there are a number of users with different requirements in reality. The foregoing models based on one privacy requirement need establish models for each users, respectively, so their efficiency is quite low.

The foregoing two solutions cannot solve the problem that different roles access data with different attributes and accuracies. Hence, this paper puts forward AP-RBAC model.

3   Access Model for Anonymized Data

AP-RBAC model is shown in Fig. 1. The model is composed of users, roles, data, tuple and verification. Users in the model are human beings. A user is granted with a role according to her or his job nature and responsibility. Data of the model are tabular data. Each piece of record is corresponding to one individual and contains three kinds of attributes, i.e., identifiers, QA and SA. A tuple of the model is the binding information between a role and its privacy policy. It decides access ability of a role. There are three processing modes for anonymized data in the model, i.e., Hierarchy Anonymity Mode (HAM), Node Storage Mode (NSM) and Hybrid Mode (HM). The former two modes are applied for the following two Situations, respectively.

1. There are at least two roles access data with different hierarchies and privacy policies that need be satisfied;
2. There are two roles access data at least and privacy policies that need be satisfied are the same, but selection rules for needed data are different.

The third mode is a synthesized model of the forward twos.

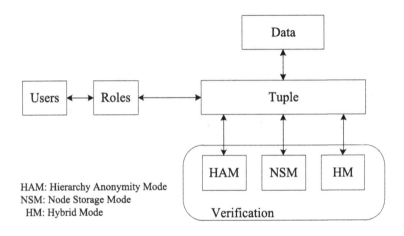

Fig. 1. The access model

3.1   Attribute Selection

All attributes of original data constitute a universal set of attributes. In any anonymity mode, the attribute that each role is authorized to access is a subset of the universal set. Based on the inclusion relation of attribute sets, the hierarchies of superior and subordinate of roles can be reflected. If the attribute that a role is authorized to access contains identifier, the role will be granted to know individuals' sensitive information that are corresponding to records. It will be no need to adopt anonymization privacy-preserving models. SA is the protected object of privacy-preserving models. For a role that has no right to access SA, it is no need to establish privacy-preserving models. For roles whose attribute sets contains SA but excludes identifier. They should comply with privacy policy and are just objects of privacy protection. Generally, data that need to be anonymized in privacy-preserving models just have QA and SA. Thus, authorized access attributes of all roles in the anonymity modes about privacy protection, which will be established in the following, do not contain identifier but contain SA, and the difference in access attribute sets is reflected by QA sets.

The universal QA set in the data sheets is Q, and $|Q|$ attributes are contained in the set. Thus, it has $2^{|Q|}$ attribute subsets for different roles' access. For instance, QA set $Q = \{a, b, c, d\}$. Hierarchy of all subsets are shown in Fig. 2. All nodes in the figure constitute power set of Q, i.e., $P(Q)$. Each set of a node in the digraph contains the set of each successive nodes on the directed path where it locates. Thus, each node contains information of all successive nodes.

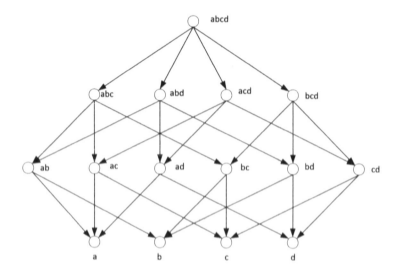

Fig. 2. Hierarchy of power set $P(Q)$ with $Q = \{a, b, c, d\}$

3.2   Three Anonymity Modes

Hierarchy Anonymity Mode. First of all, data are processed. As shown in Fig. 3, the data subset of original data, which is generated by attribute selection, is called the $0th$ data. It is supposed that $C \in P(Q)$ and the $0th$ anonymous data v_{C0} with the attribute subset C serves as input. The anonymity model M_1 is adopted for anonymization and output the $1st$ anonymous data v_{C1}. With v_{C1} as input, the anonymity model M_2 is carried out. The $2nd$ anonymous data v_{C2} is returned. Similarly, we can obtain the $m th$ anonymous data v_{Cm} that we need.

Data nodes generated in this way constitute a directed path that is called a subset path P_C. This directed tree is established according to the attribute sets that roles allow to access and their corresponding hierarchies. In accordance with generation of P_C, it satisfies monotonicity. In detail, for any two nodes v_{Ci} and v_{Cj} in the path, information of v_{Ci} contains information of v_{Cj} if $0 \leq i \leq j \leq m$.

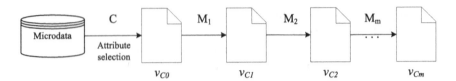

Fig. 3. Hierarchy Anonymity Mode

Hence, role hierarchies satisfy reflexivity, anti-symmetry and transmissibility. It is a partial order.

Next, each role in HAM is distributed with a 2-dimensional tuple (Attribute set, Hierarchy). The only data node that is corresponding to the tuple is the access node of the role.

Example 1. An original data owned by a hospital is shown in Table 1. According to HAM and Incognito algorithm [6], access control and privacy-preserving models are excuted. Generalization hierarchies of all QAs is shown in Fig. 4(a, b, c). We adopt models M_1: 2-anonymity [13] and M_2: 2-diversity [9] to generate the $1st$ anonymity nodes and the $2nd$ anonymity modes in accordance with the minimum of information loss (ILoss) [12,14], as shown in Fig. 5, i.e., $< A_1, G_1, Z_0 >$ and $< A_1, G_1, Z_1 >$. The anonymous data of the two nodes are shown in Table 2. For roles $R_1 : (Q, 1)$ and $R_2 : (Q, 2)$, R_1 and R_2 can access Table 2(a, b), respectively. Hierarchy of R_2 is lower than that of R_1, so Table 2(a) contains information of Table 2(b).

Table 1. Microdata

Name	Age	Gender	Zipcode	Disease
Alice	40	Female	100302	Flu
Bob	30	Male	100302	Gastritis
Carry	50	Female	100311	Flu
Daisy	40	Female	100311	Gastritis
Eric	30	Male	100313	Flu
Finn	50	Male	100313	Flu

QAs generalization hierarchy in Example 1 and the $1st$ anonymous data is shown in Figs. 5(a) and 6, respectively. For Incognito is a bottom-up algorithm, it can reduce pruning steps of the $2nd$ anonymity largely and improve efficiency, that the $1st$ anonymous data serve as input of M_2 rather than original data. Hence, HAM improve efficiency than previous models.

Node Storage Mode. The Node Storage Mode is established to satisfy Situation (2) in the beginning of Sect. 3. In traditional privacy-preserving models, one original data sheet only returns one anonymity node that satisfies a model. The NSM divides the process into two steps.

Table 2. Anonymous data for (a) R_1 and (b) R_2

Age	Gender	Zipcode	Disease
-	Person	100302	Flu
-	Person	100302	Gastritis
-	Person	100311	Flu
-	Person	100311	Gastritis
-	Person	100313	Flu
-	Person	100313	Flu

(a)

Age	Gender	Zipcode	Disease
-	Person	1003**	Flu
-	Person	1003**	Gastritis
-	Person	1003**	Flu
-	Person	1003**	Gastritis
-	Person	1003**	Flu
-	Person	1003**	Flu

(b)

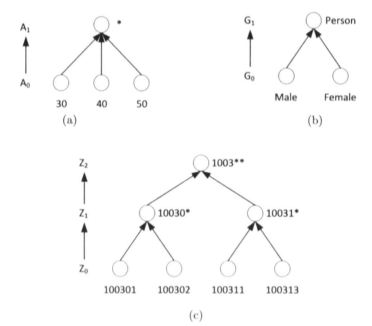

Fig. 4. Hierarchies of QAs: (a) Age, (b) Gender and (c) Zipcode

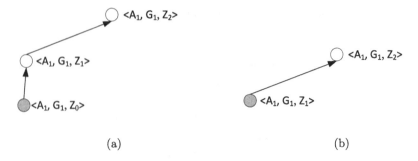

Fig. 5. Anonymization nodes of Table 1 in HAM: (a) $1st$ anonymous nodes and (b) $2nd$ anonymous nodes

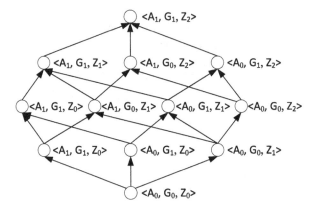

Fig. 6. The lattice of 3-QAs with $Q = \{A, G, Z\}$

Step 1. The attribute subset C generates data V_{C0}. In accordance with requirements of anonymity models, anonymization algorithm is conducted. Results of algorithm store all data nodes that satisfy privacy requirements as candidates.

Step 2. According to rules of roles, select one node from candidates, send it to the role as the access data. The selection rules involved here may be special application, for instance minimum classification loss or minimum information loss of a QA.

NSM separates the two requirements, i.e., privacy policies and a rule of node selection. Therefore, we call the first step rough model M and the second step selection rule F, respectively. Their combination are traditional model $\vec{M} = (M, F)$. In NSM, each role is endowed with a 3-dimensional tuple (Attribute set, Rough model, Rule).

Example 2. The original data is shown in Table 1. Roles R_3:(Q, 2-anonymity, minimum of ILoss) and R_4:(Q, 2-anonymity, minimum of Age ILoss) access data complying with NSM. The access subset $Q = \{A, G, Z\}$ of R_3 and R_4.

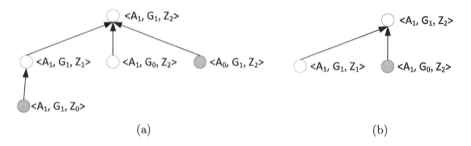

(a)   (b)

Fig. 7. Anonymization nodes: (a) 2-anonymity candidates and (b) HM: 2nd 2-diversity anonymity set

The 2-anonymous candidates are shown in Fig. 7(a). According to [12,14], the information loss of $< A_1, G_1, Z_0 >$ is the minimum. Then, send Table 2 which is generalized according to $< A_1, G_1, Z_0 >$ to R_3. Since R_4 is based on a different rule, which requires information loss of Age is the minimum. Via calculation, we obtain $< A_0, G_1, Z_2 >$ is the access node of R_4.

To sum up, for roles whose attribute sets and privacy-preserving models are the same but rules of node selection are different, we share the same candidate set that satisfies the privacy policy and then choose nodes in accordance with their different rules. A model which outputs one anonymous table once only, the algorithm need be implemented for different roles. Thus, our mode NSM saves operation time.

Hybrid Mode. The mode directs at supplement of NSM to HAM. It defines the k*th* anonymous data node in the form of candidate in NSM. In HM, the 0*th* anonymity set is defined as the original data composed of a QA subset and SA. The definition of the k*th* anonymity set is shown as follows.

Definition 1. *The kth anonymity set, denoted by D_k, is a set of nodes satisfying privacy-preserving model M_k with the (k-1)th anonymity set as input.*

D_k is just like the candidates in NSM. A naive method of the $(k+1)th$ anonymity execution is to adopt the model M_{k+1} for each node in D_k and store the results separately. Consequently, for each node in D_k, a new anonymity set is formed. Thus, several anonymity sets are generated. However, the $(k+1)th$ anonymity result of all nodes of D_k has intersection. At least, the root node satisfies all models and each $(k+1)th$ anonymity result contains it.

An improved method is to use D_k as input of M_{k+1} and combine all results satisfying M_{k+1} to D_{k+1}. In another word, D_{k+1} is obtained by pruning D_k according to whether M_{k+1} is satisfied or not. Therefore, in HM, each role is endowed with a 3-dimensional tuple: (Attribute set, Hierarchy, Rule). The anonymity set in HM satisfies monotonicity, i.e., for two anonymity sets D_i and D_j, information of v_{C_i} will contain information of v_{C_j} in case $0 \leq i \leq j \leq m$.

Example 3. We anonymize Table 1 in HM with M_1: 2-anonymity and M_2: 2-diversity. We have R_5:$(Q, 1,$ minimum of ILoss) and R_6:$(Q, 2,$ minimum of Sex ILoss). In accordance with the model M_1, we obtain $1st$ anonymity set $D_1 = \{< A_1, G_1, Z_2 >, < A_1, G_1, Z_1 >, < A_1, G_0, Z_2 >, < A_0, G_1, Z_2 >, < A_1, G_1, Z_0 >\}$. Then, $2nd$ anonymity set $D_2 = \{< A_1, G_1, Z_2 >, < A_1, G_1, Z_1 >, < A_1, G_0, Z_2 > \}$ is pruned by following M_2, as shown in Fig. 7. Thus, the node $< A_1, G_1, Z_0 >$ is an access node of R_5 and the node $< A_1, G_0, Z_2 >$ is an access node of R_6 by the selection rules of R_5 and R_6.

3.3   Access of Users

A user registers with his ID in administration center and the administrator sends a role to the user. The administrator writes down the binding information on a policy sheet which also has a tuple for each role of an anonymity mode. In accordance with the mode adopted by a role, each role is corresponding to a tuple. The tuple determines the access node for each role. A user logs in system via his or her role. After identity authentication and authorization, he or she can only read data nodes that his or her role is allowed.

3.4   Data Maintenance

With respect to role update, the administrator looks up whether there is the same attribute set in the policy sheet. In case there is the same attribute set, change attribute sets in the tuple of the role. Otherwise, add a new attribute set and generate data nodes that match with policy of the role.

   If changes in hierarchy, models or rules of a role make its access node change but access node is not added newly, we update the tuple in the policy sheet.

   If the privacy policy changes of a role cause privacy-preserving model changes, and all existing data nodes do not match the policy, then a new data node should be generated. It should comply with the attribute set, the privacy-preserving model and the rule under an anonymity mode. In HAM and HM, if the access node of a role R changes, then it may lead to access nodes whose roles with lower hierarchies than R are changed. Hence, roles with lower hierarchies need to be checked and updated.

3.5   Joint Attack

The information obtained by the role R is denoted by $E(R)$. Given two anonymous tables which is results of the privacy-preserving models $\overrightarrow{M}_1$ and the privacy-preserving models $\overrightarrow{M}_2$, additional information will disclose jointly [4,11,16]. Some articles call this sequential release and publishing do some profound research on this field [4,11,16]. Consequently, the privacy-preserving model that satisfied by combination of the two anonymous tables is weaker than $\overrightarrow{M}_1$ or $\overrightarrow{M}_2$. We denote it as $\overrightarrow{M}_1 \cap \overrightarrow{M}_2$. To recognize the privacy-preserving models satisfies different hierarchies conveniently, the "model" is added behind "Hierarchy" in HAM and HM.

In HAM, the roles $R_1 : (C_1, i, \overrightarrow{M}_1)$ and $R_2 : (C_2, j, \overrightarrow{M}_2)$ attack jointly. The attribute set that can be accessed is $C_1 \cup C_2$. According to monotonicity, we know that node on a path contain all information of successive nodes. Let $k = \min(i, j)$. Then, the information that can be obtained by jointing R_1 and R_2 does not exceed $E(C_1 \cup C_2, k, \overrightarrow{M}_1 \cap \overrightarrow{M}_2)$.

In NSM, the roles $R_1 : (C_1, M_1, F_1)$ and $R_2 : (C_2, M_2, F_2)$ attack jointly. The attribute set that can be accessed is $C_1 \cup C_2$. Let $\overrightarrow{M}_i = (M_i, F_i)$. Then, the information that can be obtained by jointing R_1 and R_2 does not exceed $E(C_1 \cup C_2, \overrightarrow{M}_1 \cap \overrightarrow{M}_2)$.

In HM, the roles $R_1 : (C_1, i, M_1, F_1)$ and $R_2 : (C_2, j, M_2, F_2)$ attack jointly. The attribute set that can be accessed is $C_1 \cup C_2$. Let $k = \min(i, j)$. Then, the information that can be obtained by jointing R_1 and R_2 does not exceed $E(C_1 \cup C_2, k, \overrightarrow{M}_1 \cap \overrightarrow{M}_2)$.

4   Conclusion

This paper proposes a model for an Anonymized data Privacy-aware Role-Based Access Control (AP-RBAC). Three anonymity modes are put forward HAM, NSM and HM. The model inherits advantages of access control, for it adopts attribute selection and carries out fine-grained access control for data security, on the one hand. On the other hand, it keeps advantages of privacy-preserving technology so that usability of data is improved while privacy is protected. The model makes anonymity operation among different privacy policies be coherent. Thus, flexibility, diversity and efficiency is improved on the premise that data security and data privacy are protected.

References

1. Abdalaal, A., Nergiz, M.E., Saygin, Y.: Privacy-preserving publishing of opinion polls. Comput. Security **37**, 143–154 (2013)
2. Bu, Y., Fu, A.W.C., Wong, R.C.W., Chen, L., Li, J.: Privacy preserving serial data publishing by role composition. Proc. VLDB Endowment **1**(1), 845–856 (2008)
3. David, F., Richard, K.: Role-based access controls. In: Proceedings of 15th NIST-NCSC National Computer Security Conference, vol. 563. NIST-NCSC, Baltimore, Maryland (1992)
4. Fung, B., Wang, K., Fu, A.W.C., Pei, J.: Anonymity for continuous data publishing. In: Proceedings of the 11th International Conference on Extending Database Technology: Advances in Database Technology, pp. 264–275. ACM (2008)
5. Huang, X., Liu, J., Han, Z., Yang, J.: A new anonymity model for privacy-preserving data publishing. China Commun. **11**(9), 47–59 (2014)
6. LeFevre, K., DeWitt, D.J., Ramakrishnan, R.: Incognito: efficient full-domain k-anonymity. In: Proceedings of the 2005 ACM SIGMOD International Conference on Management of Data. SIGMOD 2005, pp. 49–60. ACM, New York, NY, USA (2005)

7. Li, N., Li, T., Venkatasubramanian, S.: t-closeness: privacy beyond k-anonymity and l-diversity. In: IEEE 23rd International Conference on Data Engineering, 2007. ICDE 2007, pp. 106–115 (2007)
8. Li, N., Li, T., Venkatasubramanian, S.: Closeness: a new privacy measure for data publishing. IEEE Trans. Knowl. Data Eng. **22**(7), 943–956 (2010)
9. Machanavajjhala, A., Gehrke, J., Kifer, D., Venkitasubramaniam, M.: l-diversity: privacy beyond k-anonymity. In: 2013 IEEE 29th International Conference on Data Engineering (ICDE) 0, 24 (2006)
10. Ni, Q., Bertino, E., Lobo, J., Brodie, C., Karat, C.M., Karat, J., Trombeta, A.: Privacy-aware role-based access control. ACM Trans. Inf. Syst. Security (TISSEC) **13**(3), 24 (2010)
11. Shmueli, E., Tassa, T., Wasserstein, R., Shapira, B., Rokach, L.: Limiting disclosure of sensitive data in sequential releases of databases. Inf. Sci. **191**, 98–127 (2012)
12. Sun, X., Sun, L., Wang, H.: Extended k-anonymity models against sensitive attribute disclosure. Comput. Commun. **34**(4), 526–535 (2011). Special issue: Building Secure Parallel and Distributed Networks and Systems
13. Sweeney, L.: k-anonymity: a model for protecting privacy. Int. J. Uncertainty Fuzziness Knowl. Based Syst. **10**(05), 557–570 (2002)
14. Wong, R.C.W., Li, J., Fu, A.W.C., Wang, K.: (α, k)-anonymity: an enhanced k-anonymity model for privacy preserving data publishing. In: Proceedings of the 12th ACM SIGKDD International Conference on Knowledge Discovery and Data Mining. KDD 2006, pp. 754–759. ACM, New York, NY, USA (2006)
15. Xiao, X., Tao, Y.: Personalized privacy preservation. In: Proceedings of the 2006 ACM SIGMOD International Conference on Management of Data. SIGMOD 2006, pp. 229–240 (2006)
16. Xiao, X., Tao, Y.: M-invariance: towards privacy preserving re-publication of dynamic datasets. In: Proceedings of the 2007 ACM SIGMOD International Conference on Management of Data. SIGMOD 2007, pp. 689–700. ACM, New York, NY, USA (2007)

Functional Signatures
from Indistinguishability Obfuscation

Li Wang [1,2,3(✉)], Hongda Li[1,2], and Fei Tang[1,2,3]

[1] Data Assurance and Communication Security Research Center
of Chinese Academy of Sciences, Beijing, China
[2] State Key Laboratory of Information Security,
Institute of Information Engineering of Chinese Academy of Sciences, Beijing, China
[3] University of Chinese Academy of Sciences, Beijing, China
{wangli,lihongda,tangfei}@iie.ac.cn

Abstract. In PKC 2014, Boyle, Goldwasser, and Ivan introduced a cryptographic primitive called *functional signatures*. In a functional signature scheme, in addition to a master key that can be used to sign any message, there are signing keys for a function f, which allow one to sign any message in the range of f. In the same paper, Boyle et al. pointed out that in order to obtain a functional signature scheme with short signatures, we must either rely on *non-falsifiable assumptions* (as in their succinct non-interactive arguments of knowledge construction) or make use of non black-box techniques.

In this paper, we diverge from succinct non-interactive arguments of knowledge (SNARKs). We provide a construction of functional signature scheme satisfying both function privacy and succinctness under the existence of indistinguishability obfuscation for all polynomial-size circuits and one-way functions for the first time. Additionally, our scheme is under weaker assumption than $SNARK$-type assumptions for a class of functions and the size of signatures are independent of $f, f(m)$, and m.

Keywords: Functional signatures · Indistinguishability obfuscation · Non-falsifiable assumptions

1 Introduction

Indistinguishability Obfuscation. In 2001, Barak et al. [3,4] initiated the formal study of *program obfuscation*, which aims to make computer programs "unintelligible" while preserving their functionality. The obfuscator is a machine that takes as input a program, and produces a second program with same functionality while hides how the original program works. Ideally, an obfuscated program is a *virtual black-box* obfuscation which asks that an obfuscated program

---

This research is supported by the National Natural Science Foundation of China (Grant No. 60970139) and the Strategic Priority Program of Chinese Academy of Sciences (Grant No. XDA06010702).

M. Yung et al. (Eds.): INTRUST 2014, LNCS 9473, pp. 213–227, 2015.
DOI: 10.1007/978-3-319-27998-5_14

be no more useful than a black box implementing the program. Unfortunately, however, Barak et al. showed that this notion of general-purpose virtual black-box obfuscation is impossible to achieve. Motivated by this impossibility, they proposed the less intuitive, but potentially realizable, notion of *indistinguishability* obfuscation ($i\mathcal{O}$) which asks only that obfuscations of any two equal-size programs that compute the same function are computationally indistinguishable from each other. The $i\mathcal{O}$ became so important because there is a recent breakthrough result of Garg et al. [15] that put forward the first candidate construction for an efficient $i\mathcal{O}$ for general boolean circuits. The construction builds upon the multilinear map candidates of Garg et al. [14] and Coron et al. [11].

In parallel with the development of candidate obfuscation constructions, several surprising applications of indistinguishability have emerged: for instance, in the works of Garg et al. [15], Sahai and Waters [26], Hohenberger et al. [21], Boyle et al. [1], Boneh and Zhandry [10], Garg et al. [16], Bitansky et al. [2], Boyle and Pass [7]. Most notable among these is the work of Sahai and Waters [26] (and the punctured program paradigm it introduces) which shows that indistinguishability obfuscation is a powerful cryptographic primitive: it can be used to build public-key encryption from pseudorandom functions, selectively-secure short signatures, deniable encryption, and much more.

While the construction of indistinguishability obfuscation of Garg et al. [15] is based on some intractability assumptions. Actually, in the original constructive work of Garg et al. [15], the underlying explicit computational assumption encapsulated an exponential family of assumptions for each pair of circuits to be obfuscated. In the work of Pass et al. [23], the underlying assumption is a meta-assumption that also encapsulates an exponential family of assumptions, and this meta-assumption is invoked in a manner that captures the specific pair of circuits to be obfuscated. Recently, Gentry et al. [18] provide the construction of general-purpose indistinguishability obfuscation proven secure via a reduction to an instance-independent computational assumption over multilinear maps, namely, the multilinear subgroup elimination assumption.

Functional Signatures. In digital signature schemes, as defined by Diffie and Hellman [13], a signature on a message provides information which enables the receiver to verify that the message has been created by a proclaimed sender. The sender has a secret signing key, used in the signing process, and there is a corresponding verification key, which is public and can be used by anyone to verify that a signature is valid. Following Goldwasser et al. [19], the standard security requirement for signature schemes is unforgeability against chosen-message attack: an adversary that runs in probabilistic polynomial time and is allowed to request signatures for a polynomial number of messages of his choice, cannot produce a signature of any new message with non-negligible probability. While in a functional signature scheme, in addition to a *master signing key* that can be used to sign any message, there are secondary *signing keys* for functions f (called sk_f), which allow one to sign any message in the range of f. These additional keys are derived from the master signing key. To the security of functional signature, Boyle et al. [5] define the unforgeability. In addition to security, the functional signature scheme can hold two conditions which are *function privacy* and *succinctness*. In [5], Boyle et al. given three kinds of construction of a functional signature scheme. The first construction scheme, which has no function privacy and succinctness, is from any standard signature scheme. The second construction of functional signature scheme used non-interactive zero-knowledge arguments of knowledge and standard signature scheme. But it only has function privacy. In order to achieve function privacy and succinctness, Boyle et al. [5] presented the third construction scheme based on succinct non-interactive arguments of knowledge (SNARKs). However, constructing SNARKs is known to require the random oracle model [25] or non-falsifiable assumptions [20]. Boyle et al. [5] pointed out that in order to obtain a functional signature scheme with short signatures, we must either rely on *non-falsifiable assumptions* (as in their succinct non-interactive arguments of knowledge construction) or make use of non black-box techniques.

Our Motivations. In this paper, we try to diverge from SNARKs. So we should use other techniques to construct functional signature scheme which holds succinctness. There are several constructions of SNARKs known all based on non-falsifiable assumptions. A *falsifiable assumption* is an assumption that can be modeled as an interactive game between an efficient challenger and an adversary, at the conclusion of which the challenger can efficiently decide whether the adversary "won" the game. Most standard cryptographic assumptions are falsifiable. In our work, inspired by the short signature scheme proposed by Sahai and Waters [26], we use indistinguishability obfuscation to construct a new functional signature scheme. A shortcoming of our functional signature scheme can only achieve selective unforgeability, which is weaker than the adaptive unforgeability. Finally, we build a functional signature scheme which holds selective unforgeability, function privacy and succinctness using indistinguishability obfuscation and puncturable PRFs. Although the construction of indistinguishability obfuscation from falsifiable assumption is unknown, Gentry et al. [18] have constructed the indistinguishability obfuscation on a natural instance-independent sub-exponential hardness assumption over multilinear encodings which is weaker than the knowledge of exponent assumption that is used on the construction of SNARKs. Unfortunately, the primitive security definition of functional signature falls in "non-falsifiable assumptions". So if we don't set any constraints on functions, our scheme verification time is also under non-falsifiable assumptions. We constrain the functions with a property that there exists a PPT algorithm D, given message m^*, the function f and its domain, can decide whether m^* is in the range of f. This constraint makes sense. There are three reasons for that. First, this class of functions is big, all elementary functions belong to the class. Second, considering the goal of functional signatures, generally, if a party can not decide whether m^* is in the range of f, he wouldn't give the signing key sk_f to others. Because this case will cause the party which holds master key can not control the action of the party which holds signing key sk_f. Lastly, even though the same constraint is putted on the schemes which are proposed in [5], we yet can't get a functional signature scheme with short signatures under falsifiable assumptions. Because the scheme in [5] which holds succinctness uses SNARK system.

Puncturable PRFs. Recall that punctured PRFs, which is firstly introduced by Sahai and Waters [26]. Sahai and Waters [26] showed that how to use this technique to apply indistinguishability obfuscation towards cryptographic primitive. The puncturable PRFs are PRFs that can be defined on all bit strings of a certain length, except for any polynomial-size set of inputs. The next section gives a precise definition which is formulated as in [26].

1.1 Our Contributions

In our work, we provide a construction of functional signature scheme satisfying both function privacy and succinctness under the existence of indistinguishability obfuscation for all polynomial-size circuits and one-way functions for the first time. Although our scheme are only selective unforgeability, but it base on weaker assumptions. Boyle et al. [5] give three constructions of functional signature scheme. But only one construction which use SNARKs holds both succinctness and function privacy. All constructions of SNARKs known are based on non-falsifiable assumptions. Although the construction of indistinguishability obfuscation from falsifiable assumption are unknown now, Gentry et al. [18] have constructed the indistinguishability obfuscation on a natural instance-independent sub-exponential hardness assumption over multilinear encodings which is weaker than the knowledge of exponent assumption that is used on the construction of SNARKs.

Additionally, due to the primitive security definition of functional signature falling in "non-falsifiable assumptions". So if we don't set any constraints on functions, our scheme verification time is also under non-falsifiable assumptions. We set meaningful restriction on functions to make our scheme verification time under falsifiable assumption. Therefore, we think our scheme is close to the solution for the open problem that how to construct functional signatures with short (sublinear in the size of the functions supported) signatures and verification time under falsifiable assumption. And in our scheme, the size of signatures are independent of $f, f(m)$, and m.

1.2 Overview of the Paper

In Sect. 2, we describe several primitives which will be used in our constructions. In Sect. 3, we introduce the functional signature. In Sect. 4, we present how to use indistinguishability obfuscation to build a functional signature scheme.

2 Preliminaries

In this section, we define indistinguishability obfuscation, puncturable PRFs that are used in our constructions.

2.1 Indistinguishability Obufscation

Indistinguishability obfuscation was introduced in [3] and given a candidate construction in [15], and subsequently in [6,8,12].

Definition 1 (Indistinguishability obfuscation) [3]. A PPT algorithm $i\mathcal{O}$ is said to be an indistinguishability obfuscator for a class circuit $\mathcal{C}_\lambda$, if it meets the following conditions:

– Functionality preservation: For all security parameters $\lambda \in \mathbb{N}$, all $C \in \mathcal{C}_\lambda$, and all inputs x, we have:

$$\Pr[C'(x) = C(x) : C' \leftarrow i\mathcal{O}(\lambda, C)] = 1.$$

– Indistinguishability: For any (not necessarily uniform) PPT adversaries (Samp,D), there exists a negligible function α such that the following holds: if $\Pr[\forall x, C_0(x) = C_1(x) : (C_0, C_1, \delta) \leftarrow Samp(1^\lambda)] > 1 - \alpha(\lambda)$, then we have:

$$|\Pr[D(\delta, i\mathcal{O}(\lambda, C_0)) = 1 : (C_0, C_1, \delta) \leftarrow Samp(1^\lambda)]$$
$$-\Pr[D(\delta, i\mathcal{O}(\lambda, C_1)) = 1 : (C_0, C_1, \delta) \leftarrow Samp(1^\lambda)]| \le \alpha(\lambda).$$

2.2 Puncturable PRFs

Definition 2. A puncturable family of PRFs mapping is consists of a triple of algorithms Key, Pun, and F, and a pair of computable functions $n(\cdot)$ and $m(\cdot)$, satisfying the following conditions:

– Functionality preserved under puncturing: For every PPT adversary $\mathcal{A}$ such that $\mathcal{A}(1^\lambda)$ outputs a set $S \subseteq \{0,1\}^{n(\lambda)}$, then for all $x \in \{0,1\}^{n(\lambda)}$ where $x \notin S$, we have:

$$\Pr[\mathsf{F}(K, x) = \mathsf{F}(K(S), x) : K \leftarrow \mathsf{Key}(1^\lambda), K(S) \leftarrow \mathsf{Pun}(K, S)] = 1.$$

– Pseudorandom at punctured points: For every PPT adversary $(\mathcal{A}_1, \mathcal{A}_2)$ such that $\mathcal{A}_1(1^\lambda)$ outputs a set $S \subseteq \{0,1\}^{n(\lambda)}$ and state τ, consider an experiment where $K \leftarrow \mathsf{Key}(1^\lambda)$ and $K(S) \leftarrow \mathsf{Pun}(K, S)$. Then we have:

$$\Pr[\mathcal{A}_2(\tau, K(S), S, \mathsf{F}(K, S)) = 1] - \Pr[\mathcal{A}_2(\tau, K(S), S, U_{m(\lambda) \cdot |S|}) = 1] = negl(\lambda),$$

where $S = \{x_1, \ldots, x_k\}$ and $\mathsf{F}(K, S)$ denotes $\mathsf{F}(K, x_1) || \cdots || \mathsf{F}(K, x_k)$ is the concatenation of the elements of S in lexicographic order, and U_ℓ denotes the uniform distribution over ℓ bits.

The GGM tree-based construction of PRFs [17] from OWF are easily seen to yield puncturable PRFs, as realized by [5,9,22].

3   Functional Signatures

3.1   Definition

The definition of standard signature schemes is presented in Appendix A. The notion of functional signatures was introduced by Boyle et al. [5] and they have given three constructions of functional signature scheme.

Definition 3. [5] A *functional signature scheme* for a message space $\mathcal{M}$, and function family $\mathcal{F} = \{f : \mathcal{D}_f \to \mathcal{M}\}$ consists of algorithms (FS.Setup, FS.KeyGen, FS.Sign, FS.Verify):

- FS.Setup(1^k) → (msk, mvk): the setup algorithm takes as input the security parameter and outputs the master signing key and master verification key.
- FS.KeyGen(msk, f) → sk$_f$: the key generation algorithm takes as input the master signing key and a function $f \in \mathcal{F}$, and outputs a signing key for f.
- FS.Sign(f, sk$_f$, m) → ($f(m)$, σ): the signing algorithm takes as input the signing key for a function $f \in \mathcal{F}$ and an input $m \in \mathcal{D}_f$, and outputs $f(m)$ and a signature of $f(m)$.
- FS.Verify(mvk, m^*, σ) → {0, 1}: the verification algorithm takes as input the master verification key mvk, a message m^* and a signature σ, and outputs 1 if the signature is valid.

The functional signatures should hold two conditions: correctness and unforgeability. Two other conditions: function privacy, succinctness are hold optionally.

Correctness: We call a functional signature scheme correct if

$$\forall f \in \mathcal{F}, \forall m \in \mathcal{D}_f, (\text{msk}, \text{mvk}) \leftarrow \text{FS.Setup}(1^k), \text{sk}_f \leftarrow \text{FS.KeyGen}(\text{msk}, f),$$
$$(m^*, \sigma) \leftarrow \text{FS.Sign}(f, \text{sk}_f, m), \text{FS.Verify}(\text{mvk}, m^*, \sigma) = 1.$$

Unforgeability: The functional signature scheme is unforgeability if the advantage of any polynomial-time adversary $\mathcal{A}$ in the following game is negligible:

- The challenger generates (msk, mvk) ← FS.Setup(1^k), and gives mvk to the adversary $\mathcal{A}$.
- The adversary can query a key generation oracle $\mathcal{O}_{\text{key}}$, and a signing oracle $\mathcal{O}_{\text{sign}}$, that share a dictionary indexed by tuples $(f, i) \in \mathcal{F} \times \mathbb{N}$, whose entries are signing keys: sk$_f^i$ ← FS.KeyGen(msk, f). This dictionary keeps track of the keys that have been previously generated during the unforgeability game. The oracles are defined as follows:
    - $\mathcal{O}_{\text{key}}(f, i)$:
        * If there exists an entry for the key (f, i) in the dictionary, then output the corresponding value sk$_f^i$.
        * Otherwise, sample a fresh key sk$_f^i$ ← FS.KeyGen(msk, f) add an entry $(f, i) \to$ sk$_f^i$ to the dictionary, and output sk$_f^i$.
    - $\mathcal{O}_{\text{sign}}(f, i, m)$:
        * If there exists an entry for the key (f, i) in the dictionary, then generate a signature on $f(m)$ using this key: $\sigma \leftarrow$ FS.Sign(f, sk$_f^i$, m).

　*Otherwise,sample a fresh key $\mathsf{sk}_f^i \leftarrow \mathsf{FS.KeyGen}(\mathsf{msk}, f)$, add an entry $(f, i) \rightarrow \mathsf{sk}_f^i$ to the dictionary,and generate a signature on $f(m)$ using this key: $\sigma \leftarrow \mathsf{FS.Sign}(f, \mathsf{sk}_f^i, m)$.
- The adversary wins if it can produce (m^*, σ) such that
  - $\mathsf{FS.Verify}(\mathsf{mvk}, m^*, \sigma) = 1$.
  - There does not exist m such that $m^* = f(m)$ for any f which was sent as a query to the $\mathcal{O}_{\mathsf{key}}$ oracle.
  - There does not exist a (f, m) pair such that (f, m) was a query to the $\mathcal{O}_{\mathsf{sign}}$ oracle and $m^* = f(m)$.

Remark 1. One can consider assumptions where the challenger can't check in poly time whether the adversary has won or not, they are called "non-falsifiable assumptions". The security definition of functional signature falls in this category, and there are other examples in cryptography.

We define the *selective unforgeability* for functional signature schemes. The notion of *selectively unforgeable security* is as follow: a probabilistic polynomial time (PPT) adversary firstly selectively gives the challenger the message m^*, then the adversary is allowed to request signing keys for functions $f_1, \ldots, f_m$ of his choice, and signatures for messages $m_1, \ldots, m_q$ of his choice, can not produce a signature of the message m^*, which is not equal to any of the queried messages $m_1, \ldots, m_q$, and is not in the range of any queried functions $f_1, \ldots, f_m$.

Function Privacy: The functional signature scheme has the property of function privacy if the advantage of any polynomial-time adversary $\mathcal{A}$ in the following game is negligible:

- The challenger honestly generates a key pair $(\mathsf{mvk}, \mathsf{msk}) \leftarrow \mathsf{FS.Setup}(1^k)$ and gives both values to the adversary.
- The adversary chooses a function f_0 and receives an (honestly generated) secret key $\mathsf{sk}_{f_0} \leftarrow \mathsf{FS.KeyGen}(\mathsf{msk}, f_0)$.
- The adversary chooses a second function f_1 for which $|f_0| = |f_1|$(where padding can be used if there is a known upper bound) and receives an (honestly generated) secret key $\mathsf{sk}_{f_1} \leftarrow \mathsf{FS.KeyGen}(\mathsf{msk}, f_1)$.
- The adversary chooses a pair of values m_0, m_1 for which $|m_0| = |m_1|$ and $f_0(m_0) = f_1(m_1)$.
- The challenger selects a random bit $b \in \{0, 1\}$ and generates a signature on the image message $m' = f_0(m_0) = f_1(m_1)$ using secret key sk_{f_b}, and gives the resulting signature $\sigma \leftarrow \mathsf{FS.Sign}(\mathsf{sk}_{f_b}, m_b)$ to the adversary.
- The adversary outputs a bit b', and wins the game if $b' = b$.

Succinctness: We call a functional signature scheme has the property of succinctness if there exists a polynomial $s(\cdot)$ such that for every $k \in \mathbb{N}, f \in \mathcal{F}, m \in \mathcal{D}_f$,it holds with probability 1 over $(\mathsf{msk}, \mathsf{mvk}) \leftarrow \mathsf{FS.Setup}(1^k)$; $\mathsf{sk}_f \leftarrow \mathsf{FS.KeyGen}(\mathsf{msk}, f)$; $(f(m), \sigma) \leftarrow \mathsf{FS.Sign}(f, \mathsf{sk}_f, m)$ that the resulting signature on $f(m)$ has size $|\sigma| \leq s(k, |f(m)|)$. In particular, the signature size is independent of the size $|m|$ of the input to the function, and of the size $|f|$ of a description of the function f.

4   Construction

In this section, we present how to use indistinguishability obfuscation to build a functional signature scheme which holds function privacy and succinctnesss.

Our construction scheme is inspired by the short signature scheme proposed by Sahai and Waters [26]. In that construction, the signatures are essentially the classical PRF MACs. In order to make it publicly verifiable, they created an obfuscated verification program that checks the MACs. Sahai and Waters [26] showed that this construction is selectively unforgeable under the existence of indistinguishability obfuscation and one-way functions.

Our scheme consists of two obfuscated programs. The first is a sign algorithm that takes as input a message m, a signing key sk_f for a function f and the function f. It outputs $f(m)$ and a signature of $f(m)$ if $\mathsf{Sig.Verify}(\mathsf{vk}_{\mathsf{sig}}, \mathsf{sk}_f, f) = 1$. The signature is of the form of SW [26] short signature. Verification is done by an obfuscation of the algorithm Verify and is virtually identical to SW [26] short signature verification algorithm.

We construct a functional signature scheme FS = (FS.Setup, FS.KeyGen, FS.Sign, FS.Verify) as follows: Let F be a puncturable PRF that takes inputs of $\ell(k)$ bits and outputs $n(k)$ bits. Let $g(\cdot)$ be a one-way function. Let Sig = (Sig.Setup, Sig.Sign, Sig.Verify) be a signature scheme that is existentially unforgeable under chosen message attack.

- FS.Setup(1^k): The FS.Setup algorithm first chooses a puncturable pseudorandom function (PRF) key K for F. Then sample a signing and verification key pair $(\mathsf{sk}_{\mathsf{sig}}, \mathsf{vk}_{\mathsf{sig}}) \leftarrow \mathsf{Sig.Setup}(1^k)$. Next, it creates an obfuscation of the Sign of Fig. 1. The size of the program is padded to the maximum of itself and Program Sign* of Fig. 2. It also creates an obfuscation of the Verify of Fig. 3. The size of the program is padded to the maximum of itself and Program Verify* of Fig. 4. The mvk consists of the obfuscation program of Verify, the obfuscation program of Sign and $\mathsf{vk}_{\mathsf{sig}}$. The msk = $(K, \mathsf{sk}_{\mathsf{sig}})$.

- FS.KeyGen(msk, f): The FS.KeyGen algorithm generates $\mathsf{sk}_f = \mathsf{Sig.Sign}(\mathsf{sk}_{\mathsf{sig}}, f)$. The input f of algorithm FS.KeyGen is binary encoding of f.

- FS.Sign(f, sk_f, m): Run the obfuscation of Sign which takes as inputs f, sk_f, and m. The algorithm FS.Sign takes the output of the obfuscation of Sign as its output.

- FS.Verify(mvk, m', σ): Run the obfuscation of Verify, which takes as inputs m' and σ. FS.Verify takes the output of the obfuscation of Verify as its output.

Theorem 1. If our obfuscation scheme is indistingishuably secure, the signature scheme Sig is existentially unforgeable under chosen message attack, F is a secure punctured PRF, and $g(\cdot)$ be a one-way function, then our functional signature scheme is selectively unforgeable under chosen message attack.

Sign

Constants: Public key $\mathsf{vk_{sig}}$ and punctured PRF key K.
Inputs: Message m, signing key sk_f, and function f.

1. If $\mathsf{Sig.Verify}(\mathsf{vk_{sig}}, \mathsf{sk}_f, f) = 1$, then compute $m' = f(m)$, $\sigma = F(K, m')$. Output (m', σ).
2. Else output $\perp$.

Fig. 1. Program Sign

Sign*

Constants: Public key $\mathsf{vk_{sig}}$ and punctured PRF key $K(\{m^*\})$.
Inputs: Message m, signing key sk_f, and function f.

1. If $\mathsf{Sig.Verify}(\mathsf{vk_{sig}}, \mathsf{sk}_f, f) = 1$, then compute $m' = f(m)$, $\sigma = F(K, m')$. Output (m', σ).
2. Else output $\perp$.

Fig. 2. Program Sign*

Verify

Constants: Punctured PRF key K.
Inputs: Message m', signature σ.

1. Test if $g(\sigma) = g(F(K, m'))$. Output 1 if true, 0 if false.

Fig. 3. Program Verify

Verify*

Constants: Punctured PRF key $K(\{m^*\})$ and values $m^* \in \{0,1\}^{\ell(k)}, z^*$.
Inputs: Message m', signature σ.

1. If $m' = m^*$, test if $g(\sigma) = z^*$. Output 1 if true, 0 otherwise.
2. Else, test if $g(\sigma) = g(F(K, m'))$. Output 1 if true, 0 if false.

Fig. 4. Program Verify*

Proof. We describe a proof as a sequence of hybrid experiments where the first hybrid corresponds to the original functional signature security game. We prove that a PPT adversary's advantage must be negligibly close between each successive one. Then, we show that any PPT adversary in the final experiment that succeeds in forging with non-negligible probability can be used to break the security of the one-way functions.

- Hyb_0: In the first hybrid the following game is played.
    1. The adversary selectively gives the challenger the message m^*.
    2. The challenger chooses K for PRF F, then samples a signing and verification key pair $(sk_{sig}, vk_{sig}) \leftarrow Sig.Setup(1^k)$. Next, it creates an obfuscation of the Sign of Fig. 1. The size of the program is padded to the maximum of itself and Program Sign* of Fig. 2. It also creates an obfuscation of the Verify of Fig. 3. The size of the program is padded to the maximum of itself and Program Verify* of Fig. 4. The mvk consists of the obfuscation program of Verify, the obfuscation program of Sign and vk_{sig}. The challenger give mvk to the adversary.
    3. The adversary queries the key generation oracle $\mathcal{O}_{key}$, and been given sk_f. The adversary also queries the signing oracle $\mathcal{O}_{sign}$, and been given $F(K, f(m))$.
    4. The adversary sends a forgery σ and wins if
        - FS.Verify(mvk, $m^*, \sigma) = 1$.
        - there does not exist m such that $m^* = f(m)$ for any f which was sent as a query to the $\mathcal{O}_{key}$ oracle.
        - there does not exist a (f, m) pair such that (f, m) was a query to the $\mathcal{O}_{sign}$ oracle and $m^* = f(m)$.
- Hyb_1: This hybrid is the same as Hyb_0 except we replace the obfuscation of the program Sign with Sign* of Fig. 2.
- Hyb_2: This hybrid is the same as Hyb_1 except we let $z^* = g(F(K, m^*))$ and let the obfuscation of Verify be replaced by the obfuscation of the program Verify* of Fig. 4.
- Hyb_3: This hybrid is the same as Hyb_2 except $z^* = g(t)$ for t chosen uniformly at random in $\{0, 1\}^{n(k)}$.

First, we argue that the advantage for any PPT adversary in forging a signature must be negligibly close in hybrids Hyb_0 and Hyb_1.

We define the following two events:

- A: The adversary can distinguish hybrid Hyb_0 from hybrid Hyb_1.
- B: The adversary can compute a valid secret key sk_f for f with non-negligible advantage.

Then we have $Pr[A] = Pr[B] \cdot Pr[A|B] + Pr[\bar{B}] \cdot Pr[A|\bar{B}] \leq Pr[B] + Pr[A|\bar{B}]$. We now bound the probabilities of the two parts which is in the right hand of inequation.

First of all, $Pr[B] \leq negl(\lambda)$ since we have assumed that Sig is existentially unforgeable under chosen message attack.

Then we prove that $\Pr[A|\bar{B}] \leq negl(\lambda)$ by giving a reduction to the security of indistinguishability obfuscation.[1] We first observe that the input/output behavior of programs Sign and Sign* are identical. The only difference is that in Sign* the PRF at point m^* is punctured out of the constrained PRF key. However, we now consider the $\Pr[A|\bar{B}]$ that the event B will never happen. So the adversary cannot forge any valid secret key sk_f for f. Besides, there does not exist m such that $m^* = f(m)$ for any f which was sent as a query to the $\mathcal{O}_{key}$ oracle.[2] That is to say, $F(K, m^*)$ will never get called. Therefore, if there is a difference in advantage we can create an algorithm $\mathcal{B}$ that breaks indistinguishability security for obfuscation. $\mathcal{B}$ runs as the challenger. When it is to create the obfuscated program it submits both programs Sign and Sign* to the $i\mathcal{O}$ challenger. It sets the algorithm of Sign to the program returned by the challenger. If the $i\mathcal{O}$ challenger chooses the first, then we are in Hyb_0. If it chooses the second, then we are in Hyb_1. $\mathcal{B}$ will output 1 if the attacker successfully forges. Any PPT adversary with different advantages in the hybrids leads to $\mathcal{B}$ as an adversary on $i\mathcal{O}$ security.

In summary, the advantage for any PPT adversary in forging a signature must be negligibly close in hybrids Hyb_0 and Hyb_1.

Second, we argue that the advantage for any PPT adversary in forging a signature must be negligibly close in hybrids Hyb_1 and Hyb_2. We first observe that the input/output behavior of programs Verify and Verify* are identical. The only difference is that the first program computes $g(F(K, m^*))$ before applying the OWF g to it for message input m^*, whereas the second is given $g(F(K, m^*))$ as the constant z^*. Therefore, if there is a difference in advantage we can create an algorithm $\mathcal{B}$ that breaks indistinguishability security for obfuscation. $\mathcal{B}$ runs as the challenger. When it is to create the obfuscated program it submits both programs Verify and Verify* to the $i\mathcal{O}$ challenger. It sets the algorithm of FS.Verify to the program returned by the challenger. If the $i\mathcal{O}$ challenger chooses the first, then we are in Hyb_1. If it chooses the second, then we are in Hyb_2. $\mathcal{B}$ will output 1 if the adversary successfully forges. Any PPT adversary with different advantages in the hybrids leads to $\mathcal{B}$ as an adversary on $i\mathcal{O}$ security.

We now argue that the advantage for any PPT adversary in forging a signature must be negligibly close in hybrids Hyb_2 and Hyb_3. Otherwise, we can create a reduction algorithm $\mathcal{B}$ that breaks the selective security of the constrained pseudorandom function at the punctured points. $\mathcal{B}$ first gets m^* selectively from the adversary. It submits this to the constrained PRF challenger and receives the punctured PRF key $K(\{m^*\})$ and challenge a. It continues to run the experiment of Hyb_2 except it sets $z^* = g(a)$. If a is the output of the PRF at point m^*, then we are in Hyb_2. If it was chosen uniformly at random, then we are in Hyb_3. $\mathcal{B}$ will output 1 if the adversary successfully forges. Any adversary with different advantages in the hybrids leads to $\mathcal{B}$ as an adversary on the constrained PRF security. Here we were able to reduce to selective security since the attacker was defined to be selective.

---

[1] This part of proof is inspired by the soundness proof of NIZK in [26].

[2] Actually, this is a non-falsifiable assumption. Please see Remark 1 in detail.

Finally, if there is a PPT adversary in Hyb3, we can use it to break the security of the OWF. We build a reduction $\mathcal{B}$ that first takes in m^* selectively and receives y as the challenge for a OWF and sets $z^* = y$. If an adversary successfully forges on m^*, then by definition he has computed a σ such that $g(\sigma) = z^*$. Therefore, if the OWF is secure, no PPT adversary can forge with non-negligible advantage. Since the advantage of all PPT adversary's are negligibly close in each successive hybrid, this proves selective security for the functional signature scheme. $\square$

Theorem 2. Our construction of functional signature scheme holds function privacy and succinctness.

Function privacy From the signature of $f(m), \sigma = F(K, f(m))$, we can see that the signature have no information about the function f. That is if $\forall f_0, f_1, m_0, m_1$. s.t. $f_0(m_0) = f_1(m_1)$ then $\sigma_0 = F(K, f_0(m_0)) = F(K, f_1(m_1)) = \sigma_1$. So our construction of functional signature holds function privacy.

Succinctness The succinctness of our functional signature scheme follows from the structure of $\sigma = F(K, f(m))$. Namely, the signatures are essentially the classical PRF MAC. Actually, the size of a functional signatures only depend on the size of range of punctured PRF, it is independent of the size of $f, f(m)$, and m.

Theorem 3. Our functional signature scheme is under the assumptions of indistinguishability obfuscation and one-way functions. And if there exists a PPT algorithm D, given the message $m^* \in \mathcal{M}$, $f \in \mathcal{F}$ and the domain of f, D can decide whether m^* is in the range of f, then the verification time of scheme is under *falsifiable* assumptions.

The proof of Theorem 3 see Appendix B.

5   Conclusion

In this paper, we provide a construction of functional signature scheme satisfying both function privacy and succinctness under the existence of indistinguishability obfuscation for all polynomial-size circuits and one-way functions for the first time. Additionally, our scheme is under weaker assumption than $SNARK$-type assumptions for a class of functions and the size of signatures are independent of $f, f(m)$, and m.

Acknowledgements. The authors would like to thank anonymous reviewers for their helpful comments and suggestions.

A   Signature Schemes

Definition 4. A signature scheme for a message space $\mathcal{M}$ is a tuple (Gen, Sign, Verify) :

- Gen(1^k) → (sk, vk): the key generation algorithm is a probabilistic, polynomial-time algorithm which takes as input a security parameter 1^k, and outputs a signing and verification key pair (sk, vk).
- Sign(sk, m) → σ: the signing algorithm is a probabilistic polynomial time algorithm which is given the signing key sk and a message $m \in \mathcal{M}$ and outputs a string σ which we call the signature of m.
- Verify(vk, m, σ) → $\{0, 1\}$: the verification algorithm is a polynomial time algorithm which, given the verification key vk, a message m, and signature σ, return 1 or 0 indicating whether the signature is valid.

Correctness: We call a signature scheme correct if

$$\forall(\mathsf{sk}, \mathsf{vk}) \leftarrow \mathsf{Gen}(1^k), \forall m \in \mathcal{M}, \forall \sigma \leftarrow \mathsf{Sign}(\mathsf{sk}, m), \mathsf{Verify}(\mathsf{vk}, m, \sigma) \rightarrow 1$$

Unforgeability Under Chosen Message Attack: A signature scheme is unforgeable under chosen message attack if the winning probability of any probabilistic polynomial time adversary in the following game is negligible in the security parameter:

- The challenger samples a signing, verification key pair (sk, vk) ← Gen(1^k) and gives vk to the adversary.
- The adversary requests signatures from the challenger for a polynomial number of messages. In round i, the adversary chooses m_i based on $m_1, \sigma_1, \ldots, m_{i-1}, \sigma_{i-1}$, and receives $\sigma_i \leftarrow \mathsf{Sig}(\mathsf{sk}, m_i)$.
- The adversary outputs a signature σ^* and a message m^* and wins if Verify(vk, m^*, σ^*) → 1 and the adversary has not previously received a signature of m^* from the challenger.

B   The Proof of Theorem 3

Proof. In our functional signature scheme, we use indistinguishability obfuscation, signature scheme, one-way functions, puncturable PRFs. In the follow, we prove signature scheme and puncturable PRFs can be constructed if one-way function exists.

Lemma 1. [24] Under the assumption that one-way functions exist, there exists a signature scheme which is secure against existential forgery under adaptive chosen message attacks by polynomial-time algorithms.

Lemma 2. [5,9,17,22] If one-way functions exist, then for all efficiently computable functions $\ell(\lambda)$ and $n(\lambda)$, there exists a puncturable PRF family that maps $\ell(\lambda)$ bits to $n(\lambda)$ bits.

Based on lemmas 1, 2, we can conclude that our functional signature scheme is under the assumptions of indistinguishability obfuscation and one-way functions. And if exists a PPT algorithm D , $\forall f \in \mathcal{F}, m^* \in \mathcal{M}$, D can decide whether m^* is in the range of f, then in the proof of selective unforgeability, the verification time is polynomial time. Therefore, if there exists a PPT algorithm D, given the message $m^* \in \mathcal{M}$, $f \in \mathcal{F}$ and the domain of f, D can decide whether m^* is in the range of f, then our functional signature scheme verification time is under falsifiable assumptions.

References

1. Boyle, E., Chung, K.-M., Pass, R.: On extractability obfuscation. In: Lindell, Y. (ed.) TCC 2014. LNCS, vol. 8349, pp. 52–73. Springer, Heidelberg (2014)
2. Bitansky, N., Canetti, R., Paneth, O., Rosen, A.: Indistinguishability obfuscation vs. auxiliary-input extractable functions: one must fall. Technical report, Cryptology ePrint Archive, Report 2013/641 (2013)
3. Barak, B., Goldreich, O., Impagliazzo, R., Rudich, S., Sahai, A., Vadhan, S.P., Yang, K.: On the (im)possibility of obfuscating programs. In: Kilian, J. (ed.) CRYPTO 2001. LNCS, vol. 2139, p. 1. Springer, Heidelberg (2001)
4. Barak, B., Goldreich, O., Impagliazzo, R., Rudich, S., Sahai, A., Vadhan, S.P., Yang, K.: On the (im)possibility of obfuscating programs. J. ACM **59**(2), 6 (2012)
5. Boyle, E., Goldwasser, S., Ivan, I.: Functional signatures and pseudorandom functions. In: Krawczyk, H. (ed.) PKC 2014. LNCS, vol. 8383, pp. 501–519. Springer, Heidelberg (2014)
6. Barak, B., Garg, S., Kalai, Y.T., Paneth, O., Sahai, A.: Protecting obfuscation against algebraic attacks. In: Nguyen, P.Q., Oswald, E. (eds.) EUROCRYPT 2014. LNCS, vol. 8441, pp. 221–238. Springer, Heidelberg (2014)
7. Boyle, E., Pass, R.: Limits of extractability assumptions with distributional auxiliary input. IACR Cryptology ePrint Archive, p. 703 (2013)
8. Brakerski, Z., Rothblum, G.N.: Virtual black-box obfuscation for all circuits via generic graded encoding. In: Lindell, Y. (ed.) TCC 2014. LNCS, vol. 8349, pp. 1–25. Springer, Heidelberg (2014)
9. Boneh, D., Waters, B.: Constrained pseudorandom functions and their applications. In: Sako, K., Sarkar, P. (eds.) ASIACRYPT 2013, Part II. LNCS, vol. 8270, pp. 280–300. Springer, Heidelberg (2013)
10. Boneh, D., Zhandry, M.: Multiparty key exchange, efficient traitor tracing, and more from indistinguishability obfuscation. Technical report, Cryptology ePrint Archive, Report 2013/642, 2013 (2013). http://eprint.iacr.org
11. Coron, J.-S., Lepoint, T., Tibouchi, M.: Practical multilinear maps over the integers. In: Canetti, R., Garay, J.A. (eds.) CRYPTO 2013, Part I. LNCS, vol. 8042, pp. 476–493. Springer, Heidelberg (2013)
12. Canetti, R., Vaikuntanathan, V.: Obfuscating branching programs using black-box pseudo-free groups. IACR Cryptology ePrint Archive, p. 500 (2013)
13. Diffie, W., Hellman, M.E.: New directions in cryptography. IEEE Trans. Inf. Theory **22**(6), 644–654 (1976)
14. Garg, S., Gentry, C., Halevi, S., Raykova, M., Sahai, A., Waters, B.: Candidate indistinguishability obfuscation and functional encryption for all circuits. In: FOCS, pp. 40–49 (2013)

15. Garg, S., Gentry, C., Halevi, S., Sahai, A., Waters, B.: Attribute-based encryption for circuits from multilinear maps. Cryptology ePrint Archive, Report 2013/128 (2013). http://eprint.iacr.org/
16. Garg, S., Gentry, C., Halevi, S., Raykova, M.: Two-round secure MPC from indistinguishability obfuscation. In: Lindell, Y. (ed.) TCC 2014. LNCS, vol. 8349, pp. 74–94. Springer, Heidelberg (2014)
17. Goldreich, O., Goldwasser, S., Micali, S.: How to construct random functions (extended abstract). In: FOCS, pp. 464–479 (1984)
18. Gentry, C., Lewko, A., Sahai, A., Waters, B.: Indistinguishability obfuscation from the multilinear subgroup elimination assumption. Cryptology ePrint Archive, Report 2014/309 (2014)
19. Goldwasser, S., Micali, S., Rivest, R.L.: A digital signature scheme secure against adaptive chosen-message attacks. SIAM J. Comput. 17(2), 281–308 (1988)
20. Gentry, C., Wichs, D.: Separating succinct non-interactive arguments from all falsifiable assumptions. In: STOC, pp. 99–108. ACM (2011)
21. Hohenberger, S., Sahai, A., Waters, B.: Replacing a random oracle: full domain hash from indistinguishability obfuscation. Technical report, Cryptology ePrint Archive, Report 2013/509, 2013 (2013). http://eprint.iacr.org
22. Kiayias, A., Papadopoulos, S., Triandopoulos, N., Zacharias, T.: Delegatable pseudorandom functions and applications. IACRCryptology ePrint Archive, p. 379 (2013)
23. Pass, R., Seth, K., Telang, S.: Indistinguishability obfuscation from semantically-secure multilinear encodings. Cryptology ePrint Archive, Report 2013/781 (2013). http://eprint.iacr.org/
24. Rompel, J.: One-way functions are necessary and sufficient for secure signatures. In: STOC, pp 387–394 (1990)
25. Silvio, M.: Computationally sound proofs. SIAM J. Comput. 30(4), 1253–1298 (2000)
26. Sahai, A., Waters, B.: How to use indistinguishability obfuscation: deniable encryption, and more. IACR Cryptology ePrint Archive, p. 454 (2013)

Lightweight Protocol
for Trusted Spontaneous Communication

Przemysław Błaśkiewicz, Marek Klonowski, Mirosław Kutyłowski[✉], and Piotr Syga

Faculty of Fundamental Problems of Technology Department of Computer Science,
Wrocław University of Technology, Wrocław, Poland
{przemyslaw.blaskiewicz,marek.klonowski,
miroslaw.kutylowski,piotr.syga}@pwr.edu.pl

Abstract. We present a communication protocol with encryption, suitable for extremely weak devices, which communicate only by sending un-modulated, on/off signals (beeping). We assume severely constrained model with no coordination or synchronization between devices, and no mechanism for message reception acknowledgement. Under these assumptions, we present a way to handle the problem of transmissions interference (collisions) and providing message secrecy at the same time.

In order to achieve our goals in such a limited communication channel, we use special encoding and combine encryption procedure with the communication layer of the protocol. This is different from the state-of-the-art-today, where an encrypted channel is built in the highest level of the communication protocol after assigning the radio channel to one of the sender devices. We present a real-life motivations for the proposed approach as well as rigid correctness and security analysis.

Keywords: ad hoc network · Constrained device · Beeping model · Mobile device · Visible light communication

1   Introduction

In our paper we consider an ad hoc network of severely constrained mobile devices or sensors that have to transmit some information towards a selected, relatively stronger device called a *sink*. We assume that the devices are not coordinated and the communication pattern is unpredictable and possibly chaotic.

The execution time of the network is divided into units called *slots*. In a single slot, a device may only send a signal or abstain from transmitting, hence it takes several slots to transmit a complex message. Apart from limited computation capabilities of the devices, the main obstacle for successful communication are physical properties of the communication channel. We assume that each node may spontaneously start transmitting, however if more than one transmits in a given time, the messages may

The first three authors have been supported by NCN, decision number DEC-2013/08/M/ST6/00928 (Harmonia). The last author has been supported by NCN, decision number DEC-2012/07/N/ST6/02203.

© Springer International Publishing Switzerland 2015
M. Yung et al. (Eds.): INTRUST 2014, LNCS 9473, pp. 228–242, 2015.
DOI: 10.1007/978-3-319-27998-5_15

interfere, hence become unreadable. We assume that there is no prior assignment of slots to the network nodes, and nothing external to the protocol indicates the moment when they can start their transmissions. We also assume that the nodes do not listen to the channel, in order to learn if no other node is transmitting, before starting their transmission. Furthermore, the sink does not send any confirmation messages, so the nodes do not know if their messages were properly received. Lastly, we assume that all communication in the network may be eavesdropped by an external adversary.

Under these constraints we want to assure that messages sent to the sink remain private, i.e. the adversary cannot gain access to the information being sent. A natural approach to this problem is to encrypt each message, possibly after enforcing some way of authentication between sender and the receiver. Depending on the protocol in use, this may introduce a relatively heavy computation overhead and large volume of auxiliary data to be stored and/or transmitted. A lot of effort must be devoted to self-organization of the communication, as the protocol data must be delivered in a reliable way, typically in packets.

In order to avoid efficiency problems, we move away from the traditional separation of different protocol layers and by means of special encoding, we merge security and transmission layers together. Doing so, we construct a protocol suitable for constrained devices, that can be efficiently performed in the assumed channel model. Our goal is not only to allow proper transmission from the devices to the sink, but also to provide secrecy of both the content of the message as well as the identity of the sender.

Motivation. Traditionally, communication protocols are built in a layered architecture, which separates different design stages. This has many advantages, mainly concerning ease of design, flexibility and error avoidance. As data packets are transmitted, information they carry is wrapped (processed) in envelopes of each layer protocols and passed down to the device responsible for physically modifying the transmission medium and hence – transmission. Usually, cryptographic protection is implemented on specific layers (e.g. IEEE 802.1X on Data Link; IEEE 802.15.4 on MAC layer, sublayer of Data Link; IPSec on Link Layer, etc.) and pertains to this layer protection only. However, in scenarios where multi-layered architecture poses too much overhead (in hardware, memory and/or computation), this traditional and generally successful approach might be problematic.

Applicability of such constrained systems ranges from control networks in high risk industrial areas to typical military applications. As an example, consider security systems where the alert data originates from sensor devices and has to be delivered to the core system via sink nodes. In all these cases manipulations in the system may occur – (frequent) tampering with readings of the methane detectors in underground coal mines may be a good example. In case of the coal mining monitoring systems the obligatory system of methane detectors may have a shadow system that is used only for detecting manipulations on the official measurements. In this case it is crucial that the sink does not transmit anything - otherwise it can be easily detected, found and destroyed in order to remove the evidence of manipulations. One can also think of a setting similar to that of a TV set and a remote controller – it is not required for the TV set to send any information back to the remote.

Similarly, our target is also communication between vehicles in extreme conditions. This may concern road communication systems for vehicles, as well as air traffic control systems or even air defense systems. In all cases communication should be protected against sabotage (e.g. attacks by creating crashes by injecting false information) and should work in extreme conditions with unpredictable communication collisions.

Also, we may consider a sensor network based on Li-Fi infrastructures (i.e., access point and corresponding transmitting stations). While hiding the access point in this case is unfeasible, if not senseless, other means of securing the transmission could be of importance. It may be that the signals (in this case, flashes of light) intended for the access point are falsely generated by an adversarial node in order to inject false data into the system. Alternatively, such light-based communication can be prone to accidental (or purposeful) interference. We need mechanisms to overcome these problems (message trust, interference) while at the same time maintaining low requirements for hardware complexity.

The Core Problem. In order to build systems sketched above we need to design a communication protocol for the following fundamental scenario:

Information flow: The system consists of a relatively powerful node called a *sink*, and a number of nodes that have the sink in their transmission range.

Asynchronicity: Neither the sink nor the other nodes have any prior knowledge about which node will attempt to upload data to the sink at a given moment.

Shared secrets: The sink has access to a database that stores secret keys for each of the nodes deployed.

Essentially, we do not need the nodes to be synchronized with a common clock. We only assume that the clocks run at *approximately* the same speed for all devices. The transmission time is divided into slots of the same length, but we accommodate possible time drift allowing that the slot boundaries may differ. Nevertheless, for clarity of presentation from now on we assume that all devices are synchronized regarding division into the time slots.

Paper Contribution. In this paper we propose a low layer communication protocol for encrypted transmissions assuming communication channel with properties based on the beeping model. It ensures that the communication is relatively immune to the transmission collisions despite the limitations of the channel and despite overlapping of messages coming from different nodes. We achieve this by using special encoding method that we call r-sparse code. Moreover, we assume that there is no backward information channel from the sink to the nodes, and there is no synchronization nor communication between the nodes.

In Sect. 2 we present the assumed communication model. Section 3 contains description of the proposed scheme. Section 4 provides a protocol analysis with the main focus on the soundness and security. We briefly indicate related work in Sect. 5. In Sect. 6 we propose some possible improvements and provide discussion for applicability of our work.

2   Communication Model

We limit ourselves to a very simple way to send information. Namely, the sender can act on transmission medium over a specified period of time or not. The former situation corresponds to sending a logical "one", the latter – means that the station either is sending a logical "zero" or is inactive. This behavior is encompassed by the *beeping model* [1]. In this framework, a node station can perform only two actions:

1. send a "beep" – an impulse of a carrier wave,
2. sense the channel and determine if a carrier wave is present. Accordingly, the state of the radio channel is regarded as no signal (when all nodes abstain from transmitting a beep and the energy level for the beep frequency is at the noise level) and signal (if at least one node is transmitting and increases the energy level for the beep frequency).

It must be stressed that our proposal is not limited to a radio communication networks. In fact, the reader will find it plausible to apply it in Visible Light Communication (VLC) [3], where information is carried by visible light impulses. Notably, this particular communication method has been getting more attention in connection with 5G standardization (Li-Fi – see [21]).

In the following, we denote by "beep" any acting on the medium (such as sending an un-modulated carrier or flashing an LED diode). Similarly, the "carrier sensing" can refer to either typical radio wave sensing or detecting a flash of light on a CCD matrix.

Within beeping model given above we set additional limitations to construct our model as follows:

Beep length: we divide time into slots of constant length μ that is the shortest time required for the carrier sensing mechanism to determine a signal at frequency f_c. A single beep lasts for time μ.

Carrier sensing: we assume that only the *sink* can perform carrier sensing, while other nodes can only transmit "beeps". This way we extend our considerations to a purely randomized ad-hoc radio networks, where a station can commence sending its message regardless of other stations' transmissions.

Upper bound on number of senders: we assume that at most k nodes may attempt to transmit at the same time. The parameter k corresponds to a bound occurring in practice.

2.1   Discussion of Model Properties

The model we propose suits well to our supposed applications in constrained devices communicating in an ad-hoc manner to a single sink.

Un-Modulated Signals. Firstly, carrier sensing (at the sink) can be performed at a much smaller energy cost than standard listening on the radio: carrier sensing is typically done by separate circuitry: no PLL synthesizing or demodulation is required. By the same token, since the transmitter sends only an un-modulated wave, the energy costs and complexity of the transmitter are low and the protocol may be employed in simpler hardware (no need for complex modulators).

Reportedly, carrier sensing can have poor performance in a noisy environment [12]. For example, a false-positive detection can occur if a number of weak transmissions correlate and their overall power picked by the antenna exceeds a certain threshold. This problem also occurs for ASK (amplitude shift keying) modulation. As we show later, the problem of susceptibility of ASK to noise and interference can be overcome with high probability using additional computation, making it a practical, simple modulation for constrained, low-cost devices.

Efficiency. We leverage modulation similar to ASK, which has better spectral efficiency than FSK, and still improve over it by setting smallest μ possible.

In a way, the beeping model is similar to ASK, especially its variant OOK (on-off keying). In this case, the carrier wave is a sine wave of frequency f_c, and the modulating signal is a zero-one sequence of frequency f_m. Typically, $f_c = C \cdot f_m$ for some integer C. The modulated signal is obtained by multiplying carrier by modulating binary sequence, so that it is 0 when transmitting a 0, and the clear carrier when transmitting a 1. The throughput of such modulation (as well as FSK and PSK modulation schemes) for a given f_c depends on parameter C – it determines how many cycles of the carrier are necessary to encode a single bit of information. Typically, $C = 16, 32, 64, 128$. In our approach, we do not use $C \cdot f_m$ as the time base for a single character but the time constant μ, that is of length of a few periods of the carrier. This way, we approach the theoretical bandwidth efficiency of 1 bit/Hz/second, so we maximize the band utilization under ASK (or beeping model) assumptions.

Collisions and Channel Utilization. Since nodes do not perform carrier sensing, they transmit independently of each other, and, consequently, their signals may overlap. Further, we assume that a situation when two or more nodes beep in the same slot is indistinguishable by the sink from that when a single node beeps. An exemplary illustration of communication and channel state in our model is shown in Fig. 1.

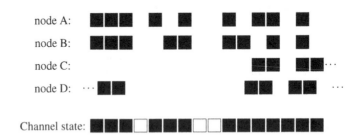

Fig. 1. Several exemplary slots out of network's lifespan for parameters $r = 2$, $m = 3$, $l = 2$, $n = 2$. Black boxes symbolize sending a beep, white boxes refer to silence. Note that each node may start transmitting anytime – some transmissions may overlap completely, some nodes may start transsmitting later on, and some may even transmit parts of two different messages, while other node transmits one.

It can be seen that differentiating between transmissions of nodes A, B, C and D is impossible if the only information available is the channel state. In the next section we describe the protocol, by which the sink can do it with the aid of additional off-line information.

2.2 Adversary Model

In our paper we consider a *passive adversary*, i.e., an outer entity that can eavesdrop radio communication, however does **not** send transmissions of its own. The adversary listens to all communications: we do not rely the security of our protocol on the fact that he fails to capture some fraction of it. Computational capabilities of the adversary are similar to those of a sink. Furthermore, the adversary knows the hash function $\mathcal{H}$ and initial identifiers of nodes, however they do not know the secrets shared between nodes and the sink. In our paper, we assume that the adversary has two main goals:

plaintext recovery: the adversary aims at learning the content of the messages transmitted to a sink,
communication linking: the adversary aims at determining if two different transmissions have been sent by the same node.

In our paper we do not assume that the adversary is able to capture nodes, however if such an ability is introduced, security guaranties of our protocol remain in effect for all the nodes that have not been compromised.

3 Protocol Description

r-sparse Coding. In our protocol we present a simple bit encoding method called *r-sparse coding*, (or r-SC, for short) suitable for channels with possible conflicts caused by transmissions of uncoordinated nodes. Its idea is loosely related to Bloom Filters [4]. This protocol turned out also to be very useful as a building block in encrypting messages for constrained devices.

In order to mitigate the interference effect between different transmissions, we encode each bit using r-SC, for a network parameter r. Each bit in r-SC is represented by r transmission slots. Two different slots, say p_0 and p_1, are distinguished: for encoding 0 there is a single beep sent in the time slot p_0, for encoding a 1 the only beep is sent in the slot p_1.

There are $r(r - 1)$ ways to choose p_0 and p_1, and therefore $r(r - 1)$ possible encodings. The protocol described below uses a pseudo-random function based on a shared secret to choose the encoding used. An example of r-codings is shown in Fig. 2.

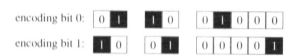

Fig. 2. All 2-sparse codes and an example of a 5-sparse code.

Setup. Before network deployment, we define a common parameter r, which determines the r-sparse coding. We choose $r > 2k$, where k is the bound on the number of nodes that may attempt to transmit at the same time.

For each node A the system generates at random a pseudonymous identifier ID_A (which can be changed during the protocol execution) and a secret $\mathcal{K}_A$ (which is constant throughout the node's lifespan). The secret $\mathcal{K}_A$ and pseudonym ID_A are stored by node A as well as in the database of the system. The length x of $\mathcal{K}_A$ should be sufficient to guarantee that a keyed hash function $\mathcal{H}$ can be modeled as a Random Oracle. The pseudonym ID_A should be long enough to avoid name collisions as well as getting a valid pseudonym used in the system when a random string is generated.

Coding. In order to encode a single bit we use r-sparse coding. The selection of the r-sparse coding out of $r(r-1)$ possibilities is done according to the value of function $\mathcal{H}(ID_A, \mathcal{K}_A, i) \mod r(r-1)$, where i is the number of current bit in a given transmission. Therefore, according to the Random Oracle Models we may assume that every single bit is encoded randomly and independently. Note however that knowing ID_A, the secret key and the number of the bit in the transmission, one can easily decode it. Figure 3 depicts an exemplary coding for $r = 4$.

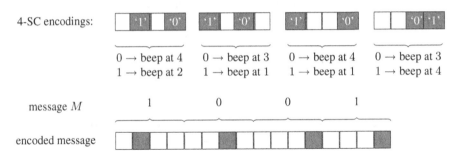

Fig. 3. Exemplary encoding of message 1001 using 4-SC. $\rightarrow$ represents choosing encoding via the function $\mathcal{H}_r(ID_A, \mathcal{K}_A, i)$. Four consecutive encodings are shown (top row), which are applied to bits of the message one at a time. Dark squares represent timeslot with a beep, they are labelled with bit they represent in this encoding. The resulting sequence of beeps and silence is shown in the last row.

Transmission. Let $\mathcal{A}$ be the set of all nodes. Each node may decide to start its transmission **at any time**. The only restriction we make is the upper bound on the number of nodes transmitting in the same time (parameter k). In order to transmit a message M with bit representation $b_1 \ldots b_n$, node $A \in \mathcal{A}$ transmits a message consisting of the following three parts.

Preamble – part responsible for signaling the start of a transmission. It consists of beeps in r consecutive slots.

Identification part – presenting the current pseudonym of the sender, enabling the sink to establish decoding method. In order to do this, the sender transmits its current ID_{sender}. Each of the m bits of ID_{sender} is sent separately encoded by r-SC. This part lasts $m \cdot r$ slots. In our protocol we take $m = 2k + \log(|\mathcal{A}|) + \lceil \log \frac{1}{\delta} \rceil$, where k is an upper bound on the number of nodes transmitting in the same time, $\mathcal{A}$ is the set of nodes in the system and δ is a parameter responsible for bounding probability of *false positives* in the identification.

Payload part – responsible for transmitting $M^{(l)}$, where $M^{(l)}$ is created from M by repeating each bit l times. Each bit of M^l is encoded separately using independently chosen encoding. For transmitting the ith bit of $M^{(l)}$ it uses the r-sparse code determined by the function $\mathcal{H}_r(ID_A, \mathcal{K}_A, i)$. We use $l = k + \lceil \log \frac{n}{\varepsilon} \rceil$, where the security parameter $\varepsilon > 0$ is an upper bound on probability that at least one bit of the original message M can be unreadable (due to collisions with other transmissions) to the sink. Note that together we need $r \cdot n \cdot l$ slots for this part.

After transmission the node updates its identifier as follows: $ID_A := \mathcal{H}(\mathcal{K}_A, ID_A)$.

Fig. 4. Example of transmitting a message 101 by the node for system parameters: $r = 2$, $m = 3$, $l = 2$. The black boxes depict beeps, the white boxes indicate silence. Note that during message repetition the encodings are different.

Decoding and Decryption. Note that in our protocol it is possible that many (however up to k) nodes start transmissions in the same slot. It is also possible that one node starts transmission while some other transmissions are in progress. In both cases we show that all messages can be decoded and decrypted.

The decoding procedure consists of three parts:

Transmission detection: the sink continuously listens to the channel. Once it detects a sequence of r slots of beeps starting in slot t, it starts the identification phase of possible transmitting node. Note that this **may** indicate that some node transmitted a preamble.

Identification phase: the sink finds in its database all A with ID_A *coherent* with slots from $t + r$ to $t + r + m \cdot r - 1$. Namely, the sink has to check if there are beeps on all respective positions with 1 in ID_A. For each such A a separated virtual channel is opened.

Decoding for virtual channel for a node A: after creating a virtual channel, the sink tries to retrieve the message bit by bit. The i-th bit (for $1 \leq i \leq n$) is encoded independently l times in slots from $(t + r + m \cdot r) + (i - 1) \cdot l \cdot r$ to $(t + r + m \cdot r) + i \cdot l \cdot r - 1$. Clearly, the sink knowing $\mathcal{K}_A$ may attempt to decode all bits. The following cases are possible:

  single bit: for only one bit value b the beeps occur at all l positions where the node ID_A is supposed to beep for value b. On the other hand, on at least one position for the bit $1 - b$ there is a silence. In this case the sink appends b to the decoded contents of the virtual channel for node ID_A.

unknown: for both $b = 0$ and $b = 1$ the beeps occur at all positions where node ID_A is supposed to beep for b. In this case the sink appends the mark "?" to the decoded contents of the virtual channel for ID_A. This corresponds to the case that the sink detects that a bit has been transmitted by ID_A, but it cannot resolve its correct value.

failure: In all other cases the sink can be sure that the node ID_A has not transmitted and can close the virtual channel corresponding to node ID_A.

After decoding n bits the sink updates the identifier according to the formula $ID_A := \mathcal{H}(\mathcal{K}_A, ID_A)$.

Let us stress that the decoding procedure can open several virtual channels starting in the same slot t. Moreover, the procedure of transmission detection is started for all consecutive r slots with beeps. In particular, if we have $w > r$ beeps in a row, then the sink takes into account $w - r + 1$ possible start positions for the preamble.

4   Protocol Analysis

Probability of Correct Message Decoding. We say that the message is *decodable* if the sink is able to retrieve the original message M from the content transmitted by a given node A. We say that a bit is *ambiguous* if it is decoded as "?" given the transmitted message and the key $\mathcal{K}_A$.

Theorem 1. *Every transmitted message is decodable by the sink with probability at least $1 - \varepsilon$ regardless other nodes' transmission starting times.*

Proof. Recall the assumption that during the transmission of any message at most $k - 1$ other nodes can transmit. Moreover, we set $r > 2k$.

First we observe that if a new station starts to transmit, this event is always detected. Indeed, a block of at least r adjacent slots of 1's appears on the channel if and only if a station transmits the initial part of its message. More precisely, the following lemma holds:

Lemma 1. *Let us assume that at most k stations are transmitting using r-SC coding. If $r > 2k$, then in each block of r consecutive slots there is at least one slot with no beep.*

Proof. Each bit of information in r-SC contains exactly one slot with a beep. Since r consecutive slots can overlap with at most two blocks representing adjacent bits (in r-SC) transmitted by a single station, the total number of 1's transmitted by all nodes is at most $2k < r$ in a block of r consecutive slots.

That is, there is at least r consecutive beeps in a row if and only if at least one node transmits its preamble. Therefore, the transmission detection sub-procedure cannot "miss" any transmission.

Next we need to prove that the identification procedure correctly determines the identity of the sending station A. It is clear that if A transmitted its ID_A, it will always be detected. Indeed, if there are beeps in all expected slots as given by r-SC encoding of ID_A, the transmissions of other nodes cannot change it (a 'beep' in the channel cannot be attenuated). In other words, there is no *false negative* (just as for the Bloom filters (cf. [4, 15])). The other error - i.e. starting decoding for a station that is not the real source of transmission is discussed latter.

Since the fact of transmission is always correctly detected and the sender is always correctly identified, the only reason why the encrypted message may be not *decodable* is that at least one bit of the message M is *ambiguous*.

Lemma 2. *Let us consider a transmission of a single bit using r-SC. If during r slots of the transmission of this bit no other node transmits its preamble, then the probability that the bit is* **not** *ambiguous is at least* $(1 - 1/r)^{k-1} > (1 - 1/r)^{r/2} \geq 1/2$.

Proof. According to the r-encoding the sink expects a beep in one of two slots determined by the key $\mathcal{K}_A$ (that is known to the sink after properly recognizing of A as the sender) and some other public parameters. If the sink observes no beep on one of these two positions, then it can be sure that beep was transmitted on the other position and the bit can be unambiguously decoded. The slot with no beep on the observed position may be covered independently by each of at most $k - 1$ other transmissions with probability at most $1/r$ each. Since choices of each transmitting stations are independent the probability of no beep is at most $(1 - 1/r)^{k-1}$. □

Lemma 3. *Let us consider an n-bit message. The probability that at least one bit of the message is ambiguous does not exceed ε.*

Proof. If there is no preamble transmitted, then a single r-SC code is successfully decoded with probability exceeding $1/2$. After l transmissions the probability that the bit is ambiguous is at most $(1/2)^{l-k} = (1/2)^{\lceil \log(\frac{\varepsilon}{n}) \rceil} \leq \frac{\varepsilon}{n}$ (since the preambles may affect at most k transmissions). Thus using union bound argument for all n bits we have that the probability that there is at least one bit that failed to be successfully transmitted does not exceed $n \cdot \frac{\varepsilon}{n} = \varepsilon$. □

Lemma 3 completes the proof of Theorem 1.

"False Positive" Sender Identification. Let us discuss the probability of the case that the protocol erroneously identifies the sender of the message. According to the model, decoding of the identification part always indicates the real sender. However, "false positives" are also possible: identification procedure may also point to some other $A' \neq A$ as a potential sender. Indeed, several beeps from different transmissions can cover the whole pattern that is a result of r-SC encoding of $ID_{A'}$. The theorem below states that such false positive error happens with a small probability for properly chosen parameters.

Theorem 2. *Probability that the decoding procedure returns a pseudonym of a node A' that has* **not** *transmitted the preamble at the considered time t is smaller than δ, provided that A' either has not transmitted its preamble at time t' where $|t' - t| \leq r$.*

Proof (Sketch). Let us recall that the identifiers of nodes of $\mathcal{A}$ are pseudorandom bits encoded by a pseudo-random r-SC. According to the properties of the hash function $\mathcal{H}$ the bits of the identifier can be regarded as stochastically independent. First note that transmitting m bits encoded with r-SC requires $r \cdot m$ slots, in which a beeps occur. A node's identifier is accepted during decoding process if and only if there is a beep on all these m positions.

We need to consider two cases. In the first case we assume that node A' has not transmitted at all. In that case messages transmitted by nodes are independent on position of m beeps in the identifier of A'. Let us fix a slot and assume that no node is transmitting a preamble. Then the probability that in this slot there is no beep from any node is $(1 - 1/r)^k > (1 - 1/r)^{r/2} \geq 1/2$ for $r \geq 2$. Now, considering possible influence of preamble transmissions, which can affect at most $2k$ (out of m) slots, one gets the upper bound on probability of a *false positive accepting* A' to be $(1/2)^{m-2k}$.

In the second case we assume that A' has transmitted, but started in the slot $t' \neq t$, where t is the time when A start its preamble. One easily proves that probability in this case cannot be greater than in the previous case. Intuitively, since $|t' - t| > r$ the problem boils down to the first case due to independence of encoding of different bits.

The number of potential identifiers that are tested is $|\mathcal{A}|$. Using union bound argument we have that the probability considered in Theorem 2 does not exceed

$$|\mathcal{A}| \cdot (1/2)^{m-2k} = (1/2)^{m-2k-\log(|\mathcal{A}|)} = (1/2)^{\lceil \log \frac{1}{\delta} \rceil} \leq \delta.$$

$\square$

In case when A and A' transmit their preambles at times $t \neq t'$ and $|t - t'| \leq r$, the situation depends on particular choice of the identifiers and encodings used. One can see that in this case probability of getting a beep in a given slot has to decrease.

Data Confidentiality. We claim that an adversary cannot retrieve message content without knowledge of the sender's secret $\mathcal{K}_{\text{sender}}$. Clearly, the information about the transmitted message can come only from the content part of the transmission. Nevertheless the message is encrypted bit-by-bit using a randomly chosen encoding, thus for a given resulting r-coding a representation of each encrypted bit is equally probable and, according to the properties of cryptographic hash functions in the Random Oracle Model can be regarded as independent of other bits. Therefore security of the protocol can be reduced to security of one-time-pad ([14]).

One can observe that if the node's pseudonym gets updated, then one cannot link the transmissions initiated by the same node before and after the update. This property may be useful for many application scenarios. Note however that an adversary can estimate quite well the number of transmissions by detecting the candidates for the preambles and observing the number of beeps in the blocks of r slots.

Transmissions Unlinkabilty. One can see that due to "refreshing of ID_A" after each transmission of A the identifier of a node is replaced and cannot be linked to the old one without the secret key $\mathcal{K}_A$. Similarly the preamble as well as independently encoded message cannot lead to logical connection of several transmissions of the same node.

Efficiency. Note that we do not restrict the duration of the protocol execution. Certainly for that reason we cannot simply use the total number of steps as a metrics for the protocol evaluation.

First note that on the side of senders the computational effort is evaluating a hash function twice per bit sent. From the point of view of encryption this is quite realistic even on weak devices, as we may expect that the hash functions are implemented on the node devices for enabling the protocols such as μTESLA (cf. [23]).

There is more computational effort on the side of the sink. However, one can prove the following:

Fact 3. *The expected number of checks for a beep presence per transmitted preamble during detection of the node pseudonym is not higher than*

$$2r|\mathcal{A}|(k+1) + m \leq 2r|\mathcal{A}|(k+1) + \log\left(\frac{2|\mathcal{A}|}{\delta}\right) + 2k,$$

for **any** *assignment of moments of starting transmissions of other nodes.*

Indeed, for the correct identifier A the number of checks is m, as each of m bits of ID_A has to be verified. However, each of $|\mathcal{A}| - 1$ other nodes has to be checked as well. One can show that for each node and each starting position we need at average less than 2 checks, provided that no preamble is sent. As at most $k - 1$ nodes may send the preamble within this time, the average number of checks does not exceed $k + 1$. The next problem is that the beginning of the preamble cannot be determined exactly. The factor $2r$ is due to this issue.

5   Related Work

Carrier sensing (CS) has been used in many scenarios regarding wireless communication, but in few cases it was used, as in our case, as a way of sending complex messages. Mainly, it is applied to avoid interference from other stations and enhance channel throughput. For example, MAC (Multiple Access Channel) protocols have been constructed [8,9,25]. In [24] the information obtained from CS mechanism (i.e., channel gain) is used to schedule a back-off time for a station, assuring that the station with the best channel conditions gets its transmission time fastest. In [5] an initialization protocol was proposed, later improved by [7], that uses CS to maintain low time complexity of the procedure. In a model similar to ours, a distributed and randomized protocol for determining dominating set with interference is proposed in [19], where dynamically adjusting CS thresholds allows reduction of message overhead. An influence of CS on performance of some basic algorithms for wireless networks has been studied in [20].

OOK modulation of visible light has been used in [22] to reach gross throughput of 230Mbps, making our solution feasible despite obvious overhead in transmitted data size.

In a standard wireless network model of distributed computing usually an *message passing* model is concerned [11]. However, as noted in [10], this assumption is often too strong for networks of constrained devices. The *beeping transmission model* was presented in [1]. The authors present interval coloring scheme, allowing stations to establish a time pattern for transmission in a decentralized way, so that neighboring stations do not interfere. Following their work, the beeping model was applied to calculating a Maximal Independent Set [2]. We combine their model with the idea of using CS as a message-carrying mechanism.

In order to provide anonymity to nodes in our scheme we leverage work on symmetric keys and their application to constrained devices. [18] gives a top-view of the problem of symmetric key management (update) for such setting. The idea of continuous modification of the key made by communicating parties is proposed in [16] and later extended in [13] to a key evolution protocol.

Our idea of OR-channel that is composed of slots where many entities can inject their bits is similar to Bloom filters [6] and our analysis follows closely that presented in [17].

6   Final Remarks

Similarity to Bloom Filters. Sending messages in the form of r-SC codes in the beeping model could be replaced by Bloom Filters (cf. [4, 15]). The main difference is that we encode each bit separately, and we use l repetitions of r-SC codes instead of putting l marks (beeps) into a single array of $l \cdot r$ positions. The reasons are the following:

- the Bloom Filters are very efficient for inserting elements and testing whether *a given* element has been inserted. However, finding what has been inserted into filter is hard and can be done only by brute force search. This cannot be improved as long a keyed cryptographic hash function $\mathcal{H}$ is used. Therefore, the blocks inserted to the filter must be short in order to avoid computationally intensive decoding. We have chosen separate encodings of single bits. Depending on the sink's capacity, encoding a few bits at once would be an option.
- Bit-wise encoding has the advantage that possible appearance of uncertain decoding is limited to single bits. In some cases, like transmitting digital images, it might be relatively easy to represent the data with the marks "?" (as missing pixels and regenerating them via interpolation). An alternative is to encode larger blocks. In this case uncertain decoding leads to difficulties in image presentation to a human user.
- One can prove that a single Bloom filter of size $l \cdot r$ is slightly more efficient in avoiding false positives (the third case where the sink decodes a bit to "?"). However, it turns out that the differences are insignificant, especially if r is relatively large and the number of senders is big. For example, if there are just two senders and $l = 2$, then the difference of false positives (the probability that the bit of the first sender will be decoded as "?") is $\frac{r-1}{r^2(2r-1)}$. This difference in probabilities decreases exponentially with the maximal number k of transmitting nodes.
- The encryption method is in fact a variant of a stream cipher. Note that the number of beeps depends only on the message length. Moreover, a potential cryptanalysis (even brute force) is made harder by the fact that many nodes may transmit at the same time. Therefore the beeps belonging to different "ciphertexts" are mixed together. Therefore it is hard to present data for cryptanalysis against a single secret K_A.

Smart Decoding. So far not all decoding possibilities have been used, as each virtual channel was considered separately. However, sometimes it is possible to learn more from the configuration of beeps. For instance, let us assume that there are two virtual channels and the beeps occur in the following time slots: $t_0, t_1, \ldots, t_{2u}$, where $t_i \in$

$[T + \frac{r}{2} \cdot i, T + \frac{r}{2} \cdot (i+1)]$ for some start moment T. We assume further that the r-SC codings for the first channel the following time slots are used by r-SC encodings: t_0 and t_1 for the first r-SC code, t_2 and t_3 for the second r-SC code, $\ldots$, t_{2u-2} and t_{2u-1} for the $u - 1$-st r-SC code. Similarly, we assume that the second channel uses t_1 and t_2 for the first r-SC code, t_3 and t_4 for the second r-SC code, $\ldots$, t_{2u-1} and t_{2u} for the $u - 1$-st r-SC code. (Note that such a situation may occur when the first and the second channel are not synchronized and the blocks of slots for r-SC codes of the second channel are shifted by $r/2$ related to the blocks used by the first channel.) According to the decoding procedure both the first and the second channel would obtain u signs "?". On the other hand, we can see the beep at t_0 must belong to the first channel, hence the beep at t_1 cannot be generated by the sender of the first channel. So this beep belongs to the second channel. But in turn this means that the beep at t_2 must belong to the sender of the first channel. We proceed in this way and assign each beep to a single channel. Thereby, one can fully decode the information corresponding to each channel. Certainly, this is not an exception situation and in many cases by a smart decoding one can eliminate many of the uncertain cases denoted by "?" signs in the decoded streams.

Multiple Systems. The proposed scheme allows more than one system to operate on the same carrier frequency. The only condition is that the sets of identifiers attributed to different systems should be disjoint. The sinks detect the initial message from all stations, and then proceed to identify the transmitting station. For a node coming from a different system the r-SC codes in the identification part and the workload part can be regarded as random and yield random noise beeps. Notably, there is no need to change the protocol neither for the nodes nor for the sinks to leverage this functionality.

References

1. Afek, Y., Alon, N., Bar-Joseph, Z., Cornejo, A., Haeupler, B., Kuhn, F.: Beeping a maximal independent set. In: Peleg, D. (ed.) Distributed Computing. LNCS, vol. 6950, pp. 32–50. Springer, Heidelberg (2011)
2. Afek, Y., Alon, N., Bar-Joseph, Z., Cornejo, A., Haeupler, B., Kuhn, F.: Beeping a maximal independent set. Distrib. Comput. 26(4), 195–208 (2013)
3. Afgani, M., Haas, H., Elgala, H., Knipp, D.: Visible light communication using OFDM. In: Proceedings of 2nd International Conference on Testbeds & Research Infrastructures for the DEvelopment of NeTworks & COMmunities, TRIDENTCOM 2006, pp. 129–134. IEEE (2006)
4. Bloom, B.H.: Space/time trade-offs in hash coding with allowable errors. Commun. ACM 13(7), 422–426 (1970)
5. Cai, Z., Lu, M., Wang, X.: Distributed initialization algorithms for single-hop ad hoc networks with minislotted carrier sensing. IEEE Trans. Parallel Distrib. Syst. 14(5), 516–528 (2003)
6. Rivest, R.L.: Chaffing and winnowing: confidentiality without encryption, May 1998. http://people.csail.mit.edu/rivest/Chaffing.txt
7. Cichoń, J., Kutyłowski, M., Zawada, M.: Adaptive initialization algorithm for ad hoc radio networks with carrier sensing. In: Nikoletseas, S.E., Rolim, J.D.P. (eds.) ALGOSENSORS 2006. LNCS, vol. 4240, pp. 35–46. Springer, Heidelberg (2006)

8. Czyzowicz, J., Gąsieniec, L., Kowalski, D.R., Pelc, A.: Consensus and mutual exclusion in a multiple access channel. In: Keidar, I. (ed.) DISC 2009. LNCS, vol. 5805, pp. 512–526. Springer, Heidelberg (2009)

9. Eisenman, S., Campbell, A.: E-CSMA: supporting enhanced CSMA performance in experimental sensor networks using per-neighbor transmission probability thresholds. In: Proceedings of INFOCOM 2007, pp. 1208–1216. IEEE (2007)

10. Emek, Y., Wattenhofer, R.: Stone age distributed computing. In: Proceedings of ACM PODC 2013, pp. 137–146. ACM, New York (2013)

11. Giaccone, P., Shah, D.: Message-passing for wireless scheduling: an experimental study. In: Proceedings of Computer Communications and Networks, IEEE ICCCN, pp. 1–6. IEEE (2010)

12. Jamieson, K., Hull, B., Miu, A., Balakrishnan, H.: Understanding the real-world performance of carrier sense. In: Proceedings of the 2005 ACM SIGCOMM Workshop on Experimental Approaches to Wireless Network Design and Analysis. E-WIND 2005, pp. 52–57. ACM, New York (2005)

13. Klonowski, M., Kutyłowski, M., Ren, M., Rybarczyk, K.: Forward-secure key evolution in wireless sensor networks. In: Bao, F., Ling, S., Okamoto, T., Wang, H., Xing, C. (eds.) CANS 2007. LNCS, vol. 4856, pp. 102–120. Springer, Heidelberg (2007)

14. Menezes, A.J., Vanstone, S.A., Oorschot, P.C.V.: Handbook of Applied Cryptography, 1st edn. CRC Press Inc., Boca Raton (1996)

15. Mitzenmacher, M.: Bloom filters. In: Liu, L., Özsu, M.T. (eds.) Encyclopedia of Database Systems, pp. 252–255. Springer, New York (2009)

16. Ren, M., Das, T.K., Zhou, J.: Diverging keys in wireless sensor networks. In: Katsikas, S.K., López, J., Backes, M., Gritzalis, S., Preneel, B. (eds.) ISC 2006. LNCS, vol. 4176, pp. 257–269. Springer, Heidelberg (2006)

17. Rivest, R.L.: All-or-nothing encryption and the package transform. In: Biham, E. (ed.) FSE 1997. LNCS, vol. 1267, pp. 210–218. Springer, Heidelberg (1997)

18. Blackburn, S.R., Martin, K.M., Paterson, M.B., Stinson, D.R.: Key refreshing in wireless sensor networks. In: Safavi-Naini, R. (ed.) ICITS 2008. LNCS, vol. 5155, pp. 156–170. Springer, Heidelberg (2008)

19. Scheideler, C., Richa, A.W., Santi, P.: An $o(\log n)$ dominating set protocol for wireless ad-hoc networks under the physical interference model. In: Jia, X., Shroff, N.B., Wan, P. (eds.) Proceedings of MobiHoc 2008, pp. 91–100. ACM Press (2008)

20. Schneider, J., Wattenhofer, R.: What is the use of collision detection (in Wireless Networks)? In: Lynch, N.A., Shvartsman, A.A. (eds.) DISC 2010. LNCS, vol. 6343, pp. 133–147. Springer, Heidelberg (2010)

21. Tsonev, D., Videv, S., Haas, H.: Light fidelity (Li-Fi): towards all-optical networking. In: SPIE Proceedings, vol. 9007, art. ID: 900702. SPIE Digital Library, pp. 900702-900702-10 (2013)

22. Vučić, J., Kottke, C., Nerreter, S., Habel, K., Buttner, A., Langer, K.D., Walewski, J.W.: 230 mbit/s via a wireless visible-light link based on OOK modulation of phosphorescent white LEDs. In: Proceedings of Optical Fiber Communication Conference, pp. 1–3. Optical Society of America, IEEE (2010)

23. Wang, M., Zhu, H., Zhao, Y., Liu, S.: Modeling and analyzing the (mu)TESLA protocol using CSP. In: Proceedings of 5th International Symposium on Theoretical Aspects of Software Engineering, TASE 2011, pp. 247–250. IEEE Computer Society (2011)

24. Zhao, Q., Tong, L.: Opportunistic carrier sensing for energy-efficient information retrieval in sensor networks. EURASIP J. Wirel. Commun. Netw. **2005**(2), 231–241 (2005)

25. Zhu, J., Guo, X., Yang, L.L., Conner, W.S., Roy, S., Hazra, M.M.: Adapting physical carrier sensing to maximize spatial reuse in 802.11 mesh networks. Wirel. Commun. Mob. Comput. J. **4**, 933–946 (2004)

Using TPM Secure Storage
in Trusted High Availability Systems

Martin Hell[1]([✉]), Linus Karlsson[1], Ben Smeets[2], and Jelena Mirosavljevic[1]

[1] Department of Electrical and Information Technology,
Lund University, P.O. Box 118, 221 00 Lund, Sweden
{martin.hell,linus.karlsson}@eit.lth.se, mat08jmi@student.lu.se
[2] Ericsson Research, Security, Mobilvägen 1, 223 62 Lund, Sweden
ben.smeets@ericsson.com

Abstract. We consider the problem of providing trusted computing functionality in high availability systems. We consider the case where data is required to be encrypted with a TPM protected key. For redundancy, and to facilitate high availability, the same TPM key is stored in multiple computational units, each one ready to take over if the main unit breaks down. This requires the TPM key to be migratable. We show how such systems can be realized using the secure storage of the TPM. Hundreds of millions TPM 1.2 chips have been shipped but with the recent introduction of TPM 2.0, more manufacturers are expected to start shipping this newer TPM. Thus, a migration from TPM 1.2 to TPM 2.0 will likely be seen in the next few years. To address this issue, we also provide an API that allows a smooth upgrade from TPM 1.2 to TPM 2.0 without having to redesign the communication protocol involving the different entities. The API has been implemented for both TPM 1.2 and TPM 2.0.

Keywords: Trusted computing · TPM · Migration · Certifiable migration key · Secure storage

1 Introduction

A High Availability System, hereafter referred to as HAS, can be used for mission critical systems like medical, trading, banking, mobile network infrastructure, and blue-light systems. Such systems often run for many years and sometimes longer than a decade. As part of high availability requirements, often such systems need trusted platform functions that guarantee that only authentic and approved system software and applications can run on them. Also, one frequently sees demands to safely store sensitive data and keys used by applications and management functions. In the types of HAS that we consider there are multiple Computational Units (CUs) that are organized so they can take over each others' tasks in the event a CU fails. To provide for trusted platform functions like authenticated boot and storage of sensitive data and keys, each CU is equipped with a TCG Trusted Platform Module (TPM). Typically, a CU is a PCB or rack

© Springer International Publishing Switzerland 2015
M. Yung et al. (Eds.): INTRUST 2014, LNCS 9473, pp. 243–258, 2015.
DOI: 10.1007/978-3-319-27998-5_16

mountable unit that can be inserted in a cabinet that hosts the HAS, and can accommodate a multitude of CUs. The use of multiple TPMs for protection in a HAS has many technical problems due to the migration problems that the use of TPM introduces.

At the same time, there are different versions of the TPM, which in some aspects are very different from each other. TPM 1.2 was introduced in 2003 and since 2006 a TPM chip has been included in many laptops. In 2012, TPM 2.0 was introduced, adding new functionality and with no backwards compatibility with TPM 1.2. Even though PCBs still come equipped with a TPM 1.2 chip, within a few years TPM 2.0 is likely to be the dominant chip on newer boards. This provides a challenge as systems utilizing trusted computing functionality may have to undergo significant, and costly, changes.

In this paper, we focus on Trusted Computing Technology, and how a CU manufacturer can offer a solution where customers have unique keys, only usable in a specific HAS, but which still utilizes generic CUs to be used as replacement boards. Moreover, we provide a general API that is independent of the TPM version used. This allows for a cost-efficient deployment of the system as it can be easily updated when TPM 2.0 gains widespread adoption.

The paper is organized as follows. Section 2 gives a brief overview of TPMs, describing some functionality relevant to the paper. In Sect. 3, we specify the use cases together with the threat model. In Sect. 4 we describe the requirements that must be met by the proposed solution, which is then described in Sect. 5. A security analysis of the proposed solution is described in Sect. 6. Section 7 describes the general API. Finally we discuss some related work in Sect. 8. Section 9 concludes the paper.

2   Overview of TPM 1.2 and TPM 2.0

TPMs have been around for more than a decade and most laptops ship with a TPM. Still, we have seen very few applications taking advantage of the functionality provided by TPMs. Microsoft's Bitlocker encryption system is the most known and widely used. A TPM enables trusted computing functionality such as authenticated boot, remote attestation and sealed storage. This section will give a short introduction to TPM 1.2 and 2.0, highlighting the differences when duplicating keys to new destinations. For a more detailed treatment we refer to the specifications [16,17].

2.1   Overview of TPM 1.2 and Certifiable Migration Keys

A TPM 1.2 provides a key hierarchy of asymmetric keys, where the private part of a child key is protected (encrypted) using the public key of the parent. Parents are of type *storage key* and are used to encrypt other keys, while leafs in the tree can be of any type, e.g., a signing key, encryption key or attestation identity key (AIK). Asymmetric keys in TPM 1.2 consist of two parts: one public part, and one private part. The public part contains data such as the public key and different flags. The private part is encrypted, and contains the private key, but also usage and migration secrets. The root of the key hierarchy is the Storage Root Key (SRK), which is created when someone takes ownership of the TPM. The TPM owner authenticates using an *owner secret* and several commands require owner authorization, e.g., commands used in migration which is the main topic of this paper. Commands that use the private part of a key are authenticated using a *usage secret* which can be unique to each key. Such commands are e.g., creation of new keys, data signing and data decryption.

The only way to have the same key protected by two TPMs is to use migratable keys. Migratable keys were introduced in TPM 1.1, offering the ability to migrate (or actually duplicate) a TPM protected key to another TPM. There are two variants of migration schemes specified, called *rewrap* and *migrate*. In the rewrap case, the private part of the migratable key is simply decrypted and re-encrypted using the destination key. In the migrate scheme, the key is instead re-encrypted using the public key of a migration authority (MA). The MA can then re-encrypt the private part with the destination public key. We will not consider the scheme using a migration authority any further in this paper.

Each key also has a *migration secret* in addition to the usage secret. Migration is only allowed if the migration secret is known. For non-migratable keys, the migration secret is *tpmproof*, a value internal to the TPM and never exposed. Also, the source TPM-owner must approve the destination, however, for any migratable key, the owner can choose any destination. Thus, if the TPM owner is not trusted, the key can end up in any TPM, or even outside a TPM if the owner migrates the key to his own keypair generated by e.g., OpenSSL.

A Certifiable Migration Key (CMK), introduced in TPM 1.2, allows for a trusted entity, called Migration Selection Authority (MSA), to be in control of destinations for each individual CMK. The MSA control is tied to each CMK by binding the CMK key to a list of MSAs at key creation time (called *MSAList*). Similar to migratable keys, there are two possible migration schemes for CMKs, *restrict_migrate* and *restrict_approve*. In restrict_approve, tickets which include both the CMK and the destination public key, are used to control the destination. Tickets are signed by the MSA and only the destination in the ticket can be used as target for migration. Then the ticket is first used to create a CMK blob encrypted with the destination SRK. Then the ticket is used again in the target TPM to convert the blob into a key in the key hierarchy. The tickets signed by the MSA are called restrictTickets. From these tickets, sigTickets are produced by letting the TPM owner approve the information in the restrictTicket. Thus, both the MSA and the TPM owner control the migration of a CMK. In the following, restrictTickets will sometimes simply be denoted "ticket" since this is the ticket that will be communicated between entities.

In the restrict_migrate scheme, the CMK is migrated directly to an MSA. No ticket is needed in this case since the key already is bound to the MSA at creation time so the MSA is trusted as destination.

Different from a migratable key, a CMK can be certified by an AIK. The certification states that the CMK key belongs to a TPM and that the private part of the key will never leave the TPM in unencrypted form (assuming the MSA enforces this). Certification of CMKs is not used in this paper and will not be considered further.

2.2 Limitations of CMKs

The TPM owner controls migratable keys in the sense that he/she can create them outside of the TPM or migrate them out from the TPM. Thus, there is no guarantee that the private key is TPM protected. While this problem is addressed by CMKs, putting an MSA in control, the CMKs have some important limitations.

- A software MSA can create CMK keys outside the TPM and migrate them into a TPM.
- When the restrict_migrate scheme is used, a software MSA can read the private CMK key.
- Each time a CMK is migrated, both out of a TPM and into a new TPM, a signed ticket from an MSA is required. Thus, from the perspective of the two TPMs, there must be communication with a third party. If tickets are created in advance this is not required, but then the destinations must be known in advance.

The last limitation above significantly restricts the use of CMKs in HAS's, because the destination CU (e.g. a replacement unit) is not known in advance. It is therefore important to find secure ways to combat this problem. This is one of the main goals in our proposed design.

2.3 Overview of TPM 2.0 and Duplication

The key hierarchy in TPM 1.2 has been replaced by an object hierarchy in TPM 2.0. Objects in the hierarchy can be both symmetric and asymmetric keys, but also data blobs. The type is determined by a combination of the binary properties *sign*, *decrypt* and *restricted*, where the last property means that the object (key) can only perform actions on data prepared by the TPM itself. This is controlled by including a specific byte sequence in these objects. Some commands can only be performed in objects with this byte sequence. Storage keys are asymmetric keys with the properties restricted and decrypt. Similar to TPM 1.2, these keys protect child keys in the hierarchy. However, the protection in TPM 2.0 is by symmetric encryption. A storage key has a unique seed in its private part, which is used to derive a symmetric encryption/decryption key. This key is derived from the seed each time a new object is created or loaded into the TPM.

In TPM 2.0 the term migration has been replaced by duplication, as it more accurately reflects the reality. Two important object attributes are used to control duplication of a key. The first, fixedTPM, controls if an object can be duplicated at all. If an object has this attribute set, the object can not be duplicated. Naturally, an object with fixedTPM set can not be below an object with fixedTPM clear in the hierarchy. The second, fixedParent, controls if an object can be explicitly duplicated (when fixedParent is clear) or if it must be implicitly duplicated (when fixedParent is set) by duplicating a parent key, which has fixedParent clear.

The notion of CMKs and migration schemes has been completely removed in TPM 2.0, and has been replaced by *policies*. A policy is a general concept that controls the actions that can be performed on an object in the hierarchy. Policies are set upon object creation time by storing a value, called authPolicy, in the public part of an object. The authPolicy is a hash value created by running several policy commands, where each command extends the authPolicy digest. This is similar to how PCR values are built by using TPM_Extend. The authPolicy can be based on e.g., time limitations on usage of the object, specific commands that are possible to execute with an object and specific parameters that can be used in a command. Before executing a command a policyDigest must be built in a policy session. This session also stores specific context values that are checked upon execution, e.g., the command code if a certain command must be executed or the fact that a certain authorization method should be used. The final policyDigest is compared to the object's authPolicy and if they match, the command is executed using the information in the context values. Policies can be combined using logical AND and OR.

The use of policies is in general optional as it is possible to authorize using HMAC, similar to authorization in TPM 1.2, or by directly providing a password. However, for duplication the use of policies is mandatory. Policy commands that are particularly interesting for key duplication are TPM2_PolicyAuthorize and TPM2_PolicyDuplicationSelect.

The TPM2_PolicyAuthorize command allows a policy to change by letting an authority sign the new policy. This is done as follows. The TPM user generates a new policy to use for an object. This policy, and the properties it represents, are evaluated by an authority. If they are acceptable, the authority signs this policy and returns the signature. The signature is verified using TPM2_VerifySignature which returns a ticket showing that the signature is valid. This ticket, together with the approved policy, is then used in the TPM2_PolicyAuthorize command. Upon executing this command with a valid ticket, the policyDigest is updated by replacing it by the hash of the name of the signature key. This hash is then the new PolicyDigest. Thus, any policy that needs to change during the lifetime of an object needs to include the TPM2_PolicyAuthorize command after all policies that are subject to change. Policies added after this command has been executed can not be changed.

The TPM2_PolicyDuplicationSelect command is used to control the destination for a duplication. The command includes both the name of the object to be duplicated and the name of the destination. The policyDigest is updated using both these names. Thus, the policy ties the object to a specific destination (or several if logic OR is used). Since the destination is typically not known when an object is created, this is typically used together with TPM2_PolicyAuthorize. This will allow an authority to verify that the destination is valid and then sign the resulting policyDigest.

2.4   Platform Configuration Registers

All TPMs, both of version 1.2 and 2.0, have a number of Platform Configuration Registers (PCRs). These registers store a hash value, which is built-up by repeatedly calling TPM_Extend or TPM2_Extend. This creates a cumulative hash, since an extend operation depends on both a new value and the previous PCR value. The PCRs are used to store measurements of the hardware configuration and software. The measured values are stored in the Stored Measurement Log (SML), outside the TPM, while the digest are secured by the TPM.

The SML can be read to ensure that the measurement values of the system are as expected, and the integrity of the SML can be verified by comparing them to the PCRs. In addition, keys in the TPM can be bound to certain PCR values, such that keys can only be used when the PCRs have the correct value, thus ensuring that keys are only used in a trusted hardware and software setting.

3   Scenario and Threat Model

The considered use case aims at building a robust infrastructure, taking the HAS life cycle into consideration. The scenarios includes four entities.

The hardware, i.e., the computational units (CUs), are produced by a **CU manufacturer**. The CU boards will include a TPM but it will not be associated with any particular, or identified, customer or end user.

A HAS is assembled by a **HAS manufacturer**. The HAS manufacturer takes two or more CUs, due to the redundancy requirements, from the CU manufacturer and assembles the HAS, also using equipment from other sources. This additional equipment is outside the scope of this work.

Customers are purchasing a HAS on which they want to store sensitive data. This data can e.g., be keys or sensitive application data of applications running on the HAS. The sensitive data is stored in *secure storage*, meaning that it resides on a hard disk in encrypted form, protected by a TPM.

A **Trusted Third Party** (TTP) is used to enable the secure migration of keys between TPMs. This is the MSA in TPM 1.2 and authority in TPM 2.0. We assume that this party keeps all keys secure, possibly, but not necessarily, with a TPM.

3.1   Threat Model

Any attacker that controls the hardware, will also be able to circumvent the protection offered by trusted computing, as the root of trust is potentially compromised. Thus, to this end it is natural to consider the CU manufacturer trusted and it can theoretically be merged with the TTP. It is also from the CU manufacturer's perspective we mainly treat the problem. Still, mounting an attack against the hardware is different from attacking the software controlling the migration on the TTP. We will therefore consider them as separate entities.

In practice, many service and operating personnel, hereafter collectively named company employees, will have access to the HAS during its lifetime. Not all company employees can be considered trusted, and this is the main reason to protect data using a TPM, as the decryption key will never leave the TPM unencrypted. Not trusting company employees will also help the customer to protect against other, potentially malicious, customers' personnel.

T1. Anyone, including customer employees, can copy data and software from drives in the HAS cabinet. They may also interact with the TPM.
T2. CU boards can be stolen, both spare boards and those already mounted in a cabinet. Boards from customer A can be used in the HAS of customer B.
T3. HAS manufacturer employees can access data in the HAS when it is being assembled, in particular data that is associated with the TPM.

The main goal is to protect stolen (encrypted) HAS data from being accessed in cleartext, while at the same time provide a system with very low downtime.

4   Requirements

Based on the scenario and threat model, we define the following requirements.

R1. **Data confidentiality.** Data stored on secondary memory, e.g., hard drives or memory cards, must always be encrypted. The key may never be stored (unencrypted) on secondary memory.
R2. **Redundancy.** The data on a HAS must at all times be accessible, even in the case of hardware failure.
R3. **Scalability.** After completed assembly by the HAS manufacturer, spare CUs can be ordered by the customer directly from the CU manufacturer. These are generic and not personalised for the specific customer. Thus, we assume that anyone will be able to buy a generic CU.
R4. **Customer lockdown.** Only TPMs initiated by the CU factory can be used as replacement boards. This will allow the CU factory to create boards that are specific for a group of customers, still allowing customers to have unique keys.
R5. **TPM Compatibility.** The API used by the different entities must be compatible with both TPM 1.2 and TPM 2.0.
R6. **Customer control.** The customer should be the owner of the TPM, allowing him to use it for other purposes such as remote attestation and key certification. This also allows the customer to reuse the hardware and TPMs in the event of a CU manufacturer going out of business.
R7. **User friendliness.** Replacing CUs in the HAS should be as easy as possible for the customer. This includes minimizing the online communication with other entities, possibly providing a completely offline solution. It also includes minimizing the HAS interaction needed by customer employees.

We return to these requirements in Sect. 6 when evaluating the security of the proposed solution.

For the sake of simplifying our expositions we assume further that the HAS uses only two CUs. Thus, the key protecting the sensitive data must be identical

in both TPMs so that the backup CU can immediately become active in case
the first CU fails. Further, when a CU breaks it should be replaced by a spare
CU from the CU manufacturer.

5   Proposed System Design

Due to the redundancy requirement (R2), one key must be associated with several TPMs. This can only be done using duplicable (migratable in TPM 1.2)
keys. We first analyze how this can be achieved in TPM 1.2. Consider the most
straightforward solution of having a plain migratable key immediately below the
SRK in the hierarchy. To migrate this key to a new SRK, the TPM owner can
simply rewrap this key with the new SRK and import it to the new TPM.

```
TPM_AuthorizeMigrationKey      //Owner authorized
TPM_CreateMigrationBlob        //On source TPM
TPM_ConvertMigrationBlob       //On destination TPM
```

The main problem with this is that the owner can rewrap the key with any key,
even one created outside the TPM. Thus, if the customer is the owner (R6) the
private part of the key is not guaranteed to be protected by the TPM at all
times (T1).

With CMK keys in TPM 1.2 and policies in TPM 2.0, the migration/duplication can be controlled by a trusted authority, even when the customer is the
TPM owner. The migration of a key then proceeds as follows.

```
TPM_CMK_ApproveMA              //On source TPM, owner authorized
TPM_CMK_CreateKey              //On source TPM
TPM_AuthorizeMigrationKey      //On source TPM, owner authorized
TPM_CMK_CreateTicket           //On source TPM, owner authorized
TPM_CMK_CreateBlob             //On source TPM
TPM_CMK_CreateTicket           //On destination TPM, owner authorized
TPM_CMK_ConvertMigration       //On destination TPM
```

The TPM_CMK_ApproveMA command lets the owner bind an MSA to the CMK. The
ticket is signed by the MSA and the key can only be migrated to a destination
given in the ticket. From this it is clear that the customer can not be owner at
the time the key is first created since he could assign any key to be an MSA
public key.

An important observation is that a TPM key, we call it K_e (e for encryption),
can be used on several TPMs provided that the parent key K_p is the same on
all TPMs. The key blob is stored on (secondary) memory and loaded into the
TPM when needed. Upon loading a key, it is decrypted by the parent key. Thus,
if the parent key is K_p it can be loaded into any TPM that has K_p in the
key hierarchy. In order to have K_p in several TPM key hierarchies, it must be
migratable and any key having a migratable key as parent key must also be
migratable. Moreover, a CMK (which is migratable) may not have a migratable
key as parent. Figure 1 summarizes these restrictions.

Thus, if we wish to be able to use K_e in several TPMs without having to
migrate it, this key must be migratable, but not a CMK. The parent key K_p
can be either a plain migratable key or a CMK. Since K_p must be explicitly

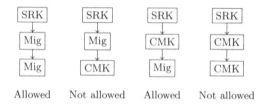

Fig. 1. Key hierarchy restrictions for migratable keys. Both the TPM_CMK_CreateKey and the TPM_CMK_ConvertMigration commands verify that the parent key is not migratable.

migrated between TPM to facilitate the use of K_e, we make use of a trusted third party that can control this migration. On a very high level, the proposed solution is given in Fig. 2 and can be summarized as follows.

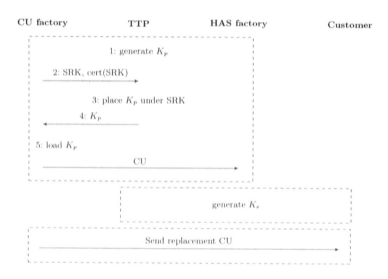

Fig. 2. Overview of the proposed system.

1. The TTP generates the CMK key K_p to be included in all new TPMs.
2. The CU manufacturer takes ownership of a new TPM and asks the TTP for K_p to be migrated under the new SRK.
3. The TTP migrates the key to the given SRK.
4. The TTP sends migrated key, encrypted with the SRK public key, to the CU factory.
5. The CU factory loads the key into the CU.

At this point, a generic board has been prepared with a unique SRK, and the K_p which is common for all boards created by the same CU. The boards are now prepared to be shipped either to a HAS factory for HAS assembly, or to a customer as replacement for a broken board. Assume it has been sent to a HAS factory. The next step is then to generate the customer specific key K_e. We consider three different alternatives for generating K_e, namely a TTP generated K_e, a HAS generated K_e, and a customer generated K_e.

Since the boards are generic, we must take two important aspects into account. First, since K_e is a migratable key in TPM 1.2, we must ensure that it can not be migrated further by a malicious customer employee (knowing the owner secret). This can be controlled by not disclosing the migration secret to untrusted users, i.e., simply to destroy it after key generation. In TPM 2.0 this can be controlled more easily by using the fixedParent attribute. Second, we must also ensure that K_e is bound to the HAS, so it can not be used by other customers. This can be done by restricting the use of the key to a given PCR setting. In TPM 1.2, the PCR settings can be directly specified in the key structure, while in TPM 2.0 this is achieved using policies.

5.1   TTP Generated K_e

If the K_e is generated by the TTP, the customer needs to send the PCR values which the new key should be bound to. Note that this requires online communication between the two entities. The new key will only be loadable under K_p, and only usable on a HAS with the correct PCR values. The steps can be described as in Fig. 3.

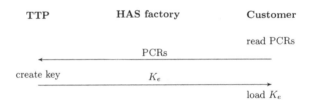

Fig. 3. TTP generated K_e.

5.2   HAS Manufacturer Generated K_e

If the HAS manufacturer generates K_e, it can be generated upon HAS assembly. The customer specific key is created on one CU, and then the blob is copied and loaded on the other CU as well. See Fig. 4 for the executed steps.

5.3   Customer Generated K_e

The customer can execute the same steps as the HAS manufacturer in the section above to generate K_e. There is no difference in commands as the hardware is assembled in the same way as when it left the HAS manufacturer. Figure 5 describes the commands.

TTP   HAS factory   Customer

                       read PCRs
                       create key
                       load key

Fig. 4. HAS factory generated K_e.

TTP   HAS factory   Customer

                                                read PCRs
                                                create key
                                                load key

Fig. 5. Customer generated K_e.

5.4   HAS Initialization

Before leaving the HAS factory, and before creating the customer-specific keys, the HAS must personalize the HAS in such a way that the PCR values are unique to every customer. This ensures that customer-specific keys can be created.

When the HAS arrives to the customer, the customer must verify that the PCR values after system startup are indeed unique to the customer. This can be done by verifying that the Stored Measurements Log (SML) includes a hash that is customer dependent. If the HAS passes this test, it is ready to be used, knowing that K_e can only be decrypted by this HAS.

6   Security Analysis and Comparison of Properties for K_e Generation

We assume that any data that resides on secondary storage on the HAS can be stolen by a malicious employee (T1). This includes the encrypted sensitive data, the encrypted sensitive part of a TPM protected key, and key usage secrets that are needed to use a key. While it could be possible to restrict the usage secret to only a small number of trusted employees, thus keeping it confidential, or to distribute it using secret sharing, we do not make such assumptions in this work. Since K_e can only be used on a customer specific HAS, the encrypted sensitive part of K_e can only be decrypted on this HAS. Thus, it is not possible to steal the encrypted data and the encrypted K_e and decrypt the data using a generic board. The sensitive data is in clear only in primary memory, when used by the HAS software.

Since the boards are generic, a stolen board will not give an attacker any additional information compared to using their own boards. This mitigates threat T2.

Threat T3 can be mitigated to different extent depending on which K_e generation alternative is used and which TPM version is used. When using a TTP for K_e generation or when K_e is generated by the customer, the HAS manufacturer employees will have no access to K_e or any information about it. If

K_e is generated by the HAS manufacturer, for TPM 1.2, the security depends on the migration secret being destroyed after the key is generated. Otherwise, this key could be leaked to a malicious customer which is able to migrate K_e outside the TPM. In TPM 2.0, K_e is created with fixedParent set, which can be verified by the customer when the HAS is being initialized. Thus, it is only for HAS manufacturer created keys in TPM 1.2 that we are not able to fully mitigate T3, but it can be noted that an attack require cooperation between HAS manufacturer and customer employees. Returning to T1, we can also note that for customer created K_e in TPM 1.2, we must ensure that the migration secret is destroyed. Thus, for TPM 1.2, higher security is achieved when K_e is generated by a trusted third party. A summary of the different properties for different cases are given in Table 1.

We note that there are three parts required to gain access to the secure information stored in the HAS: the encrypted data, the customer-specific key K_e, and the HAS itself. Thus, we cannot protect against cases where an attacker gets hold of all three of these parts. This includes a potential case where a malicious HAS employee cooperates with an malicious company employee at company A. If they have access to both stolen encrypted data, and the stolen K_e from another company B, the HAS manufacturer and employee at A may cooperate to build a HAS with the same customer-specific PCR values as customer B, thus enabling them to decrypt the stolen data.

Finally, we also note that our analysis relies on the assumption that the TTP is trusted and available.

Table 1. A summary of the properties when different entities generate the key K_e.

	TTP		HAS man.		Customer	
	1.2	2.0	1.2	2.0	1.2	2.0
No online communication with other entity needed	☐	☐	☒	☒	☒	☒
Possible to verify that K_e is bound to K_p	☒	☒	☐	☒	☐	☒

7 Unified API

We have developed a unified API for the proposed functionality, such that a move from TPM 1.2 to TPM 2.0 will be as simple as possible. By looking at the different phases of our solution, we can construct sequences of TPM commands for each of the two TPM versions, such that we get the same behaviour, abstracting away the differences between the TPM versions.

The API has been implemented and tested to ensure the correctness of the given commands, both for TPM 1.2 and TPM 2.0. To do this, two different TPM simulators and support libraries have been used, one for each TPM version.

For TPM 1.2, IBM's Software TPM version 4720 [7] has been used, which also includes libtpm, which can be used to interface with the simulator. For TPM 2.0, Microsoft's TPM2 Simulator version 1.1 [10] has been used, together with Microsoft's TPM Software Stack version 1.1 [9].

7.1   Generation and Migration of K_p

The first step in Fig. 2 is to generate K_p. The following steps are executed on the TTP:

TPM 1.2	TPM 2.0
TPM_CMK_ApproveMA TPM_CMK_CreateKey	TPM2_PolicyAuthorize TPM2_Create

In step 2, the CU factory sends the SRK to the TTP, which then in step 3 executes the following commands to create a blob which is decryptable under the given CU's SRK.

TPM 1.2	TPM 2.0
TPM_AuthorizeMigrationKey TPM_CMK_CreateTicket TPM_CMK_CreateBlob	TPM2_LoadExternal TPM2_PolicyDuplicationSelect TPM2_PolicyAuthorize TPM2_Duplicate

In step 4, the blob is sent to the CU factory, which then loads the blob into the TPM under the SRK (step 5):

TPM 1.2	TPM 2.0
TPM_CMK_CreateTicket TPM_CMK_ConvertMigration TPM_LoadKey2	TPM2_Import TPM2_Load

The CU now has K_p loaded directly beneath the SRK, and the customer-specific key K_e can be generated.

7.2   Generation of K_e

The customer-specific key K_e can be generated using any of the alternatives given in Sects. 5.1–5.3. The commands for each of the three cases are given below:

TTP Generated K_e

TPM 1.2	TPM 2.0

TPM_PcrRead   // customer
// send PCRs TTP
TPM_CreateWrapKey // TTP
// send blob to customer
TPM_LoadKey2   // customer CU1
TPM_LoadKey2   // customer CU2
	TPM2_PCR_Read // customer
	// send PCRs to TTP
	TPM2_PolicyPCR // ttp
	TPM2_Create // ttp
	// send blob to customer
	TPM2_Load // customer CU1
	TPM2_Load // customer CU2

HAS Manufacturer Generated K_e

TPM 1.2

```
TPM_PcrRead         // CU1
TPM_CreateWrapKey  // CU1
TPM_LoadKey2        // CU1
// copy blob to other CU
TPM_LoadKey2        // CU2
```

TPM 2.0

```
TPM2_PCR_Read   // CU1
TPM2_PolicyPCR // CU1
TPM2_Create     // CU1
TPM2_Load       // CU1
// copy blob to other CU
TPM2_Load       // CU2
```

Customer Generated K_e. These are the same commands as used when the HAS manufacturer generates K_e, the only difference is that they are now executed by the customer.

TPM 1.2

```
TPM_PcrRead         // CU1
TPM_CreateWrapKey  // CU1
TPM_LoadKey2        // CU1
// copy blob to other CU
TPM_LoadKey2        // CU2
```

TPM 2.0

```
TPM2_PCR_Read   // CU1
TPM2_PolicyPCR // CU1
TPM2_Create     // CU1
TPM2_Load       // CU1
// copy blob to other CU
TPM2_Load       // CU2
```

7.3 CU Failure

In the event of a CU failure, the customer will receive a new CU directly from the CU factory. This will have the key K_p loaded, as per the steps described in Sect. 7.1. The customer will however be required to load the customer-specific key K_e. Since the key is located beneath the common key K_p in the key hierarchy, the same key blob that is used on the other CU can be used directly on the new CU. Thus, the key blob of K_e is copied to the new CU, and the following commands are executed:

TPM 1.2

```
TPM_LoadKey2
```

TPM 2.0

```
TPM2_Load
```

8   Related Work

Though there are few examples of widely adopted applications taking advantage of TPM functionality, several use cases have been considered before. In [5,18], the use of TPMs to secure VANETs was proposed and studied. Using TPMs to increase the security in RFID tags and NFC communication has also been proposed in [11] and [6] respectively.

The use of Certifiable Migration Keys in the Mobile Trusted Module (MTM) for protecting secret data was proposed in [8].

Today, virtualization is a growing area, and there have been several different proposals on how to use the TPM in virtual machines. In [2] a complete virtualized TPM module is developed, which is then linked to the hardware TPM. In [3] a para-virtualized solution is discussed. [13] discusses yet another design, and also discusses migration of virtual TPMs to a large extent.

The use of TPMs in cloud computing has also been considered in recent years. In [15] secure launch and migration of VMs in the cloud is discussed in the context of trusted computing, and in [1] secure migration of virtual machines through the use of the Trusted Platform Module is further discussed.

Remote attestation has been considered in many works before [2,4,12,14]. In remote attestation, the goal is to provide the contents of PCRs to a remote party. The PCR values are signed with an AIK and the remote party can verify through the signature that the system is in a known configuration. Using an SML, the content of this, which is a set of run programs and their hashes, can be compared to the signed PCR values. In our work, it is the customer that verifies the PCRs and the SML.

9   Conclusions

We have proposed a solution for using TPMs to secure sensitive data in a high availability system. The main challenge is to create customers specific keys which can only be used in the customer's own HAS, while at the same time allowing generic computational units to be produced and shipped as replacement boards. Since employees come and go, we also do not want to trust employees. Our proposed solution relies on binding the customer specific key to a parent key which is the same on all boards, and to also bind the key to PCR values that are specific to a customer. We show that the increased functionality in TPM 2.0 allows a more secure solution in certain cases. In addition to the proposed solution we define an API such that it is possible to upgrade from TPM 1.2 to TPM 2.0 without changing the communication flow.

Acknowledgments. The authors would like to thank the anonymous reviewers for their valuable comments.

References

1. Aslam, M., Gehrmann, C., Bjorkman, M.: Security and trust preserving VM migrations in public clouds. In: Trust, Security and Privacy in Computing and Communications (TrustCom), pp. 869–876, June 2012
2. Berger, S., Cáceres, R., Goldman, K.A., Perez, R., Sailer, R., van Doorn, L.: vTPM: Virtualizing the trusted platform module. In: Proceedings of the 15th Conference on USENIX Security Symposium, USENIX-SS 2006, vol. 15. USENIX Association, Berkeley (2006). http://dl.acm.org/citation.cfm?id=1267336.1267357
3. England, P., Loeser, J.: Para-virtualized TPM sharing. In: Lipp, P., Sadeghi, A.-R., Koch, K.-M. (eds.) Trust 2008. LNCS, vol. 4968, pp. 119–132. Springer, Heidelberg (2008). http://dx.doi.org/10.1007/978-3-540-68979-9_9
4. Gu, L., Ding, X., Deng, R.H., Xie, B., Mei, H.: Remote attestation on program execution. In: Proceedings of the 3rd ACM Workshop on Scalable Trusted Computing, STC 2008, pp. 11–20. ACM, New York (2008). http://doi.acm.org/10.1145/1456455.1456458

5. Guette, G., Bryce, C.: Using TPMs to secure vehicular ad-hoc networks (VANETs). In: Onieva, J.A., Sauveron, D., Chaumette, S., Gollmann, D., Markantonakis, K. (eds.) WISTP 2008. LNCS, vol. 5019, pp. 106–116. Springer, Heidelberg (2008)
6. Hutter, M., Toegl, R.: A trusted platform module for near field communication. In: 2010 Fifth International Conference on Systems and Networks Communications (ICSNC), pp. 136–141 (2010)
7. IBM: IBM's software trusted platform module. http://ibmswtpm.sourceforge.net/
8. Kang, D.W., Jun, S.I., Lee, I.Y.: A study on migration scheme for a mobile trusted module. In: 11th International Conference on Advanced Communication Technology, 2009, ICACT 2009, vol. 3, pp. 1672–1677 (2009)
9. Microsoft: The TPM software stack from Microsoft research. https://tpm2lib. codeplex.com/
10. Microsoft: TSS.MSR v1.1 TPM2 simulator. http://research.microsoft.com/en-US/ downloads/35116857-e544-4003-8e7b-584182dc6833/default.aspx
11. Mubarak, M., Manan, J., Yahya, S.: Mutual attestation using TPM for trusted RFID protocol. In: Network Applications Protocols and Services (NETAPPS), pp. 153–158 (2010)
12. Nauman, M., Khan, S., Zhang, X., Seifert, J.-P.: Beyond kernel-level integrity measurement: enabling remote attestation for the android platform. In: Acquisti, A., Smith, S.W., Sadeghi, A.-R. (eds.) TRUST 2010. LNCS, vol. 6101, pp. 1–15. Springer, Heidelberg (2010). http://dx.doi.org/10.1007/978-3-642-13869-0_1
13. Sadeghi, A.-R., Stüble, C., Winandy, M.: Property-based TPM virtualization. In: Wu, T.-C., Lei, C.-L., Rijmen, V., Lee, D.-T. (eds.) ISC 2008. LNCS, vol. 5222, pp. 1–16. Springer, Heidelberg (2008). http://dx.doi.org/10.1007/978-3-540-85886-7_1
14. Sailer, R., Zhang, X., Jaeger, T., van Doorn, L.: Design and implementation of a TCG-based integrity measurement architecture. In: Proceedings of the 13th Conference on USENIX Security Symposium, SSYM 2004, vol. 13, p. 16. USENIX Association, Berkeley (2004). http://dl.acm.org/citation.cfm?id=1251375.1251391
15. Santos, N., Gummadi, K.P., Rodrigues, R.: Towards trusted cloud computing. In: Proceedings of the 2009 conference on Hot topics in cloud computing. USENIX Association (2009)
16. Trusted Computing Group: TPM main specification, Version 1.2, Revision 116, March 2011
17. Trusted Computing Group: Trusted Platform Module Library Specification, Family "2.0", Level 00, Revision 01.07, March 2014
18. Wagan, A., Mughal, B., Hasbullah, H.: VANET security framework for trusted grouping using TPM hardware. In: Communication Software and Networks, 2010, ICCSN 2010, pp. 309–312 (2010)

APP Vetting Based on the Consistency of Description and APK

Weili Han[1,2,3]($\boxtimes$), Wei Wang[1], Xinyi Zhang [1], Weiwei Peng[1], and Zheran Fang[1]

[1] Software School, Fudan University, Shanghai, China
[2] Key Lab of Information Network Security,
Ministry of Public Security, Shanghai, China
[3] Shanghai Key Laboratory of Data Science, Fudan University, Shanghai, China
wlhan@fudan.edu.cn

Abstract. Android has witnessed a substantial growth over the years, in the market share as well as in the number of malwares. In this paper, we proposed a novel approach to detect potentially malicious applications, based on the semantic relatedness between the applications' descriptions and the apk files. We gathered an application database of 7,570 valid applications for training and testing, finding that about 16.6 % of the tested applications exhibit a lack of relatedness between the apk files and descriptions, due to either inadequate embedded text in apk file, too short a description, unsuited description, or being a malicious application. In additions, there are 4 % of applications unjustly deemed as unrelated. Our study showed that the semantic based approach is applicable in terms of malware detection and in judging the soundness of descriptions.

Keywords: Android security · Malware · NLP · APK · Description

1 Introduction

Recent years, smart phones and mobile devices have become more and more popular. A recent report from International Data Corporation [5] showed that the number of Android smart phones reached 81.1 % in the first quarter of 2013. And the number of Android applications is rapidly growing. According to a report from AppBrain [1], the official Android Market held a total of 1,316,773 applications by July 30, 2014.

Security on Android has become a hot topic over the years, with a growing number of malicious applications threatening the privacy and financial security of users. The automated detection of malicious applications was put forward by Google in the form of Google Bouncer, which, according to RiskIQ [2], was able to detect 60 % of the malicious applications in the official android market, Google Play. However, the detection rate has drastically decreased with time, by the year 2013, it was able to detect only 23 % of the malicious applications. This shows a serious need for new methods of malware detection.

Such security problem is due to Android's open platform and unrestricted application market, using which any developer, professional, amateur or even

M. Yung et al. (Eds.): INTRUST 2014, LNCS 9473, pp. 259–277, 2015.
DOI: 10.1007/978-3-319-27998-5_17

malicious, are able to develop and sell their work to the world [7]. Such a low barrier attracts many developers unskilled in the English language, flooding the market with poorly described application, making it possible for malicious applications to hide among them.

In light of the lacking in malware detection methods and descriptions writing aids with the android market, we intend to develop a tool, being the first to utilize the relatedness between the application descriptions and the embedded text in apk files to achieve malicious application detection and to discover poorly written descriptions.

Following this innovative approach we designed a framework to analyze the relatedness between an application's description and the embedded text in the apk file. We are able to achieve a recall of 91.2% in the most tolerant case. We further analyzed the applications with its descriptions and apk file deemed unrelated, find that 77% of them falls into one of the following categories, (1) Inadequate Embedded Text, (2) Short Description, (3) Unsuited Description, (4) Malicious Application.

The rest of the paper is organized as follows. In Sect. 2, we introduce the background of the Android System and some tools used in our work, NLTK and ESA. Section 3 defines the problem and the objective of our system, while Sect. 4 outlines the framework of our system, the acquisition of data and rationalize some of the design choices. Section 5 evaluates the system and analyzed the evaluation results. We then discuss the strength and weakness of the system in Sect. 6, the related works in Sect. 7. At the end, we conclude the paper in Sect. 8.

2   Background Knowledge

2.1   Android System

Android applications are mainly written in Java, with some configurations and resources defined in XML format. Developers need to register all the Activities the application use in the Manifest file. Activity in Android is an application component that provides a screen with which users can interact in order to perform specific operations, generally, the appearance of an Activity is determined by a layout XML file (e.g. main.xml), where a hierarchy of viewable widgets is defined. Some of the texts in an Activity are statically determined, like the text on a button or some description words, these texts can be defined in three ways: (1) developer declares a string resource in string.xml, and includes the string resource in the view's text field within layout XML file, (2) developer directly assigns a string value to the text field of a view within layout XML file, (3) developer assigns a constant string to the text fields of a view's object in Java code.

By coding convention, text to be displayed to users are written mostly through the first and the second way. Thus, we can extract the displayed string from the string.xml file and layout XML file.

2.2   Text Sanitization

Natural Language Toolkit (NLTK) is a leading platform providing advanced language processing tools which are used in our system.

RegexpTokenizer Used to break sentences into words.

tag.pos_tag Used to tag words by part of speech.

RegexpParser Based on a grammar, given by regular expression, used to parse a sequence of words into a tree structure.

PorterStemmer Proposed by Porter in 1980, Porter Stemming algorithm [14] is used to stem words, by removing word suffix, uniforming tense, converting plurals and singular noun.

WordNetLemmatizer This stemmer utilize the morphy function of Word-Net [8], and preform no action if the word is not recognized by WordNet.

2.3   Explicit Relatedness Analysis (ESA)

ESA [9] is used to calculate the relatedness between text. The main idea behind ESA is to construct a multidimensional semantic space based on a given set of literature, and then put the vector to be analyzed into the semantic space, turning the semantic relatedness between two pieces on text into the cosine value of the angle between two vectors projected into the semantic space.

It is called explicit semantic analysis because each dimension in the semantic space is an explicit article in the literature. While the projection into the semantic space is based on the overlapping of words between the text.

The reason we choose ESA as the tool of semantic relatedness analysis is because that ESA has its own literature as the building blocks, greatly decreasing the size of the data needed for training. In comparison to synonym based tools like WordNet, ESA is more focused on the relatedness on a broader scale and provides a quantitative representation of the semantic relatedness.

We use `esalib` [13], the best maintained open-source ESA library available, which uses 2005 wikipedia as the literature set.

3   Problem Definition

We define the semantic vector for the embedded text of application app as below:

$$T_{app} = [v_1, v_2, v_3, ..., v_n]^T$$

$$\text{where } v_i = \begin{cases} 1 & \text{if } w_i \in MT[app] \\ 0 & \text{otherwise} \end{cases}$$

Where $MT[app]$ is the list of semantically significant phrases in the embedded text of app.

Similarly, semantic vector of *app*'s description is defined as:

$$D_{app} = [v_1, v_2, v_3, ..., v_n]^T$$

$$\text{where } v_i = \begin{cases} 1 & \text{if } w_i \in MD[app] \\ 0 & \text{otherwise} \end{cases}$$

Where $MD[app]$ is the list of semantically significant phrases in the description of *app*.

Our objective is to find a relatedness function $\hat{rel}$ approximating the actual relatedness between the description and embedded text as closely as possible.

$$\hat{rel}(T, D) = \begin{cases} 1 & \text{if } T \text{ is deemed related to } D \\ 0 & \text{otherwise} \end{cases}$$

We then need to introduce a few assumptions.

First, we suppose there is an ideal function rel.

$$rel(T, D) = \begin{cases} 1 & \text{if } T \text{ is actually related to } D \\ 0 & \text{otherwise} \end{cases}$$

Second, suppose that for most applications, its description and embedded text is related.

$$\sum_{app \in App} rel(T_{app}, D_{app}) \approx |App|$$

App here means the set of all applications in the dataset.

Third, there exist some similar applications, among which these descriptions and embedded text points to similar functionalities and thus relate to each other. Given an application, the number of applications similar to it grows linearly with the size of the dataset.

More specifically, given an X uniformly distributed on the set *App*, there exist α and β, such that the following equation is satisfied.

$$E(\sum_{app \in App} rel(T_X, D_{app}) + \sum_{app \in App} rel(T_{app}, D_X)) \approx \alpha \cdot |App| + \beta$$

The expectation is calculated as:

$$\sum_{X \in App} \sum_{app \in App} rel(T_X, D_{app}) + \sum_{X \in App} \sum_{app \in App} rel(T_{app}, D_X)$$
$$\approx \alpha \cdot |App|^2 + \beta \cdot |App|$$

$$2 \cdot \sum_{app_1 \in App} \sum_{app_2 \in App} rel(T_{app_1}, D_{app_2}) \approx \alpha \cdot |App|^2 + \beta \cdot |App|$$

$$\sum_{app_1 \in App} \sum_{app_2 \in App} rel(T_{app_1}, D_{app_2}) \approx \alpha' \cdot |App|^2 + \beta' \cdot |App|$$

We then introduce two measurements, precision and recall, to examine the effectiveness of our approximation $\hat{rel}$. To avoid the introduction of subjective error, we take the fact of whether the texts are from the same application as ground truth, which is to say, we deem a piece of embedded text and a description as matched if and only if they are from the same application. Hence, precision and recall should be in forms as below.

$$precision = \frac{\sum_{app \in App} \hat{rel}(T_{app}, D_{app})}{\sum_{app_1 \in App} \sum_{app_2 \in App} \hat{rel}(T_{app_1}, D_{app_2})}$$

$$recall = \frac{\sum_{app \in App} \hat{rel}(T_{app}, D_{app})}{|App|}$$

Recall stands for the portion of applications that have their own embedded text and description deem related, which, according to assumption two, should be approximate to 1. On the other hand, precision stands for the portion of embedded text and description pairs that actually come from the same application, among all that are deemed related, which, according to assumption number three, is approximately $\frac{1}{\alpha' \cdot |App| + \beta'}$.

In order to avoid the presence of $|App|$ in $precision$'s ideal value, we define $precision'$ as a replacement of $precision$. We define $precision'$ as the expected $precision$ when each piece of embedded text is compared with two descriptions, one from the applications itself, and another randomly chosen among all test samples, which gives,

$$precision' = \frac{\sum_{app \in App} \hat{rel}(T_{app}, D_{app})}{\sum_{app \in App}(E_X(\hat{rel}(T_{app}, D_X)) + \hat{rel}(T_{app}, D_{app}))}$$

$$\approx \frac{|App|}{\alpha' \cdot |App| + \beta' + |App|}$$

$$\approx \frac{1}{\alpha' + 1} \quad \text{when } |App| \text{ is large}$$

α' is dependent on the real life condition. Our goal is to let $recall$ be close to 1, which $precision'$ is close to $\frac{1}{\alpha'+1}$; thus the objective function is defined as,

$$F_1 = \begin{cases} 2 \cdot \frac{precision' \cdot recall}{precision' + recall} & \text{if } precision' \leq \frac{1}{1+\alpha'} \\ 2 \cdot \frac{\frac{1}{1+\alpha'} \cdot recall}{\frac{1}{1+\alpha'} + recall} & \text{otherwise} \end{cases}$$

And the problem becomes to find a $\hat{rel}$ to make F_1 as high as possible.

For example, suppose the total number of applications is 1,000, with a recall of 90 % and a precision of 10 %, then the adjusted F_1 score should be 94 % when α' is less than 1 %.

This applies when in the test set, each application has its embedded text matched against all 1,0000 descriptions. However, we now want to reduce the workload by randomly selecting 49 descriptions, in addition the one description from the applications, to matched against the embedded text. Suppose in this case, the recall is still 90 % which the precision become 12 %, the adjusted F_1 score should be 88 % with $precision'$ as 86 %.

4   System Design

4.1   Framework

Our system takes in an application and a description, after a serious of processing and decision making, decides whether these two are related semantically.

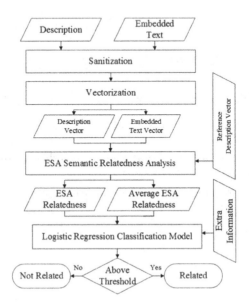

Fig. 1. System implementation outline

As shown in Fig. 1, the system is designed into the following modules.

Sanitization. In sanitization, we remove filter and clean the texts to extract strings with semantic relevance. Using the tools of NLTK, we split the sentences according to regular expression, tag words with their part of speech, combine words into phrases, stem words and remove stopwords, obtain a list of phrases in the ends.

Vectorization. Using 0 and 1 to represent the existence of a word in semantic vector. The involved vectors includes,

v_1 The vector representing the given description
v_2 The vector representing the embedded text in the given application
V_3 The reference semantic vectors generated before hand from 50 randomly chosen descriptions.

ESA Semantic Relatedness Analysis. Using v_1 and v_2 generated in Vectorization, ESA relatedness are calculated. Average ESA Relatedness are the average of the relatedness between v_2 and the 50 vectors in V_3.

Logistic Regression Classification. Using the classification function trained before hand, applying which to input vector given by the current ESA relatedness, average ESA relatedness, description, and embedded text, we obtain a result in the range of 0 to 1. Based on the threshold, the system then gives a conclusion on whether the two pieces of text are deemed related.

4.2 Data Setup

Obtaining Data. To obtain the description information of Android applications, we use an open-source third party script, android-market-api [3]. Since Google Play strictly restricts the number of requests to 500 applications per user, denying to reply useful information when the limitation is exceeded, we have to use multiple google account to accomplish the crawling of application description as well as apk file. Even then, we are only able to obtain about 10,000 applications' information. To extract semantic information from the apk file, we use the tool `apktool` [10] provided by Google to obtain string file `strings.xml`, and layout files. Since by coding convention, the text displayed in user interface would be stored in these files, we choose these as the source of embedded text information.

Data Sanitization. After gathering the application descriptions, we then sanitize them to extract key phrases for further processing.

First, we substitute the non English characters to their closest English counterparts, such as converting Loẅis to Lowis.

Second, we customize the `RegexpTokenizer` to deal with cases of abbreviation (e.g. U.S.), combined-words and percentage, currency, and numbers (e.g. 10.2 %). Then we split the sentence according to punctuation and spaces. The specific regular expressions used are as listed.

Abbreviation ([A-Z])(\.[A-Z])+\.?
Combined-words \w+(-\w+)*
Percentage, currency and numbers \$?\d+(\.\d+)?%?
Ellipse \.\.\.

Third, we use the tool `tag.pos_tag` to perform part of speech tagging.

Fourth, we use `RegexpParser` to extract phrases, focusing mainly on noun phrases, including single noun, noun+noun, adjective+noun which make up 84.8 % (need change) of the noun phrases in our dataset.

In addition, we decide to keep all the verbs that is identified. These two types of words are what we believe contains most semantic meaning.

Noun Phrases <NN.*|JJ>*<NN.*>
Verb <VB.*>

At last we stem all the extracted words, remove stopwords, duplicates or words that are too long or too short.

Embedded Text. In dealing with embedded text, we need to first remove spacing characters such as "\n" and "\t", line numbers and html tags. Texts embedded in apk file can be classified into two type, phrases and sentences. Phrases are fragments of words lacking a full grammatical structure, usually used as texts on buttons, or options; while sentences are used in introduction of the application, copyright information, documentations, etc. In this case, we roughly assume that word sequences consisting of more than four words are sentences.

Sanitization are performed to these sentences in ways similar to those in descriptions.

Data Model. The semantic information we processed in Sect. 4.2 is then stored in `json` file, with a mapping relationship as below.

$$MD[app] = [w_{i_1}, w_{i_2}, ..., w_{i_n}]$$

In the definition, *app* means the name of the given application, n means the total number of phrases in the application description. While W is the list of all phrases extracted from descriptions, with w_i meaning the ith words in W. MD stores the phrases mapping of descriptions.

Embedded text are stored in similar ways.

$$MT[app] = [w_{i_1}, w_{i_2}, ..., w_{i_n}]$$

In the case, MT stores the mapping of embedded text.

4.3 Classification Model

Since the $\hat{rel}$ function to be generated is a classification function, we use logistic regression, one of the most popular forms of binary regression [15], to derive the model.

Preliminary Analysis. We give an preliminary analysis of the dataset, which consists of 7,570 applications.

First, we analyze the difference in relatedness across various categories. According to Fig. 2, there are actually observable difference in description text relatedness across categories.

Applications in Categories like "Media and Video", "Personalization", "Productivity" and "Tools" have a higher relatedness on average. The number of applications in "Media and Video" is rather low, taking up on 52 among the total of 7,570 applications. Typical applications in this category include `hdplayer`, `videoeditor`, and `utorrent`, all require the technique of video encoding or fast decoding, some even need a high bandwidth to provide video content to the client. All these functionalities are demanding on the technical side, usually supported by well developed software company or website, with considerable number of staffs and users. It is quite natural for these kind of applications to have well written descriptions.

"Personalization" includes mobile theme, mobile font management applications, lock screens, etc. Descriptions and embedded texts in this category tend to be short, functionalities tend to be focused; thus the descriptions and texts do not diverge due to the simplicity.

"Productivity" and "Tools" are mostly used to provide technical service (such as file management and anti-virus), which are also mostly provided by established companies.

On the other hand, categories like "Books &References", "Education" and "Libraries &Demo" seems to have a lower relatedness on average. With "Libraries &Demo", applications in this categories are under development, the lack of relatedness seems quite natural. For "Education" and "Book &References", these two categories are mostly intended as a learning aid. Some are originally intended to meet the developer's own need, without a marketing team to write the description.

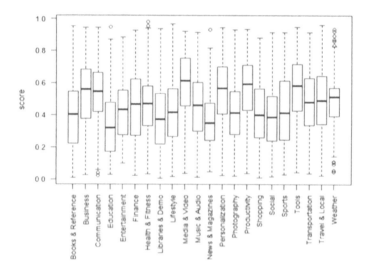

Fig. 2. ESA relatedness distribution across categories

From Fig. 3, we can reenforce our claim that "Media and Video", "Productivity" and "Tools" are mostly developed by established teams since they all have a longer description length on average; while "Libraries &Demo" have a very small average description length.

We then go on to analyze whether the relatedness between an application's embedded text and its own description is actually higher than that between others. As the reference point, we randomly selected 50 descriptions; for a given embedded text, we calculated the average relatedness between it and the selected descriptions.

Description Length Distribution

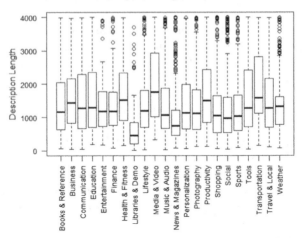

Fig. 3. Description length distribution across categories

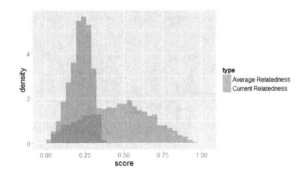

Fig. 4. Relatedness and average relatedness distribution

From Fig. 4, we can observe that texts from the same applications generally have a higher relatedness than the average. However, there is no distinct boundary between these two, which is to say, we will not able to find a threshold that could perfectly tell apart whether the texts are from one application. Say, we set the threshold to 0.2, 87.7 % of the applications can have its own embedded text and description correctly related; while 66.1 % of the average relatedness would be beyond the threshold, giving a precision' of 56.9 %.

A sampled average of relatedness is compared to the relatedness from same applications. As shown in Fig. 5, the dots in the lower triangle stands for applications whose embedded text has a relatedness to its own description lower than the average of sample texts, taking up 11.5 % of all applications. This means that using average relatedness as aid to classification can improve the recall over threshold-

Average Relatedness V.S. Relatedness

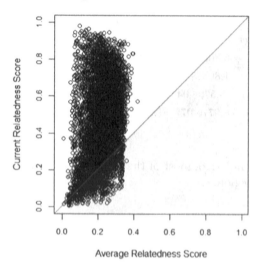

Fig. 5. Relatedness against average relatedness

based classification; however, there is still 46.6 % of the average relatedness above that of the applications own relatedness, giving a `precision'` of 65.0 %.

Beyond the usage of relatedness and average relatedness we develop a more detailed model to perform the classification.

Logistic Regression. First we examine the factors that would possibly reflect the actual relatedness.

score the ESA relatedness value between the embedded text and description to be classified

avgScore the average ESA relatedness between the given embedded text and 50 randomly chosen descriptions

descLen the number of phrases in the given description

textLen the number of phrases in the given embedded text

The effect of average relatedness had been discussed in Sect. 4.3. Considering this `avgScore` is compared against `score`, we use `reScore` which is $\frac{score}{avgScore}$ to replace it.

In addition, in `descLen` and `textLen` affect the dimensions of the semantic vector they are kept as influencing factors.

Running Logistic Regression on the factors listed gives result are shown in Table 1.

Using the model, and ground truth classifications we can then project the four factors on to the region (0, 1), making the projected value for texts from the same application close to 1 while that for texts from different applications

Table 1. Weights in logistic regression

	Estimate	Standard Err	Z value	p value
Intersect	-7.625e+00	4.214e-02	-180.948	< 2e-16
score	6.860e+00	1.427e-01	48.077	< 2e-16
reScore	1.803e+00	3.301e-02	54.628	< 2e-16
descLen	-5.570e-04	1.653e-05	-33.699	< 2e-16
textLen	6.077e-07	1.965e-07	3.092	0.00199

close to 0. As shown in Fig. 6, most of the projected value for not matched pairs are hold in a small region.

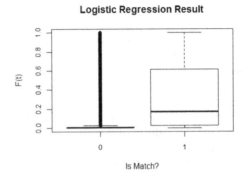

Fig. 6. Logistic regression classification

As was described in Sect. 3, given α' we want to find the threshold, such that F_1 is the highest.

In Fig. 7, we plot *recall* against *precision'*, depending on α' different threshold is selected, detailed values are listed in Table 2.

5   Evaluation

5.1   Prototype Implementation

All codes in python are written and run in Python2.7, with Windows 8 Operation System, 4-Gigabytes of memory. Functionalities implemented through python includes text sanitization and storage, handling of length information, determining threshold, and part of the graphs. ESA relatedness calculation is implemented in Java, with OpenJDK 7u51 on Operating System Ubuntu 13.10, 16-Gigabytes of memory.

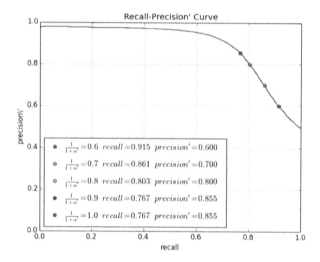

Fig. 7. *Recall* against *Precision'*

5.2   Experiment Setup

There are 7,570 valid applications obtained and used in our work, whose apk files and descriptions are crawled from Google's official Android market, Google Play. Although the default language is English, due to some intentional of unintentional error, some of the applications' descriptions are not in English, which are discarded.

In order to evaluate our system, we divide the applications into two sperate parts, performing as the training set and the test set. The training set consists of 347,428 applications while the test set consists of 38,602 applications.

Based on different choice of α' the measurements are calculated, as is shown in Table 3.

Table 2. Threshold and corresponding F_1 score

$\frac{1}{1+\alpha'}$	Threshold	Recall	*Precision'*	F_1
0.6	0.0050	0.915	0.600	0.725
0.7	0.0090	0.861	0.700	0.772
0.8	0.0143	0.803	0.800	0.801
0.9	0.0203	0.767	0.855	0.809
1.0	0.0203	0.767	0.855	0.809

We compare our method, based on Logistic Regression (annotated as LR), with the straight forward threshold on ESA relatedness approach (annotated as TH). As shown in Table 4 and illustrated in Fig. 8, logistic regression has

Table 3. Evaluation results

$\frac{1}{1+\alpha'}$	TP	FP	FN	Recall	Precision	Precision'	F_1
0.6	662	23743	64	0.912	0.0271	0.576	0.706
0.7	616	14255	110	0.848	0.0414	0.674	0.751
0.8	575	7716	151	0.792	0.0694	0.776	0.784
0.9	544	4837	182	0.749	0.101	0.835	0.790
1	544	4837	182	0.749	0.101	0.835	0.790

Table 4. Comparison with Threshold-based approach

Recall	*precision*		*precision'*		F_1	
	LR	TH	LR	TH	LR	TH
0.749	0.101	0.0605	0.835	0.752	0.790	0.750
0.792	0.0694	0.0478	0.776	0.705	0.784	0.746
0.848	0.0414	0.0351	0.674	0.637	0.751	0.728
0.912	0.0271	0.0234	0.576	0.539	0.706	0.678

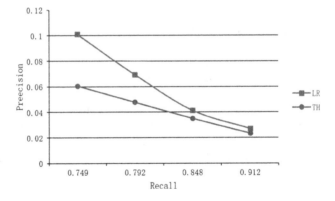

Fig. 8. Method precision comparison

made noticeable improvement in comparison to the threshold-based approach, especially when recall is not that high.

5.3 Our Findings

We examined the applications that have their own embedded text and description deemed unrelated, and classifies the reason behind such misclassification. We find that there are about 23 % applications deemed related by manual examination, containing adequate text length and indicative sentences. However, for the rest 77 % there are four reasons behind our "misclassification", the distrib-

Table 5. Distribution of reasons behind "Misclassification"

Type	Percentage(%)
Related	23
Inadequate Embedded Text	50
Short Description	26
Unsuited Description	12
Malicious Application	1

ution of which is as shown in Table 5. Note that some applications actually fall in multiple categories.

Inadequate Embedded Text. From Fig. 9, we can observe that some of the applications deemed unrelated have much less embedded text then normal ones.

The lack of embedded text is normally the result of two conditions. Either the application has very few functionalities so not much text is displayed in the user interface, a typical example of which is Muzika. This application only provides the functionality to search and download copyleft music, therefore, the only application related words in the embedded text are "search", "length", "size" and "bps", which can not set up an effective correlation with its description.

Another condition is when the developer do not follow the coding convention of putting interface text into the string.xml or layout file, but instead hardcoded it or put it in the database, leaving little useful information for our system. One typical example of this is ANT+, it has only a few entries in layout file and even fewer in string.xml, however, the magnitude of the strings hardcoded in the apk file actually approaches 72 kilo-bytes.

The inadequacy of embedded text can be somewhat mitigated by the introduction of hardcoded string as embedded text, however, it would introduce much noise to the system, considering that we should encourage the adherence to coding convention, we decide not to take in the hardcoded string into consideration.

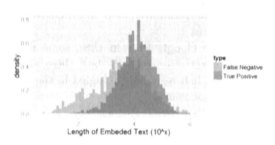

Fig. 9. Distribution of embedded text length

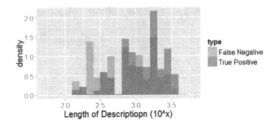

Fig. 10. Distribution of description length

Short Description. From Fig. 10, some of the applications deems unrelated has much shorter description then normal ones. This lack to description makes it harder to establish a connection between the application and the description. Typical applications that fall into this category is either made by individuals instead of a company or that it relies on introduction channel other the application market.

A typical example is `Subway Surfers News`, as a fan-made application displaying news, videos for the game `Subway Surfer`, the application is spread through fan groups of the game. Hence, its description is rather simple, only pointing out the functionality of news and video display, with no reference to functionalities implemented such as, QR-code scanning, initiating events, building albums, audio recording and coupons.

Inadequate descriptions like this could drive away potential users. Our system would warn the developers when such things happen, thus help the developers to better promote their application.

Unsuited Description. Some descriptions, although pretty long, failed to cover the actual functionalities of the application. Some elaborate on the introduction of their company (e.g. `DHgate Mobile`, some quoted many reviewers opinion (e.g. `Nexercise`), but talk little about the functionality.

All these lacks a strong correlation with the application, thus leading to a rather low relatedness. Our system can spot such problem, and suggest the developers to amend their description to a more informative state, making better impression on the potential users.

Although fetch from the Google Play in U.S., some applications still use a foreign language as its description (e.g. `Weibo`), these also pose as unsuited descriptions. Since some English words are mingled in the description they are not filtered out during preprocessing.

Malicious Application. Some applications due to reason unknown, use the application content for other, put on a different name and get into the market (e.g. `Friend Locator`, who stoles content from `360live`), which leads to difference between the application description and the embedded text.

6   Discussion

Based on the relatedness between embedded text and description, we study the state of description writing currently on the Android market. We implemented an innovative system to decided whether the description is in accordance with the application, helping the developers and Android market administrator to evaluate the quality of the description automatically.

Different from related works, our study focus on the semantic relatedness,using text shown to the user during usage to judge the description, which is an approach never attempted before.

6.1   Weaknesses

In calculating text-to-text relatedness, because the limit training set size we currently possess, we use ESA as the tool, which relies on external knowledge base in calculation. However, since new words are continually emerging in this age, external knowledge base may become outdated, failing to catch the meaning of new brand names, product names, etc. This may reduce the accuracy of our classification over time.

In terms of description aid, we are only able to alert the developers of unjust descriptions, without detailed suggestions on how it may be improved.

7   Related Work

The closest related work is AutoCog [16], which examines the consistency between an application's description on Google Play and the permissions it requires, they also leverage NPL to understand the functionality of an application, while they tried to fetch semantic features from "what the developer tells the user", i.e. the description, we get this information from "what the user actually see", i.e. the actual semantic strings, we argue that semantic strings extracted from user interfaces of applications contains more straightforward and more detailed information about what this application would do.

There are many previous works that focuses on these kinds of problems in Android permission system. Han et al. [11] proposed a framework of Collaborative policy administration, where the malicious applications can be detected due to abnormal permission configuration. The descriptions of applications may be used to measure the similarity of two applications, then help identify the abnormal permission configuration [17]. The descriptions of applications can be used to help role mining algorithms too. Enck et al. [6] observed that some specific combinations of permissions could be used as signature of malware, so they introduced a set of security rules and a tool called Kirin to check the application's permission requirements against these rules. Kirin cares about application's permission request only but does not examine whether these permission are really need by the application. Kathy et al. [4] moved one step further, they introduced

PScout to statically check the source code of Android and built a set of mappings between Android Framework APIs and the permissions each API needed. With this set of mappings at hand, one can easily verify that whether an application over claimed privileges that it would not use, but they cannot handle the permission misuse problem nor the situation where repackaged malware invoked additional APIs that are irrelevant to the functionality of the origin application to do malicious things.

8   Conclusion and Future Work

Our system takes in apk files and descriptions; uses natural language processing to normalize the text into semantic vectors; calculate the relatedness between vectors from the given text; taking into consideration size of embedded text, description length, and relatedness, decide whether the apk file and the description are related.

Assuming tested applications all have their descriptions and embedded text related, we performed a test over 7,570 applications, and achieved a recall of 91.2 % in the most tolerant scenario. Under a stricter standard, 25.1 % of the applications are misclassified, by manual examination, we find that among these 25.1 % of applications, about 77 % fall into one of the four categories: (1) Inadequate Embedded Text, (2) Short Description, (3) Unsuited Description, (4) Malicious Application.

We hope to dealt with the problem of inadequate embedded text, finding the threshold, below which the developers should be notified that the given application contains too little information. And if we detect that there is many semantic information hardcoded in the program, we should suggest the developers to perform a reconstruction.

For the to-be-outdated knowledge base, as the training set grows larger we consider using the training set itself, instead of the external knowledge base, to generate text-relatedness measurement. Online learning [12] can then be used to keep the measurement updated.

In the future we may design a better storage schema for the semantic vectors, so that pieces semantically close to the given text can be quickly located and used for description suggestion.

Acknowledgement. This paper is supported by 12th Five-Year National Development Foundation for Cryptography (MMJJ201301008), Key Lab of Information Network Security, Ministry of Public Security (C13612), Natural Science Foundation of Shanghai (12ZR1402600). We thanks anonymous reviewers for their comments.

References

1. Number of android applications. Technical report, AppBrain (2014)
2. Research also shows steady and significant drop in number of malicious apps being removed in past three years. Technical report, RiskIQ (2014)

3. An open-source api for the android market. https://code.google.com/p/android-market-api. Accessed 2014

4. Au, K.W.Y., Zhou, Y.F., Huang, Z., Lie, D.: Pscout: Analyzing the android permission specification. In: Proceedings of the 2012 ACM Conference on Computer and Communications Security, CCS 2012, pp. 217–228. ACM, New York (2012)

5. Chau, M., Reith, R., Ubrani, J.: Worldwide quarterly mobile phone tracker. Technical report, International Data Corporation (2014)

6. Enck, W., Ongtang, M., Mcdaniel, P.D.: On lightweight mobile phone application certification. In: ACM Conference on Computer and Communications Security, pp. 235–245 (2009)

7. Fang, Z., Han, W., Li, Y.: Permission based android security: issues and countermeasures. Comput. Secur. (COSE) **43**, 205–218 (2014)

8. Fellbaum, C.: WordNet An Electronic Lexical Database (1998)

9. Gabrilovich, E., Markovitch, S.: Computing semantic relatedness using wikipedia-based explicit semantic analysis. In: International Joint Conference on Artificial Intelligence, pp. 1606–1611 (2007)

10. Google. android-apktool. https://code.google.com/p/android-apktool. Accessed 2014

11. Han, W., Fang, Z., Yang, L.T., Pan, G., Wu, Z.: Collaborative policy administration. IEEE Trans. Parallel Distrib. Syst. (TPDS) **25**(2), 498–507 (2014)

12. Jordan, M.I., Jacobs, R.A.: Hierarchical mixtures of experts and the EM algorithm. In: International Symposium on Neural Networks (1993)

13. Knoth, P., Zilka, L., Zdrahal, Z.: Cross-lingual link discovery in wikipedia using explicit semantic analysis. In: The 9th NTCIR Workshop Meeting, pp. 6–9, Tokyo, Japan, December 2011. Knowledge Media Institute

14. Porter, M.: An algorithm for suffix stripping. Program-electron. Libr. Inf. Syst. **14**, 130–137 (1980)

15. Pregibon, D.: Logistic regression diagnostics. Ann. Stat. **9**, 705–724 (1981)

16. Qu, Z., Rastogi, V., Zhang, X., Chen, Y., Zhu, T., Chen, Z.: AutoCog: measuring the description-to-permission fidelity in android applications. In: ACM Conference on Computer and Communications Security (2014)

17. Zhang, X., Han, W., Fang, Z., Yin, Y., Mustafa, H.: Role mining algorithm evaluation and improvement in large volume android applications. In: Proceedings of the First International Workshop on Security in embedded systems and smartphones (SESP 2013), conjunction with ASIACCS 2013 (2013)

Traitor Tracing Based on Partially-Ordered Hierarchical Encryption

Yan Zhu$^{(\boxtimes)}$, Dandan Li, and Liguang Yang

School of Computer and Communication Engineering,
University of Science and Technology, Beijing 100083, China
zhuyan@ustb.edu.cn

Abstract. Recently, more and more enterprises and individuals have moved their data into the cloud. To meet this practical requirement, this paper addresses how to establishes a bridge between role-based access control (RBAC) and cloud storage in order to fully preserve investment in existing RBAC systems. We present a new scheme for secure migrating the resources from RBAC systems to cloud storage. This scheme takes full advantage of RBAC, which provides a well-designed and easy-to-manage approach for accessing cloud resources without user intervention. This scheme, called Partially-ordered Hierarchical Encryption (PHE), which implements the partial-order key hierarchy, similar to role hierarchy in RBAC, in public-key infrastructure. In addition, this construction provides traitor tracing to support efficient digital forensics. The performance analysis shows that our construction has following features: dynamic joining and revoking users, constant-size ciphertexts and decryption keys, and lower overloads for large-scale systems.

Keywords: Security · Encryption · Cloud storage · Partial order · Key hierarchy · Traitor tracing

1 Introduction

In recent years, more and more enterprises and individuals have moved their data, such as personal data and large archive system, into the cloud. Cloud-based storage could be particularly attractive for consumers by providing on demand capacity, low-cost service, and long-term archive. Furthermore, cloud services have brought great convenience to people's lives because consumers can access applications and data from the cloud anywhere in the world on demand.

However, there exist some obstacles for migrating the resources in information systems, especially for an amount of existing RBAC systems, into the public cloud. One of these obstacles is the security of migrated resources. Several recent surveys [1] show that 88% potential cloud consumers worry about the privacy of their data, and security is often cited as the top obstacle for cloud adoption. Unfortunately, traditional security mechanisms, such as access control technology, are not suitable for the cloud environment due to the outsourcing-service

© Springer International Publishing Switzerland 2015
M. Yung et al. (Eds.): INTRUST 2014, LNCS 9473, pp. 278–293, 2016.
DOI: 10.1007/978-3-319-27998-5_18

characteristics of cloud storage and the untrusted or honest-but-curious assumption of cloud service providers. On the other hand, the protection of the outsourced data against illegal redistribution via traitor's illegal decoders (or illegal decryption softwares) has become increasingly important due to huge potential commercial value of data stored in cloud.

In order to solve this issue, attribute-based encryption (ABE) [2–6] has been proposed in the recent years. Although ABE is a powerful tool which meets a variety of application requirements, the current ABE schemes cannot fulfill the requirements for the existing RBAC systems owing to lack of support for partial ordering relations. It is well-known that RBAC is an industry recognized and widely adopted access control model. In this model, role hierarchy (RH) is an important notion, which reflects organization's lines of authority and responsibility. Mathematically, role hierarchies are partial orders. Unfortunately, this kind of partial ordering relation still cannot be implemented in the existing ABEs. Therefore, it is necessary to develop a new RBAC-compatible encryption scheme to support the secure migration from RBAC systems into the cloud.

To construct a cryptosystem compatible with RBAC model [7], several schemes for hierarchical key management (HKM) have been designed [8,9]. These existing schemes have following common features: 1) each user has a secret-key sk_i corresponding to a role c_i in RH; 2) there exists an efficient way to derive a descendant's key sk_j from the own key sk_i in accordance with the partial order relation $c_j \preceq c_i$ in RH; and 3) key derivation can be implemented under the precondition of the existence of an one-way function.

Existing schemes can effectively derive the keys with the help of partial order structure. However, such kind of derivation process has following problems:

- A role may be assigned to multiple users who share the same secret-key. That means there is no way to distinguish those assigned users; and
- The secret-key derivation is not able to support additional function, such as the traitor tracing, in terms of digital forensics for group-oriented cryptosystem.

To address these problems, it is necessary to design a construction for hierarchical cryptosystems, considering the new features provided by some recently proposed cryptography technologies, such as IBE [2], HIBE [10] and ABE [11]. In such a construction, a user secret-key must be unique and is accompanied by the user identity. In addition, the derivation of secret-key in such a construction should be avoided. To this end, we introduce a new hierarchical key structure using the public-key settings. Our construction can achieve following functions:

- Each role is assigned with a public-key (called role-key) in RBAC, and there exists a derivation function on these public-keys in accordance with RH;
- Each user has a unique identity and private key, which retain his/her role information, but the derivation of secret-key is prohibited; and
- Such a key structure can be used to establish some important security mechanisms, such as encryption, signature, revocation, and traitor tracking.

One compelling advantage of our key structure is that it can be seamlessly integrated into the existing RBAC systems. Consequently, an RBAC system can directly use the public role-key to encrypt resources in terms of users' assigned roles, and then the users owned the senior roles can use their privacy-keys to decrypt the encrypted resources. This kind of cryptosystem can be used for secure migrating the resources from existing RBAC systems to cloud. Other potential applications of our solution include email encryption system (EES), privacy preservation for peer-to-peer (P2P) data sharing, and encrypted file system (EFS).

Table 1. Comparison of several key management methods with user management [a]

	Stateful schemes	Stateless schemes		
	LKH [12]	CS [13]	LSD [14]	Our Works
Cryptography settings	symmetric-key	symmetric-key	symmetric-key	pubic-key
User key storage	$O(\log n)$	$O(\log n)$	$O(\log^{1+\epsilon} n)^b$	$O(1)$
Encryption cost	$O(n^{1/k})^c$	$O(\log \log n)$	$O(\log n)$	$O(t + \frac{n}{m})$
Average bandwidth	$O(t \log(n/t))$	$O(t \log(n/t))$	$O(t)$	$O(t + \frac{n}{m})$ fixed
Worst case bandwidth	$\min(t \log \frac{n}{t} + t, n - t)$	$\min(t \log \frac{n}{t}, n - t)$	$\min(4t - 2, n - t)$	$t + \frac{n}{m}$ fixed
Traitor tracing	$O(\log n)$	$O(\log n)$	$O(t \log(n/t))$	$O(\log(\frac{n}{m}) + \frac{m}{t})^d$
Key-updating complexity	high	moderate	low	not modify

where, [a] n is the total number of users, t is the number of revoked devices, and m is the average number of users in a subset. [b] ϵ is any number > 0. [c] k is a parameter which mean the number of stratified subsets to obtain a reasonable computation cost, i.e., when n is less than one trillion, $n^{1/8} < \log n$. [d] references the preference evaluation.

Our Contributions. In this paper, our objective is to establishes a bridge between RBAC and secure cloud storage in order to fully preserve investment in existing RBAC systems. To meet this goal, our core task is construct an effective RBAC-compatible cryptosystem for cloud data encryption. This kind of cryptosystem takes full advantage of RBAC, which provides a well-designed and easy-to-manage approach for accessing cloud resources without user intervention. To achieve our task, we present a new cryptosystem, called as Partially-ordered Hierarchical Encryption (PHE) with traitor tracing. The major contributions of this work are summarized as follows:

- We provided a practical Partially-ordered Hierarchical Encryption (PHE) construction, which not only has semantic security and secure key hierarchy, but also supports following features: stateless receivers, dynamic granting, tight security, and a large number of users;

- We given a full security analysis of our cryptosystem, including semantical security under chosen plaintext attacks. More important, our scheme satisfied a new security definition of key management, called secure key hierarchy, against privilege attack and access attack; and
- We provided traitor tracing mechanism based on key hierarchy, which has great practical significance to preserve the integrity and validity of long-term cryptosystems and to prevent the leakage of cloud outsourced data via illegal decoders (or illegal decryption softwares).

In addition, our PHE scheme provides several new secure features, such as public user label, constant-size user key storage, $O(\log(n))$ tracing, lower computational costs and communication bandwidths.

Table 1 shows a comparison of our scheme and some broadcast encryption schemes including Logical Key Hierarchy(LKH) [12], Complete Subtree(CS) [13], Subset Difference(SD) [15], and Layered Subset Difference(LSD) [14]. Although some existing public-key schemes have adopted the hierarchical structure, this comparison does not consider them due to the reason that they do not have a unique key assigned to eash user, and therefore cannot achieve the features of traitor-tracking. From Table 1, it is obvious that the performance of our scheme is substantially better than existing methods with respect to transmission, storage, computation, and traitor tracing costs.

Organization. The rest of the paper is organized as follows. Section 2 describes the research background and the definition of key structure. In Sect. 3, we address our PHE scheme for cryptographic access control on RBAC. Section 4 describes the traitor tracing mechanism, for digital forensics. The results of security analysis is showed in Sect. 5, respectively. We summary the related workin Sect. 6. We conclude and discuss the future work in Sect. 7.

2   Background and Definition

Given a secure key hierarchy $\Psi = \langle C, E, K \rangle$ and the total number n of classes, we can define a (t, n)-Partially-ordered Hierarchical Encryption (PHE), which ensures a content provider to securely transmit a message to a subset of authorized users under the assumption of at most t collusion. More formally, a (t, n)-PHE scheme with a security parameter s is a 6-tuple of probabilistic algorithms $(Setup, Join, Encrypt, Decrypt, Trace)$ described as follows:

1. $Setup(\Omega, s, t)$: Takes as input a partial-order hierarchy Ω, a security parameter s and a maximal collusion number t. It outputs a main encryption key pk_0 as the starting point of cryptosystem, a set of public parameters P^1, and a master key mk as the manager secret.
2. $Join(P, mk, c_i \text{ or } u_{i,j})$: Includes two sub-algorithms:

---

[1] The signature of P can be generated avoid tampering.

- *Join(P, mk, c_i):* Takes as input the manager secret mk and a group identifier c_i. It generates an encryption key pk_i and some public parameters
  pp_i as the description of this class. $P = P \cup \{pp_i\}$ is made public.
- *Join($P, mk, u_{i,j}$):* Takes as input the manager secret mk and a user
  identifier $u_{i,j}$. It outputs a user key $sk_{i,j} = (lab_{i,j}, dk_{i,j})$. $P = P \cup \{lab_{i,j}\}$
  and $sk_{i,j}$ is sent to $u_{i,j}$ securely.

3. *Encrypt(P, pk_i, M):* Encrypts a message M using the public key pk_i and
   outputs a ciphertext C_i.
4. *Decrypt($P, sk_{i,j}, C_k$):* Decrypts a ciphertext C_k using a decryption keys $dk_{i,j}$
   and outputs the message M, if $u_{i,j} \in c_i$ and $c_i \preceq c_k$.
5. *Trace$^\mathcal{D}$(P, pk, mk):* Suppose an adversary uses k user keys $R =$
   $\{sk_{i_1,j_1}, \cdots, sk_{i_k,j_k}\}$ to create a decryption box $\mathcal{D}$. As an oracle algorithm
   on $\mathcal{D}$, it takes as input pk, mk, and can determine at least one key in the
   collusion R.

A tracing algorithm is said to be 'Black Box' if the decoder $\mathcal{D}$ can only be
queried as an Oracle but not opened to reveal its internal keys. The scheme is
said to be 't-resilient' if there is an effective cryptosystem with the collusion of
at most t keys. Note that, the four algorithms (*Setup, Join, Encrypt, Decrypt*)
are used to realize basic cryptographic access control under RBAC model, and
the algorithms *Trace* provide traitor tracing for digital forensics.

3   PHE Scheme for Access Control

3.1   Proposed PHE Scheme

Given a secure key hierarchy $\Omega = \langle C, E \rangle$, a security parameter s, and the maximal coalition size t. Let G_q be a group of prime order q and $\log_2 q > s$. One can
take as G_q the subgroup of $\mathbb{Z}_p^*$ of order q, where p is a large prime with $q|p - 1$.
Let $g \in_R \mathbb{Z}_p^*$ be a generator of G_q.

1. *Setup(Ω, s, t):* The manager chooses t random integers $a_1, \cdots, a_t \in$
   $\mathbb{Z}_q^*$ to construct a random polynomial $f(x) = \sum_{i=1}^t a_i x^i \pmod{q}$
   with degree t. It therefore randomly chooses t integers $x_1, \cdots, x_t$
   to generate $(x_1, f(x_1)), \cdots, (x_t, f(x_t))$. It makes the parameters $P =$
   $\{p, q, (x_1, g^{f(x_1)}), \cdots, (x_t, g^{f(x_t)})\}$ public. Without loss of generality, we
   assume that c_0 be only the senior-most class in Ω. It chooses a random integer
   $s_0 \in_R \mathbb{Z}_q$ for c_0 as its secret, so that the random polynomial $f(x)$ is replaced
   by $p_0(x) = s_0 + \sum_{i=1}^t a_i x^i \pmod{q}$. It then uses $(x_1, p_0(x_1)), \cdots, (x_t, p_0(x_t))$
   to generate an initial encryption key:

$$pk_0 = \langle g, z_{0,0}, (x_1, z_{0,1}), \cdots, (x_t, z_{0,t}), T_0 \rangle \tag{1}$$

$$= \langle g, g^{p_0(0)}, (x_1, g^{p_0(x_1)}), \cdots, (x_t, g^{p_0(x_t)}), \emptyset \rangle$$

   where, $z_{0,i} = g^{p_0(x_i)} = g^{s_0} \cdot g^{f(x_i)} \pmod{p}$ is computed from P and $T_0 =$
   $\emptyset$ denotes a null initial control domain. The system manager keeps $mk =$
   $\{s_0, a_1, \cdots, a_t\}$ secret.

2. $Join(P, mk, c_i$ or $u_{i,j})$: which includes two forms:
    (a) $Join(P, mk, c_i)$: To generate pk_i and pp_i of $c_i \in C$, the manager assigns
        the random $s_i \in \mathbb{Z}_q$ for c_i as its secret. For $\forall c_l \in C$ and $c_i \prec_d c_l$, it
        computes $t_{i,l} = g^{(s_l - s_i)} \pmod{p}$ as the public parameter of this relation,
        and then defines $pp_i = \{t_{i,l}\}_{c_i \prec_d c_l}$ as the set of all relations which directly
        dominate c_i. Finally, it appends s_i and pp_i into mk and P respectively,
        i.e., $mk = mk \cup \{s_i\}$ and $P = P \cup pp_i$.
        The encryption key pk_i in c_i can be computed from the polynomial
        $p_i(x) = s_i + f(x)$. In terms of mk and P, the manager has

$$pk_i = \langle g, z_{i,0}, (x_k, z_{i,k})_{k=1}^t, T_i \rangle \tag{2}$$
$$= \langle g, g^{p_i(0)}, (x_k, g^{p_i(x_k)})_{k=1}^t, T_i \rangle,$$

where, T_i is a set of all relations in $\uparrow c_i$, i.e., $T_i = \{t_{j,l}\}_{c_j, c_l \in \uparrow c_i, c_j \prec_d c_l}$.
    (b) $Join(P, mk, u_{i,j})$: To generate $sk_{i,j}$ of $u_{i,j}$, the manager computes the
        random polynomial $p_i(x) = s_i + f(x) \pmod{q}$ by using the secret in
        mk. It generates a new random integer $x_{i,j} \in_R \mathbb{Z}_q$ and sends $sk_{i,j} = (x_{i,j}, p_i(x_{i,j}))$ to the user via a secret channel, where $lab_{i,j} = x_{i,j}$, $dk_{i,j} = p_i(x_{i,j})$, and $P = P \cup \{lab_{i,j}\}$.
3. $Encrypt(P, pk_i, M)$: For a session key $ek \in G_q{}^2$, the user randomly chooses a
    random number $r \in_R \mathbb{Z}_q$, and then computes the ciphertext by pk_i as follows:

$$C_i = \langle h, S_i, (x_k, h_{i,k})_{k=1}^t, T_i' \rangle \tag{3}$$
$$= \langle g^r, ek \cdot z_{i,0}^r, (x_k, z_{i,k}^r)_{k=1}^t, \{t_{k_1,k_2}'\}_{t_{k_1,k_2} \in T_i} \rangle.$$

where, $h_{i,k} = z_{i,k}^r \pmod{p}$, $t_{k_1,k_2}' = t_{k_1,k_2}^r$, and $T_i' = \{t_{k_1,k_2}^r\}_{t_{k_1,k_2} \in T_i}$ denotes
a control domain which includes all relations in $\uparrow c_i$.
4. $Decrypt(P, sk_{i,j}, C_l)$:
    After receiving a cipher-text $C_l = \langle h, S_l, (x_k, h_{l,k})_{k=1}^t, \{t_{k_1,k_2}'\}_{t_{k_1,k_2} \in T_i} \rangle$, the
    user computes the following equation by the private key $sk_{i,j} = \langle x_{i,j}, y_{i,j} \rangle$ if
    we hold $u_{i,j} \in c_i$, $c_i \preceq c_l$, and

$$U_{C_l}(sk_{i,j}) = \frac{h^{y_{i,j} \cdot \lambda_0(x_{i,j})} \prod_{k=1}^t h_{l,k}^{\lambda_k(x_{i,j})}}{(\prod_{c_{k_1} \prec_d c_{k_2} \in \Delta(l,i)} t_{k_1,k_2}')^{\lambda_0(x_{i,j})}}, \tag{4}$$

where, $\lambda_k(x_{i,j}) = \prod_{l=0, l \neq k}^t \frac{x_l}{x_l - x_k} \pmod{q}$ is the coefficient of Lagrange inter-
polation polynomial[3] for $\{x_0 = x_{i,j}, x_1, \cdots, x_t\}$, and $\Delta(l,i) = \{c_{k_1} \prec_d c_{k_2} : c_{k_1}, c_{k_2} \in \Gamma(l,i)\}$ denotes the set of direct dominations on an arbitrary path
between c_i and c_l. It therefore can obtain the plaintext $ek = S_i / U_{C_l}(sk_{i,j})$.

---

[2] The plaintext (ek or M) must be converted into an element of G_q, see ElGamal
encryption system.

[3] Given a set of $t + 1$ different data points $(x_0, y_0), \cdots, (x_t, y_t)$, the language interpolation polynomial is a linear combination $L(x) = \sum_{j=0}^t y_j \lambda_j(x)$ where the coefficient
$\lambda_j(x) = \prod_{i=0, i \neq j}^t \frac{x - x_i}{x_j - x_i}$. Here, we set $x = 0$ to compute $L(0)$.

Before going further, we briefly show that the encryption scheme is valid by

$$
U_{C_l}(sk_{i,j}) = \frac{g^{p_i(x_{i,j})\cdot\lambda_0(x_{i,j})\cdot r}\prod_{k=1}^{t}g^{p_l(x_k)\cdot\lambda_k(x_{i,j})\cdot r}}{(\prod_{c_{k_1}\prec_d c_{k_2}\in\Delta(l,i)}t_{k_1,k_2}^r)^{\lambda_0(x_j)}}
$$

$$
= \frac{g^{p_i(x_{i,j})\cdot\lambda_0(x_{i,j})\cdot r}\prod_{k=1}^{t}g^{p_l(x_k)\cdot\lambda_k(x_{i,j})\cdot r}}{g^{\sum_{c_{k_1}\prec_d c_{k_2}\in\Delta(l,i)}(s_{k_2}-s_{k_1})\cdot\lambda_0(x_{i,j})\cdot r}}
$$

$$
= \frac{g^{p_i(x_{i,j})\cdot\lambda_0(x_{i,j})\cdot r}\prod_{k=1}^{t}g^{p_l(x_k)\cdot\lambda_k(x_{i,j})\cdot r}}{g^{(s_i-s_l)\cdot\lambda_0(x_{i,j})\cdot r}}
$$

$$
= g^{p_l(x_{i,j})\cdot\lambda_0(x_{i,j})\cdot r}\prod_{k=1}^{t}g^{p_l(x_k)\cdot\lambda_k(x_{i,j})\cdot r}
$$

$$
\overset{x_0=x_{i,j}}{=} g^{\sum_{k=0}^{t}p_l(x_k)\cdot\lambda_k(x_0)}\cdot r = g^{p_l(0)\cdot r} = z_{l,0}^r. \tag{5}
$$

where $s_i - s_l = \sum_{c_{k_1}\prec_d c_{k_2}\in\Delta(l,i)}(s_{k_2}-s_{k_1})$ (mod q) for an arbitrary path $\Gamma(l,i)$ between c_i and c_l[4], and $p_l(x_{i,j}) = s_l + f(x_{i,j}) = p_i(x_{i,j}) - (s_i - s_l)$ (mod q).

3.2   Further Discussion

In fact, the above process is also constructed from bottom (junior-class) to top (senior-class). In the case of many senior-most classes, the *Setup* algorithm is still available. Without loss of generality, we assume that $c_0^{(1)}, c_0^{(2)}, \cdots, c_0^{(l)}$ are l senior-most classes in Ω. Then, it chooses a random integer $s_0^{(i)} \in_R \mathbb{Z}_q$ for $c_0^{(i)}$ as the secret of this class, such that it constructs l random polynomials, $p_0^{(i)}(x) = s_0^{(i)} + \sum_{k=1}^{t}a_k x^k$, where $i \in [1,l]$. Finally, the encryption key is generated:

$$
pk_0^{(i)} = \langle g, z_{0,0}^{(i)}, (x_1, z_{0,1}^{(i)}), \cdots, (x_t, z_{0,t}^{(i)}), T_0\rangle
$$
$$
= \langle g, g^{p_0^{(i)}(0)}, (x_1, g^{p_0^{(i)}(x_1)}), \cdots, (x_t, g^{p_0^{(i)}(x_t)}), \emptyset\rangle,
$$

where, $g^{p_0^{(i)}(x_k)} = g^{s_0^{(i)}}g^{f(x_k)}$ (mod p).

In order to share information, the encryption keys pk_n of junior-most classes are usually made public, which is called the main encryption key, e.g., for a enterprise management system, if the encryption key of "Engineering Dept" class is used to send the message, all employees are able to decrypt it by their own private keys. Moreover, the storage ratio of encryption keys is also an important feature considering a number of classes in the large-scale organizations. We, of course, expect that it is as low as possible. Since $p_i(x) = (s_i - s_l) + p_l(x)$, the user can generate pk_i by using a known pk_j and public parameters P for $i \neq j$. For example, the user can compute her/his own encryption key pk_i from a junior-most encryption key pk_n by $\dot{T}_i = \prod_{c_j\prec_d c_l\in\Delta(n,i)}t_{j,l} = g^{\sum_{c_j\prec_d c_l\in\Delta(n,i)}(s_l-s_j)} = g^{s_i-s_n}$ (mod p) and

---

[4] For the different pathes, we have the same polynomial $p_i(x) = s_i + \sum_{i=1}^{t}a_i x^i$, because $p_i(x) = (s_i - s_{i-1}) + (s_{i-1} - s_{i-2}) + \cdots + (s_1 - s_l) + p_l(x)$ for any path $s_i, s_{i-1}, \cdots, s_1, s_l$.

$$pk_i = \langle g, z_{i,0}, (x_k, z_{i,k})_{k=1}^t, T_i \rangle \tag{6}$$
$$= \langle g, z_{n,0} \cdot \dot{T}_i, (x_k, z_{n,k} \cdot \dot{T}_i)_{k=1}^t, T_i \rangle,$$

where, $z_{n,k} \cdot \dot{T}_i = g^{p_n(x_k)} \cdot g^{s_i - s_n} = g^{p_i(x_k)}$ (mod p), and T_i is found from P in terms of Eq. (2). Therefore, the user only needs to store an encryption key pk_i and a private key $sk_{i,j} = (label_{i,j}, dk_{i,j})$.

The key hierarchy is saved in public parameters P, irrespective of the user private keys, so that the public parameters can be merely modified dynamically to support the change of the key hierarchy.

4   PHE Scheme for Traitor Tracing

It is very hard for the adversary to directly break a cryptosystem with provable security, but the adversary could make other means to break it. It is well-known that "the easiest way to capture a fortress is from within". Based on the same idea, the collusion attack between the adversary and some corrupted users (called traitors) is such an internal attack for group-oriented cryptosystem. In this attack, the adversary may have access to a set of legitimate user's secret keys to decrypt the ciphertext. In order to withstand such attacks, traitor tracing is introduced in the recent years. Usually, the traitor tracing algorithm is an effective detection approach to find out the corrupted users from a group of authorized users based on a found pirate decoder. We prefer that the tracing algorithm is only able to access any pirate decoder as a black box and perform the tracing based on the decoder's response on different input ciphertexts.

The traitor tracing is an efficient mechanism to support digital forensics in the existing group-oriented cryptosystems. Some tracing schemes have also been proposed via the polynomial interpolation method in the recent years. We here propose a new traitor tracing scheme for our partial-order key hierarchy on the basis of these existing schemes. This algorithm only needs to know the public label $label_{i,j}$ of users rather than their private keys. Note that, traitor tracing, as a way of digital forensics, has a precondition where the adversary cannot forge an 'unused' key to avoid tracing. We will prove that this attack is infeasible for our scheme. We now turn our attention to the tracing algorithm from the following two aspects:

4.1   Single-Key Tracing

The single-key tracing algorithm focuses on finding the traitors of collusion one by one. It is easy to find that at most t users cannot forge a new unused key in the corrupted class, such that we can find all traitors only if we search all used keys in this class. For such a collusion attack, we can use the revocation-based algorithm to construct a ciphertext, revoked by the suspicious key, into the illegal decoder. If the decoder does not work, this revocation-based key includes at least a traitor. Otherwise, we search other users in this subset. Finally, we can find all traitors. To improve the performance, we can check out t suspicious keys at the same time. Hence, the searching complexity is $O(m/t)$, where m is the total number of users in a security class or a group of users.

Many tracing algorithms [16] have noticed that a certain linear combination of $sk_1, \cdots, sk_m$ is also a 'new' private key, but in this case the adversary is not confined to the original decryption algorithm to build a decryption box. In such a case, this 'single-key' is not a new key but a linear combination of some keys. For such a decoder, we can construct an encryption key, which includes t user keys, and search all combinations among the keys in this subset. Hence, the searching complexity is $O(\binom{t}{m})$.

4.2   Hierarchical Tracing

The hierarchical tracing algorithm is a more efficient method to find the traitors in terms of partial-order key hierarchy. According to the property of threshold cryptosystem, our proposed scheme is a t-resilient encryption based on CDH assumption in the honest classes, showing that the traitors cannot collude to forge a new key outside the corrupted classes. This property gives us an advantage for constructing the tracing algorithm.

In contrast to single-key tracing, we can first go through each class c_i in a key hierarchy Ψ to locate the suspicious classes of the traitors, and then use single-key tracing algorithm to find the actual traitors in every class. In terms of this idea, given an illegal decoder, we present a black-box traitor tracing algorithm based on the key hierarchy, which involves two steps: *subtree searching* and *subset traversing*, as follows:

V1. *Subtree searching*: Given a key hierarchy Ψ, we start from $c_i \leftarrow c_n$ (the junior-most class) in C and run the following processes from bottom to top:

S1. Randomly selects t unused shares $\langle x_1, x_2, \cdots, x_t \rangle$ and constructs an enabling block:

$$\mathcal{C}_i = \langle g^r, ek \cdot g^{rp_i(0)}, (x_k, g^{rp_i(x_k)})_{k=1}^t, \{t_{j,l}^r\}_{t_{j,l} \in T_i} \rangle.$$

S2. Sends $\langle \mathcal{C}_i, E(ek, M) \rangle$ to the decoder.

S3. If the decoder can return correctly the message M, we consider c_i as a suspicious class and run V1 by $c_i \leftarrow c_j$ for $\forall c_j \prec_d c_i$, otherwise, repeat V1 by a sibling node of c_i.

V2. *Subset traversing*: Let $\langle c_1', c_2', \cdots, c_k' \rangle$ be the set of suspicious subset by $V1$, for each c_i' in this set, we run the following processes:

T1. Chooses any m user's labels in c_i' at random, $\{x_{i,1}, \cdots, x_{i,m}\}$, $m \leq t$, and then randomly selects $t - m$ unused shares, $\langle v_1, v_2, \cdots, v_{t-m} \rangle$, and constructs an enabling block:

$$\mathcal{C}_i' = \left\langle \begin{array}{c} g^r, ek \cdot g^{rp_i(0)}, (x_{i,j}, g^{rp_i(x_{i,j})})_{j=1}^m, \\ (v_k, g^{rp_i(v_k)})_{k=1}^{t-m}, \emptyset \end{array} \right\rangle.$$

T2. Sends $\langle \mathcal{C}_i', E(ek, M) \rangle$ to the pirate decoder.

T3. If the decoder does not output correctly M, we consider the set of label, $\{x_{i,1}, \cdots, x_{i,m}\}$, as a set of traitors and decrease the number of key of this set to run T1. Otherwise, repeats T1 until no more users.

Therefore, our tracing algorithm improves computation complexities and searching times as a result that key hierarchy divides the users into a large number of classes in the key hierarchy. Especially, in the worst case, the complexity of subtree searching is $O(\log n)$ time queries, where n is the number of classes.

5   Security Analysis

We define the security of PHE scheme in terms of a family of security games between a challenger and an adversary. The partial-order hierarchy Ω and system parameters P are fixed, and the adversary is allowed to depend on them. The users can be divided into two categories: the honest users and the corrupted users, so that a set of corrupted users $\mathcal{R}$ is built. The responsive classes is called as honest classes C_1 or corrupted classes C_2, in which the corrupted users can access all encrypted messages. Sometimes, there exist many honest and corrupted users in the same class. We first define a general model against collusion attacks:

1. Initial: The challenger $\mathcal{B}$ constructs an arbitrary partial-order hierarchy Ω, and then runs $Setup(\Omega, s, t)$ to generate the partial-order key hierarchy Ψ and initial public parameters P, and sends them to the adversary $\mathcal{A}$.
2. Learning: $\mathcal{A}$ adaptively issues n times queries $q_1, \cdots, q_n$ to learn the information of Ψ, where q_i is one of the following:
   – Honest class/user query ($u_{i,j} \notin \mathcal{R}$): using $Join(P, mk, c_i$ or $u_{i,j})$, $\mathcal{B}$ generates a class/user label $(pp_i, pk_i, lab_{i,j})$ and sends $lab_{i,j}$ to $\mathcal{A}$.
   – Corrupted class/user query ($u_{i,j} \in \mathcal{R}$): $\mathcal{B}$ generates a class (pp_i, pk_i) with the corrupted users, or a user label $lab_{i,j}$ and a decryption key $dk_{i,j}$, and returns $(lab_{i,j}, dk_{i,j})$ to $\mathcal{A}$.

   $\mathcal{A}$ ends up with a key hierarchy Ψ (include P, pk_i) and a collusion set $\{sk_{i,j}\}_{u_{i,j} \in \mathcal{R}}$. Note that the decryption query is unnecessary because $\mathcal{A}$ can use the corrupted key to generate it.

3. Challenge: $\mathcal{A}$ chooses two equal length plaintexts $M_0, M_1 \in \mathbb{M}$ and appoints a classes c_i on which it wishes to be challenged. $\mathcal{B}$ picks a random bit $b \in \{0,1\}$ and sends the challenge ciphertext $C_i = Encrypt(P, pk_i, M_b)$ or $Revoke(P, pk_i, M_b, R_i)$ to $\mathcal{A}$. where, R_i denotes all corrupted users in $\uparrow r_i$.
4. Guess: $\mathcal{A}$ outputs a guess $b' \in \{0,1\}$. $\mathcal{A}$ wins if $b = b'$, and otherwise it loses.

   There are several important variants for this game:

   – In a game for chosen plaintext attack (CPA), the adversary $\mathcal{A}$ may not issue the corrupted user queries and decryption queries during the learning phase.

- In a game for user's private key attack, the challenger $\mathcal{B}$ may not issue the challenge ciphertext during the challenger phase. The adversary $\mathcal{A}$ returns a forged private-key in polynomial time during the guess phase.
- In a game for unauthorized access attack, by which user can exceed its authority, we hold the above game.[5]

We denote by $\mathrm{Adv}_{\mathcal{E},\mathcal{A}}(t,n)$ the advantage of adversary $\mathcal{A}$ in winning the game:

$$\mathrm{Adv}_{\mathcal{E},\mathcal{A}}(t,n) = \frac{1}{2}\left|\Pr[\mathcal{A}_{\mathcal{E}}(\mathcal{C}_i) = \mathrm{b}] - \Pr[\mathcal{A}_{\mathcal{E}}(\mathcal{C}_i) \neq \mathrm{b}]\right|$$
$$= \left|\Pr[\mathcal{A}_{\mathcal{E}}(\mathrm{C}_i) = \mathrm{b}] - \frac{1}{2}\right|$$

We say that a PHE is (t,n)-secure if for all setup parameter P and all probabilistic polynomial-time adversaries $\mathcal{A}$, the function $\mathrm{Adv}_{\mathcal{E},\mathcal{A}}(t,n)$ is a negligible function of s.

Semantic security is a widely-used security notion in a public-key encryption scheme. Informally, it requires that it is infeasible to learn anything about the plaintext from the ciphertext. This security requirement is also fit for PHE scheme. We show that our encryption scheme is semantically secure agaist chosen plaintext attack (IND-CPA) under the Decision Diffie-Hellman (DDH) assumption as the following theorem:

Theorem 1. *The proposed (t,n)-PHE scheme is semantically secure under chosen plaintext attacks assuming the difficulty of Decisional Diffie-Hellman (DDH) problem in G_q.*

Obviously, semantic security is not enough to satisfy the security requirement of "1:n" encryption scheme. It is important to consider all types of potential attacks when we attempt to design the key hierarchy and broadcast scheme. The security of key hierarchy must assure that the adversary cannot gain any advantage by analyzing public-keys, ciphertexts, and user's private keys. There exist two strategies to attack the PHE scheme:

1. Privilege Attack: it focuses on changing the privileges of the granted users or getting the keys of the other users. This attack also involves two ways:
   - Collusion attack for corrupted classes, in which the corrupted users in $R = \{u_{i_k,j_k}\}_{k=1}^{t}$ wish to forge a (new or unused) key in $\{c_{i_1}, \cdots, c_{i_t}\}$ (called as the corrupted classes). The aim of this attack is to avoid tracing and frame the innocent users.
   - Collusion attack for honest classes, in which the corrupted users in $R = \{u_{i_k,j_k}\}_{k=1}^{t}$ wish to forge a (new or unused) key in $C \setminus \{c_{i_1}, \cdots, c_{i_t}\}$. The aim of this attack is to change the privileges in partial order hierarchy.

---

[5] This game may be more strict than the other two games.

2. Access Attack: it focuses on gaining the advantage of adversary to break the cryptosystem or extending the range of access by the collusion of corrupted users, especially gaining the advantage to break the revocation-based algorithm.

We would like to adopt appropriate technologies to prevent the above attacks, but the collusion attack is unavoidable in the way of technology because the traitor has been a granted user before s/he is not found. Thus traitor tracing is an efficient method to frighten the collusion attack. However, we must ensure that the traitors cannot forge an 'unused' key to avoid tracing but leave some 'foregone' clue of evidence to find them. We present such a definition for Secure Key Hierarchy (SKH) as follows:

Definition 1 (Secure Key hierarchy). *A (t, n)-PHE scheme $(\mathcal{S}, \mathcal{J}, \mathcal{E}, \mathcal{D})$ is said to have a secure key hierarchy $\langle C, E, K \rangle$ satisfying the following conditions:*

1. *Validity: for any member $u_{i,j}$ in $c_i \in C$, the session key ek can be efficiently computed from B_l and $sk_{i,j}$, where $c_i \prec c_l$. Then for every pair $(pk_l, sk_{i,j})$ in the range of $\mathcal{G}(1^n)$ and every sequence M_n, $|M_n| \leq poly(n)$,*

$$\Pr\left[\mathcal{D}(sk_{i,j}, \mathcal{E}(pk_l, M_n)) = M_n\right] \geq 1 - \frac{1}{|p(n)|}; \tag{7}$$

$\frac{1}{|p(n)|}$ *denotes negligible or negligibly small, which means that the absolute value is asymptotically smaller than any polynomial bound.*

2. *Privilege attack: for any set $R \subseteq \{u_{i_1,j_1}, \cdots, u_{i_m,j_m}\}$, $|R| \leq t$, it is computationally infeasible to compute $sk_{i,j}$ of a user $u_{i,j} \notin R$ and the (public) encryption key pk. Then for every probabilistic polynomial-time algorithm $\mathcal{A}$, every polynomial $p(\cdot)$, and all sufficiently large n,*

$$\Pr\left[\begin{array}{l} \mathcal{A}(pk, \{sk_{i_l,j_l}\}_{u_{i_l,j_l} \in R}) = sk_{i,j} \\ : sk_{i_l,j_l} \notin \{sk_{i,j}\}_{u_{i_l,j_l} \in R} \end{array}\right] < \frac{1}{|p(n)|}; \tag{8}$$

where, $pk = P \cup \{pk_i\}_{c_i \in C}$.

3. *Access attack: for any set $R \subseteq \{u_{i_1,j_1}, \cdots, u_{i_m,j_m}\}$ $|R| \leq t$, it is computationally infeasible to gain the advantage to break the revocation-based algorithm from the collusion set R and any ciphertexts $C_l = \mathcal{E}^R_{pk_l}(M_n)$, where $\mathcal{E}^R$ denotes revocation-based algorithm on R and M_n is a sequence with $|M_n| \leq poly(n)$. Then for every probabilistic polynomial-time algorithm $\mathcal{A}$, every pair of polynomially-bounded functions $f, h : \{0, 1\}^* \rightarrow \{0, 1\}^*$ (see [17]), every polynomial $p(\cdot)$, and all sufficiently large n,*

$$\Pr\left[A\left(\begin{array}{c} pk, h(X_n), \mathcal{E}^R_{pk_l}(X_n), \\ \{sk_{i,j}\}_{u_{i,j} \in R} \end{array}\right) = f(X_n)\right] \tag{9}$$

$$< \Pr\left[A\left(\begin{array}{c} pk, h(X_n), \\ \{sk_{i,j}\}_{u_{i,j} \in R} \end{array}\right) = f(X_n)\right] + \frac{1}{|p(n)|}.$$

Where, $f(X_n)$ denotes the information that the adversary tries to obtain from the plaintext X_n and $h(X_n)$ denotes a priori partial information about the plaintext.

In this definition, the condition 3) aims at the risk of revocation-based mechanism and puts forward this security requirement (tighter than Theorem 1), which conforms to the definition of 'semantic security' besides the additional key information $\{sk_{i,j}\}_{u_{i,j} \in R}$ for a set of revoked users R. As is well known, the encryption scheme is semantically secure if and only if it has indistinguishable encryptions (see Theorem 5.2.5 in [17]). So, we replace Eq. (9) with the following equation

$$\left| \begin{array}{l} \Pr[\mathcal{A}(\mathrm{pk}, \{\mathrm{sk}_{i,j}\}_{u_{i,j} \in R}, \mathcal{E}^R_{\mathrm{pk}_l}(X_n)) = 1] - \\ \Pr[\mathcal{A}(\mathrm{pk}, \{\mathrm{sk}_{i,j}\}_{u_{i,j} \in R}, \mathcal{E}^R_{\mathrm{pk}_l}(Y_n)) = 1] \end{array} \right| < \frac{1}{|p(n)|}, \tag{10}$$

such that it is easier than ever to prove the security of scheme against access attack. According to this definition, we can prove the following theorem.

Theorem 2. *The proposed (t,n)-PHE scheme has a secure key hierarchy satisfying Definition 1 against the privilege attack and the access attack.*

In the proof of this theorem, the security against privilege attack includes two cases: privilege attack for honest classes and one for corrupted classes. The proofs of the above-mentioned theorems were omitted due to space limitations.[6]

6   Related Work

For a large-scale group-oriented communication, broadcast encryption was first considered [18] in 1991 and, subsequently, formally defined by Fiat and Naor [19] in 1994. Since then, it has become one attractive topic in cryptography community. In symmetric-key setting, only trusted system designer can broadcast data to the receivers. However, the public-key scheme, first introduced by Boneh et al. in 1999 [20], can publish a short public key, which enables anybody to broadcast data, thus overcome the deficiency of symmetric-key setting. Also, Boneh et al. have done massive work in the development of group-oriented encryption, e.g., Boneh, Sahai, and Waters [21] propose a fully collusion resistant traitor tracing with ciphertexts of size $O(\sqrt{n})$ and private keys of size $O(1)$ in 2006, where n is the total number of users. However, these work did not take into account the hierarchy structure.

Boneh and Franklin proposed the first fully identity-based encryption (IBE) [22] in 2001, in which the public key can be an arbitrary string such as an email address. Unfortunately, IBE does not support broadcast function unless some members can share the same private-key when they hold the same identity. According to this idea, Boneh et al. provided a hierarchical identity-based encryption (HIBE) system to support an organizational hierarchy [23], but this kind of hierarchy must be a tree structure and cannot provide identity-based revocation and tracing due to the global sharing of hierarchical identity/privacy-key for all users. In addition, attribute-based encryption (ABE) is also considered

---

[6] The interesting readers may read the full proofs in the website: crypto.ustb.edu.cn.

as an effective group communication method [24], but the existing ABE schemes have not yet been able to support the hierarchical structure.

For cryptosystems on the partial order relation, Akl and Taylor put forward a simple scheme to solve multilevel security problem in 1982. In 2005, Kim [25] proposed a new key management system for multilevel security using various one-way functions. In 2008, Chung [26] proposed a method based on the elliptic curve cryptosystem and one-way hash function to solve dynamic access problems. Another related field is *hierarchical key management with time control*. For example, in 2002, Tzeng proposed a time-bound scheme based on Lucas function [27], but it is insecure against collusion attacks by Yi and Ye. Another similar schemes based on the tamper-resistant device and the hash function were proposed by Chien [28] in 2004 and Bertino *et al.* [29] in 2008. Although these work support real-time broadcast with time control rather than common access control and digital forensics, their hierarchy techniques are worth learning for hierarchy managements. In 2007, Santis *et al.* summarized and provided several provably-secure hierarchical key assignment schemes based on an existing schemes [30].

7   Conclusion and Future Work

In this paper we construct an effective RBAC-compatible cryptosystem for cloud data encryption. In our future work, we are planning to introduce a comprehensive role-based cryptosystem to support various secure mechanisms, such as encryption, signature, and authentication. Also, we would investigate a more efficient cryptosystem to realize massive-scale conditional access control systems for the practical RBAC applications of large-scale organizations.

Acknowledgments. The authors are indebted to anonymous reviewers for their valuable suggestions. This work is supported by the National 973 Program (Grant No. 2013CB329605) and National Natural Science Foundation of China (Grant Nos. 61170264 and 61472032).

References

1. F.R. Institute: Personal data in the cloud: a global survey of consumer attitudes (2010). http://www.fujitsu.com/downloads/SOL/fai/reports/fujitsu/personal-data-in-the-cloud.pdf
2. Boneh, D., Franklin, M.: Identity-based encryption from the weil pairing. In: Kilian, J. (ed.) CRYPTO 2001. LNCS, vol. 2139, pp. 213–229. Springer, Heidelberg (2001)
3. Sahai, A., Waters, B.: Fuzzy identity-based encryption. In: Cramer, R. (ed.) EURO-CRYPT 2005. LNCS, vol. 3494, pp. 457–473. Springer, Heidelberg (2005)
4. Goyal, V., Pandey, O., Sahai, A., Waters, B.: Attribute-based encryption for fine-grained access control of encrypted data. In: ACM Conference on CCS, pp. 89–98 (2006)

5. Ostrovsky, R., Sahai, A., Waters, B.: Attribute-based encryption with non-monotonic access structures. In: ACM Conference on Computer and Communications Security, pp. 195–203 (2007)

6. Nishide, T., Yoneyama, K., Ohta, K.: Attribute-based encryption with partially hidden ciphertext policies. IEICE Trans. **92–A**(1), 22–32 (2009)

7. Zhu, Y., Ahn, G.-J., Hu, H., Ma, D., Wang, S.: Role-based cryptosystem: a new cryptographic rbac system based on role-key hierarchy. IEEE Trans. Inf. Forensics Secur. **8**(12), 2138–2153 (2013)

8. Atallah, M.J., Blanton, M., Fazio, N., Frikken, K.B.: Dynamic and efficient key management for access hierarchies. ACM Trans. Inf. Syst. Secur. **12**(3), 1–43 (2009)

9. Blanton, M., Frikken, K.B.: Efficient multi-dimensional key management in broadcast services. In: Gritzalis, D., Preneel, B., Theoharidou, M. (eds.) ESORICS 2010. LNCS, vol. 6345, pp. 424–440. Springer, Heidelberg (2010)

10. Boneh, D., Boyen, X., Goh, E.-J.: Hierarchical identity based encryption with constant size ciphertext. In: Cramer, R. (ed.) EUROCRYPT 2005. LNCS, vol. 3494, pp. 440–456. Springer, Heidelberg (2005)

11. Zhu, Y., Ahn, G.-J., Hu, H., Yau, S.S., An, H.G., Hu, C.-J.: Dynamic audit services for outsourced storages in clouds. IEEE Trans. Serv. Comput. **6**(2), 227–238 (2013)

12. Wallner, D.M., Harder, E.G., Agee, R.C.: Key management for multicast: Issues and architecture. In: Internet draft draft-waller-key-arch-01.txt (1998)

13. Asano, T.: Reducing receiver's storage in CS, SD and LSD broadcast encryption schemes. IEICE Trans. Fundam. Electron. Commun. Comput. Sci. **88**(1), 203–210 (2005)

14. Halevy, D., Shamir, A.: The LSD broadcast encryption scheme. In: Yung, M. (ed.) CRYPTO 2002. LNCS, vol. 2442, pp. 47–60. Springer, Heidelberg (2002)

15. Naor, D., Naor, M., Lotspiech, J.: Revocation and tracing schemes for stateless receivers. In: Kilian, J. (ed.) CRYPTO 2001. LNCS, vol. 2139, pp. 41–62. Springer, Heidelberg (2001)

16. Tzeng, W.-G., Tzeng, Z.-J.: A public-key traitor tracing scheme with revocation using dynamic shares. In: Public Key Cryptography, pp. 207–224 (2001)

17. Goldreich, O.: Foundations of Cryptography. Basic Application, vol. II. Cambridge University Press, New York (2004)

18. Berkovits, S.: How to broadcast a secret. In: Davies, D.W. (ed.) EUROCRYPT 1991. LNCS, vol. 547, pp. 535–541. Springer, Heidelberg (1991)

19. Fiat, A., Naor, M.: Broadcast encryption. In: Stinson, D.R. (ed.) CRYPTO 1993. LNCS, vol. 773, pp. 480–491. Springer, Heidelberg (1994)

20. Boneh, D., Franklin, M.K.: An efficient public key traitor scheme (extended abstract). In: Wiener, M. (ed.) CRYPTO 1999. LNCS, vol. 1666, pp. 338–353. Springer, Heidelberg (1999)

21. Boneh, D., Sahai, A., Waters, B.: Fully collusion resistant traitor tracing with short ciphertexts and private keys. In: Vaudenay, S. (ed.) EUROCRYPT 2006. LNCS, vol. 4004, pp. 573–592. Springer, Heidelberg (2006)

22. Boneh, D., Franklin, M.: Identity-based encryption from the weil pairing. In: Kilian, J. (ed.) CRYPTO 2001. LNCS, vol. 2139, pp. 213–229. Springer, Heidelberg (2001)

23. Boneh, D., Boyen, X., Goh, E.-J.: Hierarchical identity based encryption with constant size ciphertext. In: Cramer, R. (ed.) EUROCRYPT 2005. LNCS, vol. 3494, pp. 440–456. Springer, Heidelberg (2005)

24. Bethencourt, J., Sahai, A., Waters, B.: Ciphertext-policy attribute-based encryption. In: IEEE Symposium on Security and Privacy, pp. 321–334 (2007)

25. Kim, H.K., Park, B., Ha, J.C., Lee, B., Park, D.G.: New key management systems for multilevel security. In: Gervasi, O., Gavrilova, M.L., Kumar, V., Laganá, A., Lee, H.P., Mun, Y., Taniar, D., Tan, C.J.K. (eds.) ICCSA 2005. LNCS, vol. 3481, pp. 245–253. Springer, Heidelberg (2005)
26. Chung, Y.F., Lee, H.H., Lai, F., Chen, T.S.: Access control in user hierarchy based on elliptic curve cryptosystem. Inf. Sci. **178**, 230–243 (2008)
27. Tzeng, W.G.: A time-bound cryptographic key assignment scheme for access control in a hierarchy. IEEE Trans. Knowl. Data Eng. **14**(1), 182–188 (2002)
28. Chien, H.Y.: Efficient time-bound hierarchical key assignment scheme. IEEE Trans. Knowl. Data Eng. **16**(10), 1301–1304 (2004)
29. Bertino, E., Bettini, C., Ferrari, E., Samarati, P.: An access control model supporting periodicity constraints and temporal reasoning. ACM Trans. Database Syst. **23**(3), 231–285 (1998)
30. De Santis, A., Ferrara, A.L., Masucci, B.: Efficient provably-secure hierarchical key assignment schemes. In: Kučera, L., Kučera, A. (eds.) MFCS 2007. LNCS, vol. 4708, pp. 371–382. Springer, Heidelberg (2007)

SCIATool: A Tool for Analyzing SELinux Policies Based on Access Control Spaces, Information Flows and CPNs

Gaoshou Zhai[1(✉)], Tao Guo[1,2], and Jie Huang[1]

[1] School of Computer and Information Technology,
Beijing Jiaotong University, Beijing, China
{gszhai,11120437,13120394}@bjtu.edu.cn
[2] Henan Center of Patent Examination Cooperation of the Patent Office SPIO,
Henan, China

Abstract. Although security policies configuration is crucial for operating systems to constrain applications' operations and to protect the confidentiality and integrity of sensitive resources inside the systems, it is an intractable work for security administrators to accomplish correctly and consistently solely by hands. Thus policies analysis methods are becoming research hotspots. A great deal of such researches are focused on SELinux, which is a security-enhanced module of open-source and popular Linux. Among various analysis methods for SELinux policies, those based on access control spaces, information flows and colored Petri-nets (CPNs) can be thought as the three most valuable methods and they can be exploited together and complementarily. In this paper, a prototype of SELinux policies Configuration Integrated Analysis Tool, i.e. SCIA-Tool, is designed and implemented by integrating these three methods together. Test results are provided and further researches as to construct a computer-aided configuration tool for SELinux policies are discussed.

Keywords: Security policies configuration · Analysis method · Access control spaces · Information flows · Colored Petri-nets · SELinux

1 Introduction

Security of operating systems is always the research focus in the fields of information security for their irreplaceable position inside the whole information systems. As Linux is becoming popular and powerful SELinux has been embedded into the Linux kernel, they are being research hotspots in domain of security of operating systems.

SELinux can enforce mandatory access control through policies configuration based on TE, RBAC and MLS models [1, 2] to defend against local and remote attacks and to protect systematic integrity and confidentiality and thus make Linux fulfill various security requirements for most situations. However, statements and rules of policies configuration are immense and complex because there are so many programs that can become potential subjects and so many objects including processes, files, devices, sockets and other resources of sorts inside computer systems. Moreover, subjects, objects and relationships among them are complicated and confused. Thereafter, it is difficult and

© Springer International Publishing Switzerland 2015
M. Yung et al. (Eds.): INTRUST 2014, LNCS 9473, pp. 294–309, 2015.
DOI: 10.1007/978-3-319-27998-5_19

error-prone for security administrators to accomplish the correct and consistent config-uration manually without auxiliary means.

Computer-aided policies analysis is thus becoming an effective way to provide more helpful information for policies configuration. Such analysis typically includes TCB integrity analysis and validity analysis. Generally, the former is to find all rules that could potentially influence integrity of the initially specified TCB while the latter is to work out authorized or prohibited permissions as for a specified subject and/or a specified object as well as a specified information flow path so as to verify whether the policies configuration satisfies the security targets.

A lot of researches have been done as to policies analysis methods and those based upon access control spaces [3–6], information flow [5–8] and colored Petri-nets (CPNs) [9–12] reflect more practical values. In addition, the last method also depends on information flow query and verification and is essentially an information flow analysis method. Furthermore, all of these methods have their own limitation while more analysis targets or results will be more helpful for policies configuration. In details, the policy analysis method using access control method is convenient for designing security policies while the other two methods can be used to verify whether a special security goal is in practice enforced by corresponding policies configuration. In addi-tion, the latter two methods can provide more detailed security requirements for val-idation by information flow inquiries. Therefore, they are complementary and can be exploited together.

In this paper, a prototype of SELinux policies Configuration Integrated Analysis Tool, i.e. SCIATool, is designed and implemented in C language by integrating methods based on access control spaces, information flows and colored Petri-nets. The main contributions of this paper are:

- The approach presented in this paper is, to the best of our knowledge, among the first efforts on systematic integration of such three methods for SELinux policies configuration analysis. And SCIATool is the first such prototype implemented independently without other software tools.
- Integrating different methods is always a challenging problem because it is not only to put them together but also to bring them into full play in a redundancy minimized framework. An integrated architecture is put forward for SCIATool so as to make full use of different analysis methods but minimize the redundant design.
- Both the TCB integrity analysis and various validity analyses can be achieved effectively by SCIATool. Various analysis targets such as all rules that could potentially influence integrity of the initially specified TCB, authorized or prohib-ited permissions as for a specified subject and/or a specified object as well as information flow path in CPN of policy configuration as for a specified information flow path requirement can be worked out, which provides richer analysis results than many available analysis prototypes and will be more helpful for computer-aided policies configuration.

The remaining part of the paper is organized as follows: Sect. 2 introduces SELinux policies configuration, typical analysis methods and main idea about our integrated analysis method; Sect. 3 describes our design and implementation of SCIATool; Sect. 4 tests and evaluates our SCIATool prototype with a SELinux policy configuration illustration as for some student-teacher application security requirements, then compares our research in this paper with related work of others; Sect. 5 summarize the research work in this paper and discusses the limitations of SCIATool and future work as to construct a practical computer-aided configuration tool for SELinux policies.

2 Methodology

2.1 SELinux Policies Configuration

SELinux policies configuration is made up of a series of policy source files. In another word, it is the collection of statements, i.e. rules that determine allowed access for a system, and it defines the roles any SELinux user may assume, the domains a role can access, the types any process can access and how. When a process tries to gain access to a particular object, a security decision has to be made whether the access is allowed or denied depending on the security context (i.e. $<user_i, role_j, type_k>$) of the subject, the security context of the object, and the corresponding policies.

SELinux has combined three different policy models in its policies configuration, where its RBAC model associates users with roles and roles to TE domains that are authorized to specific access permissions. All roles are used to constrain the association of users with the types of processes, except that the dummy role $object\_r$ is used in security contexts for all object types. The SELinux types are classified based on the functions performed by processes and the operations performed on the different objects. Types in SELinux can be classified into domain types, security types, device types, file types, procfs types, devpts types, nfs types and network types while domain types are special for processes and can be further classified into system domains, user program domains, and user login domains. In general, Security policy description language of SELinux provide users the following top-level components such as Flask definitions, TE and RBAC declarations and rules, user declarations, constraint definitions, and security context specifications for a policy configuration.

2.2 Typical SELinux Policies Analysis Methods

Among analysis methods for SELinux security policies, those based upon access control spaces, information flows and colored Petri-nets are the three most valuable types of methods that can be exploited in practical configuration.

Method Based on Access Control Spaces. The policy analysis method based on access control spaces is put forward by Trent Jaeger et al. and is firstly used in Gokyo [3] to analyze a SELinux policy configuration for ApacheWeb server system and the example policy of the SELinux for Linux 2.4.16. And a formal model called SELAC is developed on this basis by Giorgio Zanin and Luigi Vincenzo Mancini [4] for analyzing an arbitrary security policy configuration for the SELinux system.

Access control space is the core concept for this method, which is defined as the set of all possible permission assignments of a subject (or role) and can be divided into three natural subspaces: the permissible subspace (contains the permission assignments in the current configuration), the prohibited subspace (contains the permission assignments precluded by the constraints) and the unknown subspace (contains the permission assignments that are neither permitted nor prohibited). Ideally, these three subspaces should partition the access control space without intersection and the unknown subspace should be minimal. But in practice, subspaces are not disjoint and the unknown subspace is large. In another word, sometimes the specified space conflicts with the prohibited space and the unknown space. In addition, the permission assignments within the permissible subspace can be divided into two parts: one part is explicitly expressed in the configuration and the other part is not yet specified. The former part constitutes so-called the specified subspace and it contains a subset of the permissible assignments (accordingly named by the obligated subspace) that are obligated required for correct operation of the system.

Once access control spaces are constructed for a given SELinux policies configuration, all possible integrity conflicts embodied by the conflicting subspaces for TCB subjects can then be identified and classified for solving related conflicts. So the method can be used to help security administrator to accomplish custom-made SELinux policies configuration. However, it is focused on single special analysis target and it is not appropriate for analysis on multiple targets.

Method Based on Information Flows. The policy analysis method based on information flows is firstly developed and applied in SELinux policies for the e-commerce processing system by Guttman et al. [7, 8].

Information flow is the key conception for this method. If a subject with security context $<u_1, r_1, t_1>$ can write an object with security context $<u_2, r_2, t_2>$, we say that there is a *write-like* information flow transition from the subject with security context $<u_1, r_1, t_1>$ to the object with security context $<u_2, r_2, t_2>$. Similarly, if a subject with security context $<u_2, r_2, t_2>$ can read an object with security context $<u_1, r_1, t_1>$, we say that there is a *read-like* information flow transition from the object with security context $<u_1, r_1, t_1>$ to the subject with security context $<u_2, r_2, t_2>$. Both cases can be formalized as an information flow from security context $<u_1, r_1, t_1>$ to security context $<u_2, r_2, t_2>$ through event $<c, p>$ where c, p represents corresponding class and permission for the object. Furthermore, if there is an information flow from security context $<u_i, r_i, t_i>$ to security context $<u_{i+1}, r_{i+1}, t_{i+1}>$ through event $<c_i, p_i>$ for all $0 \leq i \leq n - 1$, it can be concluded that there is an information flow from security context $<u_0, r_0, t_0>$ to security context $<u_n, r_n, t_n>$ through event sequence $\{<c_i, p_i> \mid 0 \leq i \leq n - 1\}$. In addition, an information flow can be accepted if and only if all entities passed in the flow are trusted.

Once the information flow model for a given SELinux policies configuration is built up, information flow security goal statements for the objectives that SELinux is intended to achieve can be expressed in linear temporal logic and model checking method can be used to determine whether security goals hold in the given system. So the method is focused on the whole information flow path and can be used to validate whether the SELinux policies configuration is fully in accordance with the whole desired security objectives of the system. At the same time, those expressions or inquiries for security goal statements to be validated are difficult to write and provide and thus its application is restricted to some extent. Moreover, it is not appropriate for designing or developing policies configuration directly.

Method Based on Colored Petri-Nets. The policy analysis method based on colored Petri-nets is firstly developed and applied in SELinux policies for the e-commerce processing system by Chen and Kao [9]. It is also used to model the trusted computing based secure systems [10].

This method is also developed based on information flows model thus it can be viewed as a special policy analysis method based on information flows. But it describes the SELinux policies configuration and security objectives in the way of colored Petri-nets instead of information flow graphs. It is obvious that colored Petri-net is the key conception for this method. Colored Petri-net (CPN) is formulated on the basis of traditional Petri-net concept by introducing color set. Furthermore, colored Petri-net can be formally defined by a tuple $CPN = <\Sigma,P,T,A,N_A,C_P,G_T,E_A,I_P>$ which satisfies the following requirements: (1) Σ is color set, i.e. a finite set of non-empty data types. Different color stands for different place sort, i.e. type place or permission place. (2) P is a finite set of places which are used to describe types and permissions in SELinux. Each place can hold 0, 1 or several token(s). (3) T is a finite set of transitions which are used to describe access relationships between type places and permission places. (4) A is a finite set of directed arcs which are used to link places and transitions and to describe flow directions. (5) $N_A: A \rightarrow P \times T \cup T \times P$ is a node function that associates directed arcs with two nodes (place or transition). (6) $C_P: P \rightarrow \Sigma$ is a color function that associates places with color set. (7) $G_T: T \rightarrow EXP$ is a guard function that associates transitions with expressions such that: $\forall t \in T, (type(G_T(t)) = bool) \wedge (type(var(G_T(t))) \subseteq \Sigma)$, where $type(e)$ denotes the data type of an expression e, $type(\{e1, e2, ...\})$ denotes the set of data types of expressions $e1,e2, ...$, $var(e)$ denotes the set of free variables of an expression e, and EXP denotes the set of all expression. G_T is used to describe necessary conditions for information flows. (8) $E_A: A \rightarrow EXP$ is an arc expression function that associates directed arcs with expressions such that: $\forall a \in A$, $(type(E_A(a)) = C_P(p(a))_{MS}) \wedge (type(var(E_A(a))) \subseteq \Sigma)$, where $p(a)$ is the place of $N_A(a)$, and 't_{MS}' denotes type '*multi-set of type t*'. E_A is used to describe the tokens passes the directed arcs and corresponding update modes. (9) $I_P: P \rightarrow EXP$ is an initialization function that associates places with expressions such that: $\forall p \in P$, $type (I_P(p)) = C_P(p)_{MS}$. I_P is used to set initial token values for those places corresponding with source types of inquiry statements.

In addition, two type places are generally linked by one permission place and two transitions. This means that the former type place has the authorization of permission denoted by the permission place against the latter type place. Obviously, it is consistent completely to *allow* statements in SELinux policies configuration. Thereafter, it is mainly by analysis and process of *allow* statements to construct the CPN model for policy analysis.

A token is a dynamic entity in a place and it can move from one place to another place. A transition can be initiated if and only if the value of the token in the place matches the description on the directed arc and thus pass the test of guard function associated with the transition. During the procedure of analysis as for a given inquiry, a token will record types, permissions and all other information related to inquiry in the information flow path from the place corresponding to source type of the inquiry statement to the current place, called inquiry information flow path. So it can be denoted as a tuple *<bool, queryType, typeList, permissionList>* where *bool* is a Boolean value to store the decision result whether the type in the current place matches the type in the inquiry statement; *queryType* is a char string to record the recent classification label in the inquiry statement during analysis procedure; *typeList* is a char string list to store all types on the inquiry information flow path; *permissionList* is also char string list but it is used to record all permissions on the inquiry information flow path. Values of tokens in places on the inquiry information flow path ought to be updated with the proceeding of analysis according to corresponding different classification labels.

This method has powerful analysis and verification capabilities for SELinux policies configuration. But it has also the disadvantage at difficult inquiry description as well as the information flow analysis method for its essence of information flow.

2.3 Main Idea About an Integrated Analysis Method Based on Access Control Spaces, Information Flows and Colored Petri-Nets

SELinux policies configuration is the basic foundation for systematic security enforcement and it can be viewed as the embodiment of security objectives. Thus the analysis of SELinux policies configuration ought to dedicate to validate if it faithfully supports confidentiality and integrity under mandatory access control, to check if there is any loophole that may impair the security goals, and to help security administrator to make appropriate configuration solution that accords with principles such as least privilege and separation of permission. According to the facts that each method has its own advantages and disadvantages, an integrated analysis method should be developed based upon access control spaces, information flows and colored Petri-nets. Thus different analysis method can be exploited to achieve different analysis goals so as to make full use of respective advantages.

The main idea of the integrated analysis method can be induced as follows:

1. Method based on access control spaces can be used for validity analysis (i.e. to check if policies configuration meets security goals), e.g. to sum up all objects that a specified subject can access and all subjects that can access a specified object and to decide if a specified subject can access a specified object (where subject/object specification can take the way of assigning security context and access can also be assigned as a special access mode), and it is helpful for security administrator to separate permissions, detect configuration bugs (such as undesired authorization or obligations that cannot be fulfilled because of the lack of authorizations) and make complete configuration.

2. Method based on information flows can be used along with access control spaces to analyze integrity and the integrity of trusted computing base (TCB) is the premise and foundation to ensure the security of whole system. By analyzing integrity conflicts between the TCB entities and the non-TCB entities, information can be provided to help security administrators to ameliorate SELinux policies configuration and to optimize and consummate the assignment of TCB entities.

3. Method based on colored Petri-nets can be also used for validity analysis and to find potential problems such as information flow leaks. By elaborate design to make it support both inquiries in positive description and those in negative description as well as inquiries including intermediate types, it will be convenient for security administrators to check completeness and consistency of policies configuration against security objectives.
4. All these methods ought to be implemented in a uniform architecture while modular, simple but effective design principle should be pursued and followed.

3 Prototype Design and Implementation

3.1 Architecture Design

SCIATool takes aims at validity analysis and integrity analysis for a given SELinux policies configuration.

Validity analysis is to make sure that the configuration has put expected access regulations into effect and met corresponding security goals thus inquiries ought to be processed correctly for the following cases. Firstly, if a subject is specified by security context, all objects with corresponding security context and permissions (in the form of <*class_name*, *access_mode*>) that it is authorized to access can be worked out. Secondly, if an object is specified by security context, all subjects with corresponding security context and permissions that it is authorized to be accessed can also be worked out. Thirdly, if a subject is specified by security context, all objects with corresponding security context and permissions that it is prohibited to access can be worked out. Fourthly, if an object is specified by security context, all subjects with corresponding security context and permissions that it is prohibited to be accessed can also be worked out. Fifthly, if a subject and an object are specified in the way of security context respectively, corresponding access relationships (i.e. authorized or prohibited permissions) can be figured out. Finally, if the feature of an information flow path is specified, whether it can be supported by the configuration ought to be analyzed and concluded. Except that last inquiry is processed using colored Petri-nets method, all others are analyzed based on access control spaces.

Integrity analysis in this paper is processed around the TCB. Integrity of the TCB holds if there is no type that can be written by a type outside the TCB and read by a type inside the TCB, except for special cases in which a designated trusted program sanitizes untrusted data when it enters the TCB. Thereafter, integrity analysis is to verify that subjects inside TCB are prohibited to read wrong information from non-trusted objects while sensible information inside TCB objects are protected from wrongly modified. If results show that no non-TCB subject or object can infect any TCB ones, it can be proved that the integrity of TCB is protected by the policy configuration. In fact, it is necessary that information flow from a non-TCB one into a TCB one in some cases. But such information flow ought to be audited, which can be ensured by a different authorization way of **auditallow** statement (opposite to **allow** authorization way). So that any information flow ought to be worked out if it could influence the integrity of TCB without audit. Integrity analysis is performed based upon access control spaces and information flows.

Accordingly, SCIATool can be divided into following functional parts: (1) a common module that is to extract security elements from source files for SELinux policies configuration and to store them in elaborated data structures in memory; (2) a pair of modules that are to construct security context spaces for subjects and objects respectively; (3) a group of modules that are to construct valid access control spaces for subjects and objects respectively so that analysis of authorization and prohibition for special subject and/or object denoted by security context can be figured out conveniently; (4) a group of modules that are to perform analysis of authorization and prohibition as for inquiry with specified subject and/or object; (5) a pair of modules that are to construct TCB space and to analyze integrity conflicts of TCB on premise of specifying initial TCB entities; (6) a pair of modules that are to construct colored Petri-nets for SELinux policies configuration and to perform inquiry analysis for specified information flow path. The modular architecture of SCIATool can thus be designed and illustrated as Fig. 1.

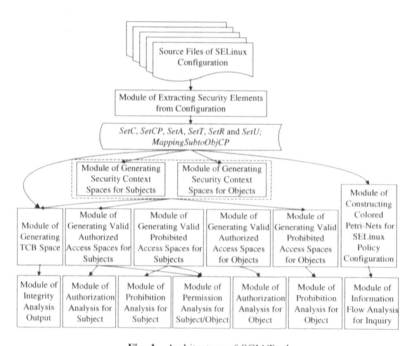

Fig. 1. Architecture of SCIATool

3.2 Main Problems and Solutions About Integration

Because we have already designed and implemented three prototypes based on the former three methods respectively, the prototype of SCIATool is implemented by the way of integrating them together.

The primary problem about integration is to start with clear and relationships among both data structures and modules whether within same prototype or across different prototypes and then to figure out a uniform framework to place those necessary modules at appropriate layer together with corresponding data structures. Just as the architecture of SCIATool we have designed finally (refer to Fig. 1), the module of extracting security elements from configuration is concluded as the fundamental module of SCIATool (thus completeness and correctness of policy information that it extracted will affect all other modules and quality for all kinds of analysis) while modules of generating security context spaces for subjects or objects and that of constructing colored Petri-nets can be placed at the second layer. In addition, all other modules can be placed at a higher layer to perform various practical and flexible analyses.

Another critical problem is that different programming styles and identifier naming habits reflected in different prototypes. So it is a rather difficult work to compose them together, especially when we select the module of extracting security elements from configuration and put it into the final prototype. The module version we have selected is more compatible with other modules placed in the final prototype but another eliminated module version have more strong extracting functions, e.g. it can process policy configuration sources files across different directories which is more closer to real application situations. Therefore, a great deal of work needs to be done to strengthen functions and to improve compatibility as for those modules of the SCIATool prototype.

The third aspect is to enrich practical analysis functions and to provide a more complete analysis result in a more convenient way for security administrators. For example, all sorts of inquiry services are provided including what kind of objects (i.e. objects in what security context) with what permissions a specified subject has been authorized or prohibited, who (i.e. subjects in what security context) is authorized or prohibited to access a specified object with what permissions and what permissions are authorized or prohibited as for a specified subject and a specified object. In addition, both input of formal inquiry statement and wizard-style inquiry input are supported by our prototype of SCIATool to provide security administrator with convenient inquiry input.

3.3 Prototype Implementation

The prototype of SCIATool is developed on Red Hat Enterprise Linux Server release 5.4 while it is written in standard C language and compiled by GCC 4.1.2 20080704 (Red Hat 4.1.2-46). Source codes of the prototype are made up of more than 6000 lines. Compared with some typical SElinux policy analysis tools such as SLAT [7, 8], SEAnalyzer [9] and etc., SCIATool is implemented completely independently and it can be executed and perform analysis tasks without any other available software tools.

4 Test Results and Discussion

The prototype is tested by using a suite of SELinux policies configuration designed for a simplified student-teacher system. And related test cases are devised around validity analysis about authorized or prohibited permissions as for a specified subject or object in the way of security context, validity analysis as for inquiry of a specified information flow path, and integrity analysis of the TCB.

4.1   Test Results and Analysis

Validity analysis about authorized and prohibited permissions is tested by having security administrator specified any subject and/or object with security context. Test results for validating prohibited permissions by specifying an object with security context *<student_u, student_r, student_t>*, and validating authorized permissions by specifying a subject with security context *<teacher_u, object_r, coursemark_t>* are illustrated in Figs. 2 and 3 respectively.

```
SCIATool - An integrated analysis tool for SELinux policies configuration based on
          access control spaces(ACS), information flows(IF) and colored Petri-nets(CPN)
Copyright© 2006-2015 Beijing Jiaotong University.

=========================[Validity Analysis Based on ACS]========================
Please enter 1~5 to perform corresponding validity analysis based on ACS:
  1. Display all permissions authorized by configuration for specified subject
  2. Display all permissions authorized by configuration to specified object
  3. Display all permissions authorized by configuration for specified subject/object pair
  4. Display all permissions prohibited by configuration for specified subject
  5. Display all permissions prohibited by configuration to specified object
  0. Return
=>4
Input security context of subject as "user role type": student_u student_r student_t

All permissions prohibited by configuration for the specified subject is as follows:
(file, write)   <student_u, object_r, coursepremark_t>
(file, write)   <college_admin_u, object_r, coursepremark_t>
(file, write)   <teacher_u, object_r, coursepremark_t>
(file, write)   <student_u, object_r, coursemark_t>
(file, write)   <college_admin_u, object_r, coursemark_t>
(file, write)   <teacher_u, object_r, coursemark_t>
=================================================================================
```

Fig. 2. Test results for prohibited permissions of subject *<student_u, student_r, student_t>*

```
=================================================================================
Please enter 1~5 to perform corresponding validity analysis based on ACS:
  1. Display all permissions authorized by configuration for specified subject
  2. Display all permissions authorized by configuration to specified object
  3. Display all permissions authorized by configuration for specified subject/object pair
  4. Display all permissions prohibited by configuration for specified subject
  5. Display all permissions prohibited by configuration to specified object
  0. Return
=>2
Input security context of object as "user role type": teacher_u, object_r, coursemark_t

All permissions authorized by configuration to the specified object is as follows:
<college_admin_u, collegeadmin_r, collegeadmin_t>              (file, read)
<student_u, student_r, student_t>              (file, read)
<college_admin_u, collegeadmin_r, accessmark_t>              (file, write)
=================================================================================
```

Fig. 3. Test results for authorized permissions of object *<teacher_u, object_r, coursemark_t>*

By careful analysis and comparison, it is confirmed that test results are consistent with the policy configuration and the configuration faithfully satisfies the original security goals of the system.

Validity analysis about inquiry of an information flow path is focused on ***allow*** rules in the policy configuration. So a series of formal description of inquiry statements are devised for its test (refer to Table 1). Because some security goals of configuration

Table 1. Inquiry statements about security configuration goals

No	Mode	Inquiry statements
1	positive	collegeadmin_t:(RL,-,!()):coursepremark_t
2	positive	collegeadmin_t:(RL,-,!()):coursemark_t
3	positive	collegeadmin_t:(RL,-,!()):coursework_t
4	positive	collegeadmin_t:(RL,-,!()):coursesourse_t
5	positive	collegeadmin_t:(RL,-,!()):coursesourse_t
6	positive	collegeadmin_t:(WL,+,!()):coursemark_t
7	positive	student_t:(RL,-,!()):coursesourse_t
8	positive	student_t:(RL,-,!()):courserecord_t
9	positive	student_t:(RL,-,!()):coursemark_t
10	positive	student_t:(WL,+,!()):coursework_t
11	positive	student_t:(WL,+,!()):courserecord_t
12	positive	teacher_t:(RL,-,!()):coursesourse_t
13	positive	teacher_t:(RL,-,!()):coursework_t
14	positive	teacher_t:(WL,+,!()):coursesourse_t
15	positive	teacher_t:(WL,+,!()):coursepremark_t
16	negative	student_t:(WL,+,!()):coursepremark_t
17	negative	student_t:(WL,+,!()):coursemark_t
18	negative	teacher_t:(WL,+,!(coursepremark_t)):coursemark_t

are not convenient to be described in positive mode (i.e. inquiry described in accordance with security goal), negative mode (i.e. inquiry described opposite to or not in accordance with configuration) is also adopted for some inquiries, e.g. inquiry statements of No. 16–18 in Table 1.

Screenshot about test for inquiry statement No. 11 in Table 1 is illustrated as Fig. 4. Note that the inquiry statement is input by a wizard-style way.

It can be seen from Fig. 4 that a subject with type *student_t* can operate firstly in the way of process transition and turn into a subject with type *accessrecord_t* and then can write the object with type *courserecord_t*.

Inquiry statements in Table 1 have been input into the running prototype and test results show that all the security targets can be met with the policy configuration.

In order to verify whether the prototype can executed effectively as for policy configuration with some bugs, some test cases are also devised and used to test the prototype. For example, it is required that a subject with type *student_t* ought to be authorized to have *read* permission as for an object with type *coursesourse_t*. Thus a new test case for inquiry statement No. 7 in Table 1 is devised for validity analysis based on CPN that the rule statement "*allow student_t coursesourse_t:file{read}*" is deleted from the configuration. Then run the prototype again with the same inquiry statement and test result shows that no information flow can be found and verifies that there is some bug about permission authorization for type *student_t* and *coursesourse_t* in the configuration. Therefore, the prototype is proved to be correct and effective from the negative aspect.

```
SCIATool - An integrated analysis tool for SELinux policies configuration based on
          access control spaces(ACS), information flows(IF) and colored Petri-nets(CPN)
Copyright® 2006-2015 Beijing Jiaotong University.

===============================[Validity Analysis Based on CPN]===========================
Please enter 1~2 to perform corresponding validity analysis based on CPN:
  1. Directly input complete query statement to describe access relationship
  2. Input query statement to describe access relationship step by step with wizard
  0. Return
=>2
======[Input query statement to describe access relationship step by step with wizard]======
Source type: student_t
Target type: courserecord_t
Enter "1" for direct access or "0" for indirect access: 0
Enter "1" for readlike access or "2" for writelike access or "0" for readlike/writelike access: 2
Input types not through flow path separated by blank space:
The query statement is as follows:
student_t:(WL,+,!()):courserecord_t
QUERY: Is there any indirect writelike information flow path from student_t to courserecord_t  acc

The query result is YES and the information flow path can be described as fllows:
student_t==>transition[process]==>accessrecord_t==>write[file]==>courserecord_t
===========================================================================================
```

Fig. 4. Screenshot about test for inquiry statement No. 11

Above results show that the prototype can not only work out all objects (or subjects) with corresponding permissions that any subject (or objects) with specified security context are authorized or prohibited according to the given SELinux policies configuration, but also can it verify correctness of configuration based upon information flow inquiry. In addition, the TCB integrity analysis can be done successfully and all rules that could potentially influence integrity of TCB subjects and objects can be detected.

4.2 Related Work and Discussion

A lot of research work has been done around SELinux policies analysis. As mentioned in the Sect. 2, Gokyo [3] and [13, 14], SELAC [4], SLAT [7, 8] and [15] and SEAnalyzer [9] provide valuable reference for our research in this paper. However, Gokyo is mainly used to check integrity of a proposed trusted computing base (i.e. to identify where untrusted data may enter the TCB) and to resolve constraint conflicts for SELinux that has multiple security goals with obviously different kinds of trust relationship, and it cannot cover all the aspects of policy violations; SELAC is focused on formalization of SELinux configuration language and to model the relationships occurring among sets of configuration rules and verification whether a given subject can access a given object in a given mode as for an arbitrary given security policy configuration; SLAT draws support from the model checker NuSMV for information flow model checking; SEAnalyzer also makes use of a software tool named CPN to aid its analysis procedure. Only one analysis method either based on access control space or based on information flow or based on colored Petri-net is used for policy analysis by each of them, and the analysis must be restricted by limitations of each method.

In addition, logic-programming approach [16, 17], deductive database approach [18], deductive spreadsheets based approach [19], model-based approach [20] and learning-based approach [21] are also used or put forward for analyzing SELinux policies. Specifically, PAL [16] uses SLAT's information flow model and creates a logic program using the XSB logic-programming system to run queries for analyzing SELinux policies; PALMS [17] is implemented in Prolog for policy analysis based on a logical specification for SELinux MLS policies; Lopol [18] normalizes and encodes SELinux policies as logical relations in conjunction with inference rules to perform a number of simple analyses and it can quickly tailor a large default policy (such as strict, targeted, or reference policy) to the specific needs of a system or a class of systems by using a logical rule set as a high-level language; XcelLog [19] is implemented based upon deductive spreadsheets as an add-in to Microsoft Excel and the XSB tabled logic programming system is used as the underlying deductive engine for security policy analysis while SELinux policies in policy.conf format are loaded into XcelLog by using a Perl script to transform the policy into comma-separated-value (.csv) format and then opening the .csv files in Excel; A model-based approach [20] is presented to analyze the dynamic proliferation of access rights in SELinux, which maps SELinux policies to an isomorphic HRU security model whose safety properties can then be analyzed by applying methods and tools available for the analysis of HRU model safety; A learning-based approach [21] is devised to analyze system call logs and to monitor an application's behavior through system calls and an application's policy within SELinux can be improved by reducing the number of Domain-Type associations, i.e. reducing SELinux application access to minimum set of types used by the application.

Compared with above research, SCIATool integrates methods based on access control spaces, information flows and colored Petri-nets organically and realizes the mutual supplement with each other's advantages. Most importantly, SCIATool is implemented independently in C language and its policy analysis doesn't require the help of other available software tools, thus it is easier to be enlarged to construct a well-functioning computer-aided SELinux configuration tool and is convenient to be integrated into other available SELinux configuration tools. However, trends of visualization and engineering [22–30] reflected in recent work to improve SELinux policy configuration ought to be considered and referenced in our future efforts to put SCIATool into practice. Furthermore, SCIATool can be improved by adopting a visualization-based policy analysis framework and it should be implemented as a policy engineering workbench encompassing the automation of engineering steps, prebuilt model patterns, integrated plausibility checks, and model analysis tools.

5 Conclusions

In this paper, we have done some exploratory research and practice for the analysis of SELinux policies configuration by making integrated use of three typical analysis methods based upon access control spaces, information flows and colored Petri-nets. And corresponding prototype, i.e. SCIATool is designed and implemented and tested. To our best of knowledge, SCIATool is the first analysis tool that integrating such three methods together.

By synthesizing complementary advantages of different methods, SCIATool can perform the TCB integrity analysis and various validity analyses effectively, as have been verified by test results. Furthermore, it can not only detect all rules that could potentially influence integrity of the initially specified TCB, but also can it figure out authorized or prohibited permissions as for a specified subject and/or a specified object as well as information flow path in CPN of policy configuration as for a specified information flow path requirement.

Specially, by using inquiry analysis based on colored Petri-nets, a whole or comparatively independent part of security goals can be described as a series of inquiry statement and then be testified by SCIATool. Therefore, SCIATool can not only perform analysis tasks of security goals (i.e. authorized and prohibited permissions) as for a single entity (i.e. subject, object or subject-object pair), but also can it perform analysis tasks of security goals as for a whole system or subsystem by in turn detecting and deciding whether there is an information flow path that satisfies each statement of security goals denoted as the specified inquiry requirement.

SCIATool is implemented completely in popular standard C language and can run independently without the help of any other available software, as make it easy to be enlarged to construct a well-functioning computer-aided SELinux configuration tool and be convenient to be integrated into other available SELinux configuration tools.

Although some non-expert oriented interface design schemes (e.g. support of wizard-style inquiry input and hierarchical display of security elements in policies configuration) have been adopted, a lot of efforts, such as comprehensive support for graphical interface and policy configuration engineering, need to be taken in order to make SCIATool achieve real practicality. At the same time, for complexity and large-scale features that the real configuration of SELinux policies has, performance of analysis and how to improve analysis efficiency ought to be considered in our future work.

Acknowledgements. The research presented in this paper was performed with the support of the Fundamental Research Funds for the Central Universities (No. 2009JBM019). This paper was also supported by the State Scholarship Fund of China Scholarship Council (File No. 201307095025).

References

1. Smalley, S., Vance, C., Salamon, W.: Implementing SELinux as a linux security module. NAI labs report #01-043 (2006)
2. Smalley, S.: Configuring the SELinux policy. NAI Labs Report #02-007 (2005)
3. Jaeger, T., Zhang, X., Edwards, A.: Policy management using access control space. ACM Trans. Inf. Syst. Secur. **6**(3), 327–364 (2003)
4. Zanin, G., Mancini, L.V.: Towards a formal model for security policies specification and validation in the SELinux system. In: Proceedings of the 9th ACM Symposium on Access Control Models and Technologies, pp. 136–145. Association for Computing Machinery (ACM), New York (2004)
5. Zhai, Gaoshou, Tong, Wu: Algorithms for automatic analysis of SELinux security policy. Int. J. Secur. Appl. **7**(1), 71–84 (2013)

6. Zhai, Gaoshou, Tong, Wu: Automatic analysis method for SELinux security policy. Int. J. Secur. Appl. **6**(2), 229–234 (2012)

7. Guttman, J.D., Herzog, A.L., Ramsdell, J.D.: Information flow in operating systems: eager formal methods. In: Workshop on Issues in the Theory of Security (WITS 2003). IFIP WG 1.7, ACM SIGPLAN and GI FoMSESS. Warsaw, Poland (2003)

8. Guttman, J.D., Herzog, A.L., Ramsdell, J.D., Skorupka, C.W.: Verifying information flow goals in security-enhanced linux. J. Comput. Secur. **13**, 115–134 (2005)

9. Chen, Y.-M., Kao, Y.-W.: Information flow query and verification for security policy of security-enhanced linux. In: Yoshiura, H., Sakurai, K., Rannenberg, K., Murayama, Y., Kawamura, S.-I. (eds.) IWSEC 2006. LNCS, vol. 4266, pp. 389–404. Springer, Heidelberg (2006)

10. Gu, L., Guo, Y., Yang, Y., Bao, F., Mei, H.: Modeling TCG-based secure systems with colored petri nets. In: Chen, L., Yung, M. (eds.) INTRUST 2010. LNCS, vol. 6802, pp. 67–86. Springer, Heidelberg (2011)

11. Ahn, G.J., Xu, W., Zhang, X.: Systematic policy analysis for high-assurance services in SELinux. In: Proceedings of 2008 IEEE Workshop on Policies for Distributed Systems and Networks, pp. 3–10. IEEE Computer Society (2008)

12. Guo, Tao, Zhai, Gaoshou: Automatic analysis of SELinux security policies based on colored petri-net (in Chinese). Inf. Secur. Technol. **4**(11), 35–40 (2013)

13. Jaeger, T., Sailer, R., Zhang, X.: Analyzing integrity protection in the SELinux example policy. In: Proceedings of the 12th USENIX Security Symposium, pp. 59–74. Washington, D.C., USA (2003)

14. Jaeger, T., Sailer, R., Zhang, X.: Resolving constraint conflicts. In: SACMAT 2004, pp. 105–114. Yorktown Heights, New York, USA (2004)

15. Guttman, J.D., Herzog, A.L., Ramsdell, J.D.: SLAT: information flow in security enhanced linux. Included in the SLAT distribution, available from http://www.nsa.gov/SELinux (2003)

16. Sarna-Starosta, B., Stoller, S.D.: Policy analysis for security-enhanced linux. In: Proceedings of the Workshop on Issues in the Theory of Security (WITS 2004), pp. 1–12. IFIP WG 1.7, ACM SIGPLAN and GI FoMSESS. Barcelona, Spain (2004)

17. Hicks, B., Rueda, S., St. Clair, L., Jaeger, T., McDaniel, P.: A logical specification and analysis for SELinux MLS policy. ACM Trans. Inf. Syst. Secur. **13**(3), 26 (2010)

18. Kissinger, A., Hale, J.C.: Lopol: a deductive database approach to policy analysis and rewriting. In: Proceedings of the Second Annual Security-enhanced Linux Symposium. Baltimore, Maryland, USA (2006)

19. Singh, A., Amakrishnan, C.R., Ramakrishnan, I.V.: Security policy analysis using deductive spreadsheets. In: FMSE 2007, pp. 42–50. Fairfax, Virginia, USA (2007)

20. Amthor, P., Kühnhauser, W.E., Pölck, A.: Model-based safety analysis of SELinux security policies. In: 2011 5th International Conference on Network and System Security (NSS), pp. 208–215. IEEE Press, New York (2011)

21. Marouf, S., Phuong, D.M., Shehab, M.: A learning-based approach for SELinux policy optimization with type mining. In: Proceedings of the Sixth Annual Workshop on Cyber Security and Information Intelligence Research (CSIIRW 2010). ACM, New York (2010)

22. Tresys Technology: SETools—policy analysis tools for SELinux. http://oss.tresys.com/projects/setools

23. Wenjuan, X., Shehab, M., Ahn, G.-J.: Visualization-based policy analysis for SELinux: framework and user study. Int. J. Inf. Secur. **12**, 155–171 (2013)

24. Clemente, P., Kaba, B., Rouzaud-Cornabas, J., Alexandre, M., Aujay, G.: SPTrack: visual analysis of information flows within SELinux policies and attack logs. In: Huang, R., Ghorbani, A.A., Pasi, G., Yamaguchi, T., Yen, N.Y., Jin, B. (eds.) AMT 2012. LNCS, vol. 7669, pp. 596–605. Springer, Heidelberg (2012)
25. Marouf, S., Shehab, M.: SEGrapher: visualization-based SELinux policy analysis. In: 2011 4th Symposium on Configuration Analytics and Automation (SAFECONFIG), pp. 1–8. Arlington, VA. IEEE Press, New York (2011)
26. Amthor, P., Kuhnhauser, W.E., Polck, A.: WorSE: a workbench for model-based security engineering. Comput. Secur. **42**, 40–55 (2014)
27. Athey, J., Ashworth, C., Mayer, F., Miner, D.: Towards Intuitive tools for managing SELinux: hiding the details but retaining the power. Tresys Technology. http://www.tresys.com/innovation/papers/Power_of_SELinux.pdf. Accessed 12 March 2007
28. MacMillan, K., Brindle, J., Mayer, F., Caplan, D., Tang, J.: Design and Implementation of the SELinux policy management server. Tresys Technology. http://www.tresys.com/innovation/papers/Design-And-Implementation-of-PMS.pdf. Accessed 1 March 2006
29. Singh, S.: Automatic verification of security policy implementations. Doctoral Dissertation in Computer Science, University of Illinois at Urbana-Champaign (2012)
30. Nakamura, Y., Sameshima, Y., Yamauchi, T.: SELinux security policy configuration system with higher level language. J. Inf. Process. **18**, 201–212 (2010)

Faster Pairing Computation on Jacobi Quartic Curves with High-Degree Twists

Fan Zhang[1] , Liangze Li[1,2], and Hongfeng Wu[3($\boxtimes$)]

[1] LMAM, School of Mathematical Sciences, Peking University, Beijing 100871, China
viczf@pku.edu.cn, liliangze2005@163.com
[2] Beijing International Center for Mathematical Research, Beijing 100871, China
[3] College of Sciences, North China University of Technology, Beijing 100144, China
whfmath@gmail.com

Abstract. In this paper, we first propose a geometric approach to explain the group law on Jacobi quartic curves which are seen as the intersection of two quadratic surfaces in space. Using the geometry interpretation we construct Miller function. Then we present explicit formulae for the addition and doubling steps in Miller's algorithm to compute the Tate pairing on Jacobi quartic curves. Our formulae on Jacobi quartic curves are better than previously proposed ones for the general case of even embedding degree. Finally, we present efficient formulas for Jacobi quartic curves with twists of degree 4 or 6. Our pairing computation on Jacobi quartic curves are faster than the pairing computation on Weierstrass curves when $j = 1728$. The addition steps of our formulae are fewer than the addition steps on Weierstrass curves when $j = 0$.

Keywords: Elliptic curve · Jacobi quartic curve · Tate pairing · Miller function · Group law

1 Introduction

In recent years, pairings on elliptic curves have become extremely useful in cryptography, and pairing-based cryptography develops rapidly. The efficient algorithms for pairing computation play a very important role in pairing-based cryptograph. The well-known method for pairing computation is Miller's algorithm. Consequently, many improvements [5,6,14] on Miller's algorithm were presented. The Weierstrass model is widely used in the early stage of elliptic curves cryptography, and many efficient formulae for pairing computation for this model can be found in [1,7,9,20,22].

One of the ideas to make improvements is to compute pairings on other elliptic curve models which may provide more efficient algorithms for the group law. Various elliptic curve models and coordinate systems show different efficiency of pairing computation. Consequently, it's necessary to carry out more research on pairing computation for different models of elliptic curves. Recently, other models, for example, Edwards curves [2,11] and twisted Edwards curves [3] are widely

© Springer International Publishing Switzerland 2015
M. Yung et al. (Eds.): INTRUST 2014, LNCS 9473, pp. 310–327, 2016.
DOI: 10.1007/978-3-319-27998-5_20

discussed. Pairing computation on twisted Edward curves was first considered by Das and Sarkar in [10] and Ionica and Joux in [18]. Then, in 2009, Arène et al. [1] developed explicit formulae for pairing computation on twisted Edwards curves. Arène et al.'s formulae for computing the Tate pairing on Edwards curves are as fast as the fastest previously published formulas on Weierstrass curves.

The use of Jacobi quartic curves in cryptology was explained in [8] and [4]. Then many other formulae for the point addition and doubling on Jacobi quartic curves were given in the literatures, see [16] for a brief introduction of the development of Jacobi quartic curves. Later, pairing computation on Jacobi quartic curves was proposed by Wang et al. [17] in 2011. A complicated geometric interpretation of Jacobi quartic curves was given in [17]. They pointed out that the doubling step of their algorithm for computing the Tate pairing was competitive with that on Weierstrass curves and Edwards curves. However the addition step of Wang et al.'s algorithm needs to be optimized.

In this paper, we present a geometric interpretation of the group law on Jacobi quartic curves. The geometric interpretation is based on the observation in [16] that Jacobi quartic curves can be seen as the intersection of two quadratic surfaces. For general elliptic curves given by intersection of two quadratic surfaces, the geometric interpretation of the group law had been discussed by Merriman et al. in [21]. And we put it into a description for Jacobi quartic curves. Using the geometric interpretation we construct Miller function and present explicit formulae for the addition steps and doubling steps in Miller's algorithm to compute the Tate pairing on Jacobi quartic curves. Miller function in this paper can reduce the cost of updating the iteration function in Miller's algorithm. So, both the addition steps and doubling steps of our formulae for pairing computation on Jacobi quartic curve are faster than that proposed by Wang et al. [17]. And our formulae on Jacobi quartic curves are better than previously proposed ones on Weierstrass curves and Edwards curves for the general case of even embedding degree.

The cost of the algorithm for pairing computation consists three parts: the cost of updating the point, the cost of updating the iteration function, and the cost of evaluating Miller function at some point Q. To reduce the cost of evaluating Miller function on Jacobi quartic curves, we employ quadratic, quartic or sextic twists to our formulae. The high-twists had been sufficiently studied by Costello, Lange and Naehrig [9] on Weierstrass curves, however, only quadratic and quartic twist had been studied for Jacobi quartic curves in [12,17]. Sextic twists of Jacobi quartic curves don't have Jacobi quartic models, this leads to some difficulties to apply sextic twists. However, the Tate pairing is performed with both P and Q on original curve, we only choose Q as the image of certain Q' on the twist curve. Thus, it's unnecessary for the twist curve to be written in Jacobi quartic form. We write the twist curve in Weierstrass form and overcome the complex derivation. Then the costs of evaluating in $j = 0$ case can reduce to a third, this result is the same as that of Weierstrass curves.

In general, when $j = 1728$, our formulae on Jacobi quartic curves are better than the formulae on Weierstrass curves. When $j = 0$, the addition steps in our formulae on Jacobi quartic curves are fewer than the addition steps on Weierstrass curves, while the doubling steps are slower. In practice, there are some famous families of paring-friendly elliptic curves, such as KSS16 [19] and TN8 [23] which have $j = 1728$. Thus, the Tate pairing computation on these curves by using Jacobi quartic model are faster than using Weierstrass curves.

The remainder of this paper is organized as follows: Sect. 2 recalls the preliminaries of the Tate pairing, Miller algorithm and the background of Jacobi quartic curves. Section 3 introduces the geometric interpretation of the group law on Jacobi quartic curves. Section 4 constructs the explicit formulae of Miller function. Section 5 proposes explicit formulae of Tate pairing on Jacobi quartic curves with quadratic, quartic or sextic twists respectively. Finally we conclude our paper and give the examples.

Note that we use **m** and **s** denote the costs of multiplication and squaring in the base field $\mathbb{F}_q$; **M** and **S** denote the costs of multiplication and squaring in the extension field $\mathbb{F}_{q^k}$; $\mathbf{m_c}$ denotes the cost of multiplying by a constant in the base field.

2 Preliminaries

In this section we briefly review the preliminaries of Tate pairing and the background of Jacobi quartic curves.

2.1 Tate Pairing

Let $p > 3$ be a prime and $\mathbb{F}_q$ be a finite field with $q = p^n$, E be an elliptic curve defined over $\mathbb{F}_q$ with neutral element denoted by O and r be a prime such that $r \mid \#E(\mathbb{F}_q)$. Let $k > 1$ denote the embedding degree, i.e. k is the smallest positive integer such that $r \mid q^k - 1$. For any point $P \in E(\mathbb{F}_q)[r]$, there exists a rational function f_P defined over $\mathbb{F}_q$ such that $\mathrm{div}(f_P) = r(P) - r(O)$, which is unique up to a non-zero scalar multiple. The group of r-th roots of unity in $\mathbb{F}_{q^k}$ is denoted by μ_r. The reduced Tate pairing is defined as follows:

$$T_r : E(\mathbb{F}_q)[r] \times E(\mathbb{F}_{q^k})/rE(\mathbb{F}_{q^k}) \to \mu_r : (P, Q) \mapsto f_P(Q)^{(q^k-1)/r}.$$

The rational function f_P can be computed in polynomial time by using Miller's algorithm ([22]). Let $r = (r_{l-1}, \cdots, r_1, r_0)_2$ be the binary representation of r, where $r_{l-1} = 1$. Let $g_{P_1,P_2} \in \mathbb{F}_q(E)$ be the rational function satisfying $\mathrm{div}(g_{P_1,P_2}) = (P_1) + (P_2) - (O) - (P_1 + P_2)$, where $P_1 + P_2$ denotes the sum of P_1 and P_2 on E, and additions of the form $(P_1) + (P_2)$ denote formal additions in the divisor group. g_{P_1,P_2} is referred as Miller function in this paper. Miller's algorithm which starts with $T = P, f = 1$ is written as Algorithm 1.

2.2 The Jacobi Quartic Curves

A Jacobi quartic curve defined over a finite field $\mathbb{F}_q$ is given by the following equation:

$$E_{a,d} : y^2 = dx^4 + 2ax^2 + 1$$

Algorithm 1. Miller's algorithm

Ensure: $r = \sum_{i=0}^{l-1} r_i 2^i$, where $r_i \in \{0,1\}$. $P \in E(\mathbb{F}_q)[r], Q \in E(\mathbb{F}_{q^k})$.

return $f_P(Q)^{(q^k-1)/r}$

1: $f \leftarrow 1, T \leftarrow P$
2: **for** $i = l - 2$ down to 0 **do**
3:     $f \leftarrow f^2 \cdot g_{T,T}(Q), T \leftarrow 2T$
4:     **if** $r_i = 1$ **then**
5:         $f \leftarrow f \cdot g_{T,P}(Q), T \leftarrow T + P$
6:     **end if**
7: **end for**
8: **return** $f^{(q^k-1)/r}$

where $d, a \in \mathbb{F}_q, d \neq 0$ and the discriminant $\triangle = 256(a^2 - d)^2 \neq 0$. In [4], O.Billet and M.Joye proved that if an elliptic curve defined over $\mathbb{F}_q$ has an $\mathbb{F}_q$-point of order 2 then E is birationally equivalent to a Jacobi quartic curve over $\mathbb{F}_q$.

The projective closure of $E_{a,d}$ in $\mathbb{P}^2$ is

$$\{(X : Y : Z) \in \mathbb{P}^2 : Y^2 Z^2 = dX^4 + 2aX^2 Z^2 + Z^4\}.$$

This curve consists of the points (x, y) on the affine curve $E_{a,d}$, embedded as usual into $\mathbb{P}^2$ by $(x, y) \mapsto (x : y : 1)$ and some extra points at infinity, i.e. the points when $Z = 0$. There is exactly one infinity point, namely $\Omega = (0 : 1 : 0)$. This point is singular.

In fact, from Hisil et al's paper [16], the Jacobi quartic curve can be seen as the intersection of two quadratic surfaces in space. That is, the Jacobi quartic curve can be written as

$$J_{a,d} : \quad 2aX^2 + Z^2 + dW^2 - Y^2 = 0, \quad X^2 - ZW = 0. \qquad (1)$$

With the projective coordinates $(X : Y : W : Z)$, the identity element is represented by the quadruplet $O = (0 : 1 : 0 : 1)$. The negative of $(X : Y : W : Z)$ is $(-X : Y : W : Z)$. The formulae of the addition steps and doubling steps on Jacobi quartic curve with projective coordinates are discussed in [16].

Point addition in $J_{a,d}$ Given $P_1 = (X_1 : Y_1 : W_1 : Z_1)$ and $P_2 = (X_2 : Y_2 : W_2 : Z_2)$ be two different points on $J_{a,d}$. Let $P_1 + P_2 = (X_3 : Y_3 : W_3 : Z_3)$, then the formulae of point addition are given in [16] as follows:

$$\begin{aligned}
X_3 &= (X_1 Y_2 - Y_1 X_2)(W_1 Z_2 - Z_1 W_2), \\
Y_3 &= (Y_1 Y_2 - 2aX_1 X_2)(W_1 Z_2 + Z_1 W_2) - 2X_1 X_2(Z_1 Z_2 + dW_1 W_2), \\
Z_3 &= (X_1 Y_2 - Y_1 X_2)^2, \\
W_3 &= (W_1 Z_2 - Z_1 W_2)^2.
\end{aligned} \qquad (2)$$

[16] points out that without any assumption on the curve constants, Y_3 can be alternatively written as:

$$Y_3 = (W_1 Z_2 + Z_1 W_2 - 2X_1 X_2)(Y_1 Y_2 - 2aX_1 X_2 + Z_1 Z_2 + dW_1 W_2) - Z_3.$$

Point doubling in $J_{a,d}$ If $P_1 = P_2$, let $2P_1 = (X_3 : Y_3 : W_3 : Z_3)$, then the formulae of point doubling are given in [16] as follows:

$$
\begin{aligned}
X_3 &= 2X_1Y_1(2Z_1{}^2 + 2aX_1{}^2 - Y_1{}^2), \\
Y_3 &= 2Y_1{}^2(Y_1{}^2 - 2aX_1{}^2) - (2Z_1{}^2 + 2aX_1{}^2 - Y_1{}^2)^2, \\
Z_3 &= (2Z_1{}^2 + 2aX_1{}^2 - Y_1{}^2)^2, \\
W_3 &= 4X_1{}^2Y_1{}^2.
\end{aligned}
\tag{3}
$$

3   Geometric Interpretation of the Group Law over $J_{a,d}$

In [17], a geometric interpretation of the group law on Jacobi quartic curves was presented. They used a cubic curve to construct Miller function. In this paper, we see the Jacobi quartic curve as the intersection of two quadratic surfaces in space as stated in [16] and adopt a standard geometric approach to explain the group law. For general elliptic curves given by intersection of two quadratic surfaces, the geometric interpretation of group law had been discussed by Merriman et al. in [21]. And we put it into a description for Jacobi quartic curves. This natural construction leads to a simpler formula for Miller function, as we will see in this section.

A projective plane is given by a homogeneous projective equation $\Pi = 0$. By abuse of notation we still use the symbol Π to denote the projective plane. Since the intersection of Π and $J_{a,d}$ is the intersection of two quadratic curves on the projective plane, any plane Π intersects $J_{a,d}$ at exactly four points, counted with appropriate multiplicities. The divisor of Π is defined as:

$$
\mathrm{div}(\Pi) = \sum_{R \in \Pi \cap J_{a,d}} n_R(R)
\tag{4}
$$

where n_R is the intersection multiplicity of Π and $J_{a,d}$ at the point R. Then the quotient of two projective planes is a well defined function which gives a principal divisor. As we will see, this divisor leads to the geometric interpretation of the group law on $J_{a,d}$.

When saying the plane Π passes three points P_1, P_2 and P_3 (not necessary distinct) which means $\mathrm{div}(\Pi) \geq (P_1) + (P_2) + (P_3)$. In fact, by Riemann-Roch theorem or by explicit discussion on multiplicity, one can know that there exists a unique plane which satisfies the above inequality. We denote this plane by Π_{P_1,P_2,P_3} in the following section.

Abel-Jacobi theorem connects the group law with principal divisor. And we can get the following lemma.

Lemma 1. *For Jacobi quartic curve $J_{a,d}$ with neutral element $O = (0 : 1 : 0 : 1)$. Then 4 points(not necessary distinct) P_1, P_2, P_3 and P_4 satisfy $P_1+P_2+P_3+P_4 = O$ if and only if there is a plane Π with $\mathrm{div}(\Pi) = (P_1) + (P_2) + (P_3) + (P_4)$.*

Proof. Firstly, it is easy to get that $\mathrm{div}(Y - Z - aW) = 4(O)$. Then the "if" part follows directly: if $\mathrm{div}(\Pi) = (P_1) + (P_2) + (P_3) + (P_4)$, the principal divisor

$\mathrm{div}(\frac{\Pi}{Y-Z-aW}) = (P_1) + (P_2) + (P_3) + (P_4) - 4(O)$ is translated to equation $P_1 + P_2 + P_3 + P_4 = O$ by the Abel-Jacobi Theorem.

For the "only if" part, suppose $P_1 + P_2 + P_3 + P_4 = O$. Consider the plane Π_{P_1,P_2,P_3}, we can assume that $\mathrm{div}(\Pi_{P_1,P_2,P_3}) = (P_1) + (P_2) + (P_3) + (P_4')$, so it derives $P_1 + P_2 + P_3 + P_4' = O$ from the "if" part. Then we get $P_4 = P_4'$, i.e. $\mathrm{div}(\Pi_{P_1,P_2,P_3}) = (P_1) + (P_2) + (P_3) + (P_4)$.

By this lemma, we can easily construct some planes to obtain a geometry approach to explain the group law on $J_{a,d}$: the fourth intersection of $\Pi_{P_1,O,O}$ and $J_{a,d}$ is $-P_1$ i.e. the negative point of P_1. The fourth intersection of $\Pi_{P_1,P_2,O}$ and $J_{a,d}$ is $-P_1 - P_2$, and its negative point is $P_1 + P_2$. Actually, this geometric interpretation is parallel with the tangent and chord law for the cubic plane curves.

4   Miller Function over $J_{a,d}$

4.1   The Construction of Miller Function

In this section we construct Miller function over $J_{a,d}$. Let P_1 and P_2 be two points on $J_{a,d}$, by Lemma 1 we can get $\mathrm{div}(\Pi_{P_1,P_2,O}) = (P_1) + (P_2) + (-P_1 - P_2) + (O)$ and $\mathrm{div}(\Pi_{P_1+P_2,O,O}) = (P_1 + P_2) + 2(O) + (-P_1 - P_2)$. Thus,

$$\mathrm{div}(\frac{\Pi_{P_1,P_2,O}}{\Pi_{P_1+P_2,O,O}}) = (P_1) + (P_2) - (P_1 + P_2) - (O).$$

In Miller's algorithm, P is always a fixed point, T is always nP for some integer n. For the addition steps, Miller function $g_{T,P}$ over $J_{a,d}$ can be given by setting $P_1 = T, P_2 = P$:

$$g_{T,P} = \frac{\Pi_{T,P,O}}{\Pi_{T+P,O,O}} \tag{5}$$

For the doubling steps, Miller function $g_{T,T}$ over $J_{a,d}$ is given by setting $P_1 = P_2 = T$ as follows:

$$g_{T,T} = \frac{\Pi_{T,T,O}}{\Pi_{2T,O,O}} \tag{6}$$

We use the equation $C_X X + C_Y Y + C_Z Z + C_W W = 0$ to denote a projective plane. Because all the planes in Formulas (5) and (6) pass through $O = (0 : 1 : 0 : 1)$, the equations of them have the form $C_X X + C_Y (Y - Z) + C_W W = 0$. Thus, to obtain the equation of the planes present in Formulas (5) and (6), we only need to compute C_X, C_Y and C_W.

To unify the notation, we use P_1, P_2 to denote the points in the Miller algorithm for both the addition and doubling steps, and consider $P_1 \neq P_2$ and $P_1 = P_2$ respectively when it is necessary. Assume that $P_1 = (X_1 : Y_1 : W_1 : Z_1), P_2 = (X_2 : Y_2 : W_2 : Z_2)$ and $P_3 = P_1 + P_2 = (X_3 : Y_3 : W_3 : Z_3)$.

4.2   The Equation of $\Pi_{P_1,P_2,O}$ with $P_1 \neq P_2$

Lemma 2. $\Pi_{P_1,P_2,O} : C_X X + C_Y(Y - Z) + C_W W = 0$ *is the plane passing through* P_1, P_2, O *where* $P_1 \neq P_2$ *and* $P_1, P_2 \neq O$ *then we have:*

$$C_X = W_1(Z_2 - Y_2) - W_2(Z_1 - Y_1), C_Y = X_2 W_1 - X_1 W_2, C_W = X_2(Z_1 - Y_1) - X_1(Z_2 - Y_2).$$

Proof. Since P_1 and P_2 are two distinct points, the coefficients are obtained by evaluating $C_X X + C_Y(Y - Z) + C_W W = 0$ at P_1 and P_2. Then we obtain two linear equations in C_X, C_Y and C_Z,

$$C_X X_1 + C_Y(Y_1 - Z_1) + C_W W_1 = 0, C_X X_2 + C_Y(Y_2 - Z_2) + C_W W_2 = 0.$$

The coefficients of the plane $\Pi_{P_1,P_2,O}$ follow from the (projective) solutions

$$C_X = \begin{vmatrix} Y_1 - Z_1 & W_1 \\ Y_2 - Z_2 & W_2 \end{vmatrix}, C_Y = \begin{vmatrix} W_1 & X_1 \\ W_2 & X_2 \end{vmatrix}, C_W = \begin{vmatrix} X_1 & Y_1 - Z_1 \\ X_2 & Y_2 - Z_2 \end{vmatrix}.$$

Thus we can get the formulae in the Lemma 2:

$$C_X = W_1(Z_2 - Y_2) - W_2(Z_1 - Y_1), C_Y = X_2 W_1 - X_1 W_2, C_W = X_2(Z_1 - Y_1) - X_1(Z_2 - Y_2).$$

4.3   The Equation of $\Pi_{P_1,P_2,O}$ with $P_1 = P_2$

Lemma 3. $\Pi_{P_1,P_2,O} : C_X X + C_Y(Y - Z) + C_W W = 0$ *is the plane passing through* P_1, P_2 *and* O, *where* $P_1 = P_2 \neq O$, *then we have:*

$$C_X = 2aX_1 W_1 + 2X_1(Z_1 - Y_1), C_Y = -Y_1 W_1, C_W = dW_1^2 - Z_1^2 + Y_1 Z_1.$$

Proof. When $P_1 = P_2 \neq O$, the tangent line to $J_{a,d}$ at P_1 is the intersection of the tangent planes to $2aX^2 + Z^2 + dW^2 - Y^2 = 0$ and $X^2 - ZW = 0$ at P_1. The tangent plane to $2aX^2 + Z^2 + dW^2 - Y^2 = 0$ at P_1 is $2aX_1 X + Z_1 Z + dW_1 W - Y_1 Y = 0$. The tangent plane to $X^2 - ZW = 0$ at P_1 is $2X_1 X - W_1 Z - Z_1 W = 0$. Then $\Pi_{P_1,P_1,O}$ has the following equation:

$$\lambda(2aX_1 X + Z_1 Z + dW_1 W - Y_1 Y) + \mu(2X_1 X - W_1 Z - Z_1 W) = 0,$$

where at least one of the constants λ, μ is non-zero. Note that $\Pi_{P_1,P_1,O}$ passes O, i.e. $\lambda(Z_1 - Y_1) - \mu W_1 = 0$. It is clear that $\lambda = W_1, \mu = Z_1 - Y_1$ satisfy the equation and at least one of them is non-zero because $P_1 \neq O$. Hence, the equation of $\Pi_{P_1,P_1,O}$ is

$$W_1(2aX_1 X + Z_1 Z + dW_1 W - Y_1 Y) + (Z_1 - Y_1)(2X_1 X - W_1 Z - Z_1 W) = 0.$$

Then we can get the coefficients of $\Pi_{P_1,P_1,O}$ as follows:

$$C_X = 2aX_1 W_1 + 2X_1(Z_1 - Y_1), C_Y = -Y_1 W_1, C_W = dW_1^2 - Z_1^2 + Y_1 Z_1.$$

4.4   The Equation of $\Pi_{P_3,O,O}$

Lemma 4. $\Pi_{P_3,O,O} : C_X X + C_Y(Y - Z) + C_W W = 0$ *is the plane passing through P_3, O and O, where $P_3 \neq O$, then we have:*

$$C_X = 0, C_Y = W_3, C_W = (Z_3 - Y_3).$$

Proof. This proof is similar to the proof of Lemma 3. Since $\Pi_{P_3,O,O}$ passes through the tangent line of $J_{a,d}$ at O, then equation of $\Pi_{P_3,O,O}$ is $\lambda(Z - Y) - \mu W = 0$. For it passes $P_3 \neq O$, we have $\lambda = -W_3, \mu = Y_3 - Z_3$ and at least one of them is non-zero. Then the equation of $\Pi_{P_3,O,O}$ is $W_3(Y - Z) + (Z_3 - Y_3)W = 0$.

4.5   The Explicit Formula of Miller Function

We summarize the above results as the following theorem:

Theorem 5. *Let* $J_{a,d} :\quad 2aX^2 + Z^2 + dW^2 - Y^2 = 0, X^2 - ZW = 0$ *be a Jacobi quartic curve,* $O = (0 : 1 : 0 : 1)$. *Let* $P_1 = (X_1 : Y_1 : W_1 : Z_1)$, $P_2 = (X_2 : Y_2 : W_2 : Z_2)$ *be two points on* $J_{a,d}$. *Let* $P_3 = P_1 + P_2 = (X_3 : Y_3 : W_3 : Z_3)$. *Then Miller function* $g_{P_1,P_2}(X, Y, W, Z)$ *which satisfies* $\mathrm{div}(g_{P_1,P_2}) = (P_1) + (P_2) - (P_3) - (O)$ *is:*

$$g_{P_1,P_2}(X, Y, W, Z) = \frac{\Pi_{P_1,P_2,O}}{\Pi_{P_3,O,O}} = \frac{C_X X + C_Y(Y - Z) + C_W W}{W_3(Y - Z) + (Z_3 - Y_3)W}.$$

In the case $P_1 \neq P_2$ and $P_1, P_2 \neq O$, the coefficients are given by

$$C_X = W_1(Z_2 - Y_2) - W_2(Z_1 - Y_1), C_Y = X_2 W_1 - X_1 W_2, C_W = X_2(Z_1 - Y_1) - X_1(Z_2 - Y_2).$$

If $P_1 = P_2 \neq O$, the coefficients are given by

$$C_X = 2aX_1 W_1 + 2X_1(Z_1 - Y_1), C_Y = -Y_1 W_1, C_W = dW_1^2 - Z_1^2 + Y_1 Z_1.$$

By the definition of Miller function, it's unique up to a factor in the base field. In practice, without changing the value of Tate pairing we can multiply Miller function by an appropriate factor in the base field to reduce the cost.

5   Tate Pairing Computation on $J_{a,d}$ Using Projective Coordinates

In this section, we analyze the formulae in Miller's algorithm explicitly. As it's shown in Algorithm 1, each addition or doubling step consists of three parts: computing the point $T + P$ or $2T$ and the function $g_{T,P}$ or $g_{T,T}$; evaluating $g_{T,P}$ or $g_{T,T}$ at Q; updating the variable f by $f \leftarrow f \cdot g_{T,P}(Q)$ or by $f \leftarrow f^2 \cdot g_{T,T}(Q)$.

The updating part costs 1M for the addition step and 1M+1S for the doubling step. It is usually the main cost but it's difficult to give further optimization for the updating part. For the evaluating part, when the embedding degree k is even, as we will introduce in the following section, some standard methods such as denominator elimination and subfield simplification can be used.

As usual, we choose $P \in J_{a,d}(\mathbb{F}_q)[r]$ and $Q \in J_{a,d}(\mathbb{F}_{q^k})$, where $k > 1$ is the embedding degree. In fact as stated in [15], Q can be chosen from a subgroup which is given by a twist of $J_{a,d}$. More precisely, for $t = \#\operatorname{Aut}(J_{a,d})$, there is a twist of $J_{a,d}$ over $\mathbb{F}_{q^{k/t}}$ with degree t denoted as E' such that there is an isomorphism $\psi : E' \to J_{a,d}$ over $\mathbb{F}_{q^k}$. Then Q is chosen as $\psi(Q')$, where Q' belongs to $E'(\mathbb{F}_{q^{k/t}})$. It is noticeable that E' is not necessary to have a Jacobi quartic model. The following theorem shows that the evaluation of Miller function can be simplified by choosing Q in the above way.

5.1 Tate Paring Computation on $J_{a,d}$ With Even Embedding Degrees

When the embedding degree k is even, one of the standard methods to cut down the expense is called denominator elimination. In this section, we can see how the denominator elimination works on $J_{a,d}$ in details.

Theorem 6. *Assume that embedding degree k is even. Let δ be a generator of $\mathbb{F}_{q^k}$ over $\mathbb{F}_{q^{k/2}}$ with $\delta^2 \in \mathbb{F}_{q^{k/2}}$. Let $\psi : J_{a\delta^2, d\delta^4} \to J_{a,d}$ be the twist isomorphism given by $(X : Y : W : Z) \mapsto (\delta X : Y : \delta^2 W : Z)$. For $Q' = (X_0 : Y_0 : W_0 : Z_0) \in J_{a\delta^2, d\delta^4}(\mathbb{F}_{q^{k/2}})$, $Q = \psi(Q') \in J_{a,d}(\mathbb{F}_{q^k})$ and $P_3 = P_1 + P_2 \neq O$, we have*

$$g_{P_1,P_2}(Q) \in (C_X \theta \delta + C_Y + C_W \eta)\mathbb{F}^*_{q^{k/2}},$$

where $\theta = \frac{X_0}{Y_0 - Z_0}$, $\eta = \frac{W_0 \delta^2}{Y_0 - Z_0}$ and C_X, C_Y, C_W are given in Theorem 5.

Proof. By Theorem 5,

$$g_{P_1,P_2}(Q) = \frac{\Pi_{P_1,P_2,O}(Q)}{\Pi_{P_3,O,O}(Q)} = \frac{C_X \delta X_0 + C_Y(Y_0 - Z_0) + C_W \delta^2 W_0}{W_3(Y_0 - Z_0) + (Z_3 - Y_3)\delta^2 W_0}$$

$$= \frac{C_X \frac{X_0}{Y_0 - Z_0}\delta + C_Y + C_W \frac{W_0 \delta^2}{Y_0 - Z_0}}{W_3 + (Z_3 - Y_3)\frac{W_0 \delta^2}{Y_0 - Z_0}} \in (C_X \theta \delta + C_Y + C_W \eta)\mathbb{F}^*_{q^{k/2}}.$$

From the above theorem, the denominator of $g_{P_1,P_2}(Q)$ can be eliminated by the final exponentiation, since it belongs to $\mathbb{F}^*_{q^{k/2}}$. Moreover, note that $\theta, \eta \in \mathbb{F}_{q^{k/2}}$ and they are fixed during the whole computation, so they can be precomputed. The coefficients C_X, C_Y and C_W are in $\mathbb{F}_q$, thus when the coefficients of the plane are given, the evaluation at Q can be computed in $k\mathbf{m}$ (the multiplications by θ and η need $\frac{k}{2}\mathbf{m}$ each).

Various tricks can be used when computing the coordinates of the points and the coefficients of the planes. We will discuss them respectively for the addition and doubling step as follows.

Addition Step. Let $P_1 = T$ and $P_2 = P$ be two distinct points. By Theorem 5 and Formula (2), the explicit formulas for computing $P_3 = T + P$ and C_X, C_Y, C_W are given as follows:

$$A = X_1 \cdot X_2; \ B = Y_1 \cdot Y_2; \ C = Z_1 \cdot Z_2; \ D = W_1 \cdot W_2;$$
$$E = (X_1 - Y_1) \cdot (X_2 + Y_2) - A + B;$$
$$F = W_1 \cdot Z_2; \ G = W_2 \cdot Z_1$$
$$H = (Y_1 - W_1) \cdot (Y_2 + W_2) - B + D;$$
$$I = (X_2 - W_2) \cdot (X_1 + W_1) - A + D;$$
$$J = (X_2 + Z_2) \cdot (X_1 - Z_1) - A + C;$$
$$Z_3 = E^2; \ X_3 = E \cdot (F - G);$$
$$Y_3 = (F + G - 2A) \cdot (B - 2aA + C + dD) - Z_3;$$
$$C_X = H + F - G; \ C_Y = I; \ C_W = E - J;$$

The coordinate W_3 is not computed in this step, because by using a trick we don't need the value of W_3 in following doubling step (see the details in the following Sect. 5.1). Then the total costs of computing $T + P$ and C_X, C_Y, C_W are $12\mathbf{m} + 1\mathbf{s} + 2\mathbf{m_c}$, where $2\mathbf{m_c}$ are multiplication by a and d. Since P is fixed during the pairing computation, we can use the mixed addition which means $Z_2 = 1$, then the costs reduce to $10\mathbf{m} + 1\mathbf{s} + 2\mathbf{m_c}$.

So the total costs of an addition step are $1\mathbf{M} + (k + 12)\mathbf{m} + 1\mathbf{s} + 2\mathbf{m_c}$, while a mixed addition step costs $1\mathbf{M} + (k + 10)\mathbf{m} + 1\mathbf{s} + 2\mathbf{m_c}$.

Doubling Step. From Theorem 5, for $P_1 = P_2 = T$, $P_3 = 2T$, we have:

$$C_X = 2aX_1W_1 + 2X_1(Z_1 - Y_1), C_Y = -Y_1W_1, C_W = dW_1^2 - Z_1^2 + Y_1Z_1$$

In order to exclude W_1, we make a trick by multiplying the above coefficients by $2Y_1Z_1$ which belongs to the base field. Instead of computing C_X, C_Y, C_W, we compute the following C'_X, C'_Y, C'_W:

$$C'_X = 2X_1Y_1(Y_1^2 - 2Y_1Z_1) + 2X_1Y_1(2Z_1^2 + 2aX_1^2 - Y_1^2)$$
$$C'_Y = -2X_1^2Y_1^2$$
$$C'_W = -2Y_1Z_1(2Z_1^2 + 2aX_1^2 - Y_1^2) + 2Y_1^2Z_1^2$$

Now the explicit formulas for computing $2P$ and C'_X, C'_Y, C'_W are given as follows:

$$A = X_1^2; \ B = Y_1^2; \ C = Z_1^2; \ D = aA; \ E = 2C + 2D - B;$$
$$F = (X_1 + Y_1)^2 - A - B; \ G = (Y_1 + Z_1)^2 - B - C;$$
$$Z_3 = E^2; \ W_3 = F^2; \ X_3 = ((E + F)^2 - Z_3 - W_3)/2;$$
$$Y_3 = 2B^2 - aW_3 - Z_3; \ C'_X = F \cdot (B - G) + X_3;$$
$$C'_Y = -W_3/2; \ C'_W = ((G - E)^2 - Z_3)/2.$$

Then the total costs are $1\mathbf{m} + 10\mathbf{s} + 2\mathbf{m_c}$, where $\mathbf{m_c}$ is the multiplication by a. Hence, a doubling step costs $1\mathbf{M} + 1\mathbf{S} + (k + 1)\mathbf{m} + 10\mathbf{s} + 2\mathbf{m_c}$.

We compare the costs of pairing computation on Jacobi quartic curves [17], Weierstrass curves [1] and twisted Edwards curves [1] in the following table.

Table 1. Costs comparison for even embedding degree($j \neq 0,1728$)

	DBL	mADD	ADD
Weierstrass [1]	$1m + 11s + 1m_c$	$6m + 6s$	$9m + 6s$
	$\approx 9.8m$	$\approx 10.8m$	$\approx 13.8m$
Weierstrass $a_4 = -3$[1]	$6m + 5s$	$6m + 6s$	$9m + 6s$
	$\approx 10m$	$\approx 10.8m$	$\approx 13.8m$
Edwards [1]	$6m + 5s + 2m_c$	$12m + 1m_c$	$14m + 1m_c$
	$\approx 10m$	$\approx 12m$	$\approx 14m$
Jacobi quartic [17]	$4m + 8s + 1m_c$	$16m + 1s + 4m_c$	$18m + 1s + 4m_c$
	$\approx 10.4m$	$\approx 16.8m$	$\approx 18.8m$
$J_{a,d}$ this paper	$1m+10s + 2m_c$	$10m + 1s + 2m_c$	$12m + 1s + 2m_c$
	$\approx 9m$	$\approx 10.8m$	$\approx 12.8m$

Each doubling step (DBL) needs $1M + km + 1S$ for the evaluation at Q and the update of f. Each mixed addition step (mADD) and addition step (ADD) needs $1M+km$ for the evaluation at Q and the update of f. In the table we do not report these expenses, since they do not depend on the chosen model. If we let $1s=0.8m$, without considering the cost of multiplication by constants, we can see the result shows that our formulae for Tate pairing computation on Jacobi quartic are effective (Table 1).

5.2   Tate Pairing on $J_{a,d}$ with quartic or sextic twists

Let $t|k$, an elliptic curve E' over $\mathbb{F}_{q^{k/t}}$ is called a twist of degree t of $E/\mathbb{F}_{q^{k/t}}$ if there is an isomorphism $\psi : E' \to E$ defined over $\mathbb{F}_{q^k}$, and this is the smallest extension of $\mathbb{F}_{q^{k/t}}$ over which ψ is defined. Depending on the j-invariant $j(E)$ of E, there exist twists of degree at most 6, since char($\mathbb{F}_q$) > 3. Elliptic curves with twists of degree higher than 2 arise from constructions with j-invariants $j(E) = 0$ and $j(E) = 1728$. Theorem 6 shows that we can reduce the cost using the twists of degree 2. In this section, we will show that twists of higher degree can further reduce the cost.

Jacobi Quartic Curve with j $= 1728$. The Jacobi quartic curve $J_{0,d} : Y^2 = dW^2 + Z^2, X^2 = ZW$ has j-invariant which is equal to 1728, hence, there exist twists of degree 4. We have the following theorem.

Theorem 7. *Assume that* $4|k$. *Let* δ *be a generator of* $\mathbb{F}_{q^k}$ *over* $\mathbb{F}_{q^{k/4}}$ *and* $\delta^4 \in \mathbb{F}_{q^{k/4}}$, *which implies* $\delta^2 \in \mathbb{F}_{q^{k/2}}{}^1$. *Then* $J_{0,d\delta^4}$ *is a twist of* $J_{0,d}$ *with degree 4 and* $\psi : (X : Y : W : Z) \mapsto (\delta X : Y : \delta^2 W : Z)$ *is the twist isomorphism. For* $Q' = (X_0 : Y_0 : W_0 : Z_0) \in J_{0,d\delta^4}(\mathbb{F}_{q^{k/4}})$, $Q = \psi(Q') \in J_{0,d}(\mathbb{F}_{q^k})$, $P_1, P_2 \neq O$ *and* $P_3 = P_1 + P_2 \neq O$, *we have*

$$g_{P_1,P_2}(Q) \in (C_X \theta \delta + C_Y + C_W \eta \delta^2)\mathbb{F}_{q^{k/2}}^*,$$

---

[1] This δ exists if and only if $\mathbb{F}_{q^{k/4}}$ contains 4th-roots of unity, i.e. $4 \mid q^{k/4} - 1$.

where $\theta = \frac{X_0}{Y_0 - Z_0}$, $\eta = \frac{W_0}{Y_0 - Z_0}$ and C_X, C_Y, C_W are given in Theorem 5.

Proof Theorem 5 shows us the explicit formulae of Miller function g_{P_1, P_2} is:

$$
\begin{aligned}
g_{P_1, P_2}(Q) &= \frac{\Pi_{P_1, P_2, O}(Q)}{\Pi_{P_3, O, O}(Q)} = \frac{C_X \delta X_0 + C_Y(Y_0 - Z_0) + C_W \delta^2 W_0}{W_3(Y_0 - Z_0) + (Z_3 - Y_3)\delta^2 W_0} \\
&= \frac{C_X \frac{X_0}{Y_0 - Z_0}\delta + C_Y + C_W \frac{W_0}{Y_0 - Z_0}\delta^2}{W_3 + (Z_3 - Y_3)\frac{W_0}{Y_0 - Z_0}\delta^2} \in (C_X \theta \delta + C_Y + C_W \eta \delta^2)\mathbb{F}^*_{q^{k/2}}.
\end{aligned}
$$

From the above theorem, the denominator of $g_{P_1, P_2}(Q)$ can be eliminated by the final exponentiation, since it belongs to $\mathbb{F}^*_{q^{k/2}}$. In addition, note that $\theta, \eta \in \mathbb{F}_{q^{k/4}}$ and they are fixed during the whole computation, so they can be precomputed. The coefficients C_X, C_Y and C_W are in $\mathbb{F}_q$, thus the evaluation at Q given the coefficients of the plane can be computed in $\frac{k}{2}\mathbf{m}$ (multiplications by θ and η need $\frac{k}{4}\mathbf{m}$ each).

Addition step: Using the algorithm in Sect. 5.1, $1\mathbf{m_c}$ can be saved for $a = 0$. Hence, an addition step in Miller's algorithm costs $1\mathbf{M} + (\frac{k}{2} + 12)\mathbf{m} + 1\mathbf{s} + 1\mathbf{m_c}$, and a mixed addition step in Miller's algorithm costs $1\mathbf{M} + (\frac{k}{2} + 10)\mathbf{m} + 1\mathbf{s} + 1\mathbf{m_c}$, where $1\mathbf{m_c}$ is multiplication by d.

Doubling step: Using the algorithm in Sect. 5.1, we compute $X_1 Y_1$ instead of computing $A = X_1^2$ and $F = (X_1 + Y_1)^2 - A - B$, since $a = 0$ leads to $D = 0$ in algorithm. Furthermore, $1\mathbf{m_c}$ can be saved for $a = 0$. Hence a doubling step in Miller's algorithm costs $1\mathbf{M} + 1\mathbf{S} + (\frac{k}{2} + 2)\mathbf{m} + 8\mathbf{s}$.

The following table shows the concrete comparison with previous results on elliptic curves with quartic twists. Each doubling step (DBL) needs $1\mathbf{M} + \frac{k}{2}\mathbf{m} + 1\mathbf{S}$ for the evaluation at Q and the update of f. Each mixed addition step (mADD) and addition step (ADD) needs $1\mathbf{M} + \frac{k}{2}\mathbf{m}$ for the evaluation at Q and the update of f. In the table we do not report these expenses, since they are the same on Weierstrass model and Jacobi model. Both the addition steps and the doubling steps in our formulae for Tate pairing computation on Jacobi quartic curves are faster than the fastest known results on Weierstrass curves and previous results on Jacobi quartic curves (Table 2).

Table 2. Costs comparison for $j = 1728$

$j = 1728$	DBL	mADD	ADD
$y^2 = x^3 + ax$ [9]	$2\mathbf{m} + 8\mathbf{s} + 1\mathbf{m_c}$	$9\mathbf{m} + 5\mathbf{s}$	$12\mathbf{m} + 7\mathbf{s}$
	$\approx 8.4\mathbf{m}$	$\approx 13\mathbf{m}$	$\approx 17.6\mathbf{m}$
$y^2 = dx^4 + 1$ [12]	$3\mathbf{m} + 7\mathbf{s} + 1\mathbf{m_c}$	$12\mathbf{m} + 7\mathbf{s} + 1\mathbf{m_c}$	$12\mathbf{m} + 11\mathbf{s} + 1\mathbf{m_c}$
	$\approx 8.6\mathbf{m}$	$\approx 17.6\mathbf{m}$	$\approx 20.8\mathbf{m}$
$J_{0,d}$ this paper	$2\mathbf{m} + 8\mathbf{s}$	$10\mathbf{m} + 1\mathbf{s} + 1\mathbf{m_c}$	$12\mathbf{m} + 1\mathbf{s} + 1\mathbf{m_c}$
	$\approx 8.4\mathbf{m}$	$\approx 10.8\mathbf{m}$	$\approx 12.8\mathbf{m}$

Jacobi Quartic Curves with $j = 0$. The Jacobi quartic curve $E_{a,d} : y^2 = dx^4 + 2ax^2 + 1$ has j-invariant $j_{a,d} = \frac{16(4a^2+12d)^3}{d(a^2-4d)^2}$. Hence, $j_{a,d} = 0$ if and only if $a^2 + 3d = 0$. Now we look into the Jacobi quartic curve

$$E_{a,-a^2/3} : y^2 = -\frac{a^2}{3}x^4 + 2ax^2 + 1$$

which has j-invariant equal to 0, hence, there exist twists of degree 6. Sextic twists of Jacobi quartic curves don't have Jacobi quartic model, this leads to some difficulties to apply sextic twists. However, the Tate pairing is performed with both P and Q on original curve, we only choose Q as the image of certain Q' on the twist curve. Thus, it's unnecessary for the twist curve to be written in Jacobi quartic form. The following lemma gives a twist of degree 6 written in Weierstrass form.

Lemma 8. *Assume that $6|k$, δ is a generator of $\mathbb{F}_{q^k}$ over $\mathbb{F}_{q^{k/6}}$ with $\delta^6 \in \mathbb{F}_{q^{k/6}}$, which implies $\delta^2 \in \mathbb{F}_{q^{k/2}}$ and $\delta^3 \in \mathbb{F}_{q^{k/3}}$.[2] Then the Weierstrass elliptic curve*

$$W_a : v^2 = u^3 + \frac{64a^3}{27}\delta^6$$

is a twist of degree 6 over $\mathbb{F}_{q^{k/6}}$ of $E_{a,-a^2/3}$. The isomorphism can be given as

$$\varphi : W_a \longrightarrow E_{a,-a^2/3}$$
$$(u,v) \longmapsto (x,y) = \left(\frac{6u\delta + 8a\delta^3}{3v}, \frac{3u - 2a\delta^2}{6\delta^2}\left(\frac{6u\delta + 8a\delta^3}{3v}\right)^2 - 1 \right).$$

Proof. Firstly, we check that φ is well defined, i.e. $\varphi(u,v) \in E_{a,-a^2/3}$. Note that

$$y = \frac{3u - 2a\delta^2}{6\delta^2}\left(\frac{6u\delta + 8a\delta^3}{3v}\right)^2 - 1 = \left(\frac{vx}{4\delta^3} - a\right)x^2 - 1$$

so

$$y + 1 + ax^2 = \frac{vx^3}{4\delta^3} = \frac{2}{v^2}\left(u + \frac{4a\delta^2}{3}\right)^3 = 2 + 2ax^2 - \frac{16a^2\delta^3}{3v}x$$

Then we have

$$(y + 1 + ax^2)(y - 1 - ax^2) = -\frac{vx^3}{4\delta^3}\frac{16a^2\delta^3}{3v}x = -\frac{4a^2}{3}x^4,$$

$$y^2 = -\frac{a^2}{3}x^4 + 2ax^2 + 1.$$

Moreover, it can be easily checked that φ is invertible and satisfies $\varphi(O) = O$, i.e. φ is an isomorphism. Besides, the minimal field that φ can be defined over is $\mathbb{F}_{q^k}$ which has degree 6 over $\mathbb{F}_{q^{k/6}}$. Hence, the twist degree is 6. In fact, $E_{a,-a^2/3}$ is isomorphic to $J_{a,-a^2/3}$ by $\iota : (x,y) \mapsto (x : y : x^2 : 1)$. Let $\psi = \iota \circ \varphi$, we get the following theorem.

---

[2] This δ exists if and only if $\mathbb{F}_{q^{k/6}}$ contains 6th-roots of unity, i.e. $6 \mid q^{k/6} - 1$.

Theorem 9. *Using the notation in Lemma 8, for* $Q' = (u, v) \in W_a(\mathbb{F}_{q^{k/6}})$, $Q = (X_Q : Y_Q : W_Q : Z_Q) = (x_Q : y_Q : x_Q^2 : 1) = \psi(Q') \in J_{a,-a^2/3}(\mathbb{F}_{q^k})$, $P_1, P_2 \neq O$ and $P_3 = P_1 + P_2 \neq O$, we have

$$g_{P_1,P_2}(Q) \in (-C_X\theta\delta - (aC_Y + C_W)\eta\delta^2 + (aC_Y - C_W)\delta^4)\mathbb{F}_{q^{k/2}}^*,$$

where $\theta = \frac{3v}{8a}$, $\eta = \frac{3u}{4a}$ *and* C_X, C_Y, C_W *are given in Theorem 5.*

Proof. Since $Q' \in W_a(\mathbb{F}_{q^{k/6}})$, we have $Q = \psi(Q') \in J_{a,-a^2/3}(\mathbb{F}_{q^k})$. One can check by substituting that:

$$\frac{X_Q}{Y_Q - Z_Q - aW_Q} = \frac{x_Q}{y_Q - 1 - ax_Q^2} = -\frac{y_Q + 1 + ax_Q^2}{\frac{4}{3}a^2x_Q^3} = -\frac{3v}{16a^2\delta^3} = -\frac{1}{2a}\theta\delta^{-3}$$

$$\frac{W_Q}{Y_Q - Z_Q - aW_Q} = \frac{x_Q^2}{y_Q - 1 - ax_Q^2} = \frac{3u}{8a^2\delta^2} - \frac{1}{2a} = -\frac{1}{2a}\eta\delta^{-2} - \frac{1}{2a}$$

$$\frac{Y_Q - Z_Q}{Y_Q - Z_Q - aW_Q} = 1 - \frac{aW_Q}{Y_Q - Z_Q - aW_Q} = \frac{3u}{8a\delta^2} + \frac{1}{2} = -\frac{1}{2}\eta\delta^{-2} + \frac{1}{2}$$

Then we get

$$g_{P_1,P_2}(Q) = \frac{\Pi_{P_1,P_2,O}(Q)}{\Pi_{P_3,O,O}(Q)} = \frac{-C_X\theta\delta - (aC_Y + C_W)\eta\delta^2 + (aC_Y - C_W)\delta^4}{(-aW_3 + Y_3 - Z_3)\eta\delta^2 + (aW_3 - Z_3 + Y_3)\delta^4}$$
$$\in (-C_X\theta\delta - (aC_Y + C_W)\eta\delta^2 + (aC_Y - C_W)\delta^4)\mathbb{F}_{q^{k/2}}^*.$$

The above theorem shows that the denominator of $g_{P_1,P_2}(Q)$ can be eliminated by the final exponentiation, since it belongs to $\mathbb{F}_{q^{k/2}}^*$. Moreover we may precompute θ and η since they are fixed during the whole computation. When $C_X, C_Y, C_W \in \mathbb{F}_q$ and $\theta, \eta \in \mathbb{F}_{q^{k/6}}$ are given, the evaluation at Q can be computed in $\frac{k}{3}\mathbf{m} + \mathbf{m}_c$, with $\frac{k}{6}\mathbf{m}$ for multiplications by θ and η respectively and a constant multiplication by a.

Special simplification has not been found for this case. Hence, we use the algorithm in Sect. 5.1. Then the total cost of the addition step using mixed addition is $1\mathbf{M} + (\frac{k}{3} + 10)\mathbf{m} + 1\mathbf{s} + 3\mathbf{m}_c$, where $3\mathbf{m}_c$ are multiplication by a, a and d, and the total cost of the doubling step is $1\mathbf{M} + 1\mathbf{S} + (\frac{k}{3} + 1)\mathbf{m} + 10\mathbf{s} + 3\mathbf{m}_c$, where $\mathbf{m}_c$ is the multiplication by a.

The following table shows the concrete comparison with previous results on elliptic curves with sextic twists. The cost of evaluating Miller function at some point Q on Weierstrass curves and Jacobi quartic curves both reach $\frac{k}{3}\mathbf{m}$. Each doubling step (DBL) needs $1\mathbf{M} + \frac{k}{3}\mathbf{m} + 1\mathbf{S}$ for the evaluation at Q and the update of f. Each mixed addition step (mADD) and addition step (ADD) needs $1\mathbf{M} + \frac{k}{3}\mathbf{m}$ for the evaluation at Q and the update of f. In the table we do not report these expenses, since they are the same on Weierstrass model and Jacobi model. The addition steps in our formulae for Tate pairing computation on Jacobi quartic curves are faster than those on Weierstrass curves, while the doubling steps are slower than those on Weierstrass curves. The Weierstrass curves gain advantages in the doubling steps that is because when $j = 0$ the formulae of the point doubling turn to much simpler. While for Jacobi quartic curves the special simplification of the point doubling has not been found when $j = 0$. Thus, if we want the Jacobi quartic curves to be competitive with Weierstrass curves in the future, we should focus on the optimization of formulae for the point doubling on Jacobi quartic curves with $j = 0$ (Table 3).

Table 3. Costs comparison for $j = 0$

$j = 0$	DBL	mADD	ADD
$y^2 = x^3 + c^2$ [9]	$3m + 5s$	$10m + 2s + 1m_c$	$14m + 2s + 1m_c$
	$\approx 7m$	$\approx 11.6m$	$\approx 15.6m$
$y^2 = x^3 + b$ [9]	$2m + 7s + 1m_c$	$10m + 2s$	$14m + 2s$
	$\approx 7.6m$	$\approx 11.6m$	$\approx 15.6m$
$J_{a,-a^2/3}$ this paper	$1m+10s + 3m_c$	$10m + 1s + 3m_c$	$12m + 1s + 3m_c$
	$\approx 9m$	$\approx 10.8m$	$\approx 12.8m$

Table 4. Costs comparison for TN8 and KSS16

	k	Security level	hw	Δ
TN8 in Appendix	8	128bits	79	$173.8m_{385}$
KSS16 in Appendix	16	192bits	148	$325.6m_{489}$

6   Conclusion and Example

Table 1 shows that Jacobi model provides efficient formulae for Tate pairing computation when the embedding degree is even and $j \neq 0, 1728$. When $j = 1728$ our formulae on Jacobi quartic curves are better than the previous results on Jacobi quartic curves as shown in Table 2; In doubling steps, the cost on Jacobi quartic curves is the same as that on Weierstrass curves; in addition steps, our formulae on Jacobi quartic curves are better than the formulae on Weierstrass curves. When $j = 0$ the addition steps of our formulae on Jacobi quartic curves are fewer than the formulae on Weierstrass curves , while the doubling steps are slower as showed in Table 3. In practice, pairing-friendly elliptic curves with $j = 0$ are more popular and gain some benefits. However, there are still some famous families of pairing-friendly elliptic curves with $j = 1728$, such as TN8 [23] and KSS16 [19]. In the following, we consider Weierstrass model and Jacobi quartic model for two examples of these families and compare the expense of Tate pairing computation. In the next table, we denote $\Delta =$ "total expense of Weierstrass model" $-$ "total expense of Jacobi model", $hw =$ the Hamming weight of r. $\mathbf{m_{385}}$ and $\mathbf{m_{489}}$ are one multiplication in a finite field $\mathbb{F}_q$ with q a prime of 385 and 489 bits respectively (Table 4).

Acknowledgment. This work was supported by National Natural Science Foundation of China (No. 11101002, No. 11271129 and No. 61370187), Beijing Natural Science Foundation (No. 1132009), and the General Program of Science and Technology Development Project of Beijing Municipal Education Commission of China.

A   Examples with $j = 1728$

Using the construction in [19] and [23] to present Jacobi quartic curves with $j = 1728$ over $\mathbb{F}_q$ for embedding degree $k = 8, 16$, we list some pairing friendly Jacobi quartic curves with $4|k$. Let q be the prime for the finite field $\mathbb{F}_q$, r be the large prime order of a subgroup in $J(\mathbb{F}_p)$, $\rho = \log(p)/\log(r)$ and hw be the Hamming weight of r (Tables 5, 6).

Table 5. An example of TN8 curve

$k = 8$, $\rho \approx 1.5$, $\log(r) = 255$ bits, $\log(q) = 383$ bits, $hw{=}79$
$r = 5789604461888328473187612216222827406174508028319419831112462543$ 8026980406793
$q = 2955150464746827054295151251415979983023077940198595723001065 28$ $4879022429106631245409593937723593546481128229278 9377$

Table 6. An example of KSS16 curve

$k = 16$, $\rho \approx 1.25$, $\log(r) = 383$ bits, $\log(q) = 488$ bits, $hw{=}148$
$r = 2093232526971065973785822292891859762190652116249515366039098 5$ $51050451808104386759445937064649846547631691040445507 3;$
$q = 1392125484899467050662229866706984519217641617553210720479 59$ $752877236736091392960342059589515891369022425188591535332 7$ $2278452287305554488470403276 17$

B   Examples with $j = 0$

Using the construction in [13] to present Jacobi quartic curves $J_{a,d} : y^2 = dx^4 + 2ax^2 + 1$ with $j = 0$ over $\mathbb{F}_q$ for embedding degree $k = 12, 24$. For each k, curves at two security levels are given. Let t be the Frobenius trace, q be the prime for the finite field $\mathbb{F}_q$, r be the large prime order of a subgroup in $J(\mathbb{F}_q)$, $n = \sharp J(\mathbb{F}_q)$, and $\rho = \log(q)/\log(r)$ (Table 7).

Table 7. $J_{-27,-9} : y^2 = -27x^4 - 18x^2 + 1$ over $\mathbb{F}_q$ for embedding degree $k = 12, 24$

$J_{d,a}$: $k = 12$, $\rho \approx 1.5$, $\log(r) = 161$ bits

$t = 1099511630726;$

$r = 1461501653010476419563824324075703470606892615001;$

$q = 5889490310694441330739011548712381814951849552463124431529211730 78632117;$

$n = 5889490310694441330739011548712381814951849552463124431529200735 67001392;$

$J_{d,a}$: $k = 12$, $\rho \approx 1.5$, $\log(r) = 257$ bits

$t = 18446744073709566686;$

$r = 1157920892373165737821551871767212460418194942614239462794724036 61265709$
$\quad$ $211401;$

$q = 1313400206546489077704631059395345592330370814691407061669418717 81698452$
$\quad$ $3607837271424913571534028427485198155447143 7;$

$n = 1313400206546489077704631059395345592330370814691407061669418717 81698452$
$\quad$ $36078372714249135715340265828107907844904752$

$J_{d,a}$: $k = 24$, $\rho \approx 1.25$, $\log(r) = 161$ bits

$t = 1048646;$

$r = 1462271190260300144437063963469081833553287590001;$

$q = 535997570850424991004603472670510699116309175557541914541557;$

$n = 535997570850424991004603472670510699116309175557541913492912$

$J_{d,a}$: $k = 12$, $\rho \approx 1.5$, $\log(r) = 257$ bits

$t = 4294970102;$

$r = 1157926942199022831048968574721142864333630419694136944823750216 16015000$
$\quad$ $100401;$

$q = 7120003282946788688767832825047892963122039770343506948090350241 49143440$
$\quad$ $4644641800571771276401 01;$

$n = 7120003282946788688767832825047892963122039770343506948090350241 49143440$
$\quad$ $46446418005717283267 0000$

References

1. Arène, C., Lange, T., Naehrig, M., Ritzenthaler, C.: Faster computation of the tate pairing. J. Number Theor. **131**, 842–857 (2011)
2. Bernstein, D.J., Lange, T.: Faster addition and doubling on elliptic curves. In: Kurosawa, K. (ed.) ASIACRYPT 2007. LNCS, vol. 4833, pp. 29–50. Springer, Heidelberg (2007)
3. Bernstein, D.J., Birkner, P., Joye, M., Lange, T., Peters, C.: Twisted edwards curves. In: Vaudenay, S. (ed.) AFRICACRYPT 2008. LNCS, vol. 5023, pp. 389–405. Springer, Heidelberg (2008)
4. Billet, O., Joye, M.: The Jacobi model of an elliptic curve and side-channel analysis. AAECC 2003. LNCS, vol. 2643, pp. 34–42. Springer, Heidelberg (2003)
5. Barreto, P.S.L.M., Kim, H.Y., Lynn, B., Scott, M.: Efficient algorithms for pairing-based cryptosystems. In: Yung, M. (ed.) CRYPTO 2002. LNCS, vol. 2442, p. 354. Springer, Heidelberg (2002)
6. Barreto, P.S.L.M., Lynn, B., Scott, M.: On the selection of pairing-friendly groups. SAC 2003. LNCS, vol. 3006, pp. 17–25. Springer, Heidelberg (2003)

7. Chatterjee, S., Sarkar, P., Barua, R.: Efficient computation of tate pairing in projective coordinate over general characteristic fields. In: Park, C., Chee, S. (eds.) ICISC 2004. LNCS, vol. 3506, pp. 168–181. Springer, Heidelberg (2005)
8. Chudnovsky, D.V., Chudnovsky, G.V.: Sequences of numbers generated by addition in formal groups and new primality and factorization tests. Adv. Appl. Math. **7**(4), 385–434 (1986)
9. Costello, C., Lange, T., Naehrig, M.: Faster pairing computations on curves with high-degree twists. In: Nguyen, P.Q., Pointcheval, D. (eds.) PKC 2010. LNCS, vol. 6056, pp. 224–242. Springer, Heidelberg (2010)
10. Das, M.P.L., Sarkar, P.: Pairing computation on twisted edwards form elliptic curves. In: Galbraith, S.D., Paterson, K.G. (eds.) Pairing 2008. LNCS, vol. 5209, pp. 192–210. Springer, Heidelberg (2008)
11. Edwards, H.M.: A normal form for elliptic curves. Bull. Am. Math. Soc. **44**(3), 393–422 (2007)
12. Duquesne, S., Fouotsa, E.: Tate pairing computation on Jacobi's elliptic curves. In: Abdalla, M., Lange, T. (eds.) Pairing 2012. LNCS, vol. 7708, pp. 254–269. Springer, Heidelberg (2013)
13. Freeman, D., Scott, M., Teske, E.: A taxonomy of pairing-friendly elliptic curves. J. Cryptology **23**(2), 224–280 (2010)
14. Galbraith, S.D., Harrison, K., Soldera, D.: Implementing the tate pairing. In: Fieker, C., Kohel, D.R. (eds.) ANTS 2002. LNCS, vol. 2369, p. 324. Springer, Heidelberg (2002)
15. Hess, F., Smart, N.P., Vercauteren, F.: The Eta pairing revisited. IEEE Trans. Inf. Theor. **52**, 4595–4602 (2006)
16. Hisil, H., Wong, K.K.-H., Carter, G., Dawson, E.: Jacobi quartic curves revisited. In: Boyd, C., González Nieto, J. (eds.) ACISP 2009. LNCS, vol. 5594, pp. 452–468. Springer, Heidelberg (2009)
17. Wang, H., Wang, K., Zhang, L., Li, B.: Pairing computation on elliptic curves of Jacobi quartic form. Chin. J. Electron. **20**(4), 655–661 (2011)
18. Ionica, S., Joux, A.: Another approach to pairing computation in edwards coordinates. In: Chowdhury, D.R., Rijmen, V., Das, A. (eds.) INDOCRYPT 2008. LNCS, vol. 5365, pp. 400–413. Springer, Heidelberg (2008)
19. Kachisa, E.J., Schaefer, E.F., Scott, M.: Constructing brezing-weng pairing-friendly elliptic curves using elements in the cyclotomic field. In: Galbraith, S.D., Paterson, K.G. (eds.) Pairing 2008. LNCS, vol. 5209, pp. 126–135. Springer, Heidelberg (2008)
20. Koblitz, N., Menezes, A.: Pairing-based cryptography at high security levels. In: Smart, N.P. (ed.) Cryptography and Coding 2005. LNCS, vol. 3796, pp. 13–36. Springer, Heidelberg (2005)
21. Merriman, J.R., Siksek, S., Smart, N.P.: Explicit 4-descents on an elliptic curve. Acta Arithmetica **77**(4), 385–404 (1996)
22. Miller, V.S.: The Weil pairing and its efficient calculation. J. Cryptol. **17**(44), 235–261 (2004)
23. Tanaka, S., Nakamula, K.: Constructing pairing-friendly elliptic curves using factorization of cyclotomic polynomials. Pairing 2008. LNCS, vol. 5209, pp. 136–145. Springer, Heidelberg (2008)

DATAEvictor: To Reduce the Leakage of Sensitive Data Targeting Multiple Memory Copies and Data Lifetimes

Min Zhu, Bibo Tu$^{(\boxtimes)}$, Ruibang You, Yanzhao Li, and Dan Meng

Institute of Information Engineering,
Chinese Academy of Sciences, Beijing, China
{zhumin, tubibo, youruibang, liyanzhao,
mengdan}@iie.ac.cn,

Abstract. In modern operating systems, when a process terminates, the data still remain in the memory for an uncertain time. In addition, encryption is insufficient because the keys may be leaked through some compulsory means. In this paper, we present a novel OS-level approach called DATAEvictor, which thoroughly and timely evicts the sensitive data not only in the user stack, heap, kernel stack, but also in page cache, kernel buffer, slab objects and virtual memory swap when the process terminates. It aims to cut short the lifetime of sensitive data in memory as early as possible, so as to reduce the possibility of these data being leaked. DATAEvictor provides a "private mode" execution for any application according to user requirements, and just needs an appropriate code extension to the Linux kernel sources. The results of performance evaluation show that the implementation of DATAEvictor only results in a reasonable system performance loss.

Keywords: Sensitive data leakage · Data encryption · Data lifetime · Memory attack · OS security

1 Introduction

Data privacy is becoming an increasing focused issue for users. In many cases, users are not willing to leave any information of their activities in the computer. For example, a user may wish to browse prohibited websites, read confidential files, or send an email without keeping a record of these activities in the computer. However, modern operating systems have no built-in mechanisms for limiting or reducing the lifetime (from creation to overwritten) of application's data [1], which results in that modern operating systems accumulate significant amounts of memories of user's activities in cleartext – even long after the corresponding activities are terminated [2].

As sensitive data lifetime increases, so does the risk of exposure. Unfortunately, as said above, most of OSes, especially Linux, have widely overlooked the issue in their design. As a result, the sensitive data of user's activities, such as confidential accounts and passwords, are pushed on the cusp of exposure, since they are often scattered widely throughout application and OS memory and remain there for indefinite periods

© Springer International Publishing Switzerland 2015
M. Yung et al. (Eds.): INTRUST 2014, LNCS 9473, pp. 328–345, 2015.
DOI: 10.1007/978-3-319-27998-5_21

[3, 4], parts of which may be swapped to an area on swap device where they could sit for days or months, even after the user discards, or otherwise dumps the swap device.

It is very difficult for user-level applications to guarantee their sensitive data not being recoverable in the presence of forensic analysis. For example, in order to automatically remove all traces of user activities from the target computer, current browsers introduce a special mode, called private browsing mode. Even in this mode, attackers are able to recover the data of emails by using the image tool LiME [5] and the analysis tool Volatility [6]. We have illustrated this in Sect. 5.1. Since most applications do not support lifetime enforcement, the problem is much harder to solve in practice [7].

One OS-level solution to avoiding leaving sensitive data in cleartext on volatile memory is to use encryption techniques, such as drive encryption systems [8], encrypted RAM [9] and encrypting virtual memory [10]. Although they are effective and realizable approaches to protect sensitive data, they do not completely address the exposure of sensitive data, especially the vulnerable encryption keys, remained in the kernel, and also these approaches did not take the physical insecurity into account. Furthermore, another drawback of these kind of solutions is that with legal or other compulsory means users may be coerced into disclosing their encryption keys, which makes the encryption approaches useless [11, 12].

Since complete exemption from hacking is kind of difficult, data erasing at OS-level is a good choice for avoiding the leakage of sensitive data. However, the PaX patch [13], and secure deallocation [14] still don't deal with many traces of user's activities in memory, including the page cache, kernel buffers, etc. Given this, it is essential to build a more effective and comprehensive data eviction approach to minimize the disaster of data leakage.

In this paper, we describe the design and implementation of DATAEvictor, an OS-level approach that protects the privacy of data by clearing all memories (including user stack and heap, kernel stack, page cache, kernel buffers, and virtual memory swap) related to a specific process timely. Inspired by the browser's "Incognito mode", DATAEvictor enables the OS to support a private execution, in which mode the user can securely execute an application to process sensitive data. During a private process execution, a per-process key and linked list are respectively used to encrypt all data written to swap device and track the kernel buffers. In addition, DATAEvictor clears the user space and slab objects when they are freed. While for the page cache and kernel stack, they are cleared while a file is closed and a system call returns, respectively. Once a private process ends, all memory becomes unrecoverable. Meanwhile, the per-process key and linked list are destroyed.

DATAEvictor does not require any explicit application supports, like application modification or recompilation, neither any supports of hardware. Nor the user experience will be changed. Unlike the encryption approaches, the sensitive data of a private process cannot be exposed through compulsory measures. We give our prototype implementation of DATAEvictor with a reasonable performance overhead through extending the existing and well-testing mechanisms of the Linux kernel. In summary, the contributions of this paper are given as follows.

- For each private process, DATAEvictor creates a temporary data recorder, a doubly linked list that assists the private process to track and clear the kernel buffers used the private execution through the recorded address and size of them.
- DATAEvictor creates a random key to each private process for its swap-pages, so that each swapped page of a private process is encrypted and decrypted only through the key of this process. The swap data become useless with the per-process key destroyed when the process exits.
- We design and implement a prototype of DATAEvictor in the Linux kernel to thoroughly and timely evict the sensitive data, which focuses not only on the user stack, heap, kernel stack, but also on the page cache, kernel buffers, slab objects and swap space. The evaluation results of security and performance show that DATAEvictor can effectively prevent the leakage of sensitive data, with a reasonable overhead that poses indistinguishable impact to the user.

The remainder of this paper is organized as follows. Section 2 provides further motivation why we present this approach, and describes the threat model we assume for this work. Section 3 describes the design and implementation of DATAEvictor. Our evaluation experiments for both the protection effectiveness and performance are shown in Sect. 4. Section 5 discusses the limitations of our approach and how to improve it in the future. At last, in Sects. 6 and 7 we respectively describe the related work and make a brief summary.

2   Approach Overview

2.1   Motivation

Today attackers are no longer dedicated to cracking crypto directly. Instead, as shown in the report [15], attackers began to focus on stealing sensitive data. Consider the severely adverse impact on a company if its business files are disclosed to its opponents; consider the damage to a person's reputation for the leakage of his crucially private information. Leakages of sensitive data are caused through many channels, such as DMA memory attack [16], core dump [16, 17] and cold boot attack [8, 18, 19]. Due to space constraints, we aren't able to provide details of these attacks again here.

Despite the best efforts of protection mechanism, there are many other attacks to obtain sensitive data. For example, software will continue to have exploitable bugs, and malicious parties will continue to gain physical access to system hardware and those raised by [20]. Due to space constraints, we will not enumerate them in details.

2.2   Thread Model

As long as the data exists, the attackers could have a chance to get them through a certain type of memory attack. Therefore, the best way to protect the data from being stolen is not to set enough security guards, but to disappear them in advance.

We assume that a private process is short-lived. In our threat model, on the base of executive status the execution of a private process can be divided into three periods:

before, during and after its execution. To specifically illustrate the threats, we presented two scenarios, one for before and during a private process executes, and another for after the private process terminates.

In the first scenario, we assume that the target computer is completely safe before running a private process. According to the concept of the "trusted computing base" (TCB), in a computer system the trusted components should be as small as possible. But it is not suitable for this case, because if malware has already existed on the system before the process running, the attacker could directly access the memory or scan the swap space through special technology. Therefore, in this case, the TCB is the entire system. During the process running, we also assume that the memory of the target computer cannot be scanned and imaged, and so also does the swap space.

In the second scenario, we assume that the attacker has the complete control of the target computer after the private process terminates, which means the attacker not only has access to the memory and swap space in the physical, but also is able to use some memory forensics tools to scan and recover the legacy data. So in this case, the TCB is empty. The entire system, including the Linux kernel, can be considered as a malware.

Our approach does not restrict communication between processes because we assume that the user clearly understands and follows that the sensitive data cannot be handled by both public and private process at the same time. The IPC buffers also can be cleared by our solution in the slab allocator. Also our approach cannot completely eliminate the data leakage from the core dump, but only little leaked.

3 Design and Implementation

3.1 System Architecture

To minimize the disclosure of sensitive data resulting from an application, DATAEvictor exploits several mechanisms to clear the data out the memory when they are no longer used. On the top of the design, DATAEvictor provides a private execution, in which the process must be subject to special principles to achieve our aim. Since DATAEvictor does not explicitly define the sensitive data, a private process must treat all data as sensitive. The system architecture of DATAEvictor is shown in Fig. 1, in which the entire sensitive data protection work are divided into six related parts: user-space, kernel stack, page cache, kernel buffers, swap space and slab objects, correspondingly as seen in the part 1–6 of Fig. 1.

3.2 Initiating the Private Process

The first requirement from DATAEvictor OS is how to create a private process. According to the user requirement, DATAEvictor must be capable of launching an application in private mode. In Linux kernel, every process has a process descriptor, represented by the structure of task_struct, which contains all the information for execution. Through the clone or fork system call, that finally invokes the function do_fork(), a new process is created and initiated.

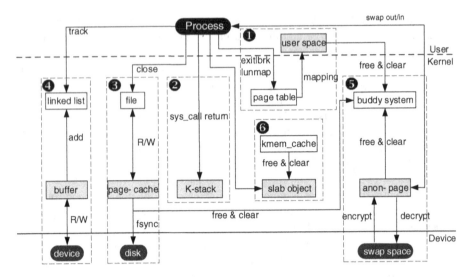

Fig. 1. The system architecture of DATAEvictor

In order to create a private process, we first extended a new process flag, PF_PRIVATE, to the flags field of task_struct for indicating that it is a private process. In our implementation, we assumed that all sub-processes or sub-threads of a private process are still in private mode by inheriting the private flags of the parent. In order to enable the communication between the users and process manager, we introduced a new flag, CLONE_PRIVATE. We modified do_fork function to set the PF_PRIVATE flag for subsequent use through checking the exist of CLONE_PRIVATE in the flags bitset. For applications to request private execution, we developed a small tool to pass the CLONE_PRIVATE flag to the clone system call and implemented the tool to the context-menu, named "private run". Thus, the user can directly launch a process of "private mode" by clicking the "private run" context-menu. The summary of all modifications to the Linux kernel described in this section is presented in Table 1.

Table 1. Changes for creating a private process

File path	Modified	Description
include/linux/sched.h	task_struct	Define PF_PRIVATE flag
	clone flags	Define CLONE_PRIVATE flag
kernel/fork.c	do_fork()	Modify it for creating a private process If the parent is a private process, its children are all private process
kernel/exit.c	do_exit()	Clear the random key
context-menu		We use nautilus-actions to create a context-menu that is used to pass the flag CLONE_PRIVATE to the clone system call to create a "private mode" process

3.3 Clearing the User-Space

In traditional OS, the expired data of user space are overwritten until the next allocation. A process's user space is subdivided into three segments: mapping segment user stack and user. In [14], a kernel thread is waked up periodically to zero pages that in polluted pool longer than a configurable amount of time. For user stack, we have two pre-selected schemes for clearing the stack pages. The first is the same as [14]. Describe no longer about it because of space cause. The second is the case that every time the process exits, all pages of the entire stack are cleared. Tests show the first has no performance advantage compared to the second, but also requires the assistance of the second in exit. Thus, DATAEvictor adopts the centralized treatment. While for heap and mapping segment, DATAEvictor clears these data immediately when the munmap and brk system-calls are called. Due to the "memory leak", DATAEvictor clears all non-freed data in exit, as illustrated in part 1 of Fig. 1.

Concretely, we extended the page's flags with PF_clear and PF_clean, which are located prior to the __NR_PAGEFLAGS in enum pageflags, and defined the operation macros of the two flags. When pages of a private process's user-space are released, the function __tlb_remove_page() is called, which we modified to support data clearing. At the __tlb_remove_page(), if the owner of the freeing page is a private process, the page flags field is marked with PF_clear, indicating that the page need to be cleared before released. And then in free_pages_check(), DATAEvictor clears the page flags, excepting the PF_clear by PAGE_FLAG_CHECK_PRIVATE, newly defined in page-flags.h. Finally, in function free_pages_prepare(), every page marked with PG_clear is cleared by clear_page(). Meanwhile the page flags is marked with PG_-cleaned, denoting the page is clean and can be reused directly.

A summary of all modifications to the Linux kernel described in this section is presented in Table 2.

3.4 Sanitizing the Kernel Stack

After the process changes to kernel mode, its execution context is stored in the kernel stack. As a result, after the system call returns or the process terminates, sensitive data put in stack-allocated variables still persist for a period of time [21]. In the solution of [14], the kernel stack is not taken into consideration. In our solution, we adopt the mechanism shown in part 2 of Fig. 1, which is zeroing the kernel stack on every return from a system call, but without causing a significant performance overhead.

To achieve this, we mainly modified two files: entry_32.S and ia32entry.S. We first created a global function named clear_kstack, in which we first checked the process mode by locating the flags filed of task_struct through thread_info. For a private process, we counted how many bytes had been written in this system call. Finally, we immediately cleared the data. We present all the modifications to the Linux kernel made in this section in Table 2.

Table 2. Changes for clearing the private process's user-space and kernel stack

File path	Modified	Description
include/linux/page-flags.h	enum pageflags	Extend two flags: PG_clear, PG_clean Define the macros of the two flags Define PAGE_FLAG_CHECK_PRIVATE
mm/memory.c	__tlb_remove_page	Mark the page structure with PG_clear for clearing the page data
mm/page_alloc.c	free_pages_prepare	Decide whether to invoke the clear_page () function
	free_pages_check	Not clear the two new flags' bit by using PAGE_FLAG_CHECK_PRIVATE
include/linux/highmem.h	clear_page	Clear pages that are marked with PG_clear and mark the pages with PG_clean
mm/page_alloc.c	prep_new_page	Cancel the page structure's PG_clean flag
arch/x86/kernel/entry_32.S	clear_kstack clear_kstack	We introduce a new function for clearing the kernel stack for int80 system call
arch/x86/ia32/ia32entry.S	clear_kstack	To clear the kernel stack for quick system call
arch/x86/kernel/asm-offsets_32.c	foo	Define THTRED_SIZE_asm, TS_flags, TS_private for clearing kernel stack

3.5 Flushing the Page Cache

In modern Linux system, in order to accelerate the speed of disk, page cache is introduced. [14] suggested that there is no need to clear the page cache. But much of our daily work is to handle a variety of files. During the process execution, all files that are not directly read/written from/to the filesystem will be catched in the page cache. These pages may exist, even rewritten to the disk, until they are reallocated.

Existing solutions such as PaX and [14] do not timely handle this issue. In order to prevent sensitive data from leakage, DATAEvictor clear the page cache when the file is closed, as shown in part 3 of Fig. 1. A file is closed by either a close system call or the termination of the private process. Though the time point is different in both scenarios, but the working process is the same.

Concretely, DATAEvictor adds a new function, clear_page_cache(), to the function filp_close(), invoked during a file closed. In Linux system, each opened file is allocated with the structure file and inode, while the inode is the sole representative of this file. In our modifications, when a file is closed, DATAEvictor firstly checks whether the process is in private mode. If so, through the reverse lookup, the structure inode and address_space are confirmed. And if the inode is dirty, the next action is to synchronize the dirty pages through the fsync function of the file operations. After that, through the f_mapping field of file structure, we can locate the address_space structure which contains the pages of the file in the page cache. Finally, all pages contained in the radix tree are released and marked with PG_clear through the function clear_inode_pages(). Also the pages are marked with PG_clean after cleared.

All changes to the Linux kernel made in this section are shown in Table 3.

Table 3. Changes for clearing the page cache and kernel buffer

File path	Modified	Description
fs/open.c	filp_close	Synchronize the data and clear them
	clear_page_cache	immediately
mm/slub.c	slab_free	Clear the freed slab objects
include/linux/sched.h	task_struct	Extend task_struct with a linked list
		Define PF_TRACK for tracking
kernel/fork.c	copy_process	Initiate the linked list
arch/x86/include/asm/uaccess.h	clear_data_node	Define a new structure to contain the kernel buffer's address and size
arch/x86/lib/usercopy_32.c	copy_to_user	Add the buffer to the linked list
	copy_from_user	Add the buffer to the linked list
fs/read_write.c	sys_read	Set the PF_TRACK
	sys_readv	Clear the buffer when read finishes
	sys_write	Set the PF_TRACK
	sys_writev	Clear the buffer when write finishes

3.6   Tracking the Kernel Level Buffers and Clearing Slab Objects

From Chow's paper [3], we learn that after a password is typed into Mozilla its journey goes through a wide range of locations, which includes keyboard queue and sk_buff etc. Contrary to the expected, PaX doesn't clean write_buf of TTY that stores the data written to the TTY device. In the solution of [14], the I/O buffers are cleared through slab buffers when they are freed. But some I/O buffers, such as the network buffer (sk_buffs) and tty buffer (write_buf) may be used for a long time at a process execution. This violates the idea of [14], clearing the data as soon as possible. In our solution, the kernel buffers are cleared immediately after every time used.

Only the private process itself knows what kernel buffers has been used during execution, so DATAEvictor creates a linked list for each private process during its creation. The private process can utilize the per-process linked list to track different kernel buffers, because the linked list connects the private process with each kernel buffer through a small node for recording the virtual address and size of buffers. Whenever a kernel buffer is used by a private process, DATAEvictor adds a new node to the linked list. When the read or write operation to kernel buffers is finished, DATAEvictor can rely on the linked list to clear the buffers immediately. The linked list as a bond connects its owner and kernel buffers as shown in part 4 of Fig. 1.

The choice of where to intercept kernel buffer in DATAEvictor requires careful consideration. In particular, in order to build a generic solution that is independent of the underlying drivers, we should leverage the drivers' common characters. If the processes read devices, it must copy the data from the kernel buffer to user-space through the function copy_to_user(). On the contrary, the function copy_from_user() is called. DATAEvictor takes advantage of these two functions to achieve its purpose. We modified the two functions to track the kernel buffers.

To achieve this, we first extended the task_struct with a field for the linked list, and extended its flags field with PF_TRACK, which indicates a private process is able to record the address of kernel buffers. To store the buffer's address and size, we defined a new structure in uaccess.h as the linked list node. The linked list is initiated in the function copy_process(). We added a new function, named track_buffers(), to the copy_to_user() and copy_from_user() to record the kernel buffers. During the private process execution, we first modified the read, readv, write and writev, etc. system calls to check whether the process is private, and if it is, the PF_TRACK flag is set to the flags field of task_struct. After all the above work is already in place, then we also modified the two system calls in order to clear the buffers' data through traversal of the linked list node. For network devices, the recvfrom, sendto, etc. system calls also need to be modified for supporting tracking.

The cleanup mechanisms described in the above sections insufficiently prevent sensitive data resulting from private execution from being leaked in the cleartext. For example, we visit a file with Vim. After Vim exits, a physical memory contains complete data of inode and dentry that may be useful for attackers or forensics.

Specifically, as shown in part 6 of Fig. 1, DATAEvictor also imposes cleanup on slab objects to prevent metadata leaks, such as mm_struct and file_struct. This is also a complementary mechanism to protect the I/O buffers. In our implementation, the slab manager is slub allocator. In kmem_cache definition, there are two fields we used: size and ctor. The size field represents the size of an object and its metadata, while the ctor field is the constructor of a slab object. To clear a slab object, we modified the objects freed function slab_free(). If the objects owner is a private process, the object data are cleared, and then a check is performed to test whether the ctor field of the kmem_cache is set. If so, the actor is invoked to initialize the object for later use. For task_struct, since it is released by its parent of the process, so when a task_struct slab object is freed the presence of PF_PRIVATE must check. If present, the task_struct is cleared. In general, since the kernel thread has no mm_struct structure and can only borrow the prior process's mm_struct. Therefore, we modified do_exit to mark the mm_struct with a flag during a private process termination. So a mm_struct is cleared if the marked flag is existence.

A summary of all modifications to the Linux kernel described in this section is presented in Table 3.

3.7   Encrypting Swap-Space and Puring the Pages by PFRA

Any data that were originally encrypted with FDE can be found as plaintext in memory when processed by a process, while if these pages are swapped out, these data will be stored as cleartext in swap space. In case of magnetic storage, it is possible for sensitive data to be contained in swap space for an indeterminate period of time even undergone several reboot and shutdown [19, 22–24].

The most radical solution is to avoid using swap. However, DATAEvictor relies on the RNG and crypto [25] API to encrypt swapped out pages and decrypt swapped in pages (see part 5 of Fig. 1). This is different from existing approaches to swap encryption, which uses a single key to encrypt the entire swap device, or use multiple rotating keys. DATAEvictor imposes a per-application partitioning of the swap through private keys and clears the pages of a private process from PFRA.

To achieve this, our first modification is during a private process creation, a random key (PEK) is generated with the RNG of the kernel, and is stored in the new field of task_struct. We also can store the random key in the TPM hardware, but for the sake of utilization we assume the task_struct is safe, protected by the OS, and never disclosed to others. In our future work, we will store the random key in TPM, and combine with the Intel TXT [26] or SGX [27–29] technology.

The exact implementation is divided into two parts. For page-in, which is handled by the kswapd, a kernel thread, we modified the pageout() function for supporting page encryption. First, based on the swapping page we can locate the mm_strcut structure through reverse mapping and then gain the task_struct structure. At this point, it is able to judge whether the page belongs to a private process. If so, the swapping page is encrypted by using the PEK. Finally, all pages that are belonged to a private process and released by kswapd thread are cleared.

While for page-out it is relatively simple. DATAEvictor first checks whether the owner of the page is a private process. If so, once the asynchronous read of the page from the swap devices has completed, DATAEvictor uses the PEK to decrypt the data prior to switching to the user mode. All the modifications to the Linux kernel made in this section are presented in Table 4.

Table 4. Changes for encrypting swap space and clearing the pages by PFRA

File path	Modified	Description
include/linux/sched.h	task_struct	Extend a filed storing the random key
mm/vmscan.c	pageout	Encrypt the swapped out page
mm/memory.c	do_swap_page	Decrypt the swapped in page
mm/vmscan.c	shrink_page_list	Mark the page for clearing

3.8 Discussion

Our design and implementation above-mentioned have reached the expected goal we set. Our approach is distributed at different stages of a private process's execution, which ensures that it is able to enforce the cleanup of sensitive data timely and guarantees the security of user's data. By its very nature, DATAEvictor is a double edged sword, which also makes the computer forensics work more difficult [30, 31].

4 Evaluation

In this section, we first demonstrate the security and feasibility of Linux's applications with DATAEvictor that are used frequently to process sensitive data. Next, we measure the performance of DATAEvictor by testing its throughput and time-delay with benchmarks. We ran all benchmarks on a Lenovo ThinkVision with Intel Core i5-3470 3.20 GHz CPU, 4 GB of RAM, running Ubuntu 12.04 desktop edition with the kernel 2.6.32.60.

4.1 Security Analysis

In this section, we firstly demonstrate the feasibility, which means any application runs normally in private mode. And next, we validate the security of DATAEvictor, namely, no traces of private processes are left in memory after they are terminated. To demonstrate these, we selected 10 popular free applications to test, which cover many kinds of categories, reported by Ubuntu Software Center [32]. Using the memory forensic methodology, we ran each application with "taint-tokens", such as a specific file, a web page data and an email, and then examined these "taint-tokens".

In our experiment, we first launched each application with DATAEvictor and checked whether they work as intended. Results of the test show that all of them have no abnormal behaviors and work correctly. Now that the applications can run as usual with DATAEvictor, we next ought to justify DATAEvictor's security character. We selected 5 representative applications that are most often used for processing a tremendous amount of sensitive data for the security test. For each test application, we used the LiME-forensic [5] to make an entire memory image in advance to ensure there were no "taint-tokens", and then ran it, and finally made a memory image again before and after the application terminated, respectively with the unmodified kernel. For comparison, we did the same experiment with PaX and DATAEvictor one hour after the computer shutdown. Finally, we utilized the image analysis tool Volatility [6] and hex editor Neo [33] to search the "taint-tokens" among these images.

From the results of experiments, we can see that without DATAEvictor, the tokens are present in many places after a process terminates. Even though with PaX, there also some tokens exist in memory. While with DATAEvictor, no tokens are found in the place specified by our prototype after the private process terminates. Let's work through an example. The test result of Gedit application shown that, on average, with PaX 5 tokens were found. While with DATAEvictor the fragments of the token were reduced to 2, about 40 % of before, and even as low as 20 %. Though we have no sufficient testimony to prove all data of a private process are clear, but this experiment is enough to justify DATAEvictor achieves the goals of our design.

4.2 Performance Evaluation

In this section, we measured the performance overhead introduced by DATAEvictor. First, we would like to measure the CPU utilization overhead with and without

DATAEvictor, respectively. Then we would like to use benchmarks to measure the filesystem I/O throughput. Finally, we would like to measure the performance of a private process at runtime in macroscopic fashion.

CPU Utilization Overhead. To evaluate raw performance overhead, we measured the CPU utilization for unmodified Linux kernel and DATAEvictor by sampling the CPU utilization of the process at 1 s interval, and computed an average score as the test results. The experiment used 6 popular applications to test the CPU utilization with and without DATAEvictor. Table 5 illustrates these results.

Table 5. CPU utilization results for six representative applications

App.	Orig. CPU (%)	DATAEvictorCPU (%)	App.	Orig. CPU(%)	DATAEvictor CPU (%)
Thunderbird	16.86	17.86	Gedit	8.26	8.93
Libreoffice	30.07	29.90	Eog	4.48	4.94
Firefox	12.30	13.20	Evince	24.71	26.20

From the "Orig. CPU" and "DATAEvictor CPU" columns of Table 5, we can see that DATAEvictor has no adverse impact (under 6 % overhead) on all 6 application.

Filesystem I/O Benchmark. Next, we used Bonnie++ [34], a widely used filesystem benchmark suite for the Linux operating system, to measure the filesystem I/O performance of DATAEvictor. Our main goal is to test the throughput of write, rewrite, and read operations by using 5 × 8 GB files. Comparably, we ran Bonnie++ with and without DATAEvictor, respectively, repeated 10 times, and calculated an average score as the final results. The results of experiments are shown in Table 6.

Table 6. Filesystem performance of DATAEvictor

Operations	Orig. Performance	Orig. Performance	Overhead
Write	82674.9 K/s (%5 cpu)	81805.75 K/s (%6 cpu)	1.06 %
Rewrite	39148 K/s (%4 cpu)	39040.13 K/s (%5 cpu)	0.28 %
Read	105324 K/s (%6 cpu)	83939.38 K/s (%6 cpu)	25.47 %

The results show that DATAEvictor has a reasonable performance with the overhead 1.06 % and 0.28 %, respectively for write and rewrite, because the time of clearing data are negligible compared to the time of writing data to disk. However, since DATAEvictor has to clear the page cache and kernel buffers but don't need to write the disk during read, so the read has a relatively high overhead.

Application's Execution Runtime Overhead. Finally, to estimate application runtime overhead, we ran the application with DATAEvictor. In these experiments, we measured the overhead with kinds of common but representative desktop and console applications. Since manual operations may cause errors, so we employ some accurate automation frameworks and testing tools.

We firstly measured the entire application's runtime overhead with DATAEvictor. We selected 8 popular Linux applications. They are audio and video players, email client, web browser, picture processing, document processing applications and a system tool (full details in the following). While for these applications we utilized the shortcut key provided by the applications themselves, and Xdotool [35], an X automation tool that can simulate keyboard, mouse and window management events. Similar to the prior experiments, we repeated each test for10 times and calculated an average of the runtimes. In a summary, we present all the results in Fig. 3.

- LibreOffice: LibreOffice is a powerful office suite. We used it to open word documents, repeated 10 times with 10 different documents (each about 5 M) that contain images, forms and text.
- Gedit: Gedit is a common text editor. We used it to open 20 different text files (each about 50 K), repeated 10 times.
- Evince: Evince is a document viewer for multiple document formats. We used it to open PDF format documents, repeated 10 times with 10 different documents (each about 550 K).
- SMPlayer: SMPlayer is a built-in codecs desktop media player. We used it to open MP4 format media, repeated 10 times with 10 different files (each about 135 M).
- Audacious: Audacious is a very friendly audio player. We configured it to open 25 MP3 files in sequence through the playlist, repeated 10 times.
- Eog: Eog is a photo viewer in a variety of formats. We utilized it to open 10 JPEG format images, repeated 10 times.
- Thunderbird: Thunderbird is a popular graphical email client. We configured it to open 20 email messages, that some with attachment, repeated 10 times.
- Firefox: Firefox is the Ubuntu built-in web browser. We used it to visit local websites, which contain images and tables, repeated 10 times with 10 different websites.

In Fig. 2, experiments results make it clear to understand that the overhead of DATAEvictor is under 20 % in most test cases, which indicates that running with DATAEvictor the runtime overhead of the applications cannot be obviously perceived by the users. From Fig. 3, we know that the overall performance of an application also depends on how much data is handled by the private process, and how often the data are released during its execution, that is why Libreoffice's overhead is higher than others. The pink and green curves of Fig. 3 show that most of the overhead is introduced by page cache cleanup, because during the execution of a private process dozens of files are opened and closed. Our test results also show that DATAEvictor spends 64.31 μs in total clearing 100 pages, on average 0.643 μs per page, which also explains the reason why the overhead is high to dispose big files.

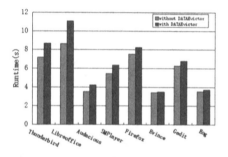
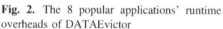

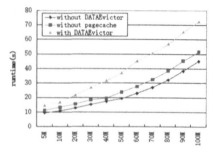

Fig. 2. The 8 popular applications' runtime overheads of DATAEvictor

Fig. 3. As the file size increasing the Libreoffice' overheads

5 Limitations and Future Work

While our prototype aims at offering an effective approach to protect user's sensitive data. Although our approach is able to effectively clear sensitive data in memory on most points, our current prototype has several limitations.

First, in our approach, we cannot control all the internal data of device drivers if there exist multiple copies, due to we can only track the data in drivers copied from the process user-space without making modifications to drivers. As a result, a private process may leave a very small amount of sensitive data in the peripheral drivers.

Second, in a Linux system, graphics process's display is controlled by the X server, which we did not take into account. The memory of the X server may hold a few sensitive data from a private process. Due to the code of the X server is notorious for its size and complexity, directly erasing graphical output is impractical.

The last is not a limitation but improvement. For performance reason, we are like to modify pages allocation strategy of the buddy system. We intend to divide the buddy system into two parts for a separate purpose, one for storing pages uncleared, and the other one as the storage of the cleared pages. It is obvious understand that when the system need zeroed pages, the buddy system directly allocates pages from the cleared part, otherwise from the uncleared part. In this way, we can achieve a higher execution performance for different page requirements.

As future work, one avenue we plan to investigate is whether we can take advantage of the hardware features (like X86 TXT and SGX) to enhance the privacy of application's data [36]. If necessary, we would like to extend the architecture to deal with the privacy issue [37–41].

6 Related Work

Operating system frequently process data that requires protection from attackers. In many case, the OS cannot against protean attacks. There are a variety of existing methods proposed to mitigate or address data leakage. Chow et al. proposed a secure deallocation, which aims at clearing the process user data at different levels [14].

Due to the design in multiple levels, thus the secure deallocation has relatively large code changes. Its first requirement is programmer's cooperation [42]. Furthermore, it needs to modify the library functions, and requires an idle process for clearing the process's stack regularly. Owing to secure deallocation limitation, the security deallocation implementation did not put page caches, device buffers into consideration.

While for encryption system, Peter et al. introduced CryptKeeper [9], in which the RAM is divided into two regions, a small cleartext working set and an encrypted area. Thus ensures only a small amount of cleartext may be exposed to the attacker. Although to some extent the encryption system protects the sensitive data of user processes, but it does not guarantee that after the process exits the related data absolutely become unrecovered, and the key management is also a problem. CleanOS [46] encrypts sensitive data after a period of non-use, and then evict the key to clouds to prevent data from disclosing. While how to define the sensitive data is a challenge. And it does not prevent leaks through the OS and I/O.

To against physical attack, two tiers of cryptographic mechanisms are proposed by Dorrendorf [8]. First, every opened file is bound with a separate key that is encrypted with the master key of this system. When the system change to unattended state, such as log out, the master key is cleared, that ensures even an attacker control the computer it is impossible to extract the master key.

Application's data may disclose via the swap, so encrypt virtual memory [10] is proposed that used a key to encrypt the page that is swapped out to the swap. Inspired by the browser's "ncognito mode", Kaan et al. proposed the PRIVEXEC in 2013 [11]. While SandBox, such as Vx32 [43] and Native Client [44], is another type of private execution. SandBox is able to isolate the memory access path of malicious code and delete the trace of the application running in the SandBox. Storage Capsules [45] can achieve this by using the snapshot and encryption. However, they didn't fully consider the legacy data in memory.

In 2012, Dunn et al. proposed Lacuna [2] which utilizes the ephemeral channels, VM and PaX to protect sensitive data in the process memory. First, using the virtual machine to isolate the space and then making use of the ephemeral channels to translate the encrypted data. Ultimately, the VM's memory is cleared by PaX. However, it also exposes a shortcoming, the data in hardware drivers is out of its control. Also to be noted that the use of virtual machine performance loss greatly and need to modify the appropriate hardware drivers.

7 Conclusion

Data privacy is of paramount importance for many users, companies and organizes. Indisputably, there is a huge demand from users for privacy-enabling technologies [11]. In this paper, we have described the design, implementation and evaluation of DATAEvictor, a novel OS approach that can throughly (targeting multiple memory copies) and timely (targeting data lifetimes) clear a private process's sensitive data before it completely terminates. We have implemented a prototype of this approach as a modification to the Linux kernel on Ubuntu. We have demonstrated that DATAEvictor is an effective approach to sensitive data protection from security analysis and performance evaluation.

References

1. Lyman, J.: Security: TaintBochs testing highlights the persistence of OS memory. http://archive09.linux.com/feature/36916. Accessed 22 June 2004
2. Dunn, A.M., Lee, M.Z., Jana, S., Kim, S., Silberstein, M., Xu, Y., Shmatikov, V., Witchel, E.: Eternal sunshine of the spotless machine: protecting privacy with ephemeral channels, In: OSDI 2012 (2012)
3. Chow, J., Pfaff, B., Garfinkel, T., Christopher, K., Rosenblum, M.: Understanding data lifetime via whole system simulation. In: Proceedings of the 13th Conference on USENIX Security Symposium, 09–13 August 2004
4. Czeskis, A., Hilaire, D.J.S., Koscher, K., Gribble, S.D., Kohno, T., Schneier, B.: Defeating encrypted and deniable file systems: TrueCrypt v5.1a and the case of the tattling OS and applications. In: Proceedings of the 3rd Conference on Hot Topics in Security, 29 July 2008 (2008)
5. Google Project Hosting. LiME-Linux memory extractor. http://code.google.com/p/lime-forensics/
6. The Volatility Framework. https://code.google.com/p/volatility/
7. Kannan, J., Altekar, G., Maniatis, P., Chun, B.-G.: Making programs forget: enforcing lifetime for sensitive data. In: Proceedings of the 13th USENIX Conference on Hot Topics in Operating Systems, 09–11 May 2011
8. Dorrendorf, L.: Protecting Drive Encryption Systems Against Memory Attacks. IACR Cryptology ePrint Archive (2011)
9. Peterson, P.A.H.: Cryptkeeper: improving security with encrypted RAM. In: Proceedings of the IEEE International Conference on Technologies for Homeland Security (2010)
10. Provos, N.: Encrypting virtual memory. In: Proceedings of the 9th Conference on USENIX Security Symposium, p. 3, 14–17 August 2000
11. Onarlioglu, K., Mulliner, C., Robertson, W., Kirda, E.: PRIVEXEC: private execution as an operating system service. In IEEE Symposium on S&P (2013)
12. Thing, V.L.L., Ying, H.-M.: A novel time-memory trade-off method for password recovery. In: Proceedings of the Ninth Annual DFRWS Conference, vol. 6, Supplement, pp. S114–S120, September 2009
13. Homepage of the PaX team. http://pax.grsecurity.net
14. Chow, J., Pfaff, B., Garfinkel, T., Rosenblum, M.: Shredding your garbage: reducing data lifetime through secure deallocation. In: Proceedings of the 14th Conference on USENIX Security Symposium, 31 July–05 August 2005
15. A new type of attack (2005). http://tech.163.com/05/1228/13/262HR1J000091KUL.html
16. Gubanovis, Y., Afonin, O.: Catching the Ghost: How to Discover Ephemeral Evidence through Live RAM Analysis (2013). http://forensic.belkasoft.com/download/info/Live_RAM_-Analysis_in_Digital_Forensics.pdf
17. Garfinkel, T., Pfaff, B., Chow, J., Rosenblum, M.: Data lifetime is a systems problem. In: ACM SIGOPS European Workshop, 19–22 September 2004

18. Halderman, J.A, Schoen, S.D., Heninger, N., Clarkson, W., Paul, W., Calandrino, J.A., Feldman, A.J., Appelbaum, J., Felten, E.W.: Lest we remember: cold boot attack on encryption keys. In: USENIX Security Symposium (2008)
19. Di Crescenzo, G., Ferguson, N., Impagliazzo, R., Jakobsson, M.: How to forget a secret. In: Meinel, C., Tison, S. (eds.) STACS 1999. LNCS, vol. 1563, pp. 500–509. Springer, Heidelberg (1999)
20. Harrison, K., Xu, S.: Protecting cryptographic keys from memory disclosure attacks. In: IEEE/IFIP International Conference on DSN (2007)
21. Oberheide, J., Rosenberg, D.: Stackjacking your way to grsecurity/PaX bypass (2011). https://jon.oberheide.org/files/stackjacking-hes11.pdf
22. Gutmann, P.: Secure deletion of data from magnetic and solid-state memory. In: Proceedings of the 6th USENIX Security Symposium (1996)
23. Hamilton, T.: 'Error' sends bank files to eBay. Toronto Star, 15 September 2003 (2003)
24. Perlman, R.: File system design with assured delete. In: Proceedings of the Third IEEE International Security in Storage Workshop, pp. 83–88 (2005)
25. Crypto Introduction: http://www.gnu.org/software/gnu-crypto/
26. Evolution of Integrity Checking with Intel® Trusted Execution Technology: an Intel IT Perspective. http://www.intel.cn/content/www/cn/zh/pc-security/intel-it-security-trusted-execution-technology-paper.html
27. McKeen, F., Alexandrovich, I., Berenzon, A., Rozas, C.V., Shafi, H., Shanbhogue, V., Savagaonkar, U.R.: Innovative instructions and software model for isolated execution. In: HASP, 2013, vol. 13, p. 10 (2013)
28. Hoekstra, M., Lal, R., Pappachan, P., Phegade, V., Del Cuvillo, J.: Using innovative instructions to create trustworthy software solutions. In: Proceedings of the 2nd International Workshop on Hardware and Architectural Support for Security and Privacy. ACM (2013)
29. Anati, I., Gueron, S., Johnson, S., Scarlata, V.: Innovative technology for cpu based attestation and sealing. In: Proceedings of the 2nd International Workshop on Hardware and Architectural Support for Security and Privacy, HASP (2013)
30. Graziano, M., Lanzi, A., Balzarotti, D.: Hypervisor memory forensics. In: Stolfo, S.J., Stavrou, A., Wright, C.V. (eds.) RAID 2013. LNCS, vol. 8145, pp. 21–40. Springer, Heidelberg (2013)
31. Petroni, N.L., Walters, A., Fraser, T., Arbaugh, W.A.: FATKit: a framework for the extraction and analysis of digital forensic data from volatile system memory. Digital Invest. 3(4), 197–210 (2006)
32. Ubuntu Software Center:http://www.ubuntu.org.cn/ubuntu/features/ubuntu-software-centre
33. HHD Software Ltd. Free Hex Editor Neo. http://www.hhdsoftware.com/free-hex-editor
34. Bonnie++. http://www.coker.com.au/bonnie++/
35. Sissel, J. (a hacker): Xdotool - fake keyboard/mouse input, window management, and more. http://www.semicomplete.com/projects/xdotool/. Posted Sun, 21 July 2013
36. Baumann, A., Peinado, M., Hunt, G.: Shielding applications from an untrusted cloud with Haven. In: Proceedings of the 11th USENIX Conference on Operating Systems Design and Implementation. USENIX Association (2014)
37. Suh, G.E., Clarke, D., Gassend, B., Van Dijk, M., Devadas, S.: AEGIS: architecture for tamper-evident and tamper-resistant processing. In: Proceedings of the 17th Annual International Conference on Supercomputing. ACM (2003)
38. Suh, G.E., Clarke, D., Gassend, B., Dijk, M.V., Devadas, S.: Efficient memory integrity verification and encryption for secure processors. In: Proceedings of the 36th Annual IEEE/ACM International Symposium on Microarchitecture. IEEE Computer Society (2003)
39. Lie, D., Thekkath, C.A., Horowitz, M.: Implementing an untrusted operating system on trusted hardware. In: ACM SIGOPS Operating Systems Review. ACM (2003)

40. Champagne, D., Lee, R.B.: Scalable architectural support for trusted software. In: 2010 IEEE 16th International Symposium on High Performance Computer Architecture (HPCA). IEEE (2010)
41. Chhabra, S., Rogers, B., Solihin, Y., Prvulovic, M.: Secureme: a hardware-software approach to full system security. In: Proceedings of the International Conference on Supercomputing. ACM (2011)
42. Viega, J.: Protecting sensitive data in memory (2001). http://www.ibm.com/developerworks/library/s-data.html?n-s-311
43. Ford, B., Cox, R.: Vx32: lightweight, user-level sandboxing on the x86. In: USENIX Annual Technical Conference (2008)
44. Yee, B., Sehr, D., Dardyk, G., Chen, J.B., Muth, R., Ormandy, T., Okasaka, S., Narula, N., Fullagar, N.: Native client: a sandbox for portable, untrusted x86 native code. In: IEEE Symposium on Security and Privacy (2009)
45. Borders, K., Vander Weele, E., Lau, B., Prakash, A.: Protecting confidential data on personal computers with storage capsules. In: USENIX Security Symposium (2009)
46. Tang, Y., Ames, P., Bhamidipati, S., Bijlani, A., Geambasu, R., Sarda, N.: CleanOS: limiting mobile data exposure with idle eviction. In: USENIX Conference on Operating Systems Design and Implementation (2012)

Template Attacks Based on Priori Knowledge

Guangjun Fan[1]([✉]) , Yongbin Zhou[2] , Hailong Zhang[2], and Dengguo Feng[1]

[1] State Key Laboratory of Computer Science, Institute of Software,
Chinese Academy of Sciences, Beijing, China
guangjunfan@163.com, feng@tca.iscas.ac.cn
[2] State Key Laboratory of Information Security, Institute of Information Engineering
Chinese Academy of Sciences, Beijing, China
{zhouyongbin,zhanghailong}@iie.ac.cn

Abstract. Template attacks are widely accepted as the *strongest* side-channel attacks from the information theoretic point of view, and they can be used as a very *powerful* tool to evaluate the physical security of cryptographic devices. Template attacks consist of two stages, the profiling stage and the extraction stage. In the profiling stage, the attacker is assumed to have a large number of power traces measured from the reference device, using which he can accurately characterize signals and noises in different points. However, in practice, the number of profiling power traces may not be sufficient. In this case, signals and noises are not accurately characterized, and the key-recovery efficiency of template attacks is significantly influenced. We show that, the attacker can still make template attacks powerfully enough in practice as long as the priori knowledge about the reference device be obtained. We note that, the priori knowledge is just a prior distribution of the signal component of the instantaneous power consumption, which the attacker can easily obtain from his previous experience of conducting template attacks, from Internet and many other possible ways. Evaluation results show that, the priori knowledge, even if not accurate, can still help increase the power of template attacks, which poses a serious threat to the physical security of cryptographic devices.

Keywords: Side-channel attacks · Power analysis attacks · Template attacks · Priori knowledge

1 Introduction

Template attacks were proposed by Chari et al. in 2002 [1], which consist of two stages, i.e. the profiling stage and the extraction stage. In the profiling stage, the attacker has a reference device identical or similar to the target device, and he can use the reference device to characterize the leakage of the target device. In the extraction stage, the attacker can exploit a small number of power traces measured from the target device to recover the correct (sub)key. In order to make template attacks powerfully enough, the attacker needs to use a large number of power traces to accurately characterize signals and noises in different

© Springer International Publishing Switzerland 2015
M. Yung et al. (Eds.): INTRUST 2014, LNCS 9473, pp. 346–363, 2015.
DOI: 10.1007/978-3-319-27998-5_22

interesting points. However, in practice, the number of profiling traces may be limited. For example, a common countermeasure is to limit the operation times of the reference device, or the key used by the reference device will be refreshed after being used several times. In these scenarios, the attacker can only obtain a limited number of power traces in the profiling stage, and signals and noises are not accurately characterized, which significantly influences the key-recovery efficiency of template attacks.

1.1 Motivations

A natural question is whether or not it is possible to further increase the power of template attacks even if the number of profiling traces is limited? We anticipate that using the priori knowledge about the reference device may be a possible way. The priori knowledge is just a kind of prior distribution of the actual value of the signal component in the instantaneous power consumption. There are many ways that the attacker can obtain the priori knowledge in practice. We show three typical examples here.

Example 1: Assume that the attacker has characterized the power leakages of some cryptographic devices whose leakage characterizations are similar to the reference device. Then, he may obtain the priori knowledge about the reference device. For example, noises in different interesting points are usually assumed to follow the normal distribution. If the attacker can estimate the mean value and the variance of the normal distribution using power traces measured from previous cryptographic devices, then the priori knowledge about the reference device can be obtained.

Example 2: From Internet (e.g. [18, 19]), the attacker may obtain some power traces or other potential useful information (e.g. Signal-to-Noise Ratio) of different devices which are similar to the reference device, using which he can infer the priori knowledge of the reference device (similarly to Example 1).

Example 3: For a sophisticated attacker, after obtaining power traces from the reference device in the profiling stage, he can use the power traces to obtain an interval estimation of the actual value of the signal component and roughly infer the prior distribution is a kind of distribution (e.g. normal distribution or uniform distribution) over the interval.

To sum up, for a seasoned attacker, it is not only reasonable but also realistic for him to possess the priori knowledge about the reference device from a practical point of view. Therefore, we need to consider the power of template attacks when the attacker can not obtain enough power traces from the reference device in the profiling stage *but* has the priori knowledge about the reference device. Specifically, two questions need to be answered. The first question is how can the attacker exploit the priori knowledge during the profiling stage in a theoretically correct and practically feasible way to make template attacks more powerful (i.e. achieve better classification performance)? The second question is whether or not the priori knowledge (even if it may not be very accurate) will make template attacks more powerful really?

Of course, one may ask such question: Why not the attacker exploits the power traces obtained from the similar devices (from his previous experience of conducting template attacks or from Internet) together with the power traces obtained from the reference device to build the templates to make template attacks more powerful? In fact, if one *directly* exploits power traces from the similar devices and the reference device to build the templates, the classification performance of template attacks will be decreased [21]. The reason is that the acquisition campaigns about the devices are different[1] even if the leakage distributions of the similar devices and the reference device are similar [21].

If we can give positive answers to the above two important questions, then in order to make template attacks more powerful in the above scenarios, the attacker can first *extract* the priori knowledge from the power traces obtained from the different but similar devices and then conduct template attacks with the priori knowledge as well as the limited power traces obtained from the reference device. From this point of view, these two questions are worth researching.

1.2 Contributions

Main contributions of our work are two-folds. Firstly, based on the method of Bayes estimation [13], we give a theoretically correct and practically feasible way of exploiting the priori knowledge when the attacker conducts template attacks with limited power traces obtained from the reference device in the profiling stage.

Secondly, we verify our way of exploiting the priori knowledge using both simulated and practical experiments. Evaluation results show that, template attacks will be more powerful if the attacker can possess accurate priori knowledge. Additionally, the more accurate the priori knowledge is, the more powerful template attacks will be. Therefore, with the priori knowledge we can further increase the power of template attacks.

1.3 Related Work

Answers to some practical issues of template attacks were provided by [2], such as how to choose interesting points in an efficient way and how to preprocess noisy data. Choudary et al. proposed efficient methods to avoid possible numerical obstacles when implementing template attacks in [4]. In [10], Hanley et al. presented a variant of template attacks which can be applied to block ciphers when the plaintext and ciphertext are unknown. In [7], template attacks were used to attack a masked implementation. Recently, a simple pre-processing technique of template attacks, normalizing the sample values using the means and variances was evaluated [6]. Standaert et al. [20] showed how to best evaluate profiling and extraction of profiled attacks by using the information theoretic metric and the security metric. Principal Component Analysis (PCA)-based template attacks

---

[1] For example, there exist offsets in the different acquisition campaigns.

were investigated in [3]. However, this kind of template attacks may not improve the classification performance [6]. Therefore, PCA-based template attacks are not widely used in practice. Linear Discriminant Analysis (LDA)-based template attacks were introduced in [8] and depend on the condition of equal covariances [4], which does not hold in most settings. Therefore, it is not a better choice compared with PCA-based template attacks [4]. Up to now, no previous work considered our important questions.

1.4 Organization of This Paper

The rest of this paper is organized as follows. In Sect. 2, we review the concept of template attacks and Bayes estimation. In Sect. 3, we give a reasonable way of exploiting the priori knowledge to make template attacks more powerful. In Sect. 4, we verify the way of exploiting the priori knowledge by both simulated and practical experiments. In Sect. 5, we conclude the whole paper.

2 Preliminaries

Template attacks mainly include: classical template attacks [1] and reduced template attacks (p. 108 in [9]). In this section, we briefly review these two kinds of template attacks and the method of Bayes estimation.

2.1 Classical Template Attacks

We will introduce the two stages of classical template attacks: the profiling stage and the extraction stage.

The Profiling Stage. Assume that there exist K different (sub)keys $key_i, i = 0, 1, \ldots, K - 1$ which need to be classified. Also, there exist K different key-dependent operations O_i, $i = 0, 1, \ldots, K - 1$. Usually, one will generate K templates, one for each key-dependent operation O_i. One can exploit some methods to choose N interesting points $(P_0, P_1, \ldots, P_{N-1})$. The interesting points are those time samples that contain the most information about the characterized key-dependent operations. Each template is composed of a mean vector and a covariance matrix. The mean vector is used to estimate the signal component of side-channel leakages. It is the average signal vector $M_i = (M_i[P_0], \ldots, M_i[P_{N-1}])$ for each one of the key-dependent operations. The covariance matrix is used to estimate the probability density of the noise component at different interesting points. It is assumed that noises at different interesting points approximately follow the multivariate normal distribution. A N dimensional noise vector $n_i(S)$ is extracted from each actual power trace $S = (S[P_0], \ldots, S[P_{N-1}])$ representing the template's key dependency O_i as $n_i(S) = (S[P_0] - M_i[P_0], \ldots, S[P_{N-1}] - M_i[P_{N-1}])$. One computes the $(N \times N)$ covariance matrix C_i from these noise vectors. The probability density of the noises occurring under key-dependent operation O_i is given by the N dimensional multivariate Gaussian distribution $p_i(\cdot)$, where the probability of observing a noise vector $\mathbf{n}_i(\mathbf{S})$ is:

$$p_i(\mathbf{n}_i(\mathbf{S})) = \frac{1}{\sqrt{(2\pi)^N |\mathbf{C}_i|}} exp\Big(-\frac{1}{2}\mathbf{n}_i(\mathbf{S})\mathbf{C}_i^{-1}\mathbf{n}_i(\mathbf{S})^T \Big) \quad \mathbf{n}_i(\mathbf{S}) \in \mathbb{R}^N. \quad (1)$$

In Eq. (1), the symbol $|\mathbf{C}_i|$ denotes the determinant of $\mathbf{C}_i$ and the symbol $\mathbf{C}_i^{-1}$ denotes its inverse.

The Extraction Stage. Assume that one obtains t power traces (denoted by $\mathbf{S}_1, \mathbf{S}_2, \ldots, \mathbf{S}_t$) from the target device in the extraction stage. When the power traces are statistically independent, one will apply maximum likelihood approach on the product of conditional probabilities (p. 156 in [9]), i.e.

$$key_{ck} := argmax_{key_i}\Big\{ \prod_{j=1}^{t} \mathsf{Pr}(\mathbf{S}_j|key_i), i = 0, 1, \ldots, K-1 \Big\},$$

where $\mathsf{Pr}(\mathbf{S}_j|key_i) = p_{f(\mathbf{S}_j, key_i)}(n_{f(\mathbf{S}_j, key_i)}(\mathbf{S}_j))$. The key_{ck} is considered to be the correct (sub)key. The output of the function $f(\mathbf{S}_j, key_i)$ is the index of a key-dependent operation.

2.2 Reduced Template Attacks

In order to avoid numerical obstacles with the inversion of the covariance matrix $\mathbf{C}_i$, one can set the covariance matrix equal to the identity matrix. This essentially means that one does not take the covariances between different interesting points into consideration. A template that only consists of a mean vector is called a *reduced template* (p. 108 in [9]). Correspondingly, template attacks based on reduced templates are called as reduced template attacks. In reduced template attacks, the probability density of the noises occurring under key-dependent operation O_i is given by the distribution $p_i'(\cdot)$, where the probability of observing a noise vector $\mathbf{n}_i(\mathbf{S})$ is:

$$p_i'(\mathbf{n}_i(\mathbf{S})) = \frac{1}{\sqrt{(2\pi)^N}} exp\Big(-\frac{1}{2}\mathbf{n}_i(\mathbf{S})\mathbf{n}_i(\mathbf{S})^T \Big) \quad \mathbf{n}_i(\mathbf{S}) \in \mathbb{R}^N.$$

2.3 Bayes Estimation

In the following, we briefly introduce the method of Bayes estimation [13]. We firstly introduce the definition of Bayes estimators. Then, we introduce how to compute a Bayes estimator.

Suppose an unknown parameter θ is known to have a prior distribution Λ (The prior distribution can be discrete or continuous distribution. In this paper, we only assume the prior distribution is continuous.). Quite generally, suppose that the consequences of estimating $g(\theta)$ by a value $\delta(X)$ (based on some measurements X) are measured by $L(\theta, \delta(X))$. As of the *loss function* L, we shall assume that

$$L(\theta, \delta(X)) \geq 0 \text{ for all } \theta \text{ and } \delta(X),$$

and $L[\theta, g(\theta)] = 0$ for all θ, so that the loss is zero when the correct value is estimated. The accuracy, or rather inaccuracy, of an estimator δ is then measured by the *risk function*

$$R(\theta, \delta) = E_\theta\{L[\theta, \delta(X)]\},$$

the long-term average loss resulting from the use of $\delta(X)$. This defines the risk function as a function of $\delta(X)$. An estimator $\delta(X)$ minimizing

$$r(\Lambda, \delta) = \int R(\theta, \delta) d\Lambda(\theta)$$

is called a *Bayes estimator* with respect to the prior distribution Λ. Note that, the prior distribution Λ is a probability distribution of the parameter θ, that is,

$$\int d\Lambda(\theta) = 1.$$

Now, we will introduce how to compute a Bayes estimator of an unknown parameter θ. Let $\lambda(\theta)$ denote the prior probability density of the parameter θ. The prior probability density of the population (or discrete probability function) is denoted by $f(X; \theta)$. If one extracts n samples $(X_1, X_2, \ldots, X_n)$ from the population, then the probability density of this group of samples is

$$f(X_1; \theta)f(X_2; \theta) \cdots f(X_n; \theta).$$

Thereby, we can compute the marginal density

$$p(X_1, X_2, \ldots, X_n) = \int \lambda(\theta)f(X_1; \theta)f(X_2; \theta) \cdots f(X_n; \theta)d\theta.$$

Then, the following posterior probability density is computed:

$$\lambda(\theta|X_1, \ldots, X_n) = \lambda(\theta)f(X_1; \theta) \cdots f(X_n; \theta)/p(X_1, X_2, \ldots, X_n). \qquad (2)$$

In general, the Bayes estimator of the parameter θ is set to be the mean value of $\lambda(\theta|X_1, \ldots, X_n)$.

3   Using Priori Knowledge to Improve Template Attacks

In this section, we introduce how to use the priori knowledge about the reference device for template attacks. The usage of the priori knowledge for template attacks is the same for both classical template attacks and reduced template attacks.

It is well known that the instantaneous power consumption PC_{total} can be modeled as the sum of an operation-dependent component PC_{op}, a data-dependent component PC_{data}, the electronic noise $PC_{el.noise}$, and a constant component PC_{const} (pp. 62–65 in [9]), i.e.

$$PC_{total} = PC_{op} + PC_{data} + PC_{el.noise} + PC_{const}.$$

The value $PC_{op} + PC_{data}$ (or $PC_{op} + PC_{data} + PC_{const}$) can be viewed as the signal component and the value $PC_{el.noise}$ can be viewed as the noise component. Usually, for each point P_j in an actual power trace, when the operation and the data are all fixed, its power consumption PC_{total} follows a normal distribution $\mathcal{N}(\mu_j, \sigma_j^2)$ and the electronic noise $PC_{el.noise}$ follows the normal distribution $\mathcal{N}(0, \sigma_j^2)$ (pp. 62–65 in [9]). For fixed operation on fixed data, due to $Var(PC_{op}) = Var(PC_{data}) = Var(PC_{const}) = 0$, we have $PC_{op} + PC_{data} + PC_{const} = \mu_j$. The priori knowledge is a kind of prior distribution of the actual value of the signal component μ_j. Due to the existence of the electronic noise, we can reasonably assume the prior distribution of the actual value of μ_j obtained by the attacker is a normal distribution.

There are many ways that the attacker can obtain the prior distribution and we just give out a specific one of them. Considering Example 1 in Sect. 1, for the same position about the target intermediate value, the attacker obtains n samples (For convenience, the samples are denoted by $X_1, \ldots, X_n$.) from power traces obtained from his previous experience of conducting template attacks against different devices which are similar to the reference device. Then, by computing

$$\theta_1 = \frac{1}{n} \cdot \sum_{i=1}^{n} X_i, \quad \theta_2^2 = \frac{1}{n-1} \cdot \sum_{i=1}^{n} (X_i - \theta_1)^2,$$

the attacker can easily obtain the prior distribution which is the normal distribution $\mathcal{N}(\theta_1, \theta_2^2)$. Because the leakage distributions of the devices are very similar to that of the reference device, the prior distribution can be used for the interesting points correspond to the same position about the target intermediate value for the reference device. We note that, compared with traditional template attacks, the computational price of obtaining the priori knowledge about the reference device is very small. This implies that the attacker can obtain the prior distribution easily in practice.

The more accurate the signal component (the value of μ_j) is estimated, the more accurate the noise component (the value $PC_{total} - \mu_j$) will be estimated. For an interesting point, if the signal component and the noise component are accurately estimated, accurate templates (reduced templates) will be built and template attacks (both classical template attacks and reduced template attacks) will be more powerful. In the classical way of building templates (reduced templates), for an interesting point, the attacker computes the mean value of the samples to estimate the actual value of the signal component μ_j. Specifically, for the key-dependent operation O_i, the point P_j is an interesting point and the attacker obtains n power traces $(\mathbf{S}_1, \mathbf{S}_2, \ldots, \mathbf{S}_n)$ from the reference device in the profiling stage. Therefore, the attacker obtains n values of the power consumption of the point P_j, one from each power trace. The n values are denoted by $S_1[P_j], S_2[P_j], \ldots, S_n[P_j]$. The actual value of μ_j is estimated by μ'_j:

$$\mu'_j = M_i[P_j] = \sum_{k=1}^{n} S_k[P_j]/n.$$

However, in our scenario, the attacker not only has n power traces (The power traces are obtained from the reference device. However, the number of the power traces is limited.), but also possesses the priori knowledge about the reference device which can be used to estimate the actual value of μ_j more accurately. Let's consider the most common case. Assume that the attacker knows that the actual value of μ_j follows the normal distribution $\mathcal{N}(\theta_1, \theta_2^2)$ from priori knowledge[1] but does not know what the actual value of μ_j accurately is. The attacker can use the method of Bayes estimation to estimate the actual value of μ_j with the priori knowledge $\mathcal{N}(\theta_1, \theta_2^2)$ in the profiling stage as follows: The attacker computes the probability density of the actual value of the signal component μ_j from priori knowledge as

$$\lambda(\mu_j) = (\sqrt{2\pi}\theta_2)^{-1} exp\left[-\frac{1}{2\theta_2^2}(\mu_j - \theta_1)^2 \right].$$

Moreover, the power consumption of the point P_j satisfies the following probability density function:

$$f(x; \mu_j, \sigma_j) = (\sqrt{2\pi}\sigma_j)^{-1} exp\left[-\frac{1}{2\sigma_j^2}(x - \mu_j)^2 \right].$$

From Eq. (2), the attacker computes the posterior probability density:

$$\lambda(\mu_j|S_1[P_j], \ldots, S_n[P_j]) = C_1 exp\left[-\frac{1}{2\theta_2^2}(\mu_j - \theta_1)^2 - \frac{1}{2\sigma_j^2}\sum_{k=1}^{n}(S_k[P_j] - \mu_j)^2 \right],$$

the constant C_1 only has relation with $\theta_1, \theta_2, \sigma_j, S_1[P_j], \ldots, S_n[P_j]$ and has no relation with μ_j. It has that

$$-\frac{1}{2\theta_2^2}(\mu_j - \theta_1)^2 - \frac{1}{2\sigma_j^2}\sum_{k=1}^{n}(S_k[P_j] - \mu_j)^2 = -\frac{1}{2A^2}(\mu_j - B)^2 + C_2,$$

where $A^2 = \sigma_j^2\theta_2^2/(\sigma_j^2 + n\theta_2^2)$, $B = (nM_i[P_j] + \sigma_j^2\theta_1/\theta_2^2)/(n + \sigma_j^2/\theta_2^2)$, and C_2 has no relation with μ_j. Furthermore, the attacker can obtain

$$\lambda(\mu_j|S_1[P_j], \ldots, S_n[P_j]) = C_3 exp\left[-\frac{1}{2A^2}(\mu_j - B)^2 \right],$$

---

[1] Note that, the normal distribution $\mathcal{N}(\theta_1, \theta_2^2)$ itself may not be very accurate. However, from the priori knowledge, the parameters θ_1, θ_2^2 are all known to the attacker.

where $C_3 = C_1 e^{C_2}$. Because it has that

$$\int_{-\infty}^{+\infty} \lambda(\mu_j | S_1[P_j], \ldots, S_n[P_j]) d\mu_j = 1,$$

hence $C_3 = (\sqrt{2\pi}A)^{-1}$. Up to now, the attacker obtains the Bayes estimator of the actual value of μ_j as

$$\mu_j'' = \frac{n}{n + \sigma_j^2/\theta_2^2} \left(\frac{\sum_{k=1}^n S_k[P_j]}{n} \right) + \frac{\sigma_j^2/\theta_2^2}{n + \sigma_j^2/\theta_2^2} \theta_1. \tag{3}$$

The Eq. (3) shows that if the attacker does not have the priori knowledge (i.e. the prior distribution $\mathcal{N}(\theta_1, \theta_2^2)$), he can only use $\sum_{k=1}^n S_k[P_j]/n$ to estimate the actual value of μ_j. If the attacker does not have power traces obtained from the reference device, he can only use the priori knowledge (i.e. the value θ_1) to estimate the actual value of μ_j. If the attacker has power traces obtained from the reference device as well as the priori knowledge, by equation (3), he will use the weighted average of $\sum_{k=1}^n S_k[P_j]/n$ and θ_1 to estimate the actual value of μ_j under the ratio $n : \sigma_j^2/\theta_2^2$ in the profiling stage. This ratio is reasonable and the relevant reasons are as follows. On one hand, when more power traces are obtained from the reference device by the attacker, the proportion of $\sum_{k=1}^n S_k[P_j]/n$ should be larger. On the other hand, when the value θ_2^2 is smaller (This implies that the prior distribution of the actual value of μ_j is more accurate.), the proportion of θ_1 should be larger. Although the attacker may not know the actual value of σ_j^2 in practice, the Bayes estimator about the actual value of μ_j can still be computed. The reason is that the attacker can reasonably assume that the actual value of σ_j^2 equals to a constant value. Of course, when the attacker knows the actual value of σ_j^2, more accurate Bayes estimation about μ_j can be obtained.

Other details of building templates (reduced templates) remain unchanged. Our way only exploits the priori knowledge to estimate the actual value of the signal component more accurately. We note that, due to the computational price of obtaining and exploiting the priori knowledge is very small, the priori knowledge can easily be used by practical attackers.

4   Experimental Evaluations

For the implementation of a cryptographic algorithm with countermeasures, one usually tries his best to use some approaches to delete the countermeasures from power traces at first. If the countermeasures can be deleted, then one tries to recover the correct (sub)key using some attacks against unprotected implementation. For example, if one has power traces with random delays [11], he may first use the approach proposed in [12] to remove the random delays from power traces and then uses some attacks to recover the correct (sub)key. The approaches of deleting countermeasures from power traces are beyond the scope of this paper. Moreover, considering power traces without any countermeasures shows the upper bound of the physical security of the target cryptographic device. Therefore, we take unprotected AES-128 implementation as an example.

We verified both classical template attacks and reduced template attacks by conducting simulated and practical experiments. In both simulated and practical experiments, we tried to attack the outputs of the S-boxes in the 1st round of AES-128. Before introducing the specific experiment details, we first introduce how to get the prior distribution of the actual value of the signal component for every interesting point for both simulated and practical experiments.

The work [17] showed that reduced template attacks are more powerful compared with classical template attacks when the number of power traces used in the profiling stage is limited. Therefore, we mainly exploit reduced template attacks to exhibit our discoveries (Note that, our method can be used for both classical template attacks and reduced template attacks.).

For simplicity, for both simulated and practical experiments, let n_p denote the number of traces used in the profiling stage and let n_e denote the number of traces used in the extraction stage. In this paper, we use the typical metric *Guessing Entropy* [5] as the metric about the classification performance of template attacks (Many other papers also used Guessing Entropy (e.g. [4,14,15]).).

4.1 How to Get the Priori Knowledge

In order to get the priori knowledge, we simulated the cases that the attacker can obtain the priori knowledge from his previous experience of conducting template attacks against a device similar to the reference device.

For both simulated and practical experiments, we get the prior distribution of the actual value of the signal component for every interesting point using the traces which were generated in the same way as those were used in the two stages of template attacks. In this way, we can clearly give out an upper bound of how powerful template attacks will become by exploiting the priori knowledge.

In both simulated and practical experiments, for each key-dependent operation O_i and each interesting point P_j, we considered the prior distribution under four different levels of accuracy and assumed the prior distribution is a normal distribution $\mathcal{N}(\theta_1, \theta_2^2)$ (For different interesting points, the corresponding prior distributions are different.).

For each key-dependent operation O_i, we generated 400 traces (simulated traces or actual power traces). The 400 traces were used to estimate the prior distributions for every interesting point as follows. We repeated the following process 300 times. Every time, we chose m traces (denoted by $S_1, \ldots, S_m$) from the 400 traces uniformly at random and computed $\sum_{k=1}^{m} S_k[P_j]/m$. Therefore, there were 300 different values about $\sum_{k=1}^{m} S_k[P_j]/m$. The mean value of the 300 different values was set to be θ_1 and the variance of the 300 different values was set to be θ_2^2. In this way, the prior distribution $\mathcal{N}(\theta_1, \theta_2^2)$ was got. Note that, in practice, the attacker has many ways to get the prior distribution $\mathcal{N}(\theta_1, \theta_2^2)$. Our method which were used in this paper is just one of them. We respectively let $m = 16, 32, 64, 128$ and obtained four different estimation of the prior distribution.

Clearly, when the value m is larger, the estimation of θ_1 and θ_2^2 is more accurate. Therefore, we obtained four different prior distributions under different levels of accuracy, which represent the priori knowledge that the attacker can possess in practical attack scenarios.

We considered many kinds of template attacks and define the following symbols to denote them. In all the experiments, we let the symbol "CTA" denotes the classical template attacks without any priori knowledge. The symbol "CTA-16" denotes classical template attacks based on priori knowledge which is obtained when the value m equals to 16. Similarly, we define the symbols "CTA-32", "CTA-64", and "CTA-128" to denote the cases that the value m equals to 32, 64, and 128 respectively. We let the symbol "RTA" denotes the reduced template attacks without any priori knowledge. The symbol "RTA-16" denotes reduced template attacks based on priori knowledge which is obtained when the value m equals to 16. Similarly, we define the symbols "RTA-32", "RTA-64", and "RTA-128" to denote the cases that the value m equals to 32, 64, and 128.

4.2   Simulated Experiments

In simulated experiments, we chose 4 interesting points and the typical Hamming-Weight power model (pp. 40–41 in [9]) was adopted to describe the power consumption. The standard deviation of simulated Gaussian noise is denoted by σ. We employed three different noise levels to test the influence of noises on the classification performance of template attacks. The standard deviations of simulated Gaussian noise for the three noise levels were 2, 3, and 4.

For each noise level, we respectively used 2,000 and 4,000 simulated traces to build the 256 reduced templates in the profiling stage for the five kinds of reduced template attacks (RTA, RTA-16, RTA-32, RTA-64, and RTA-128). This means that the attacker respectively obtained 2,000 and 4,000 traces from the reference device in the profiling stage. The simulated traces used in the profiling stage were generated with a fixed subkey and random plaintext inputs. We generated additional 100,000 simulated traces with another fixed subkey and random plaintext inputs under each noise level. The 100,000 simulated traces were used in the extraction stage. For fixed n_p and σ, we tested the Guessing Entropy of the five kinds of reduced template attacks when the attacker could use n_e simulated traces in the extraction stage as follows. We respectively repeated the five kinds of reduced template attacks 1,000 times. For each time, we chose n_e simulated traces from the 100,000 simulated traces uniformly at random and the five kinds of reduced template attacks were conducted with the same n_e simulated traces. We respectively computed the Guessing Entropy of the five kinds of reduced template attacks with the results of the 1,000 times attacks. The Guessing Entropy of the five kinds of reduced template attacks for different values of n_p and σ is shown in Fig. 1.

The Guessing Entropy of the five kinds of reduced template attacks for the case $\{n_p = 2,000, n_e = 20, \sigma = 4\}$ is shown in Table 1. From Fig. 1 and Table 1, we find that the classification performance of reduced template attacks with accurate priori knowledge will be obvious better than that of reduced template

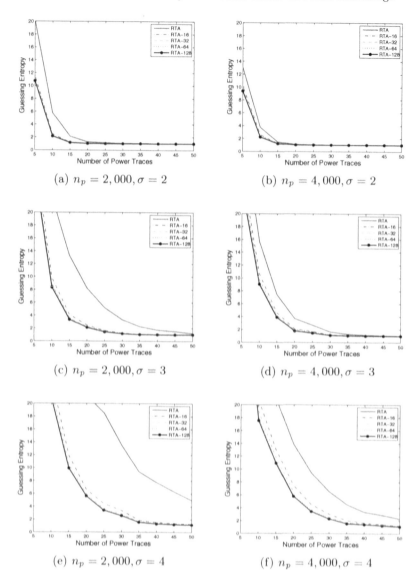

Fig. 1. The simulated experiment results

Table 1. The simulated experiment results for the case $n_p = 2,000, n_e = 20, \sigma = 4$

RTA	RTA-16	RTA-32	RTA-64	RTA-128
21.22	6.66	5.97	5.68	5.61

attacks without priori knowledge. For example, in Table 1, the Guessing Entropy of RTA equals to 21.22, while the Guessing Entropy of RTA-128 equals to 5.61. Moreover, if the priori knowledge is more accurate, the classification performance of reduced template attacks with priori knowledge will be better. For example, in Table 1, the Guessing Entropy of RTA-16 equals to 6.66, while the Guessing Entropy of RTA-128 obviously reduces to 5.61.

Table 2. The simulated experiment results for different levels of noises

$n_p = 2,000, n_e = 20$	$\sigma = 2$	$\sigma = 3$	$\sigma = 4$
RTA	1.27	8.23	21.22
RTA-128	1.03	2.11	5.61

Table 2 shows the Guessing Entropy of RTA and RTA-128 for different levels of noises when n_p is fixed to 2,000 and n_e is fixed to 20. From Fig. 1 and Table 2, we further find that, when the noise level is higher, reduced template attacks with priori knowledge will achieve larger advantage over reduced template attacks without priori knowledge. For example, in Table 2, the Guessing Entropy of RTA and RTA-128 is almost equal when σ equals to 2 (1.27 and 1.03). However, when σ equals to 4, the Guessing Entropy of RTA-128 (5.61) is much lower than that of RTA (21.22).

When more simulated traces can be obtained from the reference device (e.g. $n_p = 4,000$) in the profiling stage, the advantages of reduced template attacks with priori knowledge over template attacks without priori knowledge will be smaller. For classical template attacks, we computed the Guessing Entropy of the five kinds of classical template attacks (CTA, CTA-16, CTA-32, CTA-64, and CTA-128) similarly. The simulated experiment results show that classical template attacks with accurate priori knowledge have advantages over classical template attacks without priori knowledge.

4.3   Practical Experiments

We tried to attack the outputs of all the S-boxes in the 1st round of an unprotected AES-128 software implementation on an typical 8-bit microcontroller STC89C58RD+ whose operating frequency is 11MHz. The actual power traces were acquired with a sampling rate of 50MS/s. The average number of actual power traces during the sampling process was 10 times. For our device, the condition of equal covariances [4] does not hold.

We generated two sets of actual power traces, Set A and Set B. The Set A captured 10,000 power traces which were generated with a fixed main key and random plaintext inputs. The Set B captured 100,000 power traces which were generated with another fixed main key and random plaintext inputs. The power traces in Set A were used in the profiling stage and the power traces in Set B were used in the extraction stage. The device that was used to generate the two sets of actual power traces is the same as that was used to get the prior distribution in Sect. 4.1, which provides a good setting for the focuses of our research. In this way, we can show the actual and the greatest threats caused by the priori knowledge. For each S-box of the unprotected AES-128 software implementation, we chose 4 interesting points in 4 continual clock cycles, one in each clock cycle[1]. Both classical template attacks and reduced template attacks were conducted based on the same 4 interesting points. We only show the practical experiment results of the 1st and the 2nd S-box in this paper. For other S-boxes in the 1st round, similar evaluation results were obtained by us. In all practical experiments, we reasonably assumed that the actual value of σ_j^2 equals to a constant value for each interesting point and each target intermediate value.

For reduced template attacks, we respectively chose 2,000 and 4,000 different power traces from Set A to build the 256 templates for the five kinds of reduced template attacks (RTA, RTA-16, RTA-32, RTA-64, and RTA-128). The 100,000 power traces of Set B were used in the extraction stage for the five kinds of reduced template attacks. For fixed n_p, we tested the Guessing Entropy of the five kinds of reduced template attacks when one uses n_e power traces in the extraction stage similarly to that of the simulated experiments but used actual power traces. The Guessing Entropy of the five kinds of reduced template attacks for the 1st S-box are shown in Fig. 2. The Guessing Entropy of the five kinds of reduced template attacks for the 1st S-box when n_p is fixed to 2,000 and n_e is fixed to 20 is shown in Table 3.

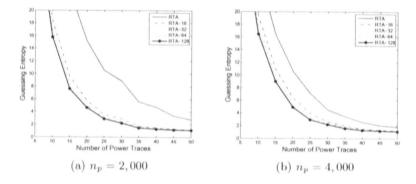

(a) $n_p = 2,000$ (b) $n_p = 4,000$

Fig. 2. The experiment results of reduced template attacks for the 1st S-box

From Fig. 2 and Table 3, we find that the classification performance of reduced template attacks with accurate priori knowledge will be obvious better than that of reduced template attacks without priori knowledge. For example, in Table 3, the Guessing Entropy of RTA equals to 15.16, while the Guessing Entropy of RTA-16 reduces to 5.78.

For classical template attacks, in order to avoid numerical obstacles with the inversion of the covariance matrix, we respectively chose 5,000 and 10,000 different power traces from Set A to build the 256 templates for the five kinds of classical template attacks (CTA, CTA-16, CTA-32, CTA-64, and CTA-128).

---

[1] In our device, the target intermediate values only continue 4 clock cycles.

Table 3. The experiment results of reduced template attacks for the 1st S-box

$n_p = 2,000$	RTA	RTA-16	RTA-32	RTA-64	RTA-128
$n_e = 20$	15.16	5.78	5.03	4.73	4.65

Moreover, using power traces from Set B, we computed the Guessing Entropy of the five kinds of classical template attacks when one uses n_e power traces in the extraction stage similarly. The Guessing Entropy of the five kinds of classical template attacks for the 1st S-box are shown in Fig. 3 in Appendix A.

For the 2nd S-box, we also used the actual power traces in Set A and Set B to compute the Guessing Entropy of the five kinds of reduced template attacks and the five kinds of classical template attacks similarly. The practical experiment results for the 2nd S-box which can also verify our discoveries are shown in Figs. 4, 5 and Table 4 in Appendix B.

The practical experiment results show that, for both reduced template attacks and classical template attacks, if the priori knowledge is more accurate, the classification performance will be better. For example, in Table 3, the Guessing Entropy of RTA-16 equals to 5.78, while the Guessing Entropy of RTA-128 reduces to 4.65. When more power traces can be obtained from the reference device, the advantages of template attacks with priori knowledge over template attacks without priori knowledge will be smaller.

5   Conclusion and Future Work

In this paper, we show that leaking the priori knowledge about the reference device poses serious threat to the physical security of cryptographic devices. Therefore, we suggest that the designers of a cryptographic device should take the priori knowledge into consideration when he uses template attacks to evaluate the physical security of the cryptographic device. The future work is as follows. First, our discoveries show that the approach to infer (estimate) the priori knowledge as accurately as possible is crucial and is worth being researched from the attacker's point of view. Second, it would be interesting to research how to prevent the attacker to obtain the priori knowledge (Using countermeasures such as the random delays [11] may be a good choice.). We should also concern on how to exploit the priori knowledge to make other profiled side-channel attacks (such as stochastic model based attacks [16], PCA-based template attacks etc.) become more powerful in a reasonable way. It is also necessary to further verify our discoveries in other devices such as ASIC and FPGA.

Acknowledgments. This work was supported by the National Basic Research Program of China (No.2013CB338003), the National Natural Science Foundation of China (Nos.61472416, 61272478), and the National Key Scientific and Technological Project (No.2014ZX01032401-001).

Appendix A: Practical Experiments for the 1st S-box

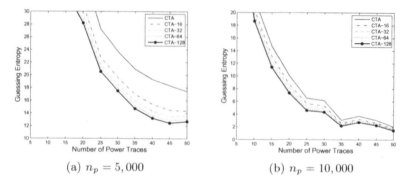

(a) $n_p = 5,000$ (b) $n_p = 10,000$

Fig. 3. The experiment results of classical template attacks for the 1st S-box

Appendix B: Practical Experiments for the 2nd S-box

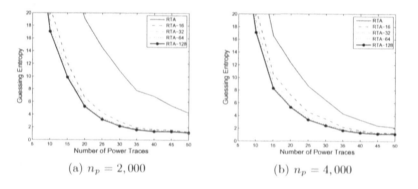

(a) $n_p = 2,000$ (b) $n_p = 4,000$

Fig. 4. The experiment results of reduced template attacks for the 2nd S-box

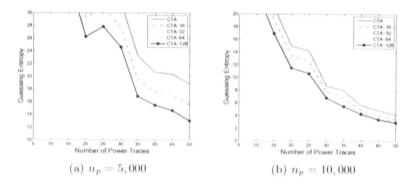

(a) $n_p = 5,000$ (b) $n_p = 10,000$

Fig. 5. The experiment results of classical template attacks for the 2nd S-box

Table 4. The experiment results of reduced template attacks for the 2nd S-box

$n_p = 2,000$	RTA	RTA-16	RTA-32	RTA-64	RTA-128
$n_e = 20$	19.05	6.64	5.76	5.34	5.25

References

1. Chari, S., Rao, J.R., Rohatgi, P.: Template attacks. In: Kaliski Jr., B.S., Koç, Ç.K., Paar, C. (eds.) CHES 2002. LNCS, vol. 2523, pp. 13–28. Springer, Heidelberg (2003)
2. Rechberger, C., Oswald, E.: Practical template attacks. In: Lim, C.H., Yung, M. (eds.) WISA 2004. LNCS, vol. 3325, pp. 440–456. Springer, Heidelberg (2005)
3. Archambeau, C., Peeters, E., Standaert, F.-X., Quisquater, J.-J.: Template attacks in principal subspaces. In: Goubin, L., Matsui, M. (eds.) CHES 2006. LNCS, vol. 4249, pp. 1–14. Springer, Heidelberg (2006)
4. Choudary, O., Kuhn, M.G.: Efficient template attacks. In: Francillon, A., Rohatgi, P. (eds.) CARDIS 2013. LNCS, vol. 8419, pp. 253–270. Springer, Heidelberg (2014)
5. Standaert, F.-X., Malkin, T.G., Yung, M.: A unified framework for the analysis of side-channel key recovery attacks. In: Joux, A. (ed.) EUROCRYPT 2009. LNCS, vol. 5479, pp. 443–461. Springer, Heidelberg (2009)
6. Montminy, D.P., Baldwin, R.O., Temple, M.A., Laspe, E.D.: Improving cross-device attacks using zero-mean unit-variance mormalization. J. Cryptographic Eng. **3**(2), 99–110 (2013)
7. Oswald, E., Mangard, S.: Template attacks on masking—resistance is futile. In: Abe, M. (ed.) CT-RSA 2007. LNCS, vol. 4377, pp. 243–256. Springer, Heidelberg (2006)
8. Standaert, F.-X., Archambeau, C.: Using subspace-based template attacks to compare and combine power and electromagnetic information leakages. In: Oswald, E., Rohatgi, P. (eds.) CHES 2008. LNCS, vol. 5154, pp. 411–425. Springer, Heidelberg (2008)
9. Mangard, S., Oswald, E., Popp, T.: Power Analysis Attacks: Revealing the Secrets of Smart Cards. Springer, New York (2007)
10. Hanley, N., Tunstall, M., Marnane, W.P.: Unknown plaintext template attacks. In: Youm, H.Y., Yung, M. (eds.) WISA 2009. LNCS, vol. 5932, pp. 148–162. Springer, Heidelberg (2009)
11. Coron, J.-S., Kizhvatov, I.: Analysis and improvement of the random delay countermeasure of CHES 2009. In: Mangard, S., Standaert, F.-X. (eds.) CHES 2010. LNCS, vol. 6225, pp. 95–109. Springer, Heidelberg (2010)
12. Durvaux, F., Renauld, M., Standaert, F.-X., van Oldeneel tot Oldenzeel, L., Veyrat-Charvillon, N.: Efficient removal of random delays from embedded software implementations using hidden markov models. In: Mangard, S. (ed.) CARDIS 2012. LNCS, vol. 7771, pp. 123–140. Springer, Heidelberg (2013)
13. Lehmann, E.L., Casella, G.: Theory of Point Estimation, 2nd edn. Springer, New York. ISBN 978-0-387-98502-6
14. Standaert, F.-X., Gierlichs, B., Verbauwhede, I.: Partition vs. comparison side-channel distinguishers: an empirical evaluation of statistical tests for univariate side-channel attacks against two unprotected CMOS devices. In: Lee, P.J., Cheon, J.H. (eds.) ICISC 2008. LNCS, vol. 5461, pp. 253–267. Springer, Heidelberg (2009)

15. Medwed, M., Standaert, F.-X., Joux, A.: Towards super-exponential side-channel security with efficient leakage-resilient PRFs. In: Prouff, E., Schaumont, P. (eds.) CHES 2012. LNCS, vol. 7428, pp. 193–212. Springer, Heidelberg (2012)
16. Schindler, W., Lemke, K., Paar, C.: A stochastic model for differential side channel cryptanalysis. In: Rao, J.R., Sunar, B. (eds.) CHES 2005. LNCS, vol. 3659, pp. 30–46. Springer, Heidelberg (2005)
17. Ye, X., Eisenbarth, T.: Wide collisions in practice. In: Bao, F., Samarati, P., Zhou, J. (eds.) ACNS 2012. LNCS, vol. 7341, pp. 329–343. Springer, Heidelberg (2012)
18. The DPA Contest. http://www.dpacontest.org/home/
19. Power analysis attacks-revealing the secrets of smartcards. http://dpabook.org/
20. Standaert, F.-X., Koeune, F., Schindler, W.: How to compare profiled side-channel attacks? In: Abdalla, M., Pointcheval, D., Fouque, P.-A., Vergnaud, D. (eds.) ACNS 2009. LNCS, vol. 5536, pp. 485–498. Springer, Heidelberg (2009)
21. Choudary, O., Kuhn, M.G.: Template attacks on different devices. In: Prouff, E. (ed.) COSADE 2014. LNCS, vol. 8622, pp. 179–198. Springer, Heidelberg (2014)

Some Observations on the Lightweight Block Cipher Piccolo-80

Wenying Zhang[1,2(✉)], Jiaqi Zhang[1], and Xiangqian Zheng[1]

[1] School of Information Science and Engineering,
Shandong Normal University, Jinan 250014, China
[2] Science and Technology on Information Assume Laboratory, Beijing 100072, China
wenyingzh@sohu.com

Abstract. Piccolo is a 64-bit lightweight block cipher proposed by SONY corporation to be used in the constrained environments such as wireless sensor net work environments. In this paper, by algebraic analysis, we give some observations on Piccolo, including the linear analysis of the F-function, and a weakness of key scheduling. We found that the F-function could be matched with linear permutation with high probability. We revealed the statistical character of the F- function, which gives the attackers chance to distinguish piccolo from random permutation. We attack two rounds Piccolo-80 with the computational complexity 2^{17} two rounds Piccolo-80 encryptions. We found that the subkeys in last two rounds of Piccolo-80 do not play the roles of hide information of internal states well, 16 bits of cipher text can be represented by the state of last but one round.

Keywords: Block cipher · Piccolo · Linear analysis · Key scheduling weakness

1 Introduction

Cryptographic techniques move into applications like sensor nodes, radio frequency identification tags and integrated circuit (IC) printing. The ever increasing demand for security and privacy in these very constrained environments requires new cryptographic primitives, like low cost, tiny and efficient ciphers. As a result, some new block ciphers were presented. Such as Lblock [1], PRINTcipher [2], LED [3] and Piccolo [4].

A 64-bit block cipher Piccolo proposed in CHES 2011 by SONY corporation supports 80- and 128- bit secret keys [4]. According to the length of the secret key, they are denoted by Piccolo-80 and Piccolo-128, respectively. The number of rounds of Piccolo-80 and Piccolo-128 is 25 and 31, respectively. The iterative

Funded by the National Science Foundation of China (No. 61272434), the Natural Science Foundation of Shandong Province (No. ZR2012FM004, ZR2013FQ021) and the Foundation of Science and Technology on Information Assume Laboratory (No. KJ-13-004).

M. Yung et al. (Eds.): INTRUST 2014, LNCS 9473, pp. 364–373, 2015.
DOI: 10.1007/978-3-319-27998-5_23

structure of Piccolo is a variant of generalized Feistel network. Until now, several cryptanalytic results on them were proposed. In [4], the strength of Piccolo against some well-known attacks were examined by the designers, including differential cryptanalysis, linear cryptanalysis, impossible differential cryptanalysis, truncated differential cryptanalysis, related-key cryptanalysis, Boomerang cryptanalysis and some other well-known attacks. The best result of actual single-key attack is 3-Subset Meet-in-the-Middle(MITM)attacks on a 14-round reduced Piccolo- 80 and a 21-round reduced Piccolo-128 without whitening keys [5]. Wang et al. introduced a biclique cryptanalysis of the full round Piccolo-80 without post whitening keys and a 28-round Piccolo-128 without prewhitening keys [6]. These attacks are respectively with data complexity of 2^{48} and 2^{24} chosen cipher texts, and with time complexity of $2^{78.95}$ and $2^{126.79}$ encryptions. In [7], the authors give differential fault analysis on Piccolo. In [8], the authors evaluate the security of Piccolo, their attacks on Piccolo-80/128 require computational complexities of $2^{79.13}$ and $2^{127.35}$, respectively.

In this paper, we give some reports on Piccolo that the F-function could be matched with linear permutation with high probability and certain weakness of the key schedule. By linear analysis, we attack the two rounds Piccolo with the computational complexity of 2^{17} two rounds Piccolo-80 encryptions. The weakness of the key schedule of Piccolo 80 is the whitening key and the round key are canceled on part of the cipher, and the cipher text leak 16 bits information of the internal state of round 23. We give a proposal to improve the key schedule. We also evaluate the key scheduling of Piccolo-128.

This paper is organized as follows. In Sect. 2, we briefly introduce the structure of Piccolo and give the inverse function of F. The linear approximation representation of the round function of Piccolo-80 is given in Sect. 3. Our attacks on two rounds Piccolo based on linear analysis are presented in Sect. 4. In Sect. 5, a weakness of the key schedule is presented. Finally, we give our conclusion in Sect. 6.

2   Description of Piccolo

2.1   Notations

$a|b$ or $(a|b)$: Concatenation.
$a \leftarrow b$: Updating a value of a by a value of b.
a^t : Transposition of a vector or a matrix a.
a^L: The left half bits of the string, a^R: the right half bits of the string.
i^L: The left half bits of k_i, i^R: the right half bits of k_i, $0 \le i < 5$.
$S = (y_0, y_1, y_2, y_3)$: a 64-bit intermediate state. y_i^j the j-th bit of y_i.
$C = (C_0, C_1, C_2, C_3)$: a 64-bit ciphertext.

2.2   Data Processing Part

Piccolo-80 is a 64-bit block cipher supporting 80-bit keys proposed by Shibutani et al. in 2011. Piccolo-80 consist of 25 rounds of a variant of a generalized Feistel network. The encryption function is defined as follows:

$X_0 \mid X_1 \mid X_2 \mid X_3 \leftarrow$ Plain text

$$X_0 \leftarrow X_0 \oplus wk_0, X_2 \leftarrow X_2 \oplus wk_1,$$

for $i \leftarrow 0$ to $r\text{-}2$ do

$$X_1 \leftarrow X_1 \oplus F(X_0) \oplus rk_{2i}, X_3 \leftarrow X_3 \oplus F(X_2) \oplus rk_{2i+1}$$
$$X_1^L \mid X_3^R \mid X_2^L \mid X_0^R \mid X_3^L \mid X_1^R X_0^L \mid X_2^R$$
$$X_1 \leftarrow X_1 \oplus F(X_0) \oplus rk_{2r-2}, X_3 \leftarrow X_3 \oplus F(X_2) \oplus rk_{2r-1}$$
$$X_0 \leftarrow X_0 \oplus wk_2, X_2 \leftarrow X_2 \oplus wk_3,$$

Cipher text $\leftarrow X_0 \mid X_1 \mid X_2 \mid X_3$.

Where F is a 16-bit F-function defined in the following:

F-function $F : \{0,1\}^{16} \rightarrow \{0,1\}^{16}$ consists of two S-box layers separated by a diffusion matrix M. The S-box layer consists of four 4-bit bijective S-boxes S. S is given by S = Array(14, 4, 11, 2, 3, 8, 0, 9, 1, 10, 7, 15, 6, 12, 5, 13), and updates a 16-bit data X by $(x_0, x_1, x_2, x_3) \leftarrow (S(x_0), S(x_1), S(x_2), S(x_3))$.

The diffusion matrix for M is

$$M = \begin{pmatrix} 2 & 3 & 1 & 1 \\ 1 & 2 & 3 & 1 \\ 1 & 1 & 2 & 3 \\ 3 & 1 & 1 & 2 \end{pmatrix}.$$

Where the multiplications between matrix and vectors are performed over $GF(2^4)$ defined by an irreducible polynomial $x^4 + x + 1$. The diffusion function updates a 16-bit data X as follows: $(x_0, x_1, x_2, x_3)^t \leftarrow M(x_0, x_1, x_2, x_3)^t$.

Constant Values. The constants con_i used in the key scheduling is a 16 bits binary string which is generated as follows:

$$Con_{2i} \mid Con_{2i+1} = (i+1)|00000|(i+1)|00|(i+1)|00000|(i+1) \oplus (0f1e2d3c),$$

where $i+1$ is a 5-bit representation of $i+1$, e.g., $12 = 01100$.

The key scheduling function for the 80-bit key mode divides an 80-bit key K into five 16-bit subkeys $k_i, (0 \leq i < 5)$ and provides $wk_i (0 \leq i < 4)$ and $rk_j (0 \leq j < 2r)$ as follows:

Whitening key.

$$wk_0 \leftarrow k_0^L \mid k_1^R, wk_1 \leftarrow k_1^L \mid k_0^R, wk_2 \leftarrow k_4^L \mid k_3^R, wk_3 \leftarrow k_3^L \mid k_4^R$$

Round key. For $i = 0$ to $(r-1)$ do

$$(rk_{2i}, rk_{2i+1}) \leftarrow (con_{2i}, con_{2i+1}) \oplus \begin{cases} (k_2, k_3) \ (if \ i \ mod \ 5 \ = 0 \ or \ 2) \\ (k_0, k_1) \ (if \ i \ mod \ 5 = 1 \ or \ 4) \\ (k_4, k_4) \ (if \ i \ mod \ 5 = 3) \end{cases}$$

From the round key scheduling we see that the last two round keys are (k_4, k_4) for round 23, and (k_0, k_1) for round 24, and the whitening key for the last round is $wk_2 \leftarrow k_4^L \mid k_3^R, wk_3 \leftarrow k_3^L \mid k_4^R$. Among the last used six 16-bit round keys (whitening keys), k_4 appears three times, this is the vulnerability of Piccolo-80.

2.3   The Inverse Function of F

In this subsection, we study the inverse function of F. $F^{-1} : \{0,1\}^{16} \to \{0,1\}^{16}$ is also consists of two S^{-1}-box layers separated by a diffusion matrix M^{-1}(See Fig. 1). By Gauss-Jordan elimination method, we get the inverse of the matrix M.

Theorem 1. $M^{-1} = \begin{pmatrix} 14 & 11 & 13 & 9 \\ 9 & 14 & 11 & 13 \\ 13 & 9 & 14 & 11 \\ 11 & 13 & 9 & 14 \end{pmatrix}.$

Proof. $2 \times 14 + 3 \times 9 + 13 + 11 = x \cdot x^{11} + x^4 \cdot x^{14} + 13 + 11 = 15 + 8 + 13 + 11 = 1$

$2 \times 11 + 3 \times 14 + 9 + 13 = x \cdot x^7 + x^4 \cdot x^{11} + 9 + 13 = 5 + 1 + 9 + 13 = 0,$

and the other inner products of the rows of M and the columns of M^{-1} can be verified.

It is obvious that $S^{-1} = Array(6, 8, 3, 4, 1, 14, 12, 10, 5, 7, 9, 2, 13, 15, 0, 11)$. And hence $F^{-1} = S^{-1} \circ M^{-1} \circ S^{-1}$

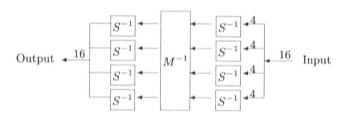

Fig. 1. The F^{-1} function

3   The Linear Approximation Representation of the Round Function

Let n be any positive integer, and let F_2 be the Galois field of two elements, i.e., $F_2 = \{0,1\}$. An n-variable Boolean function is a mapping $f : \{0,1\}^n \to \{0,1\}$.
    An n-variable Boolean function $f(x_1, \cdots, x_n)$ can be represented as a multivariate polynomial over F_2 uniquely, called *algebraic normal form* (ANF),

$$a_0 + \sum_{i=1}^{n} a_i x_i + \sum_{1 \le i < j \le n} a_{ij} x_i x_j + \cdots + a_{12\cdots n} x_1 x_2 \cdots x_n,$$

where the coefficients $a_0, a_i, a_{ij}, \cdots, a_{12\cdots n} \in F_2$, the addition and multiplication operations are in F_2.

The ANF of S-box in Piccolo is $S(x_0, x_1, x_2, x_3) = ((x_0+1)(x_1+1)+x_3, (x_1+1)(x_2+1)+x_0, x_0+x_3+x_0x_1+x_0x_2+x_1x_2+x_2x_3+x_0x_1x_2, x_0+x_1+x_2+x_0x_2+x_0x_3+x_1x_3+x_2x_3+x_0x_1x_2+x_1x_2x_3) = (S_0, S_1, S_2, S_3) = ((x_0+1)(x_1+1)+x_3, (x_1+1)(x_2+1)+x_0, (S_0+1)(x_2+1)+x_1+1, (S_0+1)(S_1+1)+x_2)).$

Note that the last two components of the S-box can be simply represented by the first two components functions, this is a novel and interesting character of the S boxes in the new proposed block ciphers. Although the ANF of S-box is very complex, S box formulation $((x_0+1)(x_1+1)+x_3, (x_1+1)(x_2+1)+x_0, (S_0+1)(x_2+1)+x_1+1, (S_0+1)(S_1+1)+x_2))$ in its first two output components is very simply. This simply formulation made it to be efficiency implemented both for software and hardware.

It is not unique, but has its counterpart. The light weight block cipher PRIDE designed by Albrecht et al., presented in CRYPTO2014 also employs the similar S box. The designers claim that their constructing method is good both in security and efficiency. The formulation of S-box in PRIDE is $A = c \oplus ab, B = d \oplus bc, C = a \oplus AB, D = b \oplus BC$, where a, b, c, d and A, B, C, D are the four components of the inputs and outputs of the S-box respectively.

From the ANF of S-box, (x_3, x_0, x_1+1, x_2) is the best linear approximation representation of S-box. In fact, with the value table of S box and the linear mapping $L(x_0, x_1, x_2, x_3) = (x_3, x_0, x_1+1, x_2)$,

x	0	1	2	3	4	5	6	7	8	9	A	B	C	D	E	F
$S(x)$	E	4	B	2	3	8	0	9	1	A	7	F	6	C	5	D
$L(x)$	2	A	3	B	0	8	1	9	6	E	7	F	4	C	5	D

it can be observed that the S box matches with linear permutation with probability 7/16.

The ANF of x multiplied by 2 or 3 over finite field F_2^4 are:
$2(x_0, x_1, x_2, x_3) = (x_1, x_2, x_0+x_3, x_0),$
$3(x_0, x_1, x_2, x_3) = (x_0+x_1, x_1+x_2, x_0+x_2+x_3, x_0+x_3).$

After multiply matrix layer, the first four bits of outputs can be linearly approximated as following $2(x_0, x_1, x_2, x_3)+3(x_4, x_5, x_6, x_7)+(x_8, x_9, x_{10}, x_{11})+(x_{12}, x_{13}, x_{14}, x_{15}) = (x_1, x_2, x_0+x_3, x_3) + (x_4+x_5, x_5+x_6, x_4+x_6+x_7, x_4+x_7) + (x_8, x_9, x_{10}, x_{11}) + (x_{12}, x_{13}, x_{14}, x_{15}) = (x_1+x_4+x_5+x_8+x_{12}, x_2+x_5+x_6+x_9+x_{13}, x_0+x_3+x_4+x_6+x_7+x_{10}+x_{14}, x_3+x_4+x_7+x_{11}+x_{12}).$ After the second S box the 16 bits output of F-function can be linearly approximated as following:

$$v_0 = x_3 + x_6 + x_7 + x_{10} + x_{14}, \quad 33792$$
$$v_1 = x_0 + x_4 + x_7 + x_{11} + x_{15}, \quad 33728$$
$$v_2 = x_1 + x_4 + x_5 + x_8 + x_{12} + 1, \quad 33856$$
$$v_3 = x_2 + x_3 + x_5 + x_6 + x_7 + x_9 + x_{13} + 1,$$
$$v_4 = x_2 + x_7 + x_{10} + x_{11} + x_{14},$$
$$v_5 = x_3 + x_4 + x_8 + x_{11} + x_{15},$$
$$v_6 = x_0 + x_5 + x_8 + x_9 + x_{12} + 1,$$
$$v_7 = x_1 + x_6 + x_7 + x_9 + x_{10} + x_{11} + x_{13} + 1,$$
$$v_8 = x_2 + x_6 + x_{11} + x_{14} + x_{15},$$
$$v_9 = x_3 + x_7 + x_8 + x_{12} + x_{15},$$
$$v_{10} = x_0 + x_4 + x_9 + x_{12} + x_{13} + 1,$$
$$v_{11} = x_1 + x_5 + x_{10} + x_{11} + x_{13} + x_{14} + x_{15} + 1,$$
$$v_{12} = x_2 + x_3 + x_6 + x_{10} + x_{15},$$
$$v_{13} = x_0 + x_3 + x_7 + x_{11} + x_{12},$$
$$v_{14} = x_0 + x_1 + x_4 + x_8 + x_{13} + 1,$$
$$v_{15} = x_1 + x_2 + x_3 + x_5 + x_9 + x_{14} + x_{15} + 1.$$

Where $(x_0, x_1, \cdots, x_{15})$ are the 16 bits input of F. By computation on a computer, the first component of F equals to the boolean function v_0 at 33792 vectors in $F^{16}(2)$, which is greater than 32768. Here 33728 and 33856 are the numbers of i'th $i = 1, 2$ component of F match with v_i among the 65536 vectors. Hence we get a linear approximation representation $(v_0, v_1, \cdots, v_{15})$ of F-function. This can be applied to distinguish the round function of Piccolo from random permutation.

4   The Linear Cryptanlysis of Two Rounds Piccolo

We give the algebraic representation of internal states firstly.

4.1   The Algebraic Representation of Internal States

Denote the 64 bits plain text of Piccolo by

$$P = (P_0, P_1, P_2, P_3) = (y_0, y_1, y_2, y_3) \in F_2^{64}, \tag{1}$$

where $y_i, i = 0, 1, 2, 3$ are 16 bits binary strings. After the F-layer the internal state is

$$y_0, F(y_0) + y_1 + wk_0 + k_2, y_2, F(y_2) + y_3 + wk_1 + k_3. \tag{2}$$

After the round permutation, the output of round 0 is

$$F(y_0)^L + y_1^L + k_2^L, F(y_2)^R + y_3^R + k_3^R, y_2^L, y_0^R, F(y_2)^L + y_3^L + k_3^L, F(y_0)^R + y_1^R + k_3^R, y_0^L, y_2^R.$$

After the F-function, the output of first two rounds is

$$(C_0, C_1, C_2, C_3) = ([F(y_0 + wk_0) + y_1 + rk_0]^L, [F(y_2 + wk_1) + y_3 + rk_1]^R,$$
$$F\{[F(y_0 + wk_0) + y_1 + rk_0]^L, [F(y_2 + wk_1) + y_3 + rk_1]^R\} + y_2{}^L y_0^R + rk_2,$$
$$[F(y_2 + wk_1) + y_3 + rk_1]^L, [F(y_0 + wk_0) + y_1 + rk_0]^R,$$
$$F\{[F(y_2 + wk_1) + y_3 + rk_1]^L, [F(y_0 + wk_0) + y_1 + rk_0]^R\} + y_0^L y_2^R + rk_3).$$ By the
algebraic representation, we give attack of two rounds Piccolo.

4.2   The Attack

Step 1. Assumed that $y_0 = y_2 = 0$ and let (y_1^L, y_3^R) runs from 0 to 65535, hence

$$([F(wk_0) + y_1 + rk_0]^L, [F(wk_1) + y_3 + rk_1]^R)$$

runs from 0 to 65535. Denote the first component of

$$F\{[F(y_0 + wk_0) + y_1 + rk_0]^L, [F(y_2 + wk_1) + y_3 + rk_1]^R\} = C_1 + rk_2$$

by f_0, let $v_0 = y_0^3 + y_0^6 + y_0^7 + y_0^{10} + y_0^{14}$, which is the linear combination of the
components of the plain text P_0.
Step 2. Recover rk_2.
    Count the numbers of (y_1^L, y_3^R) satisfy $f_0 + v_0 = 0$. If the number is 33792,
then the first component of rk_2 is equal to the first component of C_1. If the
number is 65536-33792=31744, the two bits are complementary each other.
Step 3. Similarly get the rest 15 bits of rk_2 and all 16 bits of rk_3. Hence we
recover $rk_2 = k_0 + con, rk_3 = k_1 + con$. Since $wk_0 \leftarrow k_0^L \mid k_1^R, wk_1 \leftarrow k_1^L \mid k_0^R$,
the wk_0, wk_1 are known.
Step 4. Recover $rk_0 = k_2 + con, rk_1 = k_3 + con$ by

$$rk_0^L = [C_0 + F(wk_0) + y_1]^L, rk_0^R = [C_2 + F(wk_0) + y_1]^R, \qquad (3)$$

$$rk_1^L = [C_2 + F(wk_2) + y_3]^L, rk_1^R = [C_0 + F(wk_1) + y_3]^R. \qquad (4)$$

where (C_0, C_1, C_2, C_3) is the corresponding output of round 1 with the input
$y_0, y_2 = 0$ and y_1, y_3 selected randomly.
Step 5. At last, exhaustively test k_4 by the encryption oracle until the correct
key is found.
    The computational complexity of the attack is $2^{16} + 2^{16} = 2^{17}$ two rounds
Piccolo-80 encryptions. The data complexity is 2^{17} pairs of known plaintext-
ciphertext.
    In [4], The authors recognized that the figures for the maximum differential
probability and the maximum linear probability of the F-function are not optimal
for a 16-bit bijective function. However, they said it is sufficient for their design,
since Piccolo has enough differentially and linearly active F-functions over a
certain number of rounds. We think the statistical character of the F-function
matches with linear permutation will gives the attackers a chance to distinguish
piccolo from random permutation.

5 A Weakness of the Key Schedule

We start the research at the algebraic representation of the last two rounds. Denote the round number of Piccolo-80 varied from 0 to 24. Denote the 64 bits initial state of round 23 by

$$S = (y_0, y_1, y_2, y_3) \in F_2^{64}, \tag{5}$$

where $y_i, i = 0, 1, 2, 3$ are 16 bits binary string. After the F-layer the internal state is

$$y_0, F(y_0) + y_1 + k_4, y_2, F(y_2) + y_3 + k_4. \tag{6}$$

After the round permutation, the output of round 23 is

$$F(y_0)^L + y_1^L + k_4^L, F(y_2)^R + y_3^R + k_4^R, y_2^L, y_0^R, F(y_2)^L + y_3^L + k_4^L, F(y_0)^R + y_1^R + k_4^R, y_0^L, y_2^R.$$

which is the initial state of round 24. The cipher text is

$$F(y_0)^L + y_1^L + k_4^L + k_4^L, F(y_2)^R + y_3^R + k_4^R + k_3^R,$$
$$F(F(y_0)^L + y_1^L + k_4^L, F(y_2)^R + y_3^R + k_4^R) + (y_2^L, y_0^R) + k_0,$$
$$F(y_2)^L + y_3^L + k_4^L + k_3^L, F(y_0)^R + y_1^R + k_4^R + k_4^R,$$
$$F(F(y_2)^L + y_3^L + k_4^L, F(y_0)^R + y_1^R + k_4^R) + (y_0^L, y_2^R) + k_1.$$

Let us focus on the first part and the fifth part of the cipher text, the round keys are canceled in the encryption process. Hence, the first part and the fifth part of the cipher text are $F(y_0)^L + y_1^L, F(y_0)^R + y_1^R$. This shows that the round keys in round 23 and the whiten keys are both equal to k_4, which invokes that the subkeys do not play the roles of hide information of internal states. That is to say, there are 16 bits of cipher text equal to $F(y_0) + y_1$ exactly, which is only related to the internal states of the last but one round without secret key. Hence there are 16 bits of internal state information exposed to the attacker.

Figure 2 shows the way on which the round key diffuse vividly, where the red blocks are the intermediate states or round keys related to k_4^L, the yellow blocks are the intermediate states or round keys related to k_4^R, and the green blocks are the cipher text which are not affected by k_4. In order to emphasize our idea, we did not colored the blocks indirectly related to our analysis.

6 Concluding Remarks and Proposals

In this paper, we give some observations on the lightweight block Piccolo, including the inverse function of F-function, the linear analysis of the F-function, and the weakness of the key scheduling of lightweight block cipher Piccolo-80. We do not agree with the designers on the security of nonlinear F-function, since the statistical character of the F-function matches with linear permutation, which gives the attackers chance to distinguish piccolo from random permutation.

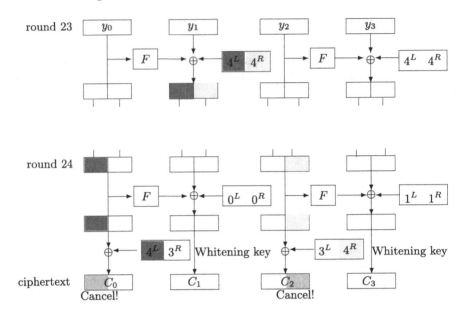

Fig. 2. The key scheduling are canceled in the last two rounds of Piccolo-80 (Color figure online)

To avoid the above weakness, the key scheduling should be modified. We suggest that $wk_2 \leftarrow k_2^L \mid k_3^R, wk_3 \leftarrow k_3^L \mid k_2^R$ instead of the current key schedule. There are five 16-bit string subkeys $k_i, i = 0, 1, 2, 3, 4$, and six k_i needed in the last two rounds including the whitening layer. According to the Pigeonhole Principle, there must be two of the subkeys are the same. We can use the same subkey one in right and the other in the left to avoid cancel out each other.

As for the Piccolo-128, the last two round keys are k_6, k_3 for round 29, k_2, k_5 for round 30 and the whiten keys are $wk_2 = k_4^L k_7^R, wk_3 = k_7^L k_4^R$. The round keys for last two rounds and the whiten keys has no intersection, hence Piccolo-128 does have this kind of weakness.

References

1. Wu, W., Zhang, L.: LBlock: a lightweight block cipher. In: Lopez, J., Tsudik, G. (eds.) ACNS 2011. LNCS, vol. 6715, pp. 327–344. Springer, Heidelberg (2011)
2. Knudsen, L., Leander, G., Poschmann, A., Robshaw, M.J.B.: PRINTCIPHER: a block cipher for IC-printing. In: Mangard, S., Standaert, F.-X. (eds.) CHES 2010. LNCS, vol. 6225, pp. 16–32. Springer, Heidelberg (2010)
3. Guo, J., Peyrin, T., Poschmann, A., Robshaw, M.: The LED block cipher. In: Preneel, B., Takagi, T. (eds.) CHES 2011. LNCS, vol. 6917, pp. 326–341. Springer, Heidelberg (2011)
4. Shibutani, K., Isobe, T., Hiwatari, H., Mitsuda, A., Akishita, T., Shirai, T.: *Piccolo*: An ultra-lightweight blockcipher. In: Preneel, B., Takagi, T. (eds.) CHES 2011. LNCS, vol. 6917, pp. 342–357. Springer, Heidelberg (2011)

5. Isobe, T., Shibutani, K.: Security analysis of the lightweight block ciphers XTEA, LED and Piccolo. In: Susilo, W., Mu, Y., Seberry, J. (eds.) ACISP 2012. LNCS, vol. 7372, pp. 71–86. Springer, Heidelberg (2012)
6. Wang, Y., Wu, W., Yu, X.: Biclique cryptanalysis of reduced-round Piccolo block cipher. In: Ryan, M.D., Smyth, B., Wang, G. (eds.) ISPEC 2012. LNCS, vol. 7232, pp. 337–352. Springer, Heidelberg (2012)
7. Kitae, J.: Security analysis of block cipher Piccolo suitable for wireless sensor networks. Peer-to-Peer Networking Appl. 7(4), 636–644 (2014)
8. Kitae, J., Hyung, C.K., Changhoon, L., Jaechul, S., Seokhie, H.: Biclique cryptanalysis of lightweight block ciphers PRESENT, Piccolo and LED. IACR Cryptology ePrint Archive 2012: 621 (2012)
9. Albrecht, M.R., Driessen, B., Kavun, E.B., Leander, G., Paar, C., Yalçın, T.: Block ciphers – focus on the linear layer (feat. PRIDE). In: Garay, J.A., Gennaro, R. (eds.) CRYPTO 2014, Part I. LNCS, vol. 8616, pp. 57–76. Springer, Heidelberg (2014)

A Memory Efficient Variant of an Implementation of the F_4 Algorithm for Computing Gröbner Bases

Yun-Ju Huang[1]([⊠]), Wei-Chih Hong[2], Chen-Mou Cheng[3],
Jiun-Ming Chen[4], and Bo-Yin Yang[5]

[1] Graduate School of Mathematics, Kyushu University, Fukuoka, Japan
y-huang@math.kyushu-u.ac.jp
[2] Department of Information Engineering and Computer Science,
Feng Chia University, Taichung, Taiwan
[3] Institute of Mathematics for Industry, Kyushu University, Fukuoka, Japan
[4] Department of Mathematics, National Taiwan University, Taipei, Taiwan
[5] Institute of Information Science, Academia Sinica, Taipei, Taiwan

Abstract. Solving multivariate systems of polynomial equations is an important problem both as a subroutine in many problems and in its own right. Currently, the most efficient solvers are the Gröbner-basis solvers, which include the XL algorithm [6], as well as F_4 [9] and F_5 [10] algorithms. The F_4 is an advanced algorithm for computing Gröbner bases. However, the algorithm has exponential space complexity and does not provide much flexibility in terms of controlling memory usage. This poses a serious challenge when we want to use it to solve instances of sizes of practical interest.

In this paper, we address the issue of memory usage by proposing a variant of F_4 algorithm called YAGS (Yet Another Gröbner-basis Solver). YAGS uses less memory than the original algorithm and runs at comparable speed with F_4. Furthermore, YAGS runs even faster than F_4 when solving dense polynomial systems. In other words, the proposed algorithm can reach better time-memory compromise via deliberately designed techniques to control its memory usage and efficiency. We have implemented a prototype of YAGS and conducted an extensive set of experiments with it. The experiment results demonstrate that the proposed modification does achieve lower time-memory products than the original F_4 over a broad set of parameters and problem sizes.

Keywords: Gröbner basis · F_4 algorithm · Time-memory trade-off

1 Introduction

The method of Gröbner bases, originated by Buchberger in 1965 [2], has been gathering more and more interest due to its widespread applications in various areas of computer science and engineering, including algorithmic algebra [7], cryptanalysis [8,11,12,18], model checking [1,5,20], coding theory [13,17,19],

© Springer International Publishing Switzerland 2015
M. Yung et al. (Eds.): INTRUST 2014, LNCS 9473, pp. 374–393, 2015.
DOI: 10.1007/978-3-319-27998-5_24

multidimensional signal processing and systems theory [15], etc. The power of the Gröbner bases method, including its theory and algorithms, lies in its ability to transform an arbitrary finite set of multivariate polynomials into another equivalent, algorithmically solvable basis. Gröbner bases can be viewed as a generalization of the Euclidean Division algorithm from the univariate polynomials to the multivariate cases. Therefore, it can be applied to solving or analyzing any problem that can be modeled by a multivariate polynomial system.

The original Buchberger algorithm is not optimized and often fails to compute the Gröbner bases for large problems owing to its inefficiency in pair selection and redundancy reduction, as well as its exponential complexity. However, real systems usually have a relatively large number of variables. For example, the algebraic attack in cryptography is done by modeling the encryption of a plaintext block $\mathbf{x}$ as a set of multivariate polynomials $(f_1(\mathbf{x}), f_2(\mathbf{x}), \ldots, f_m(\mathbf{x}))$, and solving the system using computational techniques. The number of variables in such systems grows linearly with the size of the plaintext $\mathbf{x}$, which is often rather large in cryptographic systems.

Along with the applications of Gröbner bases theory, there were also research efforts on improving the efficiency of their computation. Some of the improving techniques have been proposed to rapidly eliminate unnecessary S-polynomials by examining the corresponding polynomial pairs directly since a major part of the execution time of Buchberger algorithm is spent on checking S-polynomials. For example, Buchberger further improved his algorithm by presenting a number of criteria to remove useless S-polynomials in [3,4].

On the other hand, the F_4 algorithm proposed by Faugère in 1999 [9] focused on the pair selection strategy. It substantially elevated the speed of computing the Gröbner bases by dealing with multiple pairs simultaneously and exploiting the techniques in sparse linear algebra. The F_4 was shown to be at least one order of magnitude faster than its predecessors and able to solve previously intractable problems like Cyclic-9. Many researches recently works on the variant of F_4 algorithm focus on avoiding all reductions to zero [10,14]. However, the algorithm still has exponential space complexity, which makes it rather challenging to solve systems of sizes of practical interest. For example, if we are going to solve a multivariate polynomial system of 40 equations in 40 variables, then most of today's computers will run out of memory before the execution of the algorithm finishes, let alone to implement the algorithm in an FPGA.

In this paper, we set out to address this shortcoming by starting with the following questions about F_4's memory consumption.

1. Can F_4, or any variant of it, be executed under any memory limitation?
2. If not, at least how much memory is necessary for F_4 to be successfully executed?
3. Can we modify F_4 algorithm to run faster when given more memory?

Throughout the process of answering these questions, we observe the memory usage in each part of the F_4 algorithm, based on which we propose an improved F_4 algorithm called YAGS (Yet Another Gröbner-basis Solver). YAGS uses less memory than the original F_4. More importantly, YAGS computes Gröbner basis with comparable speed to F_4 and provides better performance considering both its time and memory usage. In certain cases, YAGS even runs faster than F_4 while using smaller amount of memory.

YAGS controls its memory consumption by dividing the work into chunks of smaller working sets and executing them one at a time. In the first place, such design intends to trade time for memory since some computation might have to be carried out repeatedly. However, due to the suboptimality of the pair selection strategy of F_4, reducing the number of pairs selected in each round of computation might eliminate some unnecessary work and speed up the procedure as well. We will show that such design makes sense in terms of time-memory product and is extremely flexible by showing the following.

1. On average, YAGS yields smaller time-memory products than the original F_4 algorithm in most simulated cases.
2. YAGS allows the Gröbner basis be computed using an arbitrary amount of memory as long as it is above the minimum amount of memory required to solve the instance.
3. When dealing with dense polynomial systems, YAGS even runs a bit faster than F_4.

The rest of this paper is structured as follows. Section 2 introduces the background knowledge about the Gröbner bases theory as well as the computing algorithms, including Buchberger and F_4. Section 3 illustrates the proposed YAGS algorithm and its proof of correctness. We then demonstrate the performance comparisons of YAGS via experiment results in Sect. 4 and conclude this paper in Sect. 5.

2   Background

In this section, we give a brief introduction to the background knowledge of this paper, including the basic concept of Gröbner basis, the necessary definitions and notations, and a couple of the previously proposed construction algorithms, e.g. the Buchberger and the F_4 algorithms.

2.1   Preliminary Definitions

We define $K[\mathbf{x}]$ the multivariate polynomial ring over the finite field K, and $\mathbf{x} = (x_1, \ldots, x_n)$ is the n-tuple of the variables. Let $\boldsymbol{\alpha} = (\alpha_1, \ldots, \alpha_n)$ be an n-tuple of non-negative integers, a monomial of multidegree $\boldsymbol{\alpha}$ is denoted by $\mathbf{x}^{\boldsymbol{\alpha}} = \prod_{i=1}^{n} x_i^{\alpha_i}$, and a polynomial is a finite linear combination of monomials with coefficients in K.

Let $F = \{f_i | i = 1, \ldots, m; f_i \in K[\mathbf{x}]\}$ be a set of m polynomials. The ideal I generated by F is defined as

$$I(F) = \left\{ \sum_{i=1}^{m} h_i f_i \,\middle|\, h_i \in K[\mathbf{x}], i = 1, \ldots, m \right\}. \tag{1}$$

Alternatively, we can write $I = \langle F \rangle = \langle f_1, \ldots, f_m \rangle$ to represent that I is the ideal spanned by the basis F.

Before introducing the Gröbner basis, we need to define an ordering of the monomials in $K[\mathbf{x}]$ so that we can arrange the terms of each polynomial. There are three commonly used orders in a multivariate polynomial ring. They are defined as follows.

1. **Lexicographic Order($>_{\text{lex}}$)**
   Given two monomials $\mathbf{x}_1^{\alpha_1}$ and $\mathbf{x}_2^{\alpha_2}$, we say $\mathbf{x}_1^{\alpha_1} >_{\text{lex}} \mathbf{x}_2^{\alpha_2}$ if the *leftmost* nonzero element of the difference vector $(\alpha_1 - \alpha_2)$ is positive.
2. **Graded Lex Order($>_{\text{grlex}}$)**
   The total degree of a monomial $\mathbf{x}^{\alpha}$ is defined by

$$\deg(\mathbf{x}^{\alpha}) = |\alpha| = \sum_{i=1}^{n} \alpha_i. \tag{2}$$

   In graded lex order, we say $\mathbf{x}_1^{\alpha_1} >_{\text{grlex}} \mathbf{x}_2^{\alpha_2}$ if $|\alpha_1| > |\alpha_2|$, or $|\alpha_1| = |\alpha_2|$ and $\mathbf{x}_1^{\alpha_1} >_{\text{lex}} \mathbf{x}_2^{\alpha_2}$.
3. **Graded Reverse Lex Order($>_{\text{grvlex}}$)**
   Like the graded lex order, this order compares the total degree first, but breaks ties according to a reversed lex order, which checks the *rightmost* nonzero element of $(\alpha_1 - \alpha_2)$ instead.

With regard to a chosen ordering, we can arrange the terms of a polynomial f in a descending order and denote the leftmost term with $\text{HT}(f)$, i.e. the head term of f. The order of two polynomials f_1 and f_2 can be determined by comparing their terms sequentially until the tie is broken.

It is noteworthy that we do not take the coefficients into account when ordering the monomials or polynomials.

2.2   Gröbner Basis

Definition 1 (Gröbner Basis). *For a chosen monomial order, a finite subset G of $I = \langle f_1, \ldots, f_m \rangle$ is said to be a Gröbner basis if and only if $\forall f \in I, \exists g \in G$, such that $\text{HT}(f)$ is divisible by $\text{HT}(g)$, denoted as $g \mid_{\text{HT}} f$.*

The definition above indicates that the head terms of the polynomials in a Gröbner basis must span an ideal containing all the head terms of the member polynomials of I. It has been shown that every ideal $I \subset K[\mathbf{x}]$ other than $\{0\}$ has a Gröbner basis. Furthermore, thanks to some good properties of the Gröbner basis, it serves as a powerful tool in solving multivariate polynomial systems.

However, since it is impossible to examine all the arithmetic combinations of $f_1, \ldots, f_m$, such definition does not provide an efficient way to verify whether a set $G \subset K[\mathbf{x}]$ is a Gröbner basis, nor does it illustrate how to construct a Gröbner basis from an arbitrarily given set of polynomials $F \subset K[\mathbf{x}]$. The following section presents the definition and theorem introduced by Buchberger for working with these problems.

2.3   Buchberger Algorithm

Definition 2 (S-polynomial and pre-S-polynomial). *Let* $\mathrm{lcm_{HT}}(f_i, f_j)$ *be the least common multiple (LCM) of the head terms of* $f_i, f_j \in K[\mathbf{x}]$, *and* $c_i = \frac{\mathrm{lcm_{HT}}(f_i,f_j)}{\mathrm{HT}(f_i)}$, $c_j = \frac{\mathrm{lcm_{HT}}(f_i,f_j)}{\mathrm{HT}(f_j)}$. *The S-polynomial of* (f_i, f_j), *denoted by* $\mathrm{Spoly}(f_i, f_j)$, *is defined as*

$$\mathrm{Spoly}(f_i, f_j) = c_i f_i - c_j f_j. \tag{3}$$

In addition, $c_i f_i$ *and* $c_j f_j$ *are the two pre-S-polynomials of the pair, denoted as* $\mathrm{PSpoly}_i(f_i, f_j)$ *and* $\mathrm{PSpoly}_j(f_i, f_j)$ *respectively.*

Theorem 1 (Buchberger Theorem). *Let* $G = \{g_1, \ldots, g_t\}$ *be a basis of* I, *then* G *is a Gröbner basis of* I *if and only if* $\forall g_i, g_j \in G$, $\mathrm{Spoly}(g_i, g_j)$ *can be reduced to zero by* G, *i.e. there exist polynomials* $h_i \in K[\mathbf{x}]$ *such that*

$$\mathrm{Spoly}(g_i, g_j) = \sum_{k=1}^{t} h_k g_k, \tag{4}$$

where $\mathrm{HT}(h_k g_k) \leq \mathrm{HT}(\mathrm{Spoly}(g_i, g_j))$ *for all* k.

This theorem reduces the seemingly impossible work of checking all the head terms of the polynomials in I into a reasonable job of examining the S-polynomials of all the pairs (g_i, g_j) in G. Based on the theorem, Buchberger devised an algorithm for the construction of Gröbner basis. The idea of this algorithm is simple. It repeatedly checks the S-polynomial of a selected pair of polynomials by multivariately dividing it with G. If the remainder is not equal to zero, it adds the remainder into the set G and checks the next pair. The selection strategy is unspecified. One can adopt any strategy as long as all the pairs will be checked eventually. Nevertheless, the strategy is very important to the efficiency of the implemented program since bad selections can incur a large amount of redundant work. It was suggested by Buchberger that the pairs with the least LCM degree should be selected first. Algorithm 1 shows the pseudocode of the Buchberger Algorithm.

Additionally, Buchberger proposed two more criteria to speed up the process of checking the S-polynomials.

Criterion 1 (Buchberger's First Criterion). *Let* $\gcd_{\mathrm{HT}}(f_i, f_j)$ *be the greatest common divisor of the head terms of* $f_i, f_j \in I$. *If* $\gcd_{\mathrm{HT}}(f_i, f_j) = 1$, *then* $\mathrm{Spoly}(f_i, f_j)$ *can be reduced to zero by* $\{f_i, f_j\}$.

Criterion 2 (Buchberger's Second Criterion). *Let* $f_i, f_j, f_k \in I$ *and* $\mathrm{lcm_{HT}}(f_i, f_j) \mid_{\mathrm{HT}} \mathrm{lcm_{HT}}(f_j, f_k)$, $\mathrm{lcm_{HT}}(f_i, f_k) \mid_{\mathrm{HT}} \mathrm{lcm_{HT}}(f_j, f_k)$. *If both* $\mathrm{Spoly}(f_i, f_j)$ *and* $\mathrm{Spoly}(f_i, f_k)$ *can be reduced to zero by* G, *then so can* $\mathrm{Spoly}(f_j, f_k)$.

Algorithm 1. Buchberger Algorithm

Input: F
1 $F \longleftarrow$ ReducedRowEchelon(F)
2 $G, P \longleftarrow$ UpdateGP(G, P, F)
3 **while** $P \neq \emptyset$ **do**
4      $spoly, P \longleftarrow$ SelectPair(P)
5      $r \longleftarrow$ MultivariateDivision$(spoly, G)$
6   **if** $r \neq 0$ **then**
7 | $G, P \longleftarrow$ UpdateGP$(G, P, \{r\})$
8   **end**
9 **end**
  Output: G

Algorithm 2. UpdateGP function

Input: G, P, F
1 **forall the** $f \in F$ **do**
2      $P \longleftarrow P \cup \{$getPair$(f, g_i) \mid g_i \in G\}$
3      $P \longleftarrow$ BuchbergerCriteria(P)
4      $G \longleftarrow G \cup \{f\} \setminus \{g_i \mid g_i \in G, f \mid_{\mathrm{HT}} g_i\}$
5 **end**
  Output: G, P

Both criteria are implemented in the **UpdateGP** function to check the necessity of adding each polynomial while updating G and P with the newly generated remainders.

It has been proved that this algorithm terminates deterministically and correctly produces a Gröbner basis of the ideal spanned by the input F. However, the computing complexity grows drastically as the number of variables and polynomials becomes large. It is not hard to notice that the set P expands quickly as we keep adding new remainders into G. Checking those pairs one by one is time consuming and quite inefficient.

Besides, a large portion of the computing work in the Buchberger algorithm is redundant. For example, if both $f_i, f_j \in G$ contain the same monomial m, which can be divided by another polynomial $g \in G$, then in Buchberger algorithm, g has to be extended to m twice to do the reduction for f_i and f_j respectively. Such redundancy calls for a smarter design of algorithm to improve the efficiency.

2.4 F_4 Algorithm

The F_4 algorithm is based on the ideas of the Buchberger algorithm and greatly increases the pair-checking throughput by simultaneously selecting a set of pairs and doing the reduction of their S-polynomials in matrix form. It is designed

---

Algorithm 3. F_4 Algorithm

Input: F
1 $F \longleftarrow$ ReducedRowEchelon(F)
2 $G, P \longleftarrow$ UpdateGP(G, P, F)
3 **while** $P \neq \emptyset$ **do**
4 $\quad$ $Poly, P \longleftarrow$ SelectPairs(P)
5 $\quad$ $M \longleftarrow$ Monomials$(Poly)$
6 $\quad$ $Done \longleftarrow$ Headterms$(Poly)$
7 $\quad$ **while** $\exists\, m \in M \setminus Done,\ \exists\, g \in G,\ g \mid_{\mathrm{HT}} m$ **do**
8 $\quad\quad$ $Done \longleftarrow Done \cup \{m\}$
9 $\quad\quad$ $Poly \longleftarrow Poly \cup \{g * \frac{m}{\mathrm{HT}(g)}\}$
10 $\quad\quad$ $M \longleftarrow$ Monomials$(Poly)$
11 $\quad$ **end**
12 $\quad$ $Poly' \longleftarrow$ ReducedRowEchelon$(Poly)$
13 $\quad$ $R \longleftarrow \{p \mid p \in Poly',\, \mathrm{HT}(p) \notin \mathrm{HT}(Poly)\}$
14 $\quad$ $G, P \longleftarrow$ UpdateGP(G, P, R)
15 **end**
Output: G

Algorithm 4. SelectPairs function in F_4

Input: P
1 $d \longleftarrow \min\{\deg(\mathrm{lcm}_{\mathrm{HT}}(p)) \mid p \in P\}$
2 $Pair \longleftarrow \{p \in P \mid \deg(\mathrm{lcm}_{\mathrm{HT}}(p)) = d\}$
3 $P \longleftarrow P \setminus Pair$
4 $Poly \longleftarrow \{\mathrm{PSpoly}_i(f_i, f_j), \mathrm{PSpoly}_j(f_i, f_j) \mid (f_i, f_j) \in Pair\}$
Output: $Poly, P$

---

to efficiently compute the Gröbner bases using the graded reverse lex order. Algorithm 3 gives the pseudocode of the F_4 algorithm.

Unlike the Buchberger algorithm, F_4 adopts a newly designed `SelectPairs` function to select multiple pairs for one round of reduction. The selection strategy suggested in F_4 is called the *normal strategy*. It selects all the pairs with the least head term LCM degree and returns the pre-S-polynomials of the selected pairs (see Algorithm 4). The reason for keeping the pre-S-polynomials in the matrix is to preserve the monomials that might be canceled out during the computation of the S-polynomials, thus increasing the probability of reducing the redundant work.

After generating the pre-S-polynomials, F_4 adds as many useful polynomials into *Poly* as possible in order to further improve the computing efficiency. This is done by examining each non-heading monomial of the polynomials in *Poly*. If there exists a non-heading monomial m and a $g \in G$ such that $g \mid_{\mathrm{HT}} m$, it extends g and adds the result into the set *Poly*. This step ensures that after performing the Gaussian elimination in the `ReducedRowEchelon` function, there will not be any $r \in R$ reducible by G.

$$F = \left\{ \begin{array}{c} 3x^2 + 2xy + y^2 + 3 \\ 2x^2 + 4xy + y^2 + 1 \end{array} \right\} \Longleftrightarrow M = \begin{bmatrix} x^2 & xy & y^2 & 1 \\ 3 & 2 & 1 & 3 \\ 2 & 4 & 1 & 1 \end{bmatrix}$$

Fig. 1. Example of polynomial-matrix transformation.

The ReducedRowEchelon function transforms the set $Poly$ into matrix form, applies the Gaussian elimination, and then transforms the resultant matrix back to a set of polynomials $Poly'$. Figure 1 gives a simple example of the transformation between polynomials and matrix. Such transformation and batch processing are crucial to the F_4 algorithm. The combination of these tricks greatly increases the throughput of the reduction step and makes the computing procedure amazingly fast.

The final step of an iteration is to pick out the useful remainders with newly "exposed" head terms and add them into G using the UpdateGP function.

3 Memory Control of F_4 Algorithm

Although the F_4 is proved to be at least one order of magnitude faster than all previous algorithms, its memory consumption is not well-controlled. In Sect. 4, it will be shown by experiments that the space complexity of F_4 grows exponentially. Therefore, most computers will run out of memory when solving large multivariate polynomial systems. This is because the size of G and P grows quickly during the execution of the algorithm, thus resulting in excessively large matrices in the reduction stage.

An intuitive idea of controlling the memory usage might be simply adjusting the pair selection strategy to cut down the number of selected polynomials. However, the reduction step of F_4 algorithm may still include a large number of extended polynomials into $Poly$ and produce a huge matrix. Moreover, selecting too few pairs in each iteration would cut the claws of F_4 algorithm and prolong the whole computing process.

As a result, we need a redesigned version of F_4 in order to achieve better trade-off between time and memory usages. The goal of the new design is to limit the memory usage while maintaining satisfactory speed. To this end, our YAGS algorithm takes a new input parameter $size$ as its memory usage constraint and makes the following three major modifications to the original F_4:

1. A new SelectPairs function to select pairs under a limitation based on the memory usage parameter $size$.
2. A modified reduction step that avoids including all the extended member polynomials of G at the same time, but splits the work into smaller chunks instead.
3. A newly added ReduceG function which decreases the memory usage of G by reducing the non-heading terms of its members.

Algorithm 5. YAGS algorithm

Input: $F, size$
1 $F \longleftarrow$ ReducedRowEchelon(F)
2 $G, P \longleftarrow$ UpdateGP(G, P, F)
3 **while** $P \neq \emptyset$ **do**
4 | $mat\_size \longleftarrow size - G\_size - P\_size$
5 | $Poly, P \longleftarrow$ SelectPairs(P, mat_size)
6 | **while** $Poly \neq \emptyset$ **do**
7 | | $Poly' \longleftarrow$ ReducedRowEchelon$(Poly)$
8 | | $Poly'' \longleftarrow Poly'' \cup \{p \mid p \in Poly', \text{HT}(p) \notin \text{HT}(Poly)\}$
9 | | **while** $Poly'' \neq \emptyset$ **do**
10 | | | $p \longleftarrow \min\{Poly''\}$
11 | | | $Poly'' \longleftarrow Poly'' \setminus \{p\}$
12 | | | **if** $\exists\, g \in G \cup R, g \mid_{\text{HT}} p$ **then**
13 | | | | $Poly \longleftarrow Poly \cup \{p, h^*g \mid \text{HT}(p) = \text{HT}(h^*g)\}$
14 | | | **else**
15 | | | | $R \longleftarrow R \cup \{p\}$
16 | | | **end**
17 | | | **if** $\#\text{Monomials}(Poly)^* \#Poly \geq mat\_size$ **then**
18 | | | | break
19 | | | **end**
20 | | **end**
21 | **end**
22 | $G, P \longleftarrow$ UpdateGP(G, P, R)
23 | $G \longleftarrow$ ReduceG(G)
24 **end**
Output: G

Algorithms 5–7 gives the pseudocode of the proposed YAGS algorithm. As will be explained in the following subsections, the compromises made in YAGS do not merely improve the computing efficiency of the algorithm, but could bring some benefits as well. Experiment results in Sect. 4 will give proof to this fact by showing some cases when YAGS computes Gröbner bases faster than F_4 using less memory.

3.1 New Pair Selection Function

The first step of controlling the memory usage is to revise the pair selection algorithm. Instead of selecting an uncertain number of polynomials in each iteration, we modify the SelectPairs function so that the total amount of memory usage is limited.

Let $size$ be the total memory usable to the program; G_size and P_size be the memory used to store G and P respectively. The SelectPairs function would select as many pairs as possible and fill up the remaining space, $mat\_size = size - G\_size - P\_size$, with their pre-S-polynomials. Also, the selection strategy

---

Algorithm 6. SelectPairs function in YAGS

Input: P, mat_size
1 **while** $P \neq \emptyset$ **do**
2       $pair \longleftarrow \min\{\mathrm{lcm}_{\mathrm{HT}}(p) \mid p \in P\}$
3       $P \longleftarrow P \setminus pair$
4       $Poly \longleftarrow \{\mathrm{PSpoly}_i(f_i, f_j), \mathrm{PSpoly}_j(f_i, f_j) \mid (f_i, f_j) \in pair\}$
5   **if** $\#\mathtt{Monomials}(Poly)^* \#Poly \geq mat\_size$ **then**
6   break
7   **end**
8 **end**
    Output: $Poly, P$

Algorithm 7. ReduceG function in YAGS

Input: G
1 **for** $g \in G$ **do**
2   **if** $length\ of\ g \geq \frac{1}{2}max\_length$ **then**
3          $Poly \longleftarrow \{g\}$
4   **for** $m \in \mathtt{Monomials}(g)$ **do**
5             $Poly \longleftarrow Poly \cup \{h^*g' \mid g' \in G,\ m = \mathrm{HT}(h^*g')\}$
6   **end**
7          $Poly' \longleftarrow \mathtt{ReducedRowEchelon}(Poly)$
8          $g \longleftarrow g' \in Poly', \mathrm{HT}(g') = \mathrm{HT}(g)$
9   **end**
10 **end**
    Output: G

---

needs a little adjustment. Unlike F₄'s normal strategy, which selects all pairs with least head term LCM degree, YAGS compares the head term LCM's according to the monomial order and selects the "minimum" one at a time before the memory limit is reached (see Algorithm 6). In most cases, when the memory is limited, the number of selected pairs is much smaller than those recommended by the F₄ as the size of G and P grows very quickly.

Such modification would confine the reduction throughput of each iteration, but it might save some duplicated calculations as well. This is because at the end of each iteration, there will be new basis elements added into G and some pairs might be removed by the UpdateGP function according to the Buchberger Criteria. It will be shown in Sect. 4 that the computing speed is not slowed down much.

3.2 S-Polynomial Reduction Step Modifications

In the S-polynomial reduction step of F₄, it includes all the possibly useful polynomials in G in addition to the pre-S-polynomials and generates a single huge matrix (see lines 7–11 in Algorithm 3). Contrarily, in our new SelectPairs function, we load the matrix with as many pre-S-polynomials as possible, and hence there is no free space for including any member of G. As a result, we need to redesign the reduction step in order to perform similar operations to those in F_4.

Lines 6–21 of Algorithm 5 illustrate our modified procedure for S-polynomial reduction in YAGS. Since we do not include any member of G into the matrix in the beginning of the reduction step, we do the Gaussian elimination first and pick out a set $Poly''$ containing the polynomials with newly exposed head terms. Next we check the members of $Poly''$ for their divisibility to the members of G. If a $p \in Poly''$ can not be divided by any $g \in G$, it is a new remainder and will be stored in R temporarily for later update of G. Otherwise, if p is divisible by some $g \in G$, we put both p and the extended g into a newly constructed $Poly$ for further reduction.

Our approach here can be seen as a hybrid of F_4 and the Buchberger algorithm. Owing to the limitation on the memory usage, we apply Gaussian elimination to a submatrix of the huge matrix that would be generated by F_4 at a time. Meanwhile, unlike F_4, we only add the polynomials that can reduce the new head terms in each loop, and this strategy can limit the size of $Poly$ that need to be processed in the next round. Accordingly, we might have to include the same $g \in G$ several times to finish the reduction of one set of pairs selected by SelectPairs.

Intuitively, limiting the available memory space makes the efficiency of the modified S-polynomial reduction procedure worse than F_4. Splitting the huge matrix into smaller submatrices introduces more overhead of transforming back and forth between the polynomials and the matrix representations. However, we may save some operations on the other hand. For example, in line 12 of Algorithm 5, the inclusion of R when checking the S-polynomials helps discover the divisibility among the newly exposed head terms. In F_4, such discovery can not be done in the same iteration and redundant new pairs will be generated in the UpdateGP function, thus resulting in unnecessary computation and memory usage.

After the reduction of S-polynomials is done, the new remainders in R are added into G by calling the same UpdateGP function as in F_4, and the primary operations of an iteration are finished.

3.3   Reduction of G

The size of G becomes a new issue in our scheme since it occupies a segment of the total available memory and it grows as the execution of the program. As we select a smaller set of pairs and merely inspect the exposed head terms, the new remainders added into G would be much longer in length than those produced by F_4. This effect makes it more memory-consuming to store G and lessens the number of pairs that can be processed in one iteration, thus resulting in the decline in reduction throughput. In the worst case, G and P might use up all available memory and the program can no longer produce anything. Actually, this phenomenon will impose a lower bound on the memory requirement of the YAGS program. We have to deal with this issue if more memory efficiency is required.

Accordingly, every time new members are added, the function ReduceG will be called to shrink the space occupied by G through shortening the length of the polynomials in it. In Algorithm 7, we examine the members of G which are longer than $\frac{1}{2}max\_length$, where max_length is the maximum length of $g \in G$. We then try to shorten their length by reducing the non-heading monomials with other polynomials in G.

However, it is noteworthy that the operations in ReduceG do not necessarily shorten the polynomials. For example, if we reduce $x^3 + x^2$ with $x^2 + x + 1$, the result will be $x^3 - x - 1$ and its length is longer than the original polynomial. This is one of the reasons why we reduce only the polynomials longer than $\frac{1}{2}max\_length$. Another reason is for the efficiency of the program. Checking all members of G would cost too much time.

3.4   Correctness of YAGS

In this subsection, we prove the correctness of the proposed YAGS algorithm by showing that it terminates deterministically and generates a Gröbner basis of the ideal spanned by the input F.

First, we claim that the S-polynomial reduction loop in YAGS terminates and produces the corresponding remainders by proving the following proposition.

Proposition 1. *For a set of finitely selected pairs, the S-polynomial reduction loop in YAGS terminates deterministically and generates a set of new remainders corresponding to the S-polynomials.*

Proof. According to Algorithm 6, the pre-S-polynomials of the selected pairs will be placed in the set *Poly*. As a result, the S-polynomials should be generated after the first iteration of the reduction loop. In the following iterations, the divisibilities of the new head terms of the S-polynomials to the elements in G are checked, and the indivisible remainders will be placed in R, as described in Subsect. 3.2. Since there are only finitely many S-polynomials each with finite number of terms, such checking and reducing loop will terminate eventually.

In addition, for each selected pair $f_i, f_j \in G$, there will be a reduced S-polynomial p when the reduction loop terminates. If $p = 0$, then this pair will be discarded after this reduction. If $p \neq 0$, then we claim that $p \in R$ according to the following arguments.

Assume that there exists a $p \neq 0$ and $p \notin R$, then p should be in *Poly* and its head term should be divisible by some $g \in G$. In this case, the reduction loop should not have terminated.

Basically, for the same set of selected pairs, the output set R of the reduction step of YAGS will contain the same head terms as those produced by F_4. Notice that R could be empty, and in that case, no new remainders will be added into G and no new pairs will be generated in the subsequent UpdateGP function. However, should R contain any new remainder, the newly updated G will span a larger head term ideal than the original set.

Next, we prove the following theorem to justify the termination and correctness of the YAGS algorithm.

Theorem 2. *The YAGS algorithm terminates deterministically and computes a Gröbner basis G for any input finite set of polynomials F.*

Proof. Termination: Let R_t be the output set of new remainders, G_t be the temporary result basis after the tth iteration of main loop and $G_0 = F$.

Let I_t denote the ideal spanned by the head terms of polynomials in G_t, i.e. $I_t = I(\text{HT}(G_t))$. Thus we have

$$\begin{cases} I_{t-1} = I_t & \text{, if } R_t = \emptyset, \\ I_{t-1} \subsetneq I_t & \text{, otherwise,} \end{cases} \tag{5}$$

and subsequently

$$I_0 \subseteq I_1 \subseteq I_2 \subseteq \dots . \tag{6}$$

Assume for a contradiction that the main loop of YAGS does not stop. There must exist an infinite ascending sequence $d_1, d_2, \dots$ of natural numbers such that $R_{d_k} \neq \emptyset$ and

$$I_{d_1} \subsetneq I_{d_2} \subsetneq I_{d_3} \subsetneq \dots \tag{7}$$

This contradicts with the fact that the polynomial ring is noetherian. If such infinite sequence does not exist, there should be a D such that $R_t = \emptyset$ for $t \geq D$ and the main loop must terminate since no new pairs will be generated after the Dth iteration. As a result, the YAGS algorithm terminates deterministically.

Correctness: According to the pseudocode of UpdateGP function, the updated G will always contain all the member polynomials or their factors in F. In addition, all the added new remainders in each iteration are linear combinations of the members of the previous G. Therefore, we have $F \subseteq G_t \subset I(F)$ holds for all t, and this suffice to prove that all G_t's are also bases of $I(F)$.

When the algorithm terminates, P does not contain any pair, which means $\text{Spoly}(g_i, g_j)$ can be reduced to zero by G for all $g_i, g_j \in G$. Hence G is a Gröbner basis of $\langle G \rangle = I(F)$.

4   Experiment Results

In this section, we will show by experiment results that the proposed YAGS algorithm can achieve better time-memory product than the original F_4. In other words, we can save huge amount of memory without sacrificing much computing speed.

We have implemented YAGS and F_4 with C++ coded programs. Both use the same data structure and field operation codes. All experiments are performed on a machine with two Intel Xeon E5620 CPUs running at 2.40 GHz and 24 GB of main memory running at 1333 MHz.

We have run two sets of experiments. The first set is to solve randomly generated general systems of polynomials in $GF_{16}[\mathbf{x}]$ by computing their Gröbner bases using F_4 and YAGS. The number of variables in each general system is

n and the number of polynomials is m. In each run of the experiment, each input polynomial of the general system is produced by generating uniformly distributed random coefficients for each of the monomials except the constant term. Afterwards, the constant terms are obtained by evaluating the polynomial with a randomly generated solution. In this way we can ensure the input system is solvable. We run 20 repetitions and take average for each parameter setting of the first set of experiments.

The second set of experiments is to solve Katsura benchmark systems [16] over the prime field GF_{31} using F$_4$ and YAGS. A Katsura n system has $n + 1$ unknowns and is a sparse system of the form:

$$\begin{cases} \sum_{i=-n}^{n} x_i x_{m-i} = x_m, \forall m \in \{-n+1, \dots, n-1\}, \\ \sum_{i=-n}^{n} x_i = 1, \end{cases} \tag{8}$$

where $x_{-i} = x_i$ and $x_i = 0$ for $|i| > n$. Since the form is fixed, there is no randomness in the computing procedure and we run each experiment only once.

In each run of both sets of experiments, we first compute the Gröbner basis with F$_4$ and measure its memory usage. Then we run the YAGS program with various memory budgets, e.g. the conditions when only 1/3, 1/4, and 1/10 of the memory used by F$_4$ are available.

4.1   General Systems

We present the results of the randomly generated general problems in this subsection. In order to observe the performance of YAGS for ordinary as well as overdetermined systems, we run three subsets of experiments with different number of equations setups, namely $m = 2n$, $m = 1.5n$, and $m = n + 2$. Figures 2 through 4 illustrate the performance results of the three general cases respectively.

There are two interesting observations about the timing results of the general cases. First, except for some instances (e.g. when YAGS uses only 1/10 of memory and $m = 2n$, $n = 11, 12, 13$), F$_4$ runs a bit slower than the proposed

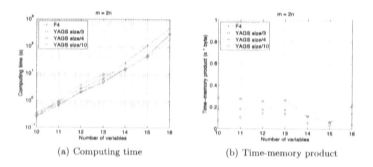

(a) Computing time   (b) Time-memory product

Fig. 2. Performance comparisons of F$_4$ and YAGS for the $m = 2n$ case.

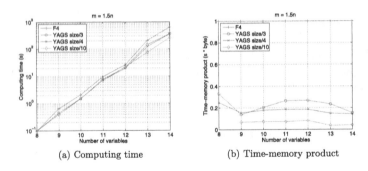

(a) Computing time   (b) Time-memory product

Fig. 3. Performance comparisons of F₄ and YAGS for the $m = 1.5n$ case.

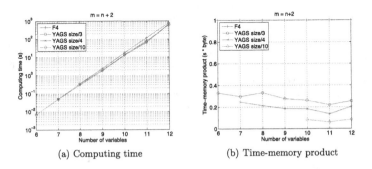

(a) Computing time   (b) Time-memory product

Fig. 4. Performance comparisons of F₄ and YAGS for the $m = n + 2$ case.

YAGS, which actually consumes less memory. Second, there is no obvious connection between the memory constraint and the running speed of YAGS. In fact, YAGS with 1/10 of memory solves generic systems faster than with 1/3 or 1/4 memory in many of the experimented cases, whereas YAGS with 1/4 memory is the fastest in other cases.

We believe such result is a combined effect of our modified tactics in pair selection and S-polynomial reduction. As described in Sect. 3.2, restricting the number of pairs processed in one iteration and reducing only the head terms may introduce repeated work and lower the reduction throughput. On the other hand, the inclusion of the remainder set R for checking and reducing S-polynomials helps discover the inter-divisibility among newly exposed head terms. Furthermore, processing a portion of the huge matrix one at a time and selecting pairs with smallest monomial order might have the effect of reducing the highest degree of the monomials need to be checked. These techniques can effectively prevent the generation of certain redundant pairs and decrease the total number of pairs we have to process.

Table 1 gives the detailed statistics of the $m = 2n$ case as an example showcase. According to our analysis, the execution time of the program is dominated by two types of multiplication operations, namely for GF numbers and monomials respectively. In this case, it is obvious that YAGS processes smaller number

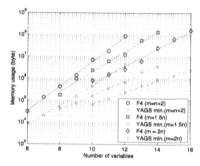

Fig. 5. Memory usage of F_4 and YAGS for different problem setups.

of pairs with lower degree and thus executing fewer GF multiplications than F_4. In fact, the less memory allocated to YAGS, the smaller number of GF multiplications it needs to execute. However, the price of using less memory is the increase in the number of monomial multiplications, which mainly comes from the repeated work due to the inability to reduce a huge matrix in one iteration. For example, for YAGS using $1/10$ memory, the numbers of monomial multiplications are relatively high when $n = 11, 12, 13$ (more than an order of magnitude larger), the savings in GF multiplications are relatively low, and the combined effect makes its computing time longer than F_4. As n becomes larger, the running speed of YAGS with $1/10$ memory turns into the fastest as the benefits of the proposed techniques grows higher.

Another interesting finding is that when the m to n ratio gets smaller, YAGS might not be able to complete the computation. Specifically, YAGS with $1/10$ memory fails to compute the Gröbner bases when $m = 1.5n$, $n = 8$, and when $m = n + 2$, $n = 6$ to 9. YAGS with $1/4$ memory is also unable to solve the system when $m = n + 2$, $n = 6$. Such cases indicate that there is a lower bound on the memory usage of the proposed YAGS. Nevertheless, according to Figs. 2(b), 3(b), and 4(b), YAGS provides better time-memory product than F_4 in solving all general cases as long as the allocated memory is above the lower bound. Note that all time-memory product values in these figures are normalized with respect to the results of F_4.

Figure 5 gives a comparison of the memory usages of F_4 as well as the minimum usages of YAGS in different experimented cases. In this figure, we also plot the regression line for each set of the data to better observe their exponents. It is obvious that both the memory usages of F_4 and YAGS min. grow exponentially with the number of variables n, but the exponents are smaller for problems with larger m to n ratio. Given the same number of variables n, an input set F with larger number of polynomials usually carries more redundant information about the spanned ideal and can be solved with less memory. As a result, given fixed amount of total memory, it is possible to solve problems with larger n when the system is more overdetermined. Moreover, the exponents of the YAGS min. memory are smaller than the corresponding usage of F_4, which means that it is

Table 1. Detailed statistics of the $m = 2n$ case.

n	Algorithm		Time (s)	Degree of regularity	Total # of pairs	GF multiplications	Monomial multiplications
10	F_4		0.35	5	1350	3.188e+07	1.103e+05
	YAGS	size/3	0.3	4	469.9	1.482e+07	3.003e+05
		size/4	0.25	4	408	1.165e+07	2.964e+05
		size/10	0.25	4	293	4.294e+06	3.746e+05
11	F_4		0.9	5	2165.5	9.117e+07	2.181e+05
	YAGS	size/3	0.75	4	616	4.238e+07	5.904e+05
		size/4	0.65	4	543	4.286e+07	5.924e+05
		size/10	0.95	4	393.6	1.816e+07	1.483e+06
12	F_4		2.75	5	3537	2.722e+08	4.156e+05
	YAGS	size/3	2.1	4	820	1.620e+08	1.200e+06
		size/4	1.95	4	722	1.368e+08	1.237e+06
		size/10	3.9	4	647.9	5.786e+07	6.627e+06
13	F_4		6.45	5	5427.85	6.253e+08	7.392e+05
	YAGS	size/3	5.2	4	1217.8	4.496e+08	2.144e+06
		size/4	4.5	4	1070	3.580e+08	2.050e+06
		size/10	9	4	876.4	2.156e+08	1.383e+07
14	F_4		24.05	5	8288.35	2.432e+09	2.444e+06
	YAGS	size/3	14.7	4	1945.7	1.378e+09	4.701e+06
		size/4	14.65	4	1945	1.376e+09	4.673e+06
		size/10	13.2	4	1491	1.081e+09	6.047e+06
15	F_4		108.85	5	12141	1.233e+10	7.273e+06
	YAGS	size/3	46.1	4	3568	4.688e+09	1.190e+07
		size/4	46.15	4	3568	4.688e+09	1.192e+07
		size/10	39.1	4	2955	3.771e+09	1.288e+07
16	F_4		463.1	6	17833	5.579e+10	1.030e+07
	YAGS	size/3	305.15	5	7557.9	2.908e+10	1.420e+08
		size/4	270.25	5	6633.1	2.454e+10	1.412e+08
		size/10	182.1	5	4538.15	1.320e+10	1.398e+08

possible for YAGS to solve systems with larger number of variables using smaller fraction of memory of F_4.

4.2   Katsura

Figure 7 shows the performance results of F_4 and the proposed YAGS solving Katsura benchmark problems with n ranging from 6 to 10. The memory usage

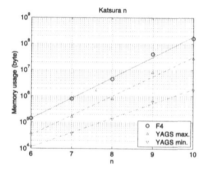

Fig. 6. Memory usage of F_4 and YAGS solving katsura benchmarks.

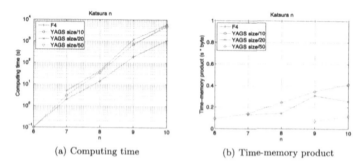

(a) Computing time   (b) Time-memory product

Fig. 7. Performance comparisons of F_4 and YAGS for the Katsura benchmark.

and execution time of both algorithms for solving sparse systems still grow exponentially with n.

In addition to F_4 and YAGS min., we have included one more set of memory usage results called YAGS max. in Fig. 6. We have found that YAGS might not consume all the allocated memory while experimenting on different memory assignments. For example, YAGS will use only a fixed amount of memory when solving Katsura systems whether 1/3 or 1/4 of the F_4 usage is allocated. In fact, the maximum memory consumption of YAGS is lower than F_4 since it tries hard to avoid the generation of redundant pairs and shorten the length of polynomials in G.

Besides, we have also found that there is larger margin for YAGS to adjust its memory allocation when solving sparse systems. In the experimented case, YAGS can solve the Katsura-10 system using as low as only 1/93 of the memory consumed by F_4. Consequently, we plot the performance results of 1/10, 1/20, 1/50 sizes in Fig. 7(a) and (b).

According to Fig. 7(a), F_4 runs much faster than YAGS in this case. This result shows the strength of F_4's strategy in solving sparse systems like Katsura, though the memory usage is rather high. When solving randomly generated dense systems, which in general consist of more dependent polynomials, F_4 tends to select too many pairs and spend substantial amount of time doing matrix reduction. However, when dealing with sparse systems, it is more efficient to include as many relevant polynomials in G as possible and reduce not only the head terms but all the monomials of the S-polynomials in one iteration.

On the contrary, by limiting the number of selected pairs and reducing the set G at the same time, YAGS can prune the redundant pairs and avoid some waste of execution time in solving dense systems. For sparse systems like Katsura benchmark, YAGS dose not have such advantage and runs slower than F_4. Therefore, we have diverse results in the comparisons of the computing speeds of YAGS and F_4. We are aware of this interesting phenomenon and are further investigating the structures that make the algorithms run faster.

Although the execution time of YAGS is longer, it still achieves much lower time-memory products thanks to its significant savings in memory consumption. In Fig. 7(b), almost all time-memory product values of YAGS are kept below 40 % of F_4. This shows that the proposed algorithm can reach better compromise between computing speed and memory usage and is better-suited for applications with space limitation.

5   Conclusions

We have successfully designed a memory efficient variant of Gröbner bases solver based on F_4 algorithm. Via a divide and conquer strategy, the proposed YAGS algorithm takes good care of its memory usage and maintains acceptable execution efficiency. It is capable of computing Gröbner bases using an arbitrary amount of memory space as long as it is above the lower bound for solving the instance. We have implemented YAGS and evaluated its performance with extensive experiments. The results show that YAGS does provide better performance than F_4 in terms of the time-memory product. In other words, the proposed algorithm can operate with as low as 2 % of memory usage while maintaining comparable execution speed with F_4. Such feature makes YAGS a better candidate for space-limited implementations of a Gröbner bases solver. Based on current design, we are working on an FPGA implementation of YAGS, as well as further boosts on the performance of the algorithm via better control on the size of intermediate result sets.

References

1. Brickenstein, M., Dreyer, A., Greuel, G.M., Wedler, M., Wienand, O.: New developments in the theory of Gröbner bases and applications to formal verification. J. Pure Appl. Algebra **213**(8), 1612–1635 (2009)
2. Buchberger, B.: An algorithm for finding the bases elements of the residue class ring modulo a zero dimensional polynomial ideal (German). Ph.D. thesis, Univ. of Innsbruck (1965)
3. Buchberger, B.: An algorithmical criterion for the solvability of algebraic systems (German). Aequationes Math. **4**(3), 374–383 (1970)

4. Buchberger, B.: Gröbner bases: an algorithmic method in polynomial ideal theory. In: Bose, N.K. (ed.) Multidimensional Systems Theory, chap. 6, pp. 184–232. Reidel Publishing Company, Dodrecht (1985)
5. Condrat, C., Kalla, P.: A Gröbner basis approach to CNF-formulae preprocessing. In: Grumberg, O., Huth, M. (eds.) TACAS 2007. LNCS, vol. 4424, pp. 618–631. Springer, Heidelberg (2007)
6. Courtois, N.T., Klimov, A.B., Patarin, J., Shamir, A.: Efficient algorithms for solving overdefined systems of multivariate polynomial equations. In: Preneel, B. (ed.) EUROCRYPT 2000. LNCS, vol. 1807, pp. 392–407. Springer, Heidelberg (2000)
7. Cox, D., Little, J., O'Shea, D.: Ideals, Varieties, and Algorithms, 3rd edn. Springer, Heidelberg (2007)
8. Faugère, J.-C., Perret, L., Petit, C., Renault, G.: Improving the complexity of index calculus algorithms in elliptic curves over binary fields. In: Pointcheval, D., Johansson, T. (eds.) EUROCRYPT 2012. LNCS, vol. 7237, pp. 27–44. Springer, Heidelberg (2012)
9. Faugère, J.C.: A new efficient algorithm for computing Gröbner bases (F₄). J. Pure Appl. Algebra 139(1–3), 61–88 (1999)
10. Faugère, J.C.: A new efficient algorithm for computing Gröbner bases without reduction to zero (F₅). In: Proceedings of the 2002 International Symposium on Symbolic and Algebraic Computation, ISSAC 2002, pp. 75–83. ACM, New York (2002)
11. Faugère, J.-C., Joux, A.: Algebraic cryptanalysis of hidden field equation (HFE) cryptosystems using Gröbner bases. In: Boneh, D. (ed.) CRYPTO 2003. LNCS, vol. 2729, pp. 44–60. Springer, Heidelberg (2003)
12. Huang, Y.-J., Petit, C., Shinohara, N., Takagi, T.: Improvement of Faugère et al.'s method to solve ECDLP. In: Sakiyama, K., Terada, M. (eds.) IWSEC 2013. LNCS, vol. 8231, pp. 115–132. Springer, Heidelberg (2013)
13. Ikegami, D., Kaji, Y.: Maximum likelihood decoding for linear block codes using Gröbner bases. IEICE Trans. Fundam. Electron. Commun. Comput. Sci. 1(3), 643–651 (2003)
14. Joux, A., Vitse, V.: A variant of the F4 algorithm. In: Kiayias, A. (ed.) CT-RSA 2011. LNCS, vol. 6558, pp. 356–375. Springer, Heidelberg (2011)
15. Lin, Z., Xu, L., Bose, N.K.: A tutorial on Gröbner bases with applications in signals and systems. IEEE Trans. Circ. Syst. 55(1), 445–461 (2008)
16. Merlet, J.P.: Polynomial systems. http://www-sop.inria.fr/coprin/logiciels/ALIAS/Benches/node1.html
17. Mora, T., Sala, M.: On the Gröbner bases of some symmetric systems and their application to coding theory. J. Symbolic Comput. 35(2), 177–194 (2003)
18. Petit, C., Quisquater, J.-J.: On polynomial systems arising from a Weil descent. In: Wang, X., Sako, K. (eds.) ASIACRYPT 2012. LNCS, vol. 7658, pp. 451–466. Springer, Heidelberg (2012)
19. Saints, K., Heegard, C.: Algebraic-geometric codes and multidimensional cyclic codes: a unified theory and algorithms for decoding using Gröbner bases. IEEE Trans. Inf. Theory 41(6), 1733–1751 (1995)
20. Wienand, O., Wedler, M., Stoffel, D., Kunz, W., Greuel, G.-M.: An algebraic approach for proving data correctness in arithmetic data paths. In: Gupta, A., Malik, S. (eds.) CAV 2008. LNCS, vol. 5123, pp. 473–486. Springer, Heidelberg (2008)

Efficient Public Key Encryption with Field-Free Conjunctive Keywords Search

Chenggen Song$^{(\boxtimes)}$, Xin Liu, and Yalong Yan

Institute of Information Security,
Beijing Electronic Science and Technology Institute,
Beijing 100070, People's Republic of China
{scg,liux,yalong}@besti.edu.cn

Abstract. In this article, we aim to the secure conjunctive keywords search problem where the keywords are field-free. Actually, many schemes have been proposed in the literature, while all the schemes need $O(k)$ pairing computing to determine a keywords set is in the ciphertext of k keywords. In this paper, we give a couple of reciprocal maps based on lagrange polynomial as basic tool to cope with this problem and we propose an efficient public key encryption scheme with field-free conjunctive keywords search(PEFCK) which reduces $O(k)$ pairing computing to $O(1)$ in once search.

Keywords: Public Key Encryption with Keywords Search · Conjunctive search · Pairing based encryption · Lagrange polynomial

1 Introduction

Public Key Encryption with Keywords Search (PEKS in short) scheme, which is also named searchable public-key encryption scheme, enables one to search encrypted documents on the untrusted server without revealing any information. Boneh et al. [13] first introduced PEKS scheme with a mail routing system in 2004. There are three entities in PEKS: data sender, receiver and server. The basic idea is as follows. User Bob (data sender) wishes to send an email to Alice (receiver). First, he encrypts the email M with keywords $w_1, w_2, \cdots, w_k$ using Alice's public key and also appends the keywords $PEKS(A_{pub}, w_1), \cdots, PEKS(A_{pub}, w_k)$. Then he sends the following ciphertext to the mail server (server):

$$C = (C_{PKE}||C_{PEKS}) = E_{A_{pub}}(M)||PEKS(A_{pub}, w_1), \cdots, PEKS(A_{pub}, w_k)$$

where A_{pub} is Alice's public key. Alice can provide the server with a certain trapdoor T_w (which is a trapdoor constructed by Alice on a keyword w) through a secure channel that enables the server to test whether the encrypted keyword associated with the message (C_{PEKS}) is equal to the keyword w selected by

This paper was supported by Institute Grant 2014GCYY02.

M. Yung et al. (Eds.): INTRUST 2014, LNCS 9473, pp. 394–406, 2015.
DOI: 10.1007/978-3-319-27998-5_25

Alice. Given $PEKS(A_{pub}, w')$ and T_w, the server can test whether $w = w'$. If $w \neq w'$, then the server learns nothing more about w'. In short, PEKS provides a mechanism that allows Alice to have the email server extract emails that contain a particular keyword by providing a trapdoor corresponding to the keyword, while the email server and other parties rather than Alice will not learn anything else about the email.

In [4], Abdalla et al. point out that we can translate any anonymous identity based encryption scheme [1,7,9,14,18] (AIBE) to a PEKS scheme. This construction was later improved by several researchers ([6,8,20]).

In 2004, Golle et al. [19] first proposed the notion of secret key encryption with conjunctive field keyword search scheme, which support Alice to retrieve the emails which contain some keywords, e.g. "urgent", "Monday" and "Marking Department". The notion of public key encryption with conjunctive field keyword search was proposed by Park et al. [22]. Byun et al. [16] and Hwang et al. [20] improved the efficiency of the conjunctive keyword search. However, all the above four schemes employ the same keyword field assumptions as: (1) The same keyword never appears in two different keyword fields in the same document. (2) Keyword field is defined for each document.

Later, many researchers improved expression to disjunctions, polynomial equations, inner products and so on [2,3,8,10,23,24].

Actually, the keyword field assumption restricts the application, keyword should be listed in any order without keyword field. Public key encryption with conjunctive field-free keyword search was studied in Boneh et al. BW scheme [8], Wang et al. WWJ scheme [26] and Zhang et al. ZZ scheme [27]. In WWJ and ZZ scheme, the basic idea is to computing the polynomial $H(w_1) + \cdots + H(w_k)$ where the k keywords included in the ciphertext are the all roots of polynomial $H(x)$. One could achieve this idea throught the attribute-hiding inner products encryption scheme OT [23]. However, all above 4 scheme cost $O(k)$ pairing computing to complete a conjunctive search in a ciphertext of k keywords.

A survey about provably secure searchable encryption was proposed in [15].

1.1   Our Contributions

The Lagrange polynomial was used in the scheme [17,25] to construct an anonymous multi-receiver identity-based encryption (AMIBE) scheme. We study the polynomial, and give a couple of reciprocal maps on $\mathbb{G}^k$, with which one could immediately index the target group elements. Using the properties of the couple of maps, we propose an efficient public key encryption with field-free conjunctive keywords search scheme. Security of our construction is based on decision linear Diffie-Hellman (DLIN) assumption in the random oracle model. To the best of our knowledge, our scheme is the most efficient searchable encryption support field-free conjunctive keywords search: we reduce the $O(k)$ pairing computing in Test function to $O(1)$.

1.2 Organization

The paper is organized as follows. In Sect. 2 we summarize notations, definitions and previous work. In Sect. 3 we study the Lagrange polynomial and give a couple of reciprocal maps on $\mathbb{G}^k$. In Sect. 4, we define security model for the public key encryption with field-free conjunctive keyword search. Proposed schemes with the security proofs are presented in Sect. 5. In Sect. 6, we give the performance analysis and comparisons. At last, we conclude our work in Sect. 7.

2 Preliminaries

In this section we summarize notations, definitions and prior work which is relevant to our result.

2.1 Bilinear Maps

We review bilinear maps, using the following standard notation [12,14]:

1. $\mathbb{G}$ and $\mathbb{G}_T$ are two (multiplicative) cyclic groups of prime order p;
2. g is a generator of $\mathbb{G}$;
3. $e : \mathbb{G} \times \mathbb{G} \to \mathbb{G}_T$ is a bilinear map.

Let $\mathbb{G}$ and $\mathbb{G}_T$ be two groups as above. A bilinear map is a map $e : \mathbb{G} \times \mathbb{G} \to \mathbb{G}_T$ with the following properties:

1. Bilinear: for all $u, v \in \mathbb{G}$ and $a, b \in \mathbb{Z}_p$, we have $e(u^a, v^b) = e(u, v)^{ab}$;
2. Non-degenerate: $e(g, g) \neq 1$.

We say that $\mathbb{G}$ is a bilinear group if the group action in $\mathbb{G}$ can be computed efficiently and there exists a group $\mathbb{G}_T$ and an efficiently computable bilinear map $e : \mathbb{G} \times \mathbb{G} \to \mathbb{G}_T$ as above. Note that $e(,)$ is symmetric since $e(g^a, g^b) = e(g, g)^{ab} = e(g^b, g^a)$.

2.2 Complexity Assumptions

The security of our schemes is based on the complexity assumption called Decision Linear Diffie-Hellman (DLIN) assumption. This assumption was introduced by Boneh et al. [11] and was used to construct a short group signature [11]. Let g_1, g_2, g_3 be random elements in group $\mathbb{G}$ and a, b, c be random elements in $\mathbb{Z}_p^*$. We define the DLIN problem in the $\mathbb{G}$ as: given 6 tuples $< g_1, g_2, g_3, g_1^a, g_2^b, g_3^c >\in \mathbb{G}^6$ as input, output 0 if $c = a + b$ and 1 otherwise. One can easily show that an algorithm for solving the DLIN problem in $\mathbb{G}$ gives an algorithm for solving DDH in $\mathbb{G}$. The converse is believed to be false. That is, it is believed that the DLIN problem is a hard problem even in the bilinear groups where DDH is easy [11,21]. We define the advantage of an algorithm $\mathcal{A}$ to solve the DLIN problem in $\mathbb{G}$ as

$$Adv_{\mathcal{A}}^{DLIN} = |Pr[\mathcal{A}(g_1, g_2, g_3, g_1^a, g_2^b, g_3^{a+b}) = 0] - Pr[\mathcal{A}(g_1, g_2, g_3, g_1^a, g_2^b, z) = 0]|$$

where the probability is over the random choice of $g_1, g_2, g_3, z \in \mathbb{G}$ and $a, b \in \mathbb{Z}_p$.

Definition 1. *We say that the (t, ϵ)-DLIN assumption holds in $\mathbb{G}$ if no t-time adversary has an advantage at least ϵ in solving the DLIN problem in $\mathbb{G}$.*

3 Lagrange Polynomial

For k distinct numbers $\bar{x} = \{x_i, i = 1, \cdots, k\}$, we have k polynomials

$$f_i(x) = \prod_{1 \leq j \neq i \leq k} \frac{x - x_j}{x_i - x_j} = a_{i,1} + a_{i,2}x + \cdots + a_{i,k}x^{k-1}$$

with degree $k - 1$. It is easy to find that

$$f_i(x_j) = \begin{cases} 1 & if \ i = j \\ 0 & if \ i \neq j \end{cases} \tag{1}$$

Those polynomials were used in Fan's [17] AMIBE scheme, we study more properties about those polynomial.

If we consider the polynomial $F_t(x) = \sum_{i=1}^{k} x_i^{t-1} f_i(x)$ for an constant $t, 1 \leq t \leq k$. It is easy to find that $F_t(x_j) = x_j^{t-1}$ for any $1 \leq j \leq k$ (for the property of Eq. (1)). However, the polynomials $F_t(x)$ and polynomials x^{t-1} have the degree at most $k - 1$, but have the same value at least k point. Thus, $F_t(x) = x^{t-1}$. So we have

$$F_t(x) = (\sum_{i=1}^{k} x_i^{t-1} a_{i,1}) + (\sum_{i=1}^{k} x_i^{t-1} a_{i,2})x + \cdots + (\sum_{i=1}^{k} x_i^{t-1} a_{i,k})x^{k-1} = x^{t-1}$$

and

$$\sum_{i=1}^{k} x_i^{t-1} a_{i,j} = \begin{cases} 1 & if \ t = j \\ 0 & if \ t \neq j \end{cases} \tag{2}$$

Let $\mathbb{G}$ be an group of order p. For k distinct numbers $\bar{x} = \{x_i \in \mathbb{Z}_p^*\}$ for $i = 1, \cdots, k$, we could construct two maps $\hat{r}_{\bar{x}}$ and $\hat{R}_{\bar{x}}$ from $\mathbb{G}^k$ to $\mathbb{G}^k$ refer to those polynomial $f_i(x)$ defined as above. That is

$$\hat{r}_{\bar{x}}: \qquad \mathbb{G}^k \ \rightarrow \ \mathbb{G}^k$$
$$(r_1, r_2, \cdots, r_k) \rightarrow (R_1, R_2, \cdots, R_k)$$
$$R_j = \prod_{i=1}^{k} r_i^{a_{i,j}}$$

$$\hat{R}_{\bar{x}}: \qquad \mathbb{G}^k \ \rightarrow \ \mathbb{G}^k$$
$$(R_1, R_2, \cdots, R_k) \rightarrow (r_1, r_2, \cdots, r_k)$$
$$r_i = \prod_{j=1}^{k} R_j^{x_i^{j-1}}$$

Lemma 1. *Let* $\mathbb{G}$, $f_i(x)$, $\hat{r}_{\overline{x}}$ *and* $\hat{R}_{\overline{x}}$ *be defined as above. Then* $\hat{R}_{\overline{x}}(\hat{r}_{\overline{x}}) = 1$ *and* $\hat{r}_{\overline{x}}(\hat{R}_{\overline{x}}) = 1$. *Thus, they are reciprocal.*

Proof. Let $(R_1, R_2, \cdots, R_k) = \hat{r}_{\overline{x}}(r_1, r_2, \cdots, r_k)$ and $(w_1, w_2, \cdots, w_k) = \hat{R}_{\overline{x}}(R_1, R_2, \cdots, R_k)$, so we have:

$$
\begin{aligned}
w_i &= \prod_{j=1}^{k} R_j^{x_i^{j-1}} = R_1 R_2^{x_i} \cdots R_k^{x_i^{k-1}} \\
&= (r_1^{a_{1,1}} r_2^{a_{2,1}} \cdots r_k^{a_{k,1}}) \cdot (r_1^{a_{1,2}x_i} r_2^{a_{2,2}x_i} \cdots r_k^{a_{k,2}x_i}) \cdots (r_1^{a_{1,k}x_i^{k-1}} r_2^{a_{2,k}x_i^{k-1}} \cdots r_k^{a_{k,k}x_i^{k-1}}) \\
&= r_1^{f_1(x_i)} r_2^{f_2(x_i)} \cdots r_k^{f_k(x_i)} \\
&= r_i
\end{aligned}
$$

the last equation comes from Eq. (1). Thus $\hat{R}_{\overline{x}}(\hat{r}_{\overline{x}}) = 1$.

Let $(r_1, r_2, \cdots, r_k) = \hat{R}_{\overline{x}}(R_1, R_2, \cdots, R_k)$ and $(W_1, W_2, \cdots, W_k) = \hat{r}_{\overline{x}}(r_1, r_2, \cdots, r_k)$, then we have:

$$
\begin{aligned}
W_j &= \prod_{i=1}^{k} r_i^{a_{i,j}} = r_1^{a_{1,j}} r_2^{a_{2,j}} \cdots r_k^{a_{k,j}} \\
&= (R_1^{a_{1,j}} R_2^{x_1 a_{1,j}} \cdots R_k^{x_1^{k-1} a_{1,j}}) \cdot (R_1^{a_{2,j}} R_2^{x_2 a_{2,j}} \cdots R_k^{x_2^{k-1} a_{2,j}}) \cdots (R_1^{a_{k,j}} R_2^{x_k a_{k,j}} \cdots R_k^{x_k^{k-1} a_{k,j}}) \\
&= R_1^{\sum_{i=1}^{k} a_{i,j}} R_2^{\sum_{i=1}^{k} x_i a_{i,j}} \cdots R_k^{\sum_{i=1}^{k} x_i^{k-1} a_{i,j}} \\
&= R_j
\end{aligned}
$$

the last equation comes from Eq. (2). Thus $\hat{r}_{\overline{x}}(\hat{R}_{\overline{x}}) = 1$. $\qquad\square$

So $\hat{r}_{\overline{x}}$ and $\hat{R}_{\overline{x}}$ can be considered as a couple of reciprocal maps. The couple of maps will be used in our scheme and the security proof. Let $\hat{R}_{\overline{x},i}$ denote the i-th component of $\hat{R}_{\overline{x}}$, that is $\hat{R}_{\overline{x},i}(R_1, R_2, \cdots, R_t) = r_i$.

4   Public Key Encryption with Field-Free Conjunctive Keyword Search Scheme

4.1   Model

We consider that a users encrypted data is outsourced in the storage of the untrusted server, such as email gateway, secure audit logs, and remote database server. In a public key model for keyword search, the server stores encrypted data collected from third parties and the user enables the server to retrieve emails containing keywords, which are the user wants to search, without leaking information. To support a conjunctive keyword search, Golle et al. [16,19,20,22] employ the same keyword fields assumptions as:

1. The same keyword never appears in two different keyword fields in the same document.
2. Every keyword field is defined for every document.

With this assumption, their scheme could immediately index the target ciphertext of keywords. Take the routing email system for example, they defined the keyword fields names as *To, From, Subject, Time* etc. However, we claim that these assumptions restrict the application. Normally, we would like to marker the email with any interesting keywords, such as *Sunshine, Seal, Homework* etc.

Actually keyword should be listed in any order without keyword fields. We propose a scheme without this keyword fields assumption.

The public key encryption with field-free conjunctive keyword search (PEFCK) consists of 4 polynomial time algorithms, (*KeyGen, PEFCK, Trapdoor, Test*) such that:

KeyGen: It takes a security parameter as input and returns *params*(system parameters) and the public/private key pair (*pk, sk*).

PEFCK: It is executed by the sender to encrypt a keyword set $\overline{W} = \{W_1, ..., W_k\}$. It produces a searchable keyword encryption S of $\overline{W}$ with the public key *pk*.

Trapdoor: It takes the secret key *sk* as input and the keyword query $\overline{Q} = \{Q_1, ..., Q_m\}$ for $m \leq k$, and returns a trapdoor $T_{\overline{Q}}$ for the conjunctive search of a given keyword query.

Test: It is executed by the server to search the documents with the keywords of a trapdoor $T_{\overline{Q}}$. It takes the public key *pk* as input, the searchable keyword encryption S, and the trapdoor $T_{\overline{Q}}$. Then output 1 if S include $\overline{Q}$ and 0 otherwise.

Actually, in the PEFCK scheme, to send a message m with keyword set $\overline{W}$, a ciphertext has a form of

$$< Enc(pk, m), PEFCK(pk, (W_1, W_2, \cdots, W_k)) >$$

where $Enc(\cdot)$ is a secure public key encryption. We concentrate on searchable encryption part.

4.2 Security Definition

We consider a semantic security against chosen keyword attacks as mentioned in the previous works [16,19,20,22]. The security games of symmetric encryptions for conjunctive keyword search was first defined in [19]. Then [16,20,22] modified the security games for public key setting. The security of the PEFCK scheme can be defined by two security games, indistinguishability of ciphertext from ciphertext (IND-CC-CKA) and indistinguishability of ciphertext from random oracle (IND-CR-CKA) against chosen keyword attacks. We briefly define them. The IND-CC-CKA game is as follows.

Setup: The challenger C takes a security parameter 1^l and runs the *Keygen* algorithm. The public key *pk* and the system parameters *params* are given to $\mathcal{A}$. The challenger C keeps the secret key *sk* to itself.

Phase 1: $\mathcal{A}$ adaptively queries a number of keyword set $\overline{Q_1}, \cdots, \overline{Q_d}$, to trapdoor oracle as follows.

    Trapdoor Queries $< \overline{Q_i} >$**.** The challenger runs Trapdoor $(sk, \overline{Q_i})$ and generates the trapdoor $T_{\overline{Q_i}}$. And then responds to $\mathcal{A}$.

Challenge: $\mathcal{A}$ selects two target keyword sets $\overline{W_0}$ and $\overline{W_1}$, and sends them to the challenger $\overline{C}$. The challenger picks a random bit $\beta \in \{0,1\}$. The only restriction is that $\overline{W_0}$ and $\overline{W_1}$ should not be distinguished by trapdoors issued in previous phases. And it sets $S_\beta = PEFCK(pk, \overline{W_\beta})$. Then send it to $\mathcal{A}$.

Phase 2: $\mathcal{A}$ additionally queries keyword sets $\overline{Q_{d+1}}, \cdots, \overline{Q_\lambda}$ to trapdoor oracle.

    Trapdoor Queries $< \overline{Q_i} \neq \overline{W_0} \text{ or } \overline{W_1} >$**.** The challenger runs Trapdoor $(sk, \overline{Q_i})$ and generates the trapdoor $T_{\overline{Q_i}}$. If $T_{\overline{Q_i}}$ cannot distinguish for $\overline{W_0}$ and $\overline{W_1}$, then it responds $T_{\overline{Q_i}}$ to $\mathcal{A}$.

Guess: Finally, $\mathcal{A}$ outputs a guess $\beta' \in \{0,1\}$. It wins the game if $\beta' = \beta$.

We define the advantage of the adversary $\mathcal{A}$ against the PEFCK scheme as the function of the security parameter 1^l: $Adv_{PEFCK,\mathcal{A}}^{IND-CC-CKA}(1^l) = |Pr[\beta' = \beta] - \frac{1}{2}|$.

We introduce another security game IND-CR-CKA which is a variant of the IND-CC-CKA game. This security game is the same as the IND-CC-CKA game except for the Challenge phase. While in the IND-CC-CKA game the adversary $\mathcal{A}$ selects two target keyword sets $\overline{W_0}$ and $\overline{W_1}$, and gives them to the challenger $\mathcal{C}$, in the IND-CR-CKA game $\mathcal{A}$ selects a target keyword set $\overline{W_0}$ and gives it to $\mathcal{C}$. Then $\mathcal{C}$ selects a random keyword set $\overline{R}$ and sets $\overline{W_1} = \overline{R}$. The IND-CR-CKA game is as follows.

Setup: The challenger $\mathcal{C}$ takes a security parameter 1^l and runs the *Keygen* algorithm. The public key pk and the system parameters *params* given to $\mathcal{A}$. The challenger $\mathcal{C}$ keeps the secret key sk to itself.

Phase 1: $\mathcal{A}$ adaptively queries a number of keyword sets $\overline{Q_1}, \cdots, \overline{Q_d}$, to trapdoor oracle as follows.

    Trapdoor Queries $< \overline{Q_i} >$**.** The challenger runs Trapdoor $(sk, \overline{Q_i})$ and generates the trapdoor $T_{\overline{Q_i}}$. And then responds to $\mathcal{A}$.

Challenge: $\mathcal{A}$ selects a keyword sets $\overline{W^*}$, and sends it to the challenger $\overline{C}$. The challenger selects a random keyword set $\overline{R}$ and picks a random bit $\beta \in \{0,1\}$. The only restriction is that $\overline{W^*}$ should not be distinguished for $\overline{R}$ from trapdoors issued in previous phases. And it sets $S_\beta = PEFCK(pk, \overline{W_\beta})$ where $\overline{W_0} = \overline{W^*}$ and $\overline{W_1} = \overline{R}$. Then sends $< S_\beta, \overline{W_0}, \overline{W_1} >$ to $\mathcal{A}$.

Phase 2: $\mathcal{A}$ additionally queries keyword sets $\overline{Q_{d+1}}, \cdots, \overline{Q_\lambda}$ to trapdoor oracle.

    Trapdoor Queries $< \overline{Q_i} \neq \overline{W_0} \text{ or } \overline{W_1} >$**.** The challenger runs Trapdoor $(sk, \overline{Q_i})$ and generates the trapdoor $T_{\overline{Q_i}}$. If $T_{\overline{Q_i}}$ cannot distinguish for $\overline{W_0}$ and $\overline{W_1}$, then it responds $T_{\overline{Q_i}}$ to $\mathcal{A}$.

Guess: Finally, $\mathcal{A}$ outputs a guess $\beta' \in \{0,1\}$. It wins the game if $\beta' = \beta$.

We define the advantage of the adversary $\mathcal{A}$ against the PEFCK scheme as the function of the security parameter 1^l: $Adv_{PEFCK,\mathcal{A}}^{IND-CR-CKA}(1^l) = |Pr[\beta' = \beta] - \frac{1}{2}|$.

The two security games, IND-CC-CKA and IND-CR-CKA, are all asymptotically equivalent [5,19].

Definition 2. *We say that a PEFCK scheme is (t, q_t, ϵ) (resp. (t, q_t, q_h, ϵ) in random oracle model)-secure if for any t-time IND-CC-CKA (resp. IND-CR-CKA) adversary $\mathcal{A}$ who makes at most q_t trapdoor queries (and q_h oracle queries), we have that $Adv_{PEFCK,\mathcal{A}}^{ATK} < \epsilon$ where ATK is IND-CC-CKA (resp. IND-CR-CKA).*

5   Our Construction

We introduce an efficient public key encryption with field-free conjunctive keyword search in which the keyword field assumption is not necessary. In addition, we provide a concrete security proof of our scheme. Our scheme is as follows.

KeyGen(1^l): Given the a security parameter 1^l, it returns $params = (\mathbb{G}, \mathbb{G}_T, e, H_1(\cdot), H_2(\cdot), H_3(\cdot), g)$ where $H_1 : \{0,1\}^* \to \mathbb{G}$, $H_2 : \{0,1\}^* \to \mathbb{G}$, $H_3 : \{0,1\}^* \to Z_p^*$ are three different collision-resistance hash functions and g is a generator of $\mathbb{G}$. And it picks a random value x in Z_p^* and computes $y = g^x$. The public/private key pair (pk, sk) is given by

$$(pk, sk) = (y, x)$$

PEFCK($pk, \overline{W}$): To generate PEFCK with keyword set $\overline{W} = \{W_1, \cdots, W_k\}$ with public key pk, the algorithm performs the following tasks:

1. computes $h_i = H_1(W_i)$, $f_i = H_2(W_i)$, $w_i = H_3(W_i)$, $i = 1, \cdots, k$.
2. constructs the maps $\hat{r}_{\overline{w}}$ from $\mathbb{G}^k$ to $\mathbb{G}^k$ refer to $\overline{w} = \{w_i\}$.
3. Picks two random values $s, r \in Z_p^*$.
4. Computes $A = g^r$, $B = y^s$.
5. For $i = 1, \cdots, k$, computes $c_i = h_i^r f_i^s$.
6. Sets $(C_1, C_2, \cdots, C_k) = \hat{r}_{\overline{w}}(c_1, c_2, \cdots, c_k)$
7. Sets $S = <A, B, C_1, \cdots, C_k>$ as the PEFCK.

Trapdoor($sk, \overline{Q}$): The algorithm first computes $h_i = H_1(Q_i)$, $f_i = H_2(Q_i)$, $q_i = H_3(Q_i)$, $i = 1, \cdots, m$, where $\overline{Q} = \{Q_1, ..., Q_m\}$. Then selects a random value $t \in Z_p^*$ and computes $D = g^t$, $E = (h_1 \cdots h_m)^t$, and $F = (f_1 \cdots f_m)^{t/x}$. At last outputs $T_{\overline{Q}} = (D, E, F, q_1, \cdots, q_m)$.

Test($pk, S, T_{\overline{Q}}$): The algorithm first computes $qc_i = \prod_{j=1}^k C_j^{q_i^{j-1}}$, $i = 1, \cdots, m$. Then checks if $e(A, E)e(B, F) = e(D, \prod_{i=1}^m qc_i)$.

Correctness: If $\overline{Q} \subseteq \overline{W}$, then $q_i = w_{u_i}$ for some u_i, so $qc_i = \prod_{j=1}^{k} C_j^{q_i^{j-1}} = \hat{R}_{\overline{w},u_i}(C_1, \cdots, C_k) = c_{u_i} = h_{u_i}^r f_{u_i}^s$, for $i = 1, \cdots, m$. So

$$e(D, \prod_{i=1}^{m} qc_i) = e(D, \prod_{i=1}^{m} c_{u_i}) = e(D, \prod_{i=1}^{m} h_{u_i}^r f_{u_i}^s)$$

$$= e(g^t, \prod_{i=1}^{m} h_{u_i}^r) e(g^t, \prod_{i=1}^{m} f_{u_i}^s)$$

$$= e(g^r, (\prod_{i=1}^{m} h_{u_i})^t) e(g^{xs}, (\prod_{i=1}^{m} f_{u_i})^{t/x})$$

$$= e(A, E) e(B, F)$$

5.1   Security

We prove that our PEFCK scheme is IND-CR-CKA secure under the DLIN assumption in the random oracle model.

Theorem 1. *Assmue that (t, ϵ)-DLIN assumption holds in $\mathbb{G}$. Then PEFCK is $(t', q_t, q_h, 4kq_t\epsilon)$-secure against IND-CR-CKA in the random oracle model for arbitrary q_t, q_h, k and $t' < t - \Theta(\tau m q_t q_h)$ where τ is the maximum time for an exponentiation in $\mathbb{G}$ and k is the maximum number of keywords in one ciphertext.*

Proof. Assume that $\mathcal{A}$ is an adversary with advantage ϵ' in breaking PEFCK against IND-CR-CKA and $H_1(\cdot)$, $H_2(\cdot)$, $H_3(\cdot)$ are modelled as random oracles. Then we can construct an adversary $\mathcal{B}$ who attacks the DLDH problem using $\mathcal{A}$ as described below.

Setup. The DLDH challenger gives the DLDH parameters $g_1, g_2, g_3, v_1, v_2, v_3$ to the adversary $\mathcal{B}$ where $v_1 = g_1^a$, $v_2 = g_2^b$ and $v_3 = g_3^{a+b}$ or z. $\mathcal{B}$ sets $g = g_1$ and $y = g_2$. At that time,the private key is regarded as α where $g_2 = g_1^\alpha$. In addition, it selects a random value η in $\mathbb{Z}_p^*$ and keeps it secret. This value is used in random oracles and Challenge phase. It gives $\mathcal{A}$ the system parameters $params = (\mathbb{G}, \mathbb{G}_T, e, H_1(\cdot), H_2(\cdot), H_3(\cdot), g)$ and the public key $pk = y$.

H_1, H_2, H_3-**queries** $\mathcal{A}$ issues at most q_h keyword queries to the random oracles. It simultaneously responds these queries. $\mathcal{B}$ maintains a list of tuples $\{W_i, cn_i; h_i, d_i; f_i, e_i; w_i\}$ called the H^{list}. If the keyword W_i is already queried, $\mathcal{B}$ returns $H_1(W_i) = h_i$, $H_2(W_i) = f_i$ and $H_3(W_i) = w_i$ in H^{list}. Otherwise, it sets w_i as a random value in $\mathbb{Z}_p^*$, and generates a random coin $cn_i \in \{0, 1\}$ so that $Pr[cn_i = 1] = 1/(kq_t)$. If $cn_i = 0$, then it selects two random values d_i and e_i in $\mathbb{Z}_p^*$, and sets $h_i = g_1^{d_i}$ and $f_i = g_2^{e_i}$. Otherwise, it selects a random value d_i in $\mathbb{Z}_p^*$ and computes $e_i = d_i/\eta$. And then it sets $h_i = g_3^{d_i}$ and $g_3^{e_i}$. It adds $\{W_i, cn_i; h_i, d_i; f_i, e_i; w_i\}$ to H^{list}, and returns h_i, f_i and w_i to $\mathcal{A}$.

Phase 1. $\mathcal{A}$ queries a number of keyword sets to trapdoor oracle.

Trapdoor Queries $\mathcal{A}$ queries a keyword set $\overline{Q_i} = \{Q_{i,1}, \cdots, Q_{i,m_i}\}$ to obtain a trapdoor $T_{\overline{Q_i}}$. $\mathcal{B}$ obtains the list such that $\{W_{i,j}, cn_{i,j}; h_{i,j}, d_{i,j}; f_{i,j}, e_{i,j}; q_{i,j}\}$ for $1 \leq j \leq m_i$ by running the above algorithm for responding to H_1, H_2, H_3-queries. If there is any $cn_{i,j} = 1$ for $1 \leq j \leq m_i$, then $\mathcal{B}$ aborts. Otherwise, it selects a random value t_i in $\mathbb{Z}_p^*$ and outputs $T_{\overline{Q_i}} = (T_{\overline{Q_{i,1}}}, T_{\overline{Q_{i,2}}}, T_{\overline{Q_{i,3}}}, q_{i,1}, \cdots, q_{i,m_i})$
where $T_{\overline{Q_{i,1}}} = g_1^{t_i}$, $T_{\overline{Q_{i,2}}} = g_1^{t_i(\sum_{j=1}^{m_i} d_{i,j})}$ and $T_{\overline{Q_{i,3}}} = g_1^{t_i(\sum_{j=1}^{m_i} e_{i,j})}$.

Challenge. $\mathcal{A}$ selects a target document $\overline{W}^*$ and sends it to $\mathcal{B}$. It selects a random document $\overline{R}$ and sets $\overline{W}_0 = \overline{W}^*$ and $\overline{W}_1 = \overline{R}$ where $\overline{W}_0 = \{W_{0,1}, \cdots, W_{0,k}\}$ and $\overline{W}_1 = \{W_{1,1}, \cdots, W_{1,k}\}$. The only restriction is that $\overline{W}_0$ and $\overline{W}_1$ should not be distinguished by trapdoors issued in previous phase. And then it selects a random bit $\beta \in \{0, 1\}$. $\mathcal{B}$ queries all keywords of $\overline{W}_\beta$ to H_1, H_2, H_3-oracles and obtains lists $\{W_{\beta,i}, cn_{\beta,i}; h_{\beta,i}, d_{\beta,i}; f_{\beta,i}, e_{\beta,i}; w_{\beta,i}\}$ for all i from oracles. If there is no $cn_{\beta,i} = 1$ for all i, it aborts. Otherwise, it computes a challenge ciphertext $S_\beta = \langle A, B, C_{\beta,1}, \cdots, C_{\beta,k} \rangle$ where $\{C_{\beta,i}\} = \hat{r}_{\overline{w_\beta}}(c_{\beta,i})$, where $\overline{w_\beta} = \{w_{\beta,1}, \cdots, w_{\beta,k}\}$ and $c_{\beta,i} = v_1^{d_{\beta,i}} v_2^{e_{\beta,i}}$ (in case that $cn_{\beta,i} = 0$) or $v_3^{d_{\beta,i}}$ (in case that $cn_{\beta,i} = 1$), $i = 1, \cdots, k$.

Phase 2. $\mathcal{A}$ continues to issue trapdoor queries which are not equal to $\overline{W}_0$ and $\overline{W}_1$. The only restriction is that it cannot issue trapdoor query distinguishing $\overline{W}_0$ for $\overline{W}_1$. $\mathcal{B}$ responds as in Phase 1.

Guess. Finally, $\mathcal{A}$ outputs a guess $\beta' \in \{0, 1\}$. If $\beta' = \beta$, then $\mathcal{B}$ outputs 1 meaning $v_3 = g_3^{a+b}$. Otherwise, it outputs 0 meaning $v_3 = z$.

The adversary $\mathcal{B}$ should not abort in **Trapdoor Queries** and **Challenge** phase for the success. The probability that it does not abort in **Trapdoor Queries** is $(1-1/kq_t)^{kq_t}$ for q_t queries. And the probability that it does not abort in **Challenge** phase is $(1-(1-1/kq_t)^k)$. Therefore, the probability that it did not abort during the simulation is greater than $1/(4 \cdot kqt)$ because $(1-1/kq_t)^{kq_t} \geq 1/4$ for any $kq_t \geq 2$ and $(1 - (1 - 1/kq_t)^l) \leq (1 - (1 - 1/kq_t)) = 1/kq_t$. If $v_3 = g_3^{a+b}$, then the challenge ciphertext is a valid encryption of the keyword set $\overline{W}_\beta$.

$$A = v_1 = g_1^a = g^a$$
$$B = v_2^\eta = g_2^{b\eta}$$
$$\{C_{\beta,i}\} = \hat{r}_{\overline{w_\beta}}(c_{\beta,i})$$
$$(if\, cn_{\beta,i} = 0)\ c_{\beta,i} = v_1^{d_{\beta,i}} v_2^{e_{\beta,i}\eta} = g_1^{d_{\beta,i}a} g_2^{e_{\beta,i}b\eta} = h_{\beta,i}^a f_{\beta,i}^{b\eta}$$
$$(if\, cn_{\beta,i} = 1)\ c_{\beta,i} = v_3^{d_{\beta,i}} = g_3^{(a+b)d_{\beta,i}} = (g_3^{d_{\beta,i}})^a (g_3^{d_{\beta,i}/\eta})^{b\eta}$$
$$= (h_{\beta,i})^a (g_3^{e_{\beta,i}})^{b\eta} = h_{\beta,i}^a f_{\beta,i}^{b\eta}$$

In this case, $\mathcal{A}$'s view is identical to its view in a real attack game and it must satisfy $|Pr[\beta' = \beta] - 1/2| \geq \epsilon'$. If $v_3 = z$ and $\mathcal{B}$ does not abort, since the map $\hat{r}_{\overline{w}}$ is invertible refer to any $\overline{w}$, then $|Pr[\beta' = \beta]| = 1/2$. Therefore, we have that

$$|Pr[\mathcal{B}(g_1, g_2, g_3, g_1^a, g_2^b, g_3^{a+b}) = 0] - Pr[\mathcal{B}(g_1, g_2, g_3, g_1^a, g_2^b, z) = 0]|$$

$$\geq \frac{1}{4lg_t}|(1/2 \pm \epsilon') - 1/2| = \frac{1}{4lg_t}$$

We complete the proof of the theorem. □

6   Performance Analysis and Comparisons

Actually, our aim is that we want to find out whether a trapdoor (include some keywords) could match with the ciphertext (matched when the keywords are the subset of keywords in the message). In order to make the comparison, the BW scheme [8], OT scheme [23], WWJ schem [26] and ZZ scheme [27] should adapt to this problem, while the scheme [20,22] do not meet the target(since the keyword field assumption). Assumption that there exist k keywords of the message and m keywords in trapdoor, we simple comparison our scheme with the BW scheme [8], OT scheme [23] and ZZ scheme [27] in Table 1.

Table 1. Performance comparisons

Scheme	Encryption	Trapdoor	Test
BW [8]	$(6k+2)G$	$(5m+2)G$	$(2k+1)e$
OT [23]	$(k+2)G$	$(2k+2)G$	$(4k+2)e$
WWJ [26]	$(k+1)G$	kG	ke
ZZ [27]	$(2k+4)G+1e$	$(3k+2)e$	$(2k+1)e$
Our scheme	$(k^2+2k+2)G$	$3G$	$kG+1e$

G: exponentiation in $\mathbb{G}$;
e: bilinear pairing mapping;
We omit the multiplication in $\mathbb{G}$ and $\mathbb{G}_T$

As other scheme use $O(k)$ pairing computing to determine whether the keywords set is in the ciphertext, our scheme reduce the $O(k)$ pairing computing in Test funciton to $O(1)$.

7   Conclusion

In this paper, we study the Lagrange polynomial which was used in the scheme [17,25] to construct an anonymous multi-receiver identity-based encryption scheme, and give a couple of reciprocal maps on $\mathbb{G}^k$, with which one could immediately index the target group elements. Using the properties of the couple of maps,we propose an efficient public key encryption with field-free conjunctive keywords search scheme. This scheme's aim is to find out whether a trapdoor (include some keywords) could match with the ciphertext (matched when the keywords are the subset of keywords in the message). Security of our construction is based on decision linear Diffie-Hellman (DLIN) assumption. To the best of our knowledge, our scheme is the most efficient searchable encryption support field-free conjunctive keywords search: we reduce the $O(k)$ pairing computing in Test to $O(1)$.

References

1. Agrawal, S., Boneh, D., Boyen, X.: Efficient lattice (H)IBE in the standard model. In: Gilbert, H. (ed.) EUROCRYPT 2010. LNCS, vol. 6110, pp. 553–572. Springer, Heidelberg (2010)
2. Agrawal, S., Freeman, D.M., Vaikuntanathan, V.: Functional encryption for inner product predicates from learning with errors. In: Lee, D.H., Wang, X. (eds.) ASIACRYPT 2011. LNCS, vol. 7073, pp. 21–40. Springer, Heidelberg (2011)
3. Agrawal, S., Gorbunov, S., Vaikuntanathan, V., Wee, H.: Functional encryption: new perspectives and lower bounds. In: Canetti, R., Garay, J.A. (eds.) CRYPTO 2013, Part II. LNCS, vol. 8043, pp. 500–518. Springer, Heidelberg (2013)
4. Kohno, T., Abdalla, M., Bellare, M., Catalano, D., Neven, G., Shi, H., Kiltz, E., Lange, T., Malone-Lee, J., Paillier, P.: Searchable encryption revisited: consistency properties, relation to anonymous IBE, and extensions. In: Shoup, V. (ed.) CRYPTO 2005. LNCS, vol. 3621, pp. 205–222. Springer, Heidelberg (2005)
5. Ballard, L., Kamara, S., Monrose, F.: Achieving efficient conjunctive keyword searches over encrypted data. In: Qing, S., Mao, W., López, J., Wang, G. (eds.) ICICS 2005. LNCS, vol. 3783, pp. 414–426. Springer, Heidelberg (2005)
6. Baek, J., Safavi-Naini, R., Susilo, W.: Public key encryption with keyword search revisited. In: Gervasi, O., Murgante, B., Laganà, A., Taniar, D., Mun, Y., Gavrilova, M.L. (eds.) ICCSA 2008, Part I. LNCS, vol. 5072, pp. 1249–1259. Springer, Heidelberg (2008)
7. Bellare, M., Kiltz, E., Peikert, C., Waters, B.: Identity-based (lossy) trapdoor functions and applications. In: Pointcheval, D., Johansson, T. (eds.) EUROCRYPT 2012. LNCS, vol. 7237, pp. 228–245. Springer, Heidelberg (2012)
8. Boneh, D., Waters, B.: Conjunctive, subset, and range queries on encrypted data. In: Vadhan, S.P. (ed.) TCC 2007. LNCS, vol. 4392, pp. 535–554. Springer, Heidelberg (2007)
9. Boyen, X., Waters, B.: Anonymous hierarchical identity-based encryption (without random oracles). In: Dwork, C. (ed.) CRYPTO 2006. LNCS, vol. 4117, pp. 290–307. Springer, Heidelberg (2006)
10. Boneh, D., Sahai, A., Waters, B.: Functional encryption: definitions and challenges. In: Ishai, Y. (ed.) TCC 2011. LNCS, vol. 6597, pp. 253–273. Springer, Heidelberg (2011)
11. Boneh, D., Boyen, X., Shacham, H.: Short group signatures. In: Franklin, M. (ed.) CRYPTO 2004. LNCS, vol. 3152, pp. 41–55. Springer, Heidelberg (2004)
12. Boneh, D., Boyen, X.: Efficient selective-ID secure identity-based encryption without random oracles. In: Cachin, C., Camenisch, J.L. (eds.) EUROCRYPT 2004. LNCS, vol. 3027, pp. 223–238. Springer, Heidelberg (2004)
13. Boneh, D., Di Crescenzo, G., Ostrovsky, R., Persiano, G.: Public key encryption with keyword search. In: Cachin, C., Camenisch, J.L. (eds.) EUROCRYPT 2004. LNCS, vol. 3027, pp. 506–522. Springer, Heidelberg (2004)

14. Boneh, D., Franklin, M.: Identity-based encryption from the Weil pairing. In: Kilian, J. (ed.) CRYPTO 2001. LNCS, vol. 2139, pp. 213–229. Springer, Heidelberg (2001)
15. Bösch, C., Hartel, P., Jonker, W., Peter, A.: A survey of provably secure searchable encryption. ACM Comput. Surv. (CSUR) **47**(2), 18 (2014)
16. Byun, J.W., Lee, D.-H., Lim, J.-I.: Efficient conjunctive keyword search on encrypted data storage system. In: Atzeni, A.S., Lioy, A. (eds.) EuroPKI 2006. LNCS, vol. 4043, pp. 184–196. Springer, Heidelberg (2006)
17. Fan, C.-I., Huang, L.-Y., Ho, P.-H.: Anonymous multireceiver identity-based encryption. IEEE Trans. Comput. **59**(9), 1239–1249 (2010)
18. Gentry, C.: Practical identity-based encryption without random oracles. In: Vaudenay, S. (ed.) EUROCRYPT 2006. LNCS, vol. 4004, pp. 445–464. Springer, Heidelberg (2006)
19. Golle, P., Staddon, J., Waters, B.: Secure conjunctive keyword search over encrypted data. In: Jakobsson, M., Yung, M., Zhou, J. (eds.) ACNS 2004. LNCS, vol. 3089, pp. 31–45. Springer, Heidelberg (2004)
20. Hwang, Y.-H., Lee, P.J.: Public key encryption with conjunctive keyword search and its extension to a multi-user system. In: Takagi, T., Okamoto, T., Okamoto, E., Okamoto, T. (eds.) Pairing 2007. LNCS, vol. 4575, pp. 2–22. Springer, Heidelberg (2007)
21. Kiltz, E.: Chosen-ciphertext security from tag-based encryption. In: Halevi, S., Rabin, T. (eds.) TCC 2006. LNCS, vol. 3876, pp. 581–600. Springer, Heidelberg (2006)
22. Park, D.J., Kim, K., Lee, P.J.: Public key encryption with conjunctive field keyword search. In: Lim, C.H., Yung, M. (eds.) WISA 2004. LNCS, vol. 3325, pp. 73–86. Springer, Heidelberg (2005)
23. Okamoto, T., Takashima, K.: Adaptively attribute-hiding (hierarchical) inner product encryption. In: Pointcheval, D., Johansson, T. (eds.) EUROCRYPT 2012. LNCS, vol. 7237, pp. 591–608. Springer, Heidelberg (2012)
24. Shen, E., Shi, E., Waters, B.: Predicate privacy in encryption systems. In: Reingold, O. (ed.) TCC 2009. LNCS, vol. 5444, pp. 457–473. Springer, Heidelberg (2009)
25. Wang, H., Zhang, Y., Xiong, H., Qin, B.: Cryptanalysis and improvements of an anonymous multi-receiver identity-based encryption scheme. IET Inf. Secur. **6**(1), 20–27 (2012)
26. Wang, P., Wang, H., Pieprzyk, J.: Keyword field-free conjunctive keyword searches on encrypted data and extension for dynamic groups. In: Franklin, M.K., Hui, L.C.K., Wong, D.S. (eds.) CANS 2008. LNCS, vol. 5339, pp. 178–195. Springer, Heidelberg (2008)
27. Zhang, B., Zhang, F.: An efficient public key encryption with conjunctive-subset keywords search. J. Netw. Comput. Appl. **34**(1), 262–267 (2011)

mOT+: An Efficient and Secure Identity-Based Diffie-Hellman Protocol over RSA Group

Baoping Tian[✉], Fushan Wei, and Chuangui Ma

State Key Laboratory of Mathematical Engineering and Advanced Computing,
Zhengzhou Information Science and Technology Institute, Zhengzhou 450001, China
baoping_tian@163.com

Abstract. In 2010, Rosario Gennaro et al. revisited the old and elegant Okamoto-Tanaka scheme and presented a variant of it called mOT. However the compromise of ephemeral private key will lead to the leakage of the session key and the user's static private key. In this paper, we propose an improved version of mOT(denoted as mOT+). Moreover, based on RSA assumption and CDH assumption we provide a tight and intuitive security reduction in the id-eCK model. Without any extra computational cost, mOT+ achieves security in the id-eCK model, and furthermore it also meets full perfect forward secrecy against active adversary.

Keywords: Public key cryptography · Diffie-Hellman · Composite modulus · id-eCK model

1 Introduction

Designing authenticated key exchange protocol that is as efficient as the original unauthentivated Diffie-Hellman protocol [1] attracts many cryptographers. In the process of pursuing this objects, many excellent protocols are born, such as HMQV [2] and MQV [3]. Efficient as they are, but when considering the verification and transmission of public key certificates, they will lose some efficiency and also result to a larger bandwidth. At the same time, the use of certificates will introduce some other troubles. Of note are the problems of certificate management, such as revocation, storage and distribution. In order to eliminate the requirement for certificates in the traditional PKI and some related issues with them, Adi Shamir [4] first introduced the concept of identity-based public key cryptography(ID-PKC) in 1984. Since then, much work has been dedicated to designing identity-based authenticated key exchange(IB-AKE) protocols [5–7], most of them are pairing-based after the pioneering work by Joux [8]. However, pairings are hard to implement in the real word and the computational cost is very expensive. Thus, pairing-free IB-AKE protocol has some advantages over pairing-based ones both in efficiency and implementation.

Among the pairing-free IB-AKE protocols, the old and elegant Okamoto-Tanaka scheme [9] catches our eyes. The exponentiation per user is very close to that in unauthentivated Diffie-Hellman protocol [1] when e takes the value

© Springer International Publishing Switzerland 2015
M. Yung et al. (Eds.): INTRUST 2014, LNCS 9473, pp. 407–421, 2015.
DOI: 10.1007/978-3-319-27998-5_26

of small exponent, e.g. $e = 3$. Masahiro Mambo and Hiroki Shizuya [10] showed that breaking the Okamoto-Tanaka scheme can be reduced to breaking the Diffie-Hellman scheme over $\mathbb{Z}_N$. Later, Seungjoo KIM et al. [11] analyzed the security of the Okamoto-Tanaka protocol against active attacks, and they showed the rigorous security of Okamoto-Tanaka scheme in their attack models. In 2010, Rosario Gennaro et al. [12] analyzed Okamoto-Tanaka scheme at length, and they showed that the original Okamoto-Tanaka scheme is susceptible to some attacks, particularly known-key and malleability attacks. After some simple modifications, they obtained a variant(denoted as mOT) of Okamoto-Tanaka scheme and provided a security proof of mOT in the Canetti-Krawczyk(CK) model [13] based on the RSA assumption. In addition, they proved full perfect forward secrecy of mOT against active attacks.

Yet, both mOT [12] and the original Okamoto-Tanaka scheme [9] suffer from ephemeral key compromise. The compromise of ephemeral key will lead to that the adversary can compute the shared secret value by itself, then followed by the leakage of session key. Furthermore, the compromise of ephemeral key can also result to the leakage of the user's long-term private key. Menezes and Ustaoglu [14] pointed out that the compromise of ephemeral key may happen in the cases of side-channel attack, the using of a weak random number generator and stored insecurely. Taking the ephemeral key compromise into account, Lamacchia et al. [15] proposed the eCK model on the foundation of the Canetti-Krawczyk(CK) model [13]. Afterwards, huang et al. [16] extended the eCK model so that it can be applied to identity-based setting, hence called id-eCK model.

In this paper, by applying the NAXOS trick [15] to mOT without incurring any computational overhead, we obtain a securer two-pass authenticated Diffie-Hellman protocol named mOT+. Further, based on RSA assumption and CDH assumption with composite modulus, we prove the security of mOT+ in the id-eCK model [16].

ORGANIZATION. The rest of this paper is organized as follows: In Sect. 2, we introduce some related preliminaries. Section 3 outlines the id-eCK model. Then, we give the protocol description and prove the security of the first one in Sect. 5. In the final section, we draw a conclusion on this paper.

2   Preliminaries

Notations. Let k be the security parameter. By $\mathcal{A}(x, y, \cdots)$, we denote a polynomial time algorithm $\mathcal{A}$ with inputs $(x, y, \cdots)$. And By $r \leftarrow_R \mathcal{A}(x, y, \cdots)$, we denote that running the polynomial time algorithm $\mathcal{A}$ with inputs $(x, y, \cdots)$, it outputs r. We denote by $z \in_R \mathbb{Z}$ choosing an element z uniformly at random from $\mathbb{Z}$.

The Group of Quadratic Residues. Let $N = PQ$ be a safe-prime RSA modulus($P = 2p + 1, Q = 2q + 1$ and p, q are also primes) and $\mathbb{Z}_N$ be the set of integers modulo N. $\mathbb{Z}_N^*$ is defined as the set of elements from $\mathbb{Z}_N$ that have an inverse modulo N. The set of quadratic residues modulo N forms a multiplicative cyclic group denoted as $\mathbb{QR}_N$. The order of which is $\frac{\phi(N)}{4} = pq$. When selecting a random element u from $\mathbb{QR}_N$, by convention we choose $v \in_R \mathbb{Z}_N^*$ and set $u = v^2$. A random element from $\mathbb{QR}_N$ is the generator of $\mathbb{QR}_N$ with overwhelming probability.

Definition 1 (Computational Diffie-Hellman(CDH) Assumption with Composite Modulus). *Let $\mathcal{IG}$ be an integer instance generator that on input 1^k, outputs (N, P, Q) where $P = 2p+1, Q = 2q+1$ and P, Q, p, q are primes. Let g be a random generator of order pq. CDH assumption holds for $\mathcal{IG}$, if for all probabilistic polynomial time adversary $\mathcal{A}$, given (g, g^u, g^v) where $u, v \in_R \mathbb{Z}_{pq}$, the probability is bounded by*

$$\Pr[g^{uv} \leftarrow_R \mathcal{A}(g, N, U = g^u, V = g^v) : (N, P, Q) \leftarrow_R \mathcal{IG}(1^k); u, v \in_R \mathbb{Z}_{pq}] < \epsilon(k)$$

where $\epsilon(k)$ is negligible in k. The probability is taken over the coin tosses of $\mathcal{A}$, uniformly random choice of u, v from $\mathbb{Z}_{pq}$ and integer instance generator $\mathcal{IG}$.

Remark. It is worth noting that if there exist an efficient algorithm to factor N, then it is left with the CDH problem modulo the factors of N, which is intractable provided P and Q are large enough. Actually, CDH assumption with composite modulus can be implied by the hardness of factoring [17].

Definition 2 (RSA Assumption). *Let $\mathcal{IG}$ be an RSA instance generator that on input 1^k outputs (N, e) such that $N = PQ$ where P, Q are safe primes and $e \in \mathbb{Z}_{\phi(N)}^*$ where ϕ is Euler's totient function.*
*   The RSA assumption holds relative to $\mathcal{IG}$ if for all probabilistic polynomial time adversary $\mathcal{A}$, the successful probability is*

$$\Pr[x \leftarrow_R \mathcal{A}(N, e, y) : (N, e) \leftarrow_R \mathcal{IG}(1^k), y \in_R \mathbb{Z}_N^*] < \epsilon(k)$$

where $x^e \equiv y \pmod{N}$ and $\epsilon(k)$ is negligible in k. The probability is taken over the coin tosses of $\mathcal{A}$, uniformly random choice of y from $\mathbb{Z}_N^*$ and RSA instance generator $\mathcal{IG}$.

Lemma 1 [18]. *Let N, e, d be the RSA parameters and b be an integer such that $\gcd(e, b) = 1$, then we can obtain $x^d \pmod{N}$ from $(x^b)^d \pmod{N}$ where $x \in \mathbb{Z}_N^*$.*

3   A Review of id-eCK Model

We model the users from the finite set $\mathcal{U}$ of size n as probabilistic polynomial time Turing machines(PPT). Each user owns an unique identifier ID_i(e.g. a binary string related to their actual name). The user obtains the static private key corresponding to her/his identity from the KGC through a secure channel.

Sessions. An instance of a protocol is called a session which can be activated by an incoming message of the forms (ID_i, ID_j) or (ID_i, ID_j, Y). If it's activated by the first form, then ID_i is called the *session initiator*, otherwise the *session responder*. The session can be uniquely identified by the *session identifier* which is in the form of $sid = (ID_i, ID_j, X, Y)$ where X and Y are the messages prepared by ID_i and ID_j respectively. For $sid = (ID_i, ID_j, X, Y)$, we call ID_i the *owner* of session sid and ID_j the *peer*. We denote by $\Pi^s_{i,j}$ the sth session executed by the owner ID_i with intended peer ID_j. If the session identifier of the tth session between ID_i and ID_j is of the form $sid' = (ID_j, ID_i, X, Y)$, then we say $\Pi^t_{j,i}$ is the *matching session* of $\Pi^s_{i,j}$ and vice versa.

Adversary. The PPT adversary owns the competence of controlling all the communications between users(e.g. intercept, modify, delay, inject its own messages, schedule sessions etc.) except that between users and KGC. The adversary can capture the leakage of private information of users via the following queries

1. *StaticKeyReveal*(ID_i): The adversary learns the static private key of ID_i.
2. *EphemeralKeyReveal*($\Pi^s_{i,j}$): The adversary learns the ephemeral secret key of $\Pi^s_{i,j}$.
3. *SessionKeyReveal*($\Pi^s_{i,j}$): The adversary learns the session key of $\Pi^s_{i,j}$ if it holds one.
4. *MasterKeyReveal*(): The adversary learns the master key of KGC.
5. *Establish*(ID_i): The adversary can register legal user on behalf of ID_i, and ID_i is completely controlled by the adversary. Users who are not revealed by this query are called *honest*.
6. *Send*($\Pi^s_{i,j}$,m): The adversary sends a message m to the session $\Pi^s_{i,j}$ and obtains a response according to the description of the protocol.
7. *Test*($\Pi^s_{i,j}$): On this query, the simulator flips a coin, and returns the session key of $\Pi^s_{i,j}$ to the adversary if it's 1, otherwise a random string under the distribution of the session key. This query can only be asked once. The goal of adversary is to distinguish the session key from a random string. If the adversary guesses the coin correctly and the test session is still *fresh*, then we say the adversary *wins*.

Definition 3 (Freshness). *Let $\Pi^s_{i,j}$ be a completed session executed between two honest users ID_i and ID_j. Let $\Pi^t_{j,i}$ be the matching session of $\Pi^s_{i,j}$ if it exists. $\Pi^s_{i,j}$ is fresh if none of the following happens:*

1. *The adversary issues **SessionKeyReveal**($\Pi^s_{i,j}$ or $\Pi^t_{j,i}$) (if $\Pi^t_{j,i}$ exists)*
2. *$\Pi^t_{j,i}$ exists and the adversary makes either of the following queries:*
   *(a) both **StaticKeyReveal**(ID_i) and **EphemeralKeyReveal**($\Pi^s_{i,j}$)*
   *(b) both **StaticKeyReveal**(ID_j) and **EphemeralKeyReveal**($\Pi^t_{j,i}$)*
3. *$\Pi^t_{j,i}$ does not exist and the adversary makes either of the following queries:*
   *(a) both **StaticKeyReveal**(ID_i) and **EphemeralKeyReveal**($\Pi^s_{i,j}$)*
   *(b) **StaticKeyReveal**(ID_j)*

Definition 4 (id-eCK Security). *The advantage of the adversary $\mathcal{M}$ is defined as*

$$Adv^{IB-AKE}_{\mathcal{M}}(k) = Pr[\mathcal{M} \; wins] - \frac{1}{2}$$

We say that an IB-AKE protocol is secure if the following conditions hold:

- If two honest users complete matching sessions then they both compute the same session key.
- For all PPT adversary, $Adv_{\mathcal{M}}^{IB-AKE}(k)$ is negligible.

4   mOT+ Protocol

In this section, we first describe mOT+ protocol, then give some related arguments. At last, we compare it with mOT [12] and Okamoto-Tanaka scheme [9]. In the rest of this paper, for simplicity we omit " (mod N)" operation when operating on $\mathbb{Z}_N^*$.

4.1   Protocol Description

Set up. Key Generation Center(KGC) chooses RSA parameters (N, e, d) properly where N is safe-prime RSA modulus(i.e. N is the product of two safe primes), a random generator g for $\mathbb{QR}_N$ and three hash functions $H_1 : \{0,1\}^* \rightarrow \mathbb{QR}_N$, $H_2 : \{0,1\}^k \times \mathbb{QR}_N \rightarrow \mathbb{S}$ and $H : \{0,1\}^* \rightarrow \{0,1\}^k$. Then, KGC publishes (N, e, g, H_1, H_2, H). The user will receive the KGC's RSA signature on his/her identity as the secret key (e.g. the secret key of user U with identifier ID_U is the value $S_U = H_1(ID_U)^d$).

Key Agreement. Let's assume that the initiator A and the responder B with identifier ID_A and ID_B respectively want to establish a shared session key.

Step1: A first chooses $x \in_R \{0,1\}^k$, then computes $\tilde{x} = H_2(x, S_A)$ and $\alpha = g^{\tilde{x}} S_A$. On finishing computing α, A destroys $\tilde{x}$ and sends (ID_A, α) to B.

Step2: On receiving the communication from A, B first checks if $\alpha \in \mathbb{Z}_N^*$. If so, B chooses $y \in_R \{0,1\}^k$ and computes $\tilde{y} = H_2(y, S_B)$, after calculating $\beta = g^{\tilde{y}} S_B$, B destroys $\tilde{y}$. Then, B sends (ID_B, β) to A. At last, B computes the shared secret value $\sigma = \left(\frac{\alpha^e}{H_1(ID_A)} \right)^{2H_2(y, S_B)}$ and sets the session key to be $H(\sigma, \alpha, \beta, ID_A, ID_B)$.

Step3: After receiving the communication from B, A first checks if $\beta \in \mathbb{Z}_N^*$. If so, A computes the shared secret value $\sigma = \left(\frac{\beta^e}{H_1(ID_B)} \right)^{2H_2(x, S_A)}$ and sets the session key to be $H(\sigma, \alpha, \beta, ID_A, ID_B)$.

In the process of key agreement, if any verification fails, then the user will abort the session. If the key agreement is successful then it's straightforward to see that A and B share the same secret value $g^{2eH_2(x, S_A)H_2(y, S_B)}$ and therefore the same session key (Fig. 1).

4.2   Some Arguments

Exponentiation with a Short Exponent Set $\mathbb{S}$. If $N = PQ$ where $P = 2p + 1, Q = 2q + 1$, normally $\mathbb{S}$ should be $\{1, \ldots, pq\}$. But the knowledge of

$$A$$
$$(S_A = H_1(ID_A)^d)$$
$$x \in_R \{0,1\}^k, \tilde{x} = H_2(x, S_A)$$
$$\alpha = g^{\tilde{x}} S_A$$
$$\text{Destroy } \tilde{x}$$

$$B$$
$$(S_B = H_1(ID_B)^d)$$

$$\xrightarrow{ID_A, \alpha}$$

$$\text{Check } \alpha \overset{?}{\in} \mathbb{Z}_N^*$$
$$y \in_R \{0,1\}^k, \tilde{y} = H_2(y, S_B)$$
$$\beta = g^{\tilde{y}} S_B$$
$$\text{Destroy } \tilde{y}$$

$$\xleftarrow{ID_B, \beta}$$

$$\text{Check } \beta \overset{?}{\in} \mathbb{Z}_N^*$$
$$\sigma = \left(\frac{\beta^e}{H_1(ID_B)}\right)^{2H_2(x, S_A)}$$

$$\sigma = \left(\frac{\alpha^e}{H_1(ID_A)}\right)^{2H_2(y, S_B)}$$

$$K = H(\sigma, \alpha, \beta, ID_A, ID_B)$$

Fig. 1. mOT+ protocol

pq yields the factoring of N. But we can use $\{1, \ldots, \lfloor N/4 \rfloor\}$ as a substitution, because the two sets $\{g^x \pmod{N}, x \in_R \{1, \ldots, pq\}\}$ and $\{g^x \pmod{N}, x \in_R \{1, \ldots, \lfloor N/4 \rfloor\}\}$ are statistically close if p, q are in the same size [19]. Using the related result in [20], we can derive that $\{g^x \pmod{N}, x \in_R \{1, \ldots, \lfloor N/4 \rfloor\}\}$ and $\{g^x \pmod{N}, x \in_R \{1, \ldots, 2^\lambda\}\}$ are computationally indistinguishable. To save time, we can implement with short exponents i.e. setting $\mathbb{S} = \{1, \ldots, 2^\lambda\}$ on condition that the discrete logarithm problem over $\mathbb{Z}_N^*$ is still hard with this setting. For example, when the modulus is of size 1024 bits, the exponent can be of length 164 bits(or 2048-bit modulus with 226-bit exponent).

Binding the Ephemeral Key with the Long-Term Private Key. In our protocol, we bind the ephemeral key with the long-term private key with a hashing operation. The merits of doing in this way is that without knowing both the ephemeral private key and the static private key of user A, no one can compute $H_2(x, S_A)$. The leakage of ephemeral private key won't put the session key and the static private key at risk. Besides, this hash value is never stored. Whenever the hash value is needed, it is computed.

The Square Operation in the Shared Value. In the original Okamoto-Tanaka scheme [9], when computing the shared value, this is no extra square operation for both users. While in mOT [12], in order to carry the simulation the authors add a square operation in computing the shared value. During the proof of mOT+ in the id-eCK model, we find that this square operation doesn't contribute to our simulation. In other words, no matter whether this square operation is added or not, we can carry the simulation smoothly. However, problem is that the user cannot determine whether the received communication belongs to the group of quadratic residues. What the user can determine is the Jacobi symbol and the membership in $\mathbb{Z}_N^*$ of the received communication. If the Jacobi symbol of the received communication is -1, then of course the received communication is outside QR_N. Otherwise, it's hard to determine[1]. In other words, the users cannot perform complete subgroup validation. In order to avoid the potential small subgroup attack [22], we keep this square operation in our protocol. It is worth noting that if we instantiate our protocol over the group of signed quadratic residues [23] in which the membership can be publicly verified, then this square operation can be removed.

4.3   Comparison

In this section, we compare our protocol with the original Okamoto-Tanaka scheme [9](denoted as OT) and mOT [12] in terms of efficiency(exponentiations per party), CK-security, eCK-security and full perfect forward security(Full-PFS). When considering the efficiency, we let $e = 3$. Then in this case, our protocol is as efficient as OT and mOT. In the original paper [9], Okamoto and Tanakam didn't provide a proof, but mOT protocol [12] is proved in the CK model [13]. Our protocol can be proved secure not only in the CK model as the mOT does[2], but also in the eCK model.

The most attractive place of mOT is that it meets the full perfect forward security against active attackers. While for implicitly-authenticated 2-message protocols, they can only achieve weak forward security against passive attackers [2]. Arming full PFS for implicitly-authenticated 2-message protocols will need an extra message or explicit signature. Thus, compared with implicitly-authenticated protocols mOT enjoys full-PFS in only two messages without additional message or explicit signature. Our modification doesn't make mOT to lose this desirable property, i.e. mOT+ achieves also full-PFS, and the proof of which is almost identical to that of mOT. Thereby, we won't repeat it in our paper (Fig. 2).

	efficiency(e=3)	CK Model	eCK Model	Full-PFS
OT [9]	2	?	×	?
mOT [12]	2	√	×	√
mOT+	2	√	√	√

Fig. 2. Protocol comparison

5   Security Proof in the id-eCK Model

In this section, we will presents a formal security proof for mOT+ in the id-eCK model.

---

[1] With Jacobi symbol 1, determining the membership in QR_N is equivalent to solving the quadratic residues assumption [21].

[2] As the simulation is almost the same with that in mOT, so we omit the proof in this paper.

Theorem 1. *If RSA assumption and CDH assumption with composite modulus hold for $\mathcal{IG}(1^k)$, and H_1, H_2, H are modeled as random oracles, then mOT+ is id-eCK secure.*

PROOF. Let $\mathcal{M}$ be a polynomially bounded adversary. We assume that there are at most $n(k)$ users and $\mathcal{M}$ activates up to $s(k)$ sessions within a user. It's straightforward to verify that if two honest users complete the matching session, then they must compute the same session key. Since the session key is computed via $H(\sigma, \alpha, \beta, ID_i, ID_j)$, while $H()$ is modeled as a random oralce, then the adversary can only distinguish the session key from a random string in the following three ways

1. **Guessing attack:** $\mathcal{M}$ guesses the session key correctly.
2. **Key replication attack:** $\mathcal{M}$ coerces two non-matching sessions to output the same session key. Then, $\mathcal{M}$ selects one of two as the test session and queries the session key of the left one.
3. **Forging attack:** $\mathcal{M}$ tries to compute σ of the test session, then queries H with $(\sigma, \alpha, \beta, ID_i, ID_j)$ to get $H(\sigma, \alpha, \beta, ID_i, ID_j)$.

As $H()$ is modeled as a random oralce, the output of which follows uniform distribution. Without querying $H()$, the probability of guessing the session key correctly for the adversary is $\mathcal{O}(\frac{1}{2^k})$, which is negligible. Since the definition of session key includes the session informatoin, two non-matching sessions cannot have the same session key except with negligible probability. Thus, guessing attack and key replication attack are excluded. So, forging attack is the only possibly feasible method for the adversary. Let M be the event that the adversary wins the distinguishing game by forging attack. Then, the following inequality holds

$$Adv_{\mathcal{M}}^{IB-AKE}(k) \le \Pr[M] + \frac{1}{2}(1 - \Pr[M]) - \frac{1}{2} \iff Adv_{\mathcal{M}}^{IB-AKE}(k) \le \frac{1}{2}\Pr[M]$$

Next, we will show that if adversary wins the distinguishing game with non-negligible advantage, this implies that M happens with non-negligible probability, then we can use $\mathcal{M}$ to construct an algorithm $\mathcal{S}$ to solve the RSA problem or the CDH problem in polynomial time with non-negligible probability.

Considering the following three sub-events

E1. There exists an honest user B such that $\mathcal{M}$ queries $H_2(*, S_B)$ before (or without) performing the $StaticKeyReveal(ID_B)$.
E2. E1 never happens and the test session has a matching session.
E3. E1 never happens and the test session has no matching session.

Let $\bar{E1}$ denotes the complementary event of E1. Then we have $\Pr[M] = \Pr[M \wedge E1] + \Pr[M \wedge \bar{E1}]$, where $\Pr[M \wedge E1]$ and $\Pr[M \wedge \bar{E1}]$ are the probabilities that event $M \wedge E1$ and event $M \wedge \bar{E1}$ happen respectively. While event $M \wedge \bar{E1}$ can then be divided into two complementary events E2 and E3 with regard to whether the matching session of test session exists. Namely, the equation $\Pr[M \wedge \bar{E}1] = \Pr[M \wedge E2] + \Pr[M \wedge E3]$ holds. At last, we have

$$\Pr[M] = \Pr[M \wedge E1] + \Pr[M \wedge E2] + \Pr[M \wedge E3]$$

If M happens with non-negligible probability, then it must be that one of the three events (event $M \wedge E1$, event $M \wedge E2$ and event $M \wedge E3$) occurs with non-negligible probability. Next, we will analysis these three events separately.

5.1 Event $E1$

Assuming that event $M \wedge E1$ happens with non-negligible probability. In this case, we will use the adversary as a subroutine to build a RSA problem solver S. The input to S is (N, e, y) where N, e are drawn from the same distribution as that in the protocol and y is chosen uniformly at random from $\mathbb{Z}_N^*$. The goal of S is to output x such that $x^e \equiv y \pmod{N}$.

First, S prepares $n(k)$ honest parties. Within these users, S selects the user B at random with probability at least $\frac{1}{n(k)}$. Then, it sets $H_1(ID_B) = y^2$. For the remaining $n(k) - 1$ users, the simulator assigns random static key pairs. The simulator S chooses $\bar{r} \in_R \mathbb{QR}_N$, and sets $r = \bar{r}^e, g = (ry)^e$.

The simulation of the session whose owner is not B follows the description of the protocol. However, there are some problems for S to simulate the session related to B. Because it doesn't have knowledge of the static private key of B, but may have to answer the $SessionKeyReveal$. In order to keep the consistency of the oracle $H()$, S maintains an extra list S^{list}. Next, we mainly simulate the sessions between B and C where C is impersonated by the adversary. Without loss of generality, let's assume that B is responder.

1. $Send(\Pi_{i,j}^s, m)$: S maintains an initially empty list S^{list} with entries of the form $(\alpha, \beta, ID_i, ID_j, K)$.
   - if $ID_i = ID_B$, for simplicity, let $ID_j = ID_C, m = \gamma$. S chooses $b \in_R \mathbb{Z}_{\lfloor N/4 \rfloor}$, then it returns $\beta = g^b/\bar{r}$ to the adversary. Afterwards, S searches the H^{list} to check if the entry $(\sigma, \beta, \gamma, ID_C, ID_B, h)$ exists.
     - if it exists in H^{list}, with the choice of $y = g^d/r, \beta = g^b/\bar{r}, r = \bar{r}^e$, we have that $\beta^e/y = g^{eb-d}$. While the shared secret value can be represented as $\sigma = CDH^2(\beta^e/y, \gamma/S_C) = (\gamma/S_C)^{2(eb-d)}$. So, to verify

       $$\sigma \stackrel{?}{=} (\gamma/S_C)^{2(eb-d)}$$

       is equivalent to

       $$\sigma^e \stackrel{?}{=} (\gamma/S_C)^{2(e^2b-1)}$$

       With the knowledge of S_C, e, b, S can carry this verification easily. If the verification passes, then S inserts the new tuple $(\beta, \gamma, ID_C, ID_B, h)$ into S^{list} where h comes from H^{list}; otherwise, it chooses $K \in_R \{0, 1\}^k$, then stores the new tuple $(\beta, \gamma, ID_C, ID_B, K)$ in S^{list}.

- else(no such entry in H^{list}) S chooses $K \in_R \{0,1\}^k$, then stores the new tuple $(\beta, \gamma, ID_C, ID_B, K)$ in S^{list}.
  - if $ID_i \neq ID_B$, S follows the description of the protocol.
2. $H_1(ID_i)$: S maintains an initially empty list H_1^{list} with entries of the form (ID_i, p_i^e).
   The action of S is as follows
   - if $ID_i \neq ID_B$, S selects $p_i \in \mathbb{QR}_N$ randomly, then inserts the tuple (ID_i, p_i^e) into H_1^{list}. The benefits of setting in this way is that the simulator knows all the users' secret keys(except B) as $S_{ID_i} = H_1(ID_i)^d = p_i^{ed} = p_i$ holds.
   - else($ID_i = ID_B$), S sets $H_1(ID_B) = y^2$ and stores the tuple (ID_B, y^2) in H_1^{list}, where y is chosen from $\mathbb{Z}_N^*$ at random.
3. $H_2(x, S_{ID_i})$: S checks $S_{ID_i}{}^e \overset{?}{=} y^2$, in which case the simulator stops and outputs $(y^2)^d = S_{ID_i}$, then by using the lemma 1 we obtain $x = y^d$ successfully; otherwise, S simulates a random oracle in the usual way.
4. $H(\sigma, \alpha, \beta, ID_i, ID_j)$: S maintains an initially empty list H^{list} with entries of the form $(\sigma, \alpha, \beta, ID_i, ID_j, h)$. S simulates a random oracle in the usual way, except for the entries with the form $(\sigma, \beta, \gamma, ID_C, ID_B)$. In which case, it simulates in the following way
   - if the entry $(\sigma, \beta, \gamma, ID_C, ID_B)$ is already in the list, then S returns the stored value.
   - else the simulator searches the entry $(\beta, \gamma, ID_C, ID_B, *)$ in the S^{list}.
     - If such entry exists in S^{list}, then S verifies the correctness of σ by checking if $\sigma^e = (\gamma/S_C)^{2(e^2b-1)}$ holds, in which case S returns the stored value K in S^{list} to the adversary and inserts the tuple $(\sigma, \beta, \gamma, ID_C, ID_B, K)$ into H^{list} where K comes from S^{list}; otherwise S selects $h \in \{0,1\}^k$ at random, and stores $(\sigma, \beta, \gamma, ID_C, ID_B, h)$ in H^{list}.
     - else(no such entry exists) S chooses $h \in \{0,1\}^k$ randomly, and stores the new tuple $(\sigma, \beta, \gamma, ID_C, ID_B, h)$ in the H^{list}.
5. $EphemeralKeyReveal(\Pi_{i,j}^s)$: S simulates this query faithfully.
6. $SessionKeyReveal(\Pi_{i,j}^s)$: S returns the stored value K in the S^{list}.
7. $StaticKeyReveal(ID_i)$: If $ID_i = ID_B$, S aborts; otherwise, responds faithfully.
8. $Establish(ID_i)$: S simulates this query faithfully.
9. $Test(\Pi_{i,j}^s)$: S simulates this query faithfully.
10. If $\mathcal{M}$ outputs a guess, S aborts.

Analysis of Event M $\wedge$ E1. The simulation of S for the adversary $\mathcal{M}$ is perfect only except with negligible probability. In the precess of simulation, if the adversary first query $(*, S_B)$ to $H_2()$ before(or without) $StaticKeyReveal(ID_B)$, then S is successful and the abortion won't happen in the step7 and step10. Let $\Pr[S]$ be the probability that S succeeds. Then, the following inequality holds

$$\Pr[S] \geq \frac{1}{n(k)} \Pr[M \wedge E1] \tag{1}$$

So, if event $M \wedge E1$ occurs with non-negligible probability, i.e. $\Pr[M \wedge E1]$ is non-negligible, then the probability of S being successful is also non-negligible.

5.2   Event $E2$

Let's assume that event $M \wedge E2$ happens with non-negligible probability. In this case, given $U = g^u$ and $V = g^v$ where $U, V \in_R QR_N$ as input, we will use $\mathcal{M}$ to construct a CDH solver $\mathcal{S}$ over QR_N. The output of $\mathcal{S}$ is $CDH(U, V) = g^{uv}$. Meanwhile, in this case, we can simulate the compromise of KGC i.e. KGC forward secrecy(KGC-fs). One thing to be noted here is that when the RSA secret key d compromises, then the adversary can factor N in polynomial time deterministically owing to the contribution by Coron and May [24], i.e.knowing the private key of RSA is equivalent deterministically to the factorization of N. If the modulus N is factored, the adversary is still left with the CDH problem modulo the factors of N. It's easy to see that any solver of CDH problem with modulus N can be used to solve the CDH problem modulo the factors of N.

  $\mathcal{S}$ prepares $n(k)$ honest users and chooses the RSA parameters $(N = PQ, e, d)$ for KGC where P, Q are safe primes. $\mathcal{S}$ distributes the secret keys of users as the description of the protocol. Then, with probability at least $\frac{2}{s(k)^2}$, $\mathcal{S}$ guesses $\mathcal{M}$ will choose two sessions one as the test session and the other as the matching session. Without loss of generality, we assume that the owners of the test session and its matching session are A and B respectively. Since $\mathcal{S}$ owns all the users' secret keys? it's easy for it to simulate all the sessions. For the test session, $\mathcal{S}$ sets $\alpha = US_A$ as the outgoing message of user A. Similarly, the simulator sets $\beta = VS_B$ to be the outgoing message of user B. By this setting, the simulator implicitly defines $H_2(x, S_A) = log_g U$ and $H_2(y, S_B) = log_g V$ where $x, y \in_R \{0, 1\}^k$. When H_2 is queried with (x, S_A) or (y, S_B), $\mathcal{S}$ aborts; otherwise simulates a random oracle in the usual way. With the knowledge of all the users' secret keys, the simulator is easy to simulate all the other queries faithfully.

Analysis of Event M ∧ E2. In the simulation of above, if the adversary queries the H_2 with (x, S_A) or (y, S_B), then the simulation fails. But in the case of $E2$, $E1$ doesn't occur which means for every honest user B, before querying H_2 with $(*, S_B)$, the adversary performs the $StaticKeyReveal(ID_B)$. Becasue x is only used in the test session, so $\mathcal{M}$ can only know it by issuing $EphemeralKeyReveal(\Pi_{i,j}^s)$. Thus, if $\mathcal{M}$ queries the H_2 with (x, S_A) in this case, this implies the adversary knows the ephemeral secret key and static secret key of user A, which violates the freshness of the test session. The same thing will happen if the adversary queries the H_2 with (y, S_B). So, the simulation of $\mathcal{S}$ for the adversary $\mathcal{M}$ is perfect only except with negligible probability. If the adversary wins the test game in this case, then it must query $H()$ with $(\sigma, \alpha, \beta, ID_A, ID_B)$ where $\sigma = CDH^{2e}(U, V)$. Let $h_3(k)$ be the maximum number of query times to $H()$ made by $\mathcal{M}$. Then, the simulator can choose one of these queries randomly to obtain the $\sigma = CDH^{2e}(U, V)$ with probability at least $\frac{1}{h_3(k)}$. To extract the instance $CDH(U, V)$ from $\sigma = CDH^{2e}(U, V)$, $\mathcal{S}$ performs $\sigma^{\frac{d}{2}} = CDH^{2e\frac{d}{2}}(U, V)$ to get $CDH(U, V) = \sigma^{\frac{d}{2}}$. The success probability of $\mathcal{S}$ is

$$\Pr[\mathcal{S}] \geq \frac{2}{s(k)^2 h_3(k)} \Pr[M \wedge E2] \tag{2}$$

5.3   Event $E3$

Suppose that event $M \wedge E3$ occurs with non-negligible probability. Then, we will construct a RSA problem solver $\mathcal{S}$. Given as input (N, e, y) where N, e are sampled from the same distribution as used in the protocol and $y \in_R \mathbb{Z}_N^*$, $\mathcal{S}$ outputs x such that $x^e \equiv y \pmod{N}$.

First, the simulator prepares $n(k)$ honest users. Then $\mathcal{S}$ guesses the adversary will selects two users one as the owner of test session and other as the peer with probability at least $\frac{2}{n(k)^2}$. Without loss of generality, we assume that the test session owner and the peer are A and B respectively. Further, with probability at least $\frac{1}{s(k)}$, $\mathcal{S}$ guesses M will choose $\Pi_{A,B}^s$ as the test session. $\mathcal{S}$ sets $H_1(ID_B) = y^2$ and assigns random key pairs for the remaining $n(k) - 1$ users. The simulator $\mathcal{S}$ chooses $\bar{r} \in_R \mathbb{QR}_N$, and sets $r = \bar{r}^e, g = (ry)^e$. The simulation of this case is almost identical to that in event $E1$, in the next we mainly describe the different part.

1. $Send(\Pi_{i,j}^s, m)$: $\mathcal{S}$ maintains an initially empty list S^{list} with entries of the form $(\alpha, \beta, ID_i, ID_j, K)$.
   - if $\Pi_{i,j}^s$ is the test session, then $\mathcal{S}$ sets $\alpha' = (ry)^f S_A$ where $f \in_R \mathbb{Z}_{\lfloor N/4 \rfloor}$ and $gcd(f, e) = 1$. Thereby, $\mathcal{S}$ defines implicitly $H_2(x', S_A) = df$ $(\bmod\ \phi(N)/4)$ where $x' \in_R \{0,1\}^k$. Obviously, the simulator cannot respond to $H_2(x', S_A)$.
   - else the simulation is similar to that in event $E1$.

2. $H_2(x, S_{ID_i})$: If $ID_i = ID_A$ and $x = x'$, then $\mathcal{S}$ aborts; otherwise simulates a random oracle in the usual way.

3. $H(\sigma, \alpha, \beta, ID_i, ID_j)$: $\mathcal{S}$ maintains an initially empty list H^{list} with entries of the form $(\sigma, \alpha, \beta, ID_i, ID_j, h)$.
   - if $(\sigma, \alpha, \beta, ID_i, ID_j)$ is in the form $(\sigma, \alpha', \beta, ID_A, ID_B)$ where β is chosen by the adversary, the right σ should be $\sigma = \left(\frac{\beta^e}{y^2}\right)^{2df}$. In order to verify the correctness of σ, $\mathcal{S}$ can check $\sigma^e \stackrel{?}{=} \left(\frac{\beta^e}{y^2}\right)^{2f}$. In which case, the simulator stops and is successful by outputting $\left(y^{4f}\right)^d = \frac{\beta^{2f}}{\sigma}$. Since $gcd(4f, e) = 1$, using lemma 1 the simulator can obtain $x = y^d$.
   - else the simulation is the same to event $E1$.

4. $Test(\Pi_{i,j}^s)$: If $\Pi_{i,j}^s$ is not the test session, $\mathcal{S}$ aborts; otherwise responds faithfully.

The simulation of other queries for the simulator is similar to that in event $E1$.

Analysis of Event M $\wedge$ E3. The simulation of $\mathcal{S}$ for the adversary is perfect except with negligible probability. During the simulation, if the adversary queries (x', S_A) to $H_2()$, then $\mathcal{S}$ fails. However, in this event, event $E1$ doesn't occur which implies for every honest user, before issuing $H_2(*, S_{ID_i})$ query, the adversary learns the static private key S_{ID_i}. While, x' is only used in the test session, hence the adversary can only know it via performing $EphemeralKeyReveal$ to the test session. So, if M queries (x', S_A) to $H_2()$, then it must learn the ephemeral and static private key of user A at the same time which is against the freshness of the test session. Thus, the abortion in Step2 won't happen. Under event $E3$, except with negligible probability, $\mathcal{S}$ is successful as described in Step3 and the abortion does not occur in Step4. If event $M \wedge E3$ happens, then the success probability of $\mathcal{S}$ is

$$\Pr[\mathcal{S}] \geq \frac{2}{n(k)^2 s(k)} \Pr[M \wedge E3] \tag{3}$$

5.4 Overall Analysis

Combining the Eqs. (1), (2) and (3), the success probability of $\mathcal{S}$ is

$$\Pr[\mathcal{S}] \geq \max\{\frac{1}{n(k)} \Pr[M \wedge E1], \frac{2}{s(k)^2 h_3(k)} \Pr[M \wedge E2], \frac{2}{n(k)^2 s(k)} \Pr[M \wedge E3]\} \tag{4}$$

Thereby if M occurs with non-negligible probability, then one of the three events($M \wedge E1, M \wedge E2, M \wedge E3$) must happen with non-negligible probability. Thus from the Eq. (4) the success probability of $\mathcal{S}$ is non-negligible. Since all the simulations are polynomially bounded, therefore $\mathcal{S}$ is a polynomial-time algorithm that can solve the RSA problem or CDH problem over $\mathbb{QR}_N$ with non-negligible probability, which contradicts the assumed security of RSA problem or CDH problem with composite modulus.

6 Conclusion

In this paper, by binding the static private key with the ephemeral secret key, we propose an improved version of mOT protocol named mOT+. Compared with mOT, mOT+ not only is secure both in CK model and eCK model, but also can resist the leakage of ephemeral key. Moreover, mOT+ also meets PFS against active adversary just as mOT. Additionally, we provide a simple and intuitive proof in the eCK model. In a word, our improved version inherits all the advantages of the original one and at the same time gets rid of the flaws of it.

Acknowledgments. The authors would like to thank the anonymous referees for their helpful comments. This work is supported by the National Natural Science Foundation of China (Nos. 61309016,61379150,61201220), Post-doctoral Science Foundation of China (No. 2014M562493), Post-doctoral Science Foundation of Shanxi Province and Key Scientific and Technological Project of Henan Province (No. 122102210126) and the National Cryptology Development Project of China (No. MMJJ201201005).

References

1. Diffie, W., Hellman, M.E.: New directions in cryptography. IEEE Trans. Inf. Theory **22**(6), 644–654 (1976)

2. Krawczyk, H.: HMQV: a high-performance secure diffie-hellman protocol. In: Shoup, V. (ed.) CRYPTO 2005. LNCS, vol. 3621, pp. 546–566. Springer, Heidelberg (2005)
3. Law, L., et al.: An efficient protocol for authenticated key agreement. Des. Codes Crypt. 28(2), 119–134 (2003)
4. Shamir, A.: Identity-based cryptosystems and signature schemes. In: Blakely, G.R., Chaum, D. (eds.) CRYPTO 1984. LNCS, vol. 196, pp. 47–53. Springer, Heidelberg (1985)
5. Chen, L., Kudla, C.: Identity based authenticated key agreement protocols from pairings. In: Proceedings of the 16th IEEE Computer Security Foundations Workshop, pp. 219–233. IEEE (2003)
6. McCullagh, N., Barreto, P.S.L.M.: A new two-party identity-based authenticated key agreement. In: Menezes, A. (ed.) CT-RSA 2005. LNCS, vol. 3376, pp. 262–274. Springer, Heidelberg (2005)
7. Smart, N.P.: Identity-based authenticated key agreement protocol based on Weil pairing. Electron. Lett. 38(13), 630–632 (2002)
8. Joux, A.: A one round protocol for tripartite Diffie-Hellman. In: Bosma, W. (ed.) ANTS-IV. LNCS, vol. 1838, pp. 385–393. Springer, Heidelberg (2000)
9. Okamoto, E., Tanaka, K.: Key distribution system based on identi cation information. IEEE J. Sel. Areas Commun. 7(4), 481–485 (1989)
10. Mambo, M., Shizuya, H.: A note on the complexity of breaking Okamoto- Tanaka ID-based key exchange scheme. IEICE Trans. Fundam. Electron. Commun. Comput. Sci. 82(1), 77–80 (1999)
11. Seungjoo, K.I.M., et al.: On the security of the Okamoto-Tanaka ID-Based Key Exchange scheme against Active attacks. IEICE Trans. Fundam. Electron. Commun. Comput. Sci. 84(1), 231–238 (2001)
12. Gennaro, R., Krawczyk, H., Rabin, T.: Okamoto-Tanaka revisited: fully authenticated Diffie-Hellman with minimal overhead. In: Yung, M., Zhou, J. (eds.) ACNS 2010. LNCS, vol. 6123, pp. 309–328. Springer, Heidelberg (2010)
13. Canetti, R., Krawczyk, H.: Analysis of key-exchange protocols and their use for building secure channels. In: Pfitzmann, B. (ed.) EUROCRYPT 2001. LNCS, vol. 2045, pp. 453–474. Springer, Heidelberg (2001)
14. Menezes, A., Ustaoglu, B.: Security arguments for the UM key agreement protocol in the NIST SP 800–56A standard. In: Proceedings of the 2008 ACM Symposium on Information, Computer and Communications Security, pp. 261–270. ACM (2008)
15. LaMacchia, B.A., Lauter, K., Mityagin, A.: Stronger security of authenticated key exchange. In: Susilo, W., Liu, J.K., Mu, Y. (eds.) ProvSec 2007. LNCS, vol. 4784, pp. 1–16. Springer, Heidelberg (2007)
16. Huang, H., Cao, Z.: An ID-based authenticated key exchange protocol based on bilinear Diffe-Hellman problem. In: Proceedings of the 4th International Symposium on Information, Computer, and Communications Security, pp. 333–342. ACM (2009)
17. Shmuely, Z.: Composite Diffie-Hellman public-key generating systems are hard to break. Technical report 356. Computer Science Department, Technion, Israel (1985)
18. Shamir, A.: On the generation of cryptographically strong pseudorandom sequences. ACM Trans. Comput. Syst. (TOCS) 1(1), 38–44 (1983)
19. De Santis, A., et al.: How to share a function securely. In: Proceedings of the Twenty- Sixth Annual ACM Symposium on Theory of Computing, pp. 522–533. ACM (1994)

20. Goldreich, O., Rosen, V.: On the security of modular exponentiation with application to the construction of pseudorandom generators. J. Crypt. **16**(2), 71–93 (2003)
21. Goldwasser, S., Micali, S.: Probabilistic encryption. J. Comput. Syst. Sci. **28**(2), 270–299 (1984)
22. Lim, C.H., Lee, P.J.: A key recovery attack on discrete log-based schemes using a prime order subgroup. In: Kaliski Jr., B.S. (ed.) CRYPTO 1997. LNCS, vol. 1294, pp. 249–263. Springer, Heidelberg (1997)
23. Hofheinz, D., Kiltz, E.: The group of signed quadratic residues and applications. In: Halevi, S. (ed.) CRYPTO 2009. LNCS, vol. 5677, pp. 637–653. Springer, Heidelberg (2009)
24. Coron, J.-S., May, A.: Deterministic polynomial-time equivalence of computing the RSA secret key and factoring. J. Crypt. **20**(1), 39–50 (2007)

Secure ($M + 1$)st-Price Auction with Automatic Tie-Break

Takashi Nishide[1]([✉]), Mitsugu Iwamoto[2], Atsushi Iwasaki[2], and Kazuo Ohta[2]

[1] University of Tsukuba, Tsukuba, Japan
nishide@risk.tsukuba.ac.jp
[2] The University of Electro-Communications, Chofu, Japan

Abstract. In auction theory, little attention has been paid to a situation where the tie-break occurs because most of auction properties are not affected by the way the tie-break is processed. Meanwhile, in secure auctions where private information should remain hidden, the information of the tie can unnecessarily reveal something that should remain hidden. Nevertheless, in most of existing secure auctions, ties are handled outside the auctions, and all the winning candidates or only the non-tied partial bidders are identified in the case of ties, assuming that a subsequent additional selection (or auction) to finalize the winners is held publicly. However, for instance, in the case of the ($M + 1$)st-price auction, the tied bidders in the ($M + 1$)st-price need to be identified for such a selection, which implies that their bids (unnecessary private information) are revealed. Hence it is desirable that secure auctions reveal neither the existence of ties nor the losing tied bidders.

To overcome these shortcomings, we propose a secure ($M + 1$)st-price auction protocol with automatic tie-breaks and no leakage of the tie information by improving the bit-slice auction circuit without increasing much overhead.

Keywords: ($M + 1$)st-Price auction · Multiparty computation · Tie-break

1 Introduction

1.1 Background

The research on secure sealed-bid auctions focuses on enhancing privacy protection for electronic auction in the network environment. That is, in the secure sealed-bid auctions, only the winning bid and identities of the winning bidders are obtained, and other unnecessary private information such as the bids of losing (and sometimes winning) bidders and the existence of *ties* should not be revealed. As one of the important applications of secure multiparty computation (MPC), there is a long line of research in this area (e.g., [FR96, NPS99, Sak00, Kik01, AS02, LAN02, KO02, JS02, SY03, BS05, Bra06, NS10, MMO11])[1]. With the

---

[1] Another line of research addresses a secure double auction [BDJ+06, BCD+09] where many sellers and bidders participate and a market clearing price is determined based on the supply and demand in the market.

© Springer International Publishing Switzerland 2015
M. Yung et al. (Eds.): INTRUST 2014, LNCS 9473, pp. 422–436, 2015.
DOI: 10.1007/978-3-319-27998-5_27

privacy-enhancing mechanism, the bidders may have more incentive to participate in the electronic auction.

In this work, we deal with $(M+1)$st-price sealed-bid auctions where M units
of a single item are auctioned, and we assume all the participants obtain the
outcome of the auction[2] (i.e., the winning bid and identities of the winning bidders). Ideally it is desirable that exactly M winning bidders should be identified
even when *tie* situations occur.

Meanwhile in most of existing secure auction protocols, tie situations are
not handled automatically, and for example it is assumed that no tie situation
occurs, or all the winning candidates are identified and a subsequent tie-breaking
selection (or auction) to finalize the winners is held outside the cryptographic
protocols. However, if the price range is not large, we will not be able to assume
easily that no tie situation occurs. Also if all the winning candidates are identified
and a subsequent additional selection (or auction) is held publicly in the case
of first/second-price auctions, it means that several winning candidates become
the losing bidders although their bids were already revealed. Furthermore in the
case of the $(M + 1)$st-price auction, the tied bidders in the $(M + 1)$st-price need
to be identified for such a selection, thus revealing their bids. Hence the bidders
may prefer the secure auctions with automatic tie-breaks revealing neither the
existence of ties nor the losing tied bidders. Also the sellers will prefer the secure
auctions that identify exactly M winning bidders to several of existing protocols
[Kik01, AS02, MMO11] that can identify only less than M winning bidders in the
case of ties. Thus, it will be worthwhile to construct secure sealed-bid auction
protocols with automatic tie-breaks.

1.2 Our Approach and Contribution

We propose an $(M+1)$st-price auction protocol where we deal with tie situations
in a more privacy-enhancing way with an automatic tie-break.

For automatic tie-breaks, we have two options: (1) random secret priority
order among bidders and (2) public priority order among bidders. With the random secret priority order, tied bidders are selected at random without revealing
the priority order among the bidders, while, with the public priority order, tied
bidders are selected based on the publicly known priority order. With a public
priority order, private information can sometimes be leaked[3], and the public priority order can be realized with the random secret priority order, so we focus on
supporting the random secret priority order in secure auctions. We note that as
Vickrey auctions [Vic61], our proposed protocol satisfies incentive compatibility
in terms of auction theory as shown in [My81, MR89], that is, truthful bidding
is still an optimal strategy even with automatic tie-breaks in our protocol.

---

[2] In [Bra06], this is called *public outcome setting* as opposed to *private outcome setting*
where losing bidders can know neither the winning bid nor winning bidders.

[3] For example, suppose we have a list of bids $(5, 2, 3, 4, 4)$ where the i-th bid belongs
to the bidder B_i, and B_i has a higher priority than B_j when $i < j$ and $M = 2$. Then
the winning $(M + 1)$st-price is 4 and the winning bidders are B_1, B_4. In this case,
B_4 can know the bid of B_5 is 4.

To this end, we have two possibilities. One possibility is (1) to adapt (the modified variants of) multiparty sorting protocols to secure auction protocols with automatic tie-breaks. As discussed in Sect. 1.3, however, several of existing implemented sorting protocols assume the honest majority setting for efficiency, and incur larger communication complexity in the dishonest majority setting in terms of the number of bids, and this may become a bottleneck for large-scale auctions. The dishonest majority setting is preferable and may be required in several situations including the case where the bidders perform the secure auction protocol by themselves (as discussed in [Bra06]). The other possibility is (2) to adapt existing secure auction protocols. In this work, we choose this approach and use the bit-slice auction circuit [KO02, MMO11] as a starting point because it can be performed efficiently also in the dishonest majority setting and a bid is represented as a binary number. Thus, our approach can lead to the less communication complexity, as opposed to a kind of unary representation of the bid used in secure auctions such as [Kik01, AS02, SY03, Bra06] (i.e., $O(2^\ell)$ communication complexity where ℓ is the bit length of a bid).

In the bit-slice circuit, the winning price and bidders are identified by handling three secret vectors to maintain the states of all the bidders without comparing the bids directly. Thus, building on the bit-slice circuit [KO02, MMO11], we improve the $(M + 1)$st-price auction of [MMO11] such that exactly M winning bidders are identified by breaking ties automatically at random without revealing the information of the tie (e.g., priority order, the existence of the tie, tied bidders, etc.). For that, we add the random secret bits[4] to the LSBs of bids such that the random secret priority order among bidders is determined. However, just adding the random secret bits does not suffice because part of these random bits are leaked as the output result in the bit-slice circuit [MMO11]. This is not a problem in [MMO11] where automatic tie-breaks are not handled, but in our case, it can lead to leaking part of the random priority order and information of the tie (see Appendix A for details). Therefore, to solve this and avoid the leakage, we hide more intermediate results in evaluating the bit-slice circuit, and modify the way to handle the three secret vectors used in [MMO11] to maintain the states of all the bidders such that the computation can proceed even with the hidden intermediate results. Our adaptation can be done without increasing much overhead, and is comparable to other secure $(M + 1)$st-price auctions based on the sorting protocols [WLG+10, Zh11, HKI+12, HIC+14] with automatic tie-breaks (see Table 2).

1.3   Related Works

1.3.1   Auction Protocols

We summarize the comparison between existing auction protocols and this work in Table 1 in terms of functionality. The auction protocols of [NPS99, JS02, Bra06, KSS09] deal with tie situations[5].

---

[4] If we want to use the public priority order among bidders, these bits can be public.

[5] Actually only the protocol of [Bra06] deals with tie situations explicitly.

Table 1. Comparison between this work and major existing auction protocols in terms of functionality.

Protocol	Auction type	Automatic tie-break	Notes
[NPS99] [JS02]	Any	Yes	Yao's garbled circuit supporting tie-break needed. Limited to the case of secure 2-party computation.
[KO02]	1st	No	More than one winners revealed in the case of tie.
[MMO10]	2nd	No	
[LAN02]	$(M + 1)$	No	More than M winners revealed in the case of tie. Bid statistics revealed to auction authority.
[Kik01] [AS02] [MMO11]	$(M + 1)$	No	Only less than M winners may be identified in the case of tie. No winner is identified in worst case where more than M winners specify same highest bid.
[SY03]	GVA[a]	No	All tie situations identified in the case of tie.
[Bra06]	1st	Partially yes	Tie-break possible only in private outcome setting.
	2nd		Tie-break based only on public priority order.
	$(M + 1)$	No	More than M winners revealed in the case of tie.
[DGK08]	1st	No	Online auction based on secure two-party comparison. More than one winners revealed in the case of tie.
[KSS09]	1st	Partially yes	Auction based on improved Yao's garbled circuit computing minimum value and its index. Tie-break based only on public priority order.
[NS10]	1st	No	More than one winners revealed in the case of tie.
This work	$(M + 1)$	Yes	Exactly M winners identified in the case of tie with both public and random secret priority order among bidders.

[a]Generalized Vickrey auction
Combinatorial clock proxy auction

In [NPS99, JS02][6], a general auction protocol based on the Yao's garbled circuit [Yao82] was proposed, and it requires two entities, the auction issuer and the auctioneer that are not supposed to collude. The auction issuer generates a garbled circuit and the auctioneer executes the circuit after receiving the private bids from the bidders and the aucton issuer by using proxy oblivious transfer. The computation is limited to the two-party case and the general auction circuit with a tie-break based on the random secret priority order could be large and inefficient in practice. The protocol of [KSS09] follows the approach of [NPS99, JS02], but it handles only the first-price auction with the tie-break based on the pubic priority order.

In [Bra06], the first-price, second-price, and $(M + 1)$st-price auctions were proposed based on threshold ElGamal encryption with a dishonest majority. The protocols in [Bra06] can support two settings: public outcome setting (as most of existing works including this work) and private outcome setting. However, in the public outcome setting of [Bra06], all the winning candidates are revealed without an automatic tie-break. In the private outcome setting of [Bra06], tie situations are handled *partially*, that is, automatic tie-breaks can work only for first/second-price auctions and only the tie-break based on the pubic priority order among bidders is supported.

1.3.2 Sorting Protocols

Besides secure auction protocols, there are secure multiparty sorting protocols such as [WLG+10, Zh11, HKI+12, HIC+14], part of which can be used to realize secure $(M + 1)$st-price auctions with the random secret priority order for the tie-break.

We show the performance comparison in Table 2. Here we assume that we add random secret bits to bids to realize automatic tie-breaks in sorting protocols as we do for the bit-slice circuit. Also we analyze the performance in terms of both honest and dishonest majority settings. We note that actually it is not necessary to sort all the bids in the context of auctions, so we use simplified sorting protocols if possible for a fair comparison.

The protocols of [HKI+12, HIC+14] are quite efficient in the honest majority setting, assuming n (the number of parties executing MPC) is small, but they depend on the resharing based shuffle protocol[7] [LWZ11] that cannot generalize to the dishonest majority setting. Therefore, if the dishonest majority setting is required, we need to use the shuffle based on permutation matrices [LWZ11], thus leading to the inefficient quadratic complexity in the number of bidders k as [Zh11] in terms of communication complexity. Also the protocol of [WLG+10] can be efficient as our protocol, but it is probabilistic (with a small error rate), and it supports only the two-party setting. We note that in the context of the

---

[6] [JS02] improved [NPS99] by replacing proxy oblivious transfer in [NPS99] with verifiable proxy oblivious transfer.

[7] One resharing corresponds to one permutation, and the complexity of one resharing corresponds to that of one multiplication. Resharing is repeated such that an adversary cannot know the whole permutation.

Table 2. Performance comparison between this work and sorting protocols to realize secure $(M + 1)$st-price auction with random priority order for tie-break.

Protocol[a]	Round Complexity	Communication Complexity (#multiplication)	Notes
Selection network[b] [WLG+10]	$O\left(\log_2 k\right)$	$O\left(\mathrm{COM} \cdot k \log_2(M+1)\right)$	Based on garbled circuit Limited to 2-party case
Oblivious keyword sort [Zh11]	$O\left(1\right)$	$O\left(\mathrm{COM} \cdot k^2\right)$	Honest & dishonest majority[d]
Modified quicksort[c] & shuffle [HKI+12]	$O\left(\frac{2^n}{\sqrt{n}} + \log_2 \frac{k}{M}\right)$	$O\left(k\frac{2^n}{\sqrt{n}} + \mathrm{COM} \cdot k\right)$	w/ honest majority
	$O\left(n + \log_2 \frac{k}{M}\right)$	$O\left(nk^2 + \mathrm{COM} \cdot k\right)$	w/ dishonest majority
Radix sort & shuffle [HIC+14]	$O\left(\frac{2^n}{\sqrt{n}}(\ell + \log_2 k)\right)$	$O\left(\frac{2^n}{\sqrt{n}}k(\ell + \log_2 k)^2\right)$	w/ honest majority
	$O\left(n(\ell + \log_2 k)\right)$	$O\left(nk^2(\ell + \log_2 k)^2\right)$	w/ dishonest majority
This work	$O\left(\ell + \log_2 k\right)$	$O\left(k(\ell + \log_2 k)\right)$	Honest & dishonest majority[d]

[a]We also assume we add random secret bits to bids to realize automatic tie-breaks in sorting protocols.
[b]A variant of sorting network to identify M largest elements
[c]Sorting all the bids is not required, so we can use the simplified quicksort, but we need to use full protocols for other protocols because of their internal structures.
[d]This means the protocol can work with both honest and dishonest majority
k: #bid(der)s
p: MPC is executed $\bmod\ p$
ℓ: bit length of bid, i.e., bid $< 2^\ell$
n: #parties executing MPC
M: M units of a single item are auctioned
COM:communication complexity of comparison protocol, typically $O(|p|)$ multiplications (with $O(1)$ round). We need to assume $\ell + O(\log_2 k) < |p|$ for automatic tie-breaks to work.

$(M + 1)$st-price auction, we need a set of M winners rather than a sorted list of M winners in the end, so actually we will need to adapt the sorting protocols by using shuffle protocols though we ignore such additional necessary processing here.

2 Preliminaries

2.1 Basic Computation Model and Notations

We assume that MPC is performed in the arithmetic black box (ABB) model formalized by [DN03] to abstract away the details of MPC implementations by following a gate evaluation technique. MPC we use in this work can be realized by the techniques such as the BGW protocol [BGW88], threshold homomorphic encryption [DN03], mix-and-match protocol [JJ00] in the honest majority setting, and SPDZ [DPSZ12] in the dishonest majority setting, and the security follows from these underlying MPC protocols with the simulation-based proof. We use the notation $[\![a]\!]$ to denote a secret handled in the ABB model. The addition and multiplication of secrets are written as, e.g., $[\![ab + c]\!] \leftarrow [\![a]\!] \cdot [\![b]\!] + [\![c]\!]$.

We can assume that the addition is essentially for free and the multiplication is costly because it needs communication. If c is a public constant and a is a secret, computing $[\![ca]\!] \leftarrow c \cdot [\![a]\!]$ and $[\![c+a]\!] \leftarrow c + [\![a]\!]$ is for free. We also use $\sum$, for example, like $\sum_{i=1}^{3} [\![a_i]\!]$ to denote $[\![a_1]\!] + [\![a_2]\!] + [\![a_3]\!]$. By a secret vector $[\![\vec{v}]\!]$, we mean that each element of the vector $\vec{v}$ is a secret (e.g., if $\vec{v} = (x, y, z)$, $[\![\vec{v}]\!] = ([\![x]\!], [\![y]\!], [\![z]\!])$). We also use the notation like $[\![t]\!] \cdot [\![\vec{v}]\!]$ meaning that each element of $\vec{v}$ is multiplied by t.

Because the multiplication is a dominant factor of the complexity, we measure the round complexity of a protocol by the number of rounds of parallel invocations of the multiplication and we also measure the communication complexity by the number of invocations of the multiplication. The round complexity relates to the time required for a protocol to be completed and the communication complexity relates to the amount of data communicated during a protocol run. For simplicity, we assume that the reveal of one secret $[\![a]\!]$ is measured as one invocation of the multiplication[8].

We will evaluate the round complexity of a protocol by performing the multiplication in parallel as much as possible.

We describe our protocols in the so-called "honest-but-curious" model, but standard techniques such as [Ped91, DN03] will be applicable to make our protocols robust.

2.2   Auction Model

We assume there are k bidders (denoted by $B_1, \ldots, B_k$ in this work) in the $(M + 1)$st-price auction and $M < k$[9] because otherwise every bidder can win with any bid. The bidders submit the private bids, and the computation of auction is performed in the ABB model, and typically this means that a set of servers $P_1, \ldots, P_n$ receive the secret inputs from $B_1, \ldots, B_k$ and perform MPC instead of the bidders (as in the client-server model in [DI05])[10].

Also we assume the auction outcome (i.e., the winning bid and identities of the winning bidders) becomes public to the seller and all the bidders after the computation is finished.

We denote the plaintext space for the ABB model by $\mathbb{Z}_p$ where p is public and will be either a prime or an RSA modulus.

In the protocol based on the bit-slice circuit, each bid is represented as a binary number and each bidder submits each shared (or encrypted) bit of the bid as in [KO02, MMO10, MMO11][11].

---

[8] In [BGW88], we note that one round for the reveal can sometimes be ignored if $[\![ra]\!]$ is revealed after the multiplication $[\![r]\!] \cdot [\![a]\!]$ with another secret $[\![r]\!]$ because the multiplication and the reveal of ra can be done simultaneously as in [BF01].

[9] We assume k and M are public information.

[10] Of course MPC can also be performed by the bidders themselves.

[11] We assume the maximum bit length of a bid is public information.

2.3 Arithmetic Black Box Techniques

RandVal: We can generate a random secret value $[\![r]\!]$ such that $r \in \mathbb{Z}_p$[12]. We assume that the complexity for this is almost the same as the complexity of 1 invocation of the multiplication as in [DFK+06][13]. We note that all the necessary random value generation can be done in advance to reduce the round complexity.

RandBit: We can generate a random bit $[\![a]\!]$ such that $a \in \{0, 1\}$. The total complexity is 2 rounds and 3 invocations ([DFK+06])[14].

Evaluating Polynomial in Constant Rounds: We can evaluate a polynomial with a secret in constant rounds by using an unbounded fan-in multiplication and the inversion protocol [BB89]. That is, given $[\![A]\!]$[15] and a public $g_t(x) = \sum_{i=0}^{t} \alpha_i x^i \bmod p$, we can compute $[\![g_t(A)]\!]$ with 3 rounds (including 2 rounds for random value generation) and $7t$ invocations [NO07].

3 $(M + 1)$st-Price Sealed-Bid Auction with Automatic Tie-Break

We propose a protocol for $(M + 1)$st-price sealed-bid auction where M units of a single item are auctioned and each of M winning bidders with M highest bids wins one unit[16] of the item by paying the uniform $(M+1)$st-price. This protocol computes only the $(M + 1)$st-price (i.e., the highest losing bid) and exactly M winning bidders by breaking ties automatically and hiding the existence of ties, while the protocols such as [AS02, Kik01, MMO11] may compute only less than M winning bidders when there are ties, and the protocol of [Bra06] may reveal more than M winning candidates and the number of the tied bids in the $(M + 1)$st-price without breaking ties automatically (see Table 1).

To deal with the ties more appropriately with automatic selection, the secret *priority order* among the bidders is determined at random such that even if there are more than M winning candidates, only M winners are identified based on the priority order without revealing the priority order. For example, if we have a sorted list of bids $(7, 7, 6, 5, 5, 5, 3)$ including ties and $M = 4$, our protocol computes the auction outcome such that the $(M+1)$st-price is 5 and the winning bidders include the bidders with bids $7, 6$ and one of the tied bidders with bid 5 selected according to the priority order by following auction theory [Kri09].

---

[12] p is usually large, so we can assume r is a non-zero value. If not, we can use the efficient technique to generate random invertible (i.e., non-zero) value from [BB89].

[13] In the setting of [BGW88], we have possibilities to realize less expensive RandVal such as non-interactive pseudo-random secret sharing [CDI05] and improved RandVal based on the use of hyper-invertible matrices [BTH08].

[14] In the setting of [DN03, JJ00], this can be computed as $a = \oplus_{i=1}^{n} b_i$ where $\oplus$ is XOR and $b_i \in_R \{0, 1\}$ is generated by P_i.

[15] We require that A is not a zero for this technique to work.

[16] That is, we assume bidders with unit-demand.

3.1 $(M+1)$st-Price Bit-Slice Auction Circuit

The $(M+1)$st-price auction circuit of [MMO11] extends the first/second-price bit-slice auctions of [KO02], and is originally based on homomorphic encryption and the mix-and-match technique, and each bid is represented as a binary number and encrypted bit by bit.

Informal Description. First we give an informal description of how the $(M+1)$st-price bit-slice auction circuit [MMO11] works because our protocol is based on it.

Now let's assume that there are 4 bidders B_1, B_2, B_3, B_4 (i.e., $k = 4$) and the bit length of prices, ℓ, is 5 and $M = 2$.

(1) [Bidding:] Each bidder B_i submits the bit representation $\vec{v}_i$ of the bid as follows:

$\vec{v}_1 = (v_{1,1}, v_{1,2}, \ldots, v_{1,5}) = (0, 0, 1, 1, 0)$ when the B_1's bid is 6

$\vec{v}_2 = (v_{2,1}, v_{2,2}, \ldots, v_{2,5}) = (0, 0, 1, 0, 0)$ when the B_2's bid is 4,

$\vec{v}_3 = (v_{3,1}, v_{3,2}, \ldots, v_{3,5}) = (0, 0, 1, 0, 1)$ when the B_3's bid is 5, and

$\vec{v}_4 = (v_{4,1}, v_{4,2}, \ldots, v_{4,5}) = (0, 1, 0, 1, 0)$ when the B_4's bid is 10.

(2) Next, we identify the M winning bidders and the $(M+1)$st-price. We can assume the $(M+1)$st-price has the bit representation $(m_1, m_2, \ldots, m_\ell)$ where $m_i \in \{0, 1\}$ and m_ℓ is the least significant bit.

(i) [Definition of $\vec{a}^i$] To do so, we consider the vectors $\vec{a}^i$ such that

$$\begin{pmatrix} \vec{v}_1 \\ \vec{v}_2 \\ \vec{v}_3 \\ \vec{v}_4 \end{pmatrix} = \begin{pmatrix} {}^t\vec{a}^1 & {}^t\vec{a}^2 & {}^t\vec{a}^3 & {}^t\vec{a}^4 & {}^t\vec{a}^5 \end{pmatrix}$$

where ${}^t\vec{a}^i$ means the transpose of $\vec{a}^i$.

We use the notation as $\vec{a}^i = (\vec{a}^i[1], \vec{a}^i[2], \ldots, \vec{a}^i[k])$.

(ii) We also handle 3 kinds of secret vectors $\vec{c} = (c_1, \ldots, c_k), \vec{w} = (w_1, \ldots, w_k), \vec{s} = (s_1, \ldots, s_k)$ corresponding to "candidates", "winners", and "survivors" respectively. Initially $\vec{c}$ is set to $(1, 1, 1, 1)$ (meaning that all the bidders are still the candidates for the winners) and $\vec{w}$ is set to $(0, 0, 0, 0)$ (meaning that no bidders are finalized to be the winners) here and $\vec{s}$ is used to store the intermediate temporary results.

(iii) [Updating vectors $\vec{c}, \vec{w}, \vec{s}$] Each time we process $\vec{a}^i$ from $i = 1$ to ℓ, we update $\vec{c}, \vec{w}, \vec{s}$ as follows: First we compute $\vec{s} \leftarrow (c_1 \cdot \vec{a}^i[1], \ldots, c_k \cdot \vec{a}^i[k])$, that is, several (possibly zero) survivors are selected from the candidates based on the bids.

(Case 1):If $\sum_{i=1}^{k}(w_i + s_i) \leq M$, then the survivors can be added to the winners, so $\vec{c}, \vec{w}$ are updated as follows:

$$\vec{c} \leftarrow \vec{c} - \vec{s}$$
$$\vec{w} \leftarrow \vec{w} + \vec{s}.$$

That is, the current survivors are finalized to be the winners and moved from the candidates to the winners. For the $(M + 1)$st-price, publicly we set $m_i = 0$.

(Case 2): If $\sum_{i=1}^{k}(w_i + s_i) > M$, we can see that the further selection of the survivors need to be done, so $\vec{c}, \vec{w}$ are updated as follows:

$$\vec{c} \leftarrow \vec{s}$$

$$\vec{w} \leftarrow \vec{w} \text{ (i.e., remains the same).}$$

That is, only the current survivors can be the candidates for the subsequent processing. For the $(M + 1)$st-price, publicly we set $m_i = 1$. Here we note that the result of whether or not $\sum_{i=1}^{k}(w_i + s_i) \leq M$ holds can be made public because it is related to the (public) $(M + 1)$st-price, and we can use $\text{ZeroIfLeq}(M, \sum_{i=1}^{k}(w_i + s_i))$ to compute this, which returns 0 if $\sum_{i=1}^{k}(w_i + s_i) \leq M$, and a non-zero random value otherwise. The detail of $\text{ZeroIfLeq}(M, \cdot)$ is given later.

ZeroIfLeq$(M, \cdot)$ We consider how to check whether $[\![a]\!]$ satisfies $0 \leq a \leq M$ and publish its result. To check this condition, we define a polynomial $g_{M+1}(x) = \prod_{i=1}^{M+1}(x - i) \bmod p$, generate a random secret $[\![r]\!]$, and compute and publish $rg_{M+1}(a + 1)$ by using the technique of Sect. 2.3. Here we use $a + 1$ instead of a to make sure that $a + 1$ is not a zero. If $rg_{M+1}(a + 1)$ is a zero, it means that $0 \leq a \leq M$, and if $rg_{M+1}(a + 1)$ is a non-zero, it means that $M + 1 \leq a$. We can realize $[\![\text{ZeroIfLeq}(M, a)]\!]$ by $[\![rg_{M+1}(a + 1)]\!]$ with fresh random r. The complexity for computing $g_{M+1}(a + 1)$ is 3 rounds and $7(M + 1)$ invocations. Therefore, the total complexity for computing $rg_{M+1}(a + 1)$ is 4 rounds and $7(M + 1) + 3$ invocations[17].

3.2 $(M + 1)$st-Price Bit-Slice Auction with Automatic Tie-Break

As already mentioned in [MMO11], the protocol of Sect. 3.1 cannot handle the tie situation appropriately. We propose how to adapt the $(M + 1)$st-price bit-slice auction circuit so that it can support the automatic tie-breaks. To do so, we (1) add random secret bits to the LSBs of the bids that determine the (secret) random priority order among bidders and (2) change how to deal with the secret vectors $\vec{c}, \vec{w}, \vec{s}$ to hide the information of the tie.

Informal Description. First we compute and add secret $O(\log_2 k)$ random bits to the LSBs of each bid by using RandBit in Sect. 2.3, and these bits determine the priority order among bidders. The processing of the non-random secret bits of the bids is done in the same way as the original protocol, but for the processing of the added random bits, we need to change how to deal with vectors $\vec{c}, \vec{w}, \vec{s}$ to hide the random bits and the information of the tie.

---

[17] We note that if M is small, $g_{M+1}(x)$ can be evaluated in $\log_2(M + 1)$ rounds with $M + 1$ multiplications.

As described in Sect. 3.1, we have Case 1 and Case 2 where $\vec{c}, \vec{w}$ are updated based on the result of $\sum_{i=1}^{k}(w_i + s_i) \leq M$. In our protocol, we need a trick to merge Case 1 and Case 2 into one step in order to hide which case happened unlike [MMO11][18], and to merge two cases we use the technique IsLeq($M, \sum_{i=1}^{k}(w_i + s_i)$) (see also Step 3 in Fig. 1), the detail of which is given later. With this technique, we can also hide the number of the tied bidders in the $(M + 1)$st-price and the number of the tied bidders to be selected as the winner in our protocol.

The formal description of our protocol is given in Fig. 1.

IsLeq($M, \cdot$) to Check the Range of a Secret Privately. We consider how to check whether $[\![a]\!]$ satisfies $0 \leq a \leq M$ or $M+1 \leq a \leq k$ without revealing which case holds. To check this condition, we construct a polynomial $f_{k,M}(x) \bmod p$ such that $f_{k,M}(x) = 1$ for $1 \leq x \leq M+1$ and $f_{k,M}(x) = 0$ for $M+2 \leq x \leq k+1$ by using Lagrange interpolation. Then we compute $[\![f_{k,M}(a + 1)]\!]$ by using the technique of Sect. 2.3. Here we use $a + 1$ instead of a to make sure that $a + 1$ is not a zero. We can realize $[\![\text{IsLeq}(M, a)]\!]$ by $[\![f_{k,M}(a + 1)]\!]$. The total complexity is 3 rounds and $7k$ invocations.

Correctness. This follows immediately from the bit-slice auction circuit [MMO11] in which the bids are scanned from the most significant bits such that all the M winning bidders are stored in $\vec{w}$ if there are no ties, and as described in Fig. 1, our adaptation for breaking ties does not affect how to compute the auction outcome.

Privacy. Besides the auction outcome, the intermediate values revealed in the protocol are the results of ZeroIfLeq($\cdot$) that are used to determine each i-th bit m_i of the $(M+1)$st-price. These results are 0 if m_i is 0 and the uniformly random values otherwise, so these can be simulated easily with the output, $(M + 1)$st-price in the simulation-based proof.

Remark 1. In the case of the second-price auction such as [KO02, MMO10], there is a trivial way to break ties. That is, after identifying the winning candidates privately, we can check whether each bidder is a winner one by one in the random priority order until we find the first winner. However, this technique does not generalize to the case where $M > 1$, and the round complexity is $O(k)$, so it may be disadvantageous when there are many bidders.

Remark 2. If the priority order is determined and public in advance[19], we note that the public $\log_2 k$ bits can be added to the LSBs of each bid according to the public priority order. That is, the random secret bits $([\![v_{i,\ell+1}]\!], \dots, [\![v_{i,\ell+L}]\!])$

---

[18] In [MMO11], the result of this conditional branch is related to the public $(M + 1)$st-price, so it does not need to be hidden. In contrast, in our protocol, the result and the random bits are related to the secret priority of the bidder whose bid is the $(M + 1)$st-price, which was used to break the tie when necessary. Therefore, they need to be hidden to avoid the leakage of the information of the tie.

[19] For example, the priority order can be determined based on the order in which the bidders sign up for the auction as mentioned in [Bra06].

for each bid in Fig. 1 are replaced with the public bits where $L = \log_2 k$, and the protocol is executed in the same way.

Acknowledgments. This work was partially supported by JSPS KAKENHI Grant Number 26330151, Kurata Grant from The Kurata Memorial Hitachi Science and Technology Foundation, and the Telecommunications Advancement Foundation.

A   Leakage of Tie Information

We describe why the leakage of the tie information can lead to unnecessary private information leakage in terms of MPC for auctions.

In our protocol as shown in Sect. 3.2, random secret bits are added to each bid as least significant bits (LSBs). For example, let's assume here that the random LSBs are 4-bit and that we obtained the $(M + 1)$st-price 87&(1111) at the end of the execution of our protocol. Here 87&(1111) means that the $(M + 1)$st-price was 87 dollars and (1111) were the 4-bit random LSBs added to the $(M + 1)$st-price. The whole 87&(1111) is obtained in public if we just use the auction protocol of [MMO11].

At this point, the bids of the winners could also be 87&(1111), but the probability that the added LSBs were (1111) also for all other winning bidders is low, so we can partially guess that the bids of the winners were > 87 although ideally we should be able to guess only that the bids of the winners were ≥ 87 in terms of MPC for auctions (i.e., the tie information led to unnecessary information leakage). The similar discussion also applies when the LSBs are large (and the LSBs (1111) are an extreme example).

Also similarly when the $(M + 1)$st-price was 87&(0000), we can guess the bids of the losing bidders except the losing bidder who specified the $(M + 1)$st-price were < 87 although ideally we should be able to guess only that the bids were ≤ 87 in terms of MPC for auctions.

If only the $(M + 1)$st-price 87 is obtained, we can avoid the above unnecessary information leakage. Although this leakage is not catastrophic, it is desirable to eliminate this if we can do so with small overhead, and that will be what ideal MPC for auctions is supposed to realize.

In our protocol, we solve the above leakage problem by adding random secret LSBs and also hiding the LSBs when the $(M + 1)$st-price is obtained in public.

B   Generating Random Bits for Tie-Break

Another possible way, by assuming we use homomorphic encryption and the mix-and-match technique as in [MMO11], is to permute a list of encryptions of $(0, 1, \ldots, k - 2, k - 1)$ by a mix protocol [JJ00] where each element in the list is in the bit representation and encrypted bit by bit. Then we can append those randomly permuted encrypted $\log_2 k$ bits to each bid that is also encrypted bit by bit in [MMO11] to realize an automatic tie-break at random.

Here we follow the notation in Sect. 3.1.

Bidding: Each bidder B_i $(1 \leq i \leq k)$ submits a bid in the bit representation, $[\vec{v_i}] = ([v_{i,1}], [v_{i,2}], \ldots, [v_{i,\ell}])$ where $v_{i,j} \in \{0,1\}$ and $v_{i,\ell}$ is the least significant bit. We add L random bits[a] to the LSBs of each bid, so now $[\vec{v_i}] = ([v_{i,1}], [v_{i,2}], \ldots, [v_{i,\ell}], [v_{i,\ell+1}], \ldots, [v_{i,\ell+L}])$.

Identifying winning bidders and $(M+1)$st price: Initially, let

$$[\vec{c}] = ([c_1], \ldots, [c_k]) \leftarrow ([1], \ldots, [1]) \text{ and}$$

$$[\vec{w}] = ([w_1], \ldots, [w_k]) \leftarrow ([0], [0], \ldots, [0]).$$

We assume the $(M+1)$st price has the bit representation $(m_1, m_2, \ldots, m_\ell)$ where $m_i \in \{0,1\}$ and m_ℓ is the least significant bit.

From $i = 1$ to ℓ, we do the following:

1. We compute $[\vec{s}] = ([s_1], \ldots, [s_k]) \leftarrow ([c_1] \cdot [\vec{a}^i[1]], \ldots, [c_k] \cdot [\vec{a}^i[k]])$.
2. We compute $[\![\text{ZeroIfLeq}(M, \sum_{i=1}^{k}(w_i + s_i))]\!]$ and publish it. If the result was 0, we update

$$[\vec{c}] \leftarrow [\vec{c}] - [\vec{s}],$$

$$[\vec{w}] \leftarrow [\vec{w}] + [\vec{s}]$$

and we set $m_i = 0$. If the result was not 0, we update

$$[\vec{c}] \leftarrow [\vec{s}]$$

and $[\vec{w}]$ remains the same and we set $m_i = 1$[b].

Breaking tie (if any) with random secret bits: From $i = \ell + 1$ to $\ell + L$,

1. We compute $[\vec{s}] = ([s_1], \ldots, [s_k]) \leftarrow ([c_1] \cdot [\vec{a}^i[1]], \ldots, [c_k] \cdot [\vec{a}^i[k]])$.
2. We compute $t' \in \{0,1\}$ such that

$$[t'] \leftarrow [\![\text{IsLeq}(M, \sum_{i=1}^{k}(w_i + s_i))]\!]$$

where for $\text{IsLeq}(M, \cdot)$ a polynomial $f_{k,M}(x)$ is defined such that $f_{k,M}(x) = 1 \bmod p$ for $1 \leq x \leq M + 1$ and $f_{k,M}(x) = 0 \bmod p$ for $M + 2 \leq x \leq k + 1$ and $\text{IsLeq}(M, x) = f_{k,M}(x + 1)$. Here $\text{IsLeq}(M, \cdot)$ is used to judge privately whether or not more than M winning bidders are identified.

3. We update $[\vec{w}], [\vec{c}]$ as follows without revealing t':

$$[\vec{c}] \leftarrow [t'] \cdot ([\vec{c}] - [\vec{s}]) + (1 - [t']) \cdot ([\vec{s}])$$
$$= [\vec{s}] + [t'] \cdot ([\vec{c}] - 2[\vec{s}])$$
$$[\vec{w}] \leftarrow [\vec{w}] + [t'] \cdot [\vec{s}]$$

The meaning of the processing here is that if $t' = 1$, the survivors can be added to the winners and the survivors are eliminated from the candidates because these survivors could become the winners. If $t' = 0$, the survivors cannot be added to the winners because the total number of the winning bidders becomes more than M, so the winners remain the same and only these survivors are selected as the new candidates with further subsequent selection.

Publishing outcome: By publishing $[\vec{w}]$, we have M winning bidders and the $(M+1)$st price is $(m_1, m_2, \ldots, m_\ell)$.

---

[a] L is $O(\log_2 k)$, and in practice, say, $2.5 \log_2 k$ will suffice based on the probabilistic analysis with the birthday paradox. Another possible way to add random bits is discussed in Appendix B.

[b] We note that at this point several bidders in $\vec{c}$ may need to be selected further to break the tie if any.

Fig. 1. Proposed $(M+1)$st-price bit-slice auction with automatic tie-break

References

[AS02] Abe, M., Suzuki, K.: $M+1$-st price auction using homomorphic encryption. In: Naccache, D., Paillier, P. (eds.) PKC 2002. LNCS, vol. 2274, pp. 115–124. Springer, Heidelberg (2002)

[BB89] Bar-Ilan, J., Beaver, D.: Non-cryptographic fault-tolerant computing in a constant number of rounds of interaction. In: Proceedings of the ACM Symposium on Principles of Distributed Computing, pp. 201–209 (1989)

[BTH08] Beerliová-Trubíniová, Z., Hirt, M.: Perfectly-secure MPC with linear communication complexity. In: Canetti, R. (ed.) TCC 2008. LNCS, vol. 4948, pp. 213–230. Springer, Heidelberg (2008)

[BGW88] Ben-Or, M., Goldwasser, S., Wigderson, A.: Completeness theorem for non-cryptographic fault-tolerant distributed computation. In: Proceedings of the 20th Annual ACM Symposium on Theory of Computing (STOC), pp. 1–10 (1988)

[BCD+09] Bogetoft, P., Christensen, D.L., Damgård, I., Geisler, M., Jakobsen, T., Krøigaard, M., Nielsen, J.D., Nielsen, J.B., Nielsen, K., Pagter, J., Schwartzbach, M., Toft, T.: Secure multiparty computation goes live. In: Dingledine, R., Golle, P. (eds.) FC 2009. LNCS, vol. 5628, pp. 325–343. Springer, Heidelberg (2009)

[BDJ+06] Bogetoft, P., Damgård, I.B., Jakobsen, T., Nielsen, K., Pagter, J.I., Toft, T.: A practical implementation of secure auctions based on multiparty integer computation. In: Di Crescenzo, G., Rubin, A. (eds.) FC 2006. LNCS, vol. 4107, pp. 142–147. Springer, Heidelberg (2006)

[BF01] Boneh, D., Franklin, M.: Efficient generation of shared RSA keys. J. ACM 48(4), 702–722 (2001)

[Bra06] Brandt, F.: How to obtain full privacy in auctions. Int. J. Inf. Sec. 5(4), 201–216 (2006)

[BS05] Brandt, F., Sandholm, T.W.: Efficient privacy-preserving protocols for multi-unit auctions. In: S. Patrick, A., Yung, M. (eds.) FC 2005. LNCS, vol. 3570, pp. 298–312. Springer, Heidelberg (2005)

[CDI05] Cramer, R., Damgård, I.B., Ishai, Y.: Share conversion, pseudorandom secret-sharing and applications to secure computation. In: Kilian, J. (ed.) TCC 2005. LNCS, vol. 3378, pp. 342–362. Springer, Heidelberg (2005)

[DFK+06] Damgård, I.B., Fitzi, M., Kiltz, E., Nielsen, J.B., Toft, T.: Unconditionally secure constant-rounds multi-party computation for equality, comparison, bits and exponentiation. In: Halevi, S., Rabin, T. (eds.) TCC 2006. LNCS, vol. 3876, pp. 285–304. Springer, Heidelberg (2006)

[DGK08] Damgård, I., Geisler, M., Krøigaard, M.: Homomorphic encryption and secure comparison. Int. J. Appl. Crypt. 1(1), 22–31 (2008)

[DI05] Damgård, I.B., Ishai, Y.: Constant-round multiparty computation using a black-box pseudorandom generator. In: Shoup, V. (ed.) CRYPTO 2005. LNCS, vol. 3621, pp. 378–394. Springer, Heidelberg (2005)

[DN03] Damgård, I.B., Nielsen, J.B.: Universally composable efficient multiparty computation from threshold homomorphic encryption. In: Boneh, D. (ed.) CRYPTO 2003. LNCS, vol. 2729, pp. 247–264. Springer, Heidelberg (2003)

[DPSZ12] Damgård, I., Pastro, V., Smart, N., Zakarias, S.: Multiparty computation from somewhat homomorphic encryption. In: Safavi-Naini, R., Canetti, R. (eds.) CRYPTO 2012. LNCS, vol. 7417, pp. 643–662. Springer, Heidelberg (2012)

436   T. Nishide et al.

[FR96]   Franklin, M.K., Reiter, M.K.: The design and implementation of a secure auction service. IEEE Trans. Software Eng. **22**(5), 302–312 (1996)

[HKI+12]   Hamada, K., Kikuchi, R., Ikarashi, D., Chida, K., Takahashi, K.: Practically efficient multi-party sorting protocols from comparison sort algorithms. In: Kwon, T., Lee, M.-K., Kwon, D. (eds.) ICISC 2012. LNCS, vol. 7839, pp. 202–216. Springer, Heidelberg (2013)

[HIC+14]   Hamada, K., Ikarashi, D., Chida, K., Takahashi, K.: Oblivious radix sort: an efficient sorting algorithm for practical secure multi-party computation. Cryptology ePrint Archive 2014/121 (2014)

[JJ00]   Jakobsson, M., Juels, A.: Mix and match: secure function evaluation via ciphertexts. In: Okamoto, T. (ed.) ASIACRYPT 2000. LNCS, vol. 1976, pp. 162–177. Springer, Heidelberg (2000)

[JS02]   Juels, A., Szydlo, M.: A two-server, sealed-bid auction protocol. In: Blaze, M. (ed.) FC 2002. LNCS, vol. 2357, pp. 72–86. Springer, Heidelberg (2002)

[Kik01]   Kikuchi, H.: (M+1)st-price auction protocol. In: Syverson, P.F. (ed.) FC 2001. LNCS, vol. 2339, pp. 341–353. Springer, Heidelberg (2002)

[KSS09]   Kolesnikov, V., Sadeghi, A.-R., Schneider, T.: Improved garbled circuit building blocks and applications to auctions and computing minima. In: Garay, J.A., Miyaji, A., Otsuka, A. (eds.) CANS 2009. LNCS, vol. 5888, pp. 1–20. Springer, Heidelberg (2009)

[Kri09]   Krishna, V.: Auction Theory, 2nd edn. Academic Press, San Diego (2009)

[KO02]   Kurosawa, K., Ogata, W.: Bit-slice auction circuit. In: Gollmann, D., Karjoth, G., Waidner, M. (eds.) ESORICS 2002. LNCS, vol. 2502, pp. 24–38. Springer, Heidelberg (2002)

[LWZ11]   Laur, S., Willemson, J., Zhang, B.: Round-efficient oblivious database manipulation. In: Lai, X., Zhou, J., Li, H. (eds.) ISC 2011. LNCS, vol. 7001, pp. 262–277. Springer, Heidelberg (2011)

[LAN02]   Lipmaa, H., Asokan, N., Niemi, V.: Secure vickrey auctions without threshold trust. In: Blaze, M. (ed.) FC 2002. LNCS, vol. 2357, pp. 87–101. Springer, Heidelberg (2002)

[MR89]   Maskin, E., Riley, J.: Optimal multi-unit auctions. In: The Economics of Missing Markets, Information, and Games, pp. 312–335. Oxford University Press (1989)

[MMO10]   Mitsunaga, T., Manabe, Y., Okamoto, T.: Efficient secure auction protocols based on the Boneh-Goh-Nissim encryption. In: Echizen, I., Kunihiro, N., Sasaki, R. (eds.) IWSEC 2010. LNCS, vol. 6434, pp. 149–163. Springer, Heidelberg (2010)

[MMO11]   Mistunaga, T., Manabe, Y., Okamoto, T.: A secure M + 1st price auction protocol based on bit slice circuits. In: Iwata, T., Nishigaki, M. (eds.) IWSEC 2011. LNCS, vol. 7038, pp. 51–64. Springer, Heidelberg (2011)

[My81]   Myerson, R.B.: Optimal auction design. Math. Oper. Res. **6**(1), 58–73 (1981)

[NPS99]   Naor, M., Pinkas, B., Sumner, R.: Privacy preserving auctions and mechanism design. In: ACM Conference on Electronic Commerce, pp. 129–139 (1999)

[NO07]   Nishide, T., Ohta, K.: Multiparty computation for interval, equality, and comparison without bit-decomposition protocol. In: Okamoto, T., Wang, X. (eds.) PKC 2007. LNCS, vol. 4450, pp. 343–360. Springer, Heidelberg (2007)

[NS10]   Nojoumian, M., Stinson, D.R.: Unconditionally secure first-price auction protocols using a multicomponent commitment scheme. In: Soriano, M., Qing, S., López, J. (eds.) ICICS 2010. LNCS, vol. 6476, pp. 266–280. Springer, Heidelberg (2010)

[Ped91] Pedersen, T.P.: Non-interactive and information-theoretic secure verifiable secret sharing. In: Feigenbaum, J. (ed.) CRYPTO 1991. LNCS, vol. 576, pp. 129–140. Springer, Heidelberg (1992)

[Sak00] Sako, K.: An auction protocol which hides bids of losers. In: Imai, H., Zheng, Y. (eds.) PKC 2000. LNCS, vol. 1751, pp. 422–432. Springer, Heidelberg (2000)

[SY03]   Suzuki, K., Yokoo, M.: Secure generalized vickrey auction using homomorphic encryption. In: Wright, R.N. (ed.) FC 2003. LNCS, vol. 2742, pp. 239–249. Springer, Heidelberg (2003)

[Vic61] Vickrey, W.: Counterspeculation, auctions, and competitive sealed tenders. J. Finance **16**(1), 8–37 (1961)

[WLG+10] Wang, G., Luo, T., Goodrich, M.T., Du, W., Zhu, Z.: Bureaucratic protocols for secure two-party sorting, selection, and permuting. In: ASIACCS, pp. 226–237. ACM (2010)

[Yao82] Yao, A.: Protocols for secure computations. In: Proceedings of the 23rd Annual Symposium on Foundations of Computer Science (FOCS), pp. 160–164 (1982)

[Zh11]   Zhang, B.: Generic constant-round oblivious sorting algorithm for MPC. In: Boyen, X., Chen, X. (eds.) ProvSec 2011. LNCS, vol. 6980, pp. 240–256. Springer, Heidelberg (2011)

Author Index

Keyword Index

Printed in the United States
By Bookmasters